Inorganic Materials

Front cover illustration

A perspective view of the X-ray structure of $SnS_2\{Co(\eta\text{-}C_5H_5)_2\}_{0.3}$.
Reprinted with permission from Verlag Chemie.
D O'Hare, J S'O Evans, C K Prout and P J Wiseman,
Angew Chemie Ind Ed Engl, **30**, 1156 (1991).

Back cover illustration

The X-ray structure of $MnCu(pbaOH)(H_2O)_3$
(pbaOH = 2-hydroxy-1,3-propylenebis(oxamato))
O Kahn, Y Pei, M Verdaguer, J P Renard and J Sletten,
J Am Chem Soc, **110**, 782 (1988)

Both illustrations were drawn using CHEM-X, the Editors would like to thank
Chemical Design Ltd, Oxford, UK for use of their graphics facilities.

Inorganic Materials

Edited by

Duncan W Bruce

Lecturer in Chemistry
University of Sheffield

and

Dermot O'Hare

University Lecturer and Fellow of Balliol College
University of Oxford

JOHN WILEY & SONS
Chichester · New York · Brisbane · Toronto · Singapore

Other Wiley Editorial Offices

John Wiley & Sons, Inc., 605 Third Avenue,
New York, NY 10158-0012, USA

Jacaranda Wiley Ltd, G.P.O. Box 859, Brisbane,
Queensland 4001, Australia

John Wiley & Sons (Canada) Ltd, 22 Worcester Road,
Rexdale, Ontario M9W 1L1, Canada

John Wiley & Sons (SEA) Pte Ltd, 37 Jalan Pemimpin #05-04,
Block B, Union Industrial Building, Singapore 2057

Library of Congress Cataloging-in-Publication Data

Inorganic materials /   edited by Duncan W. Bruce and Dermot O'Hare.
    p.   cm.
  Includes bibliographical references and index.
  ISBN 0 471 92889 5
  1. Materials.   2. Inorganic compounds.   3. Superconductors.
4. Metals.   I. Bruce, Duncan W.   II. O'Hare, Dermot.
TA403.6.I54   1992
620.1′1—dc20   92–27214
                                                       CIP

British Library Cataloguing in Publication Data

A catalogue record for this book is available from the British Library

ISBN 0 471 92889 5

Produced from camera ready copy by author
Printed and bound in Great Britain by Biddles Ltd, Guildford, Surrey

To our families, Sue, Ciarán, Anita, Katie and Annie

Preface

In recent years, Materials Chemistry has enjoyed something of a renaissance and expansion, although there are still few texts available which cover the area. This book then seeks to fill a part of that gap by considering some aspects of inorganic materials.

We have chosen a multi-author format in order to benefit from researchers who are active in their chosen fields and who can therefore give the best account of their subject. We have also used mainly younger authors as we felt that their energy, enthusiasm and relatively new entry in these areas would provide new perspectives. To try to do justice to the whole field of inorganic materials would have been impossible in a volume such as this, so we have emphasised discussions of the properties of molecular solids, as these provide the exciting possibility of controlling bulk properties by tuning molecular properties.

The book is intended to provide a well-referenced introduction to each subject, followed by an overview of the area and then concentrating on selected examples in order to emphasise best the materials under discussion. We feel that the authors have achieved this admirably and that the book will be useful for anyone wanting to start work in any of these areas, or requiring an overview of a particular field. As such, we hope that the book will be of some use in the final year of undergraduate courses, as well as to researchers in both academia and industry.

Readers may be interested to learn that the editing and production of this book have been achieved using Apple™ Macintosh computers running Microsoft Word™, ChemDraw™, Chem3D™, Cricket Graph™, Kaleidagraph™ and MacDraw™ software.

Oxford DO'H
Sheffield DWB
November 1992

Acknowledgements

The editors would like to thank their commissioning editors at John Wiley & Sons, namely Heather Bewers who initiated the project and Jenny Cossham who ably took us through to production and who were always there to answer our various technical questions regarding the production of a camera-ready manuscript. Thanks also go to the highly-skilled copy editors for spotting all the deliberate mistakes; any remaining errors are ours and not theirs. We would also like to thank Mike Manterfield (Sheffield) and the University of Sheffield Printing Unit for drawing and printing a number of the diagrams for various chapters. Chemical Design, Oxford were very helpful when we were producing a view of the structure on the front-cover and Verlag Chemie gave their permission for us to use this illustration. The final camera ready copy was produced by the National Academic Typesetting Facility at Oxford.

Finally, we would like to thank the main victims of our endeavours, namely our families, Sue and Ciarán in Oxford and Anita, Katie and Annie in Sheffield who many times have gone to sleep to the friendly chatter of a keyboard!

Contents

List of Contributors

Duncan W Bruce
Centre for Molecular Materials, Department of Chemistry,
The University, SHEFFIELD, S3 7HF, UK

Patrick Cassoux
Laboratoire de Chimie de Coordination du CNRS
205 Route de Narbonne, 31077 TOULOUSE, FRANCE

John G Gaudiello
Systems Technology Division, IBM Corporation
ENDICOTT, New York 13760-8003, USA

Yves Journaux
Laboratoire de Chimie Inorganique, Université de Paris-Sud
91405 ORSAY, FRANCE

Olivier Kahn
Laboratoire de Chimie Inorganique, Université de Paris-Sud
91405 ORSAY, FRANCE

Glen Eugene Kellogg
Department of Medicinal Chemistry and Division of Biomedical
Engineering, Virginia Commonwealth University, RICHMOND
Virginia 23298-0540, USA

Stephen Mann
School of Chemistry, University of Bath,
Claverton Down, BATH, BA2 7AY, UK

Seth R Marder
The Beckmann Institute, California Institute of Science and Technology
PASADENA, CA 91125, USA

Richard W McCabe
Department of Chemistry, University of Central Lancashire
PRESTON, PR1 2TQ, UK

Paul O'Brien
School of Chemistry, Queen Mary and Westfield College
University of London, Mile End Road, LONDON, E1 4NS, UK

Dermot O'Hare
Inorganic Chemistry Laboratory, University of Oxford,
South Parks Road, OXFORD, OX1 3QR, UK

Yu Pei
Laboratoire de Chimie Inorganique, Université de Paris-Sud
91405 ORSAY, FRANCE

Lydie Valade
Laboratoire de Chimie de Coordination du CNRS
205 Route de Narbonne, 31077 TOULOUSE, FRANCE

1 Molecular Inorganic Superconductors

Patrick Cassoux and Lydie Valade

Inorganic Materials. Edited by Duncan W Bruce and Dermot O'Hare
© 1992 John Wiley & Sons Ltd

1.1   BACKGROUND

Since the discovery of superconductivity in mercury by Kammerlingh-Onnes in 1911 until the early nineteen seventies, only elements or metal alloys were proven to exhibit this incredible property: namely that below a given temperature, called the critical temperature T_c, the material has zero resistivity. In 1973 the highest critical temperature, 25.5 K, was observed for the [Nb$_3$Al$_{0.8}$Ge$_{0.2}$] alloy.

More recently, superconducting Nobel-prize winners based on copper oxides, such as [$YBa_2Cu_3O_{6+x}$], have broken the 'wall of the liquid nitrogen temperature' ($T_c > 77$ K).

Another family of superconductors, the Chevrel phases, discovered in the seventies, consists of ternary molybdenum chalcogenides, such as [$PbMo_6S_8$] ($T_c = 15.2$ K).

However, the first true molecular superconducting compounds were obtained in 1980 by Bechgaard, and were derived from a purely organic molecule, tetramethyltetraselenafulvalene, TMTSF. Clearly, the molecular inorganic chemists were challenged.

Gratifyingly, the first molecular inorganic superconductor, [TTF][Ni(dmit)$_2$]$_2$ (TTF = tetrathiafulvalene; dmit^{2-} = 1,3-dithiol-2-thione-4,5-dithiolato) was obtained in 1986 in our group in Toulouse. The following chapter is devoted to this exciting, challenging, and finally successful, quest.

1.1.1 The KCP Complexes

The first episode of what could be called the 'Molecular Inorganic Superconductors Saga' was unconsciously written by Knop as early as 1842 [1]. He prepared 'kupferglänzenden' (copper-shining) crystals by oxidizing $K_2[Pt(CN)]_4$ with chlorine or bromine, but could not fully characterise these crystals. Knop was not at all aware that his compound was the first 'Molecular Inorganic Conductor'.

In fact, if Levy in 1912 did suggest the presence of mixed valent states in these complexes [2], later called 'KCP' from the German 'kalium tetracyanoplatinat', more than a century passed before Krogmann clarified their actual stoichiometry in 1968; $K_2[Pt(CN)_4X_{0.3}].nH_2O$ (X = Cl, Br), with all platinum atoms in the same non-integral oxidation state [3]. Their structure (Figure 1) was characterised by columns of Pt(CN)$_4$ anions stacked along the direction perpendicular to the Pt(CN)$_4$ plane, with Pt–Pt distances of 2.88 Å, i.e. slightly longer than the Pt–Pt distance in platinum metal (2.77 Å).

The suggestion made on the basis of these structural features (and, perhaps, their metallic lustre) that the KCP complexes might exhibit novel electrical properties was eventually confirmed by Zeller [4]: the room temperature conductivity along the stacking direction was ≈ 300 S cm^{-1}, with a large anisotropic ratio (the conductivity parallel to the Pt chains is 10^5 greater than the conductivity perpendicular to the chain direction), and the temperature dependence in the high temperature range is indeed consistent with a metallic band structure. In these compounds metallic behaviour arose from electron delocalisation along overlapped platinum $3d_{z^2}$-orbitals and from the formation of a partially-filled band induced by partial oxidation. Therefore, the KCP complexes could be described as the first 'One-dimensional Molecular Metals'.

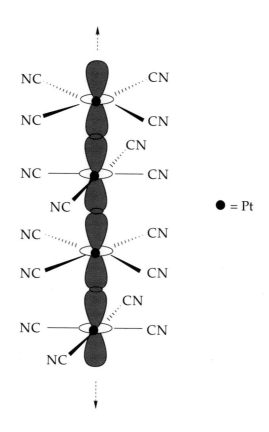

Figure 1 A chain of square-planar $Pt(CN)_4$ groups in $K_2[Pt(CN)_4]X_{0.3} \cdot nH_2O$ showing the overlapping of Pt d_{z^2}-orbitals (from [6])

Following these findings, the KCP complexes have been the subject of a large number of chemical modifications and physical and theoretical studies, which have been discussed in several review articles and books [5–8]. Whilst it appears from this extensive work that the KCP complexes may be used as textbook examples for illustrating how tight binding band theory may be successfully applied to one-dimensionally stacked molecular systems for rationalizing their physical behaviour, it is also clear that, to date, neither the studied KCP-like complexes, nor their chemical modifications such as the related partially-oxidised bis(oxalato)platinate salts [8], retain their metallic characteristics down to low temperatures (not to mention becoming superconducting). When the temperature is decreased, all these systems undergo a lattice distortion accompanied by a metal-to-insulator transition that had been predicted by Peierls [9] as an inherent instability of a one-dimensional (1-D) metal.

1.1.2   The Organic Metals and Superconductors

Nevertheless, the fire of hope was still kept burning thanks to the suggestion by Little [10] that appropriate one-dimensional systems could fulfil the criteria for high temperature superconductivity.

Given the failure of the platinum chain compounds, an alternative route towards molecular superconductivity was offered by purely organic compounds. Shortly after the first synthesis of 7,7,8,8-tetracyano-*p*-quinodimethane (TCNQ; Figure 2) in 1962 [11], it was discovered that many salts of this molecule were electrically conducting [12].

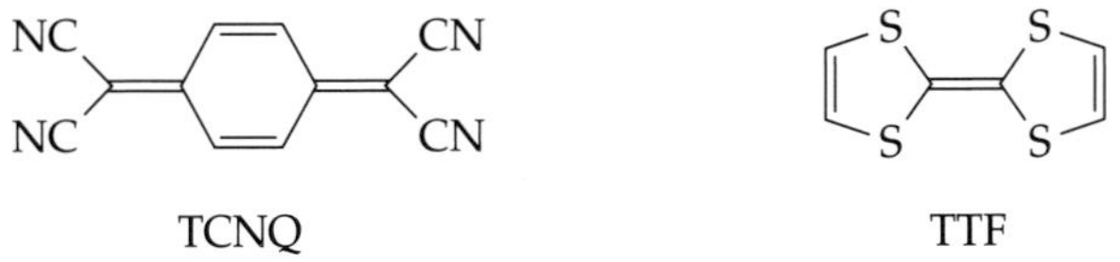

TCNQ   TTF

Figure 2 The TCNQ and TTF molecules

Starting in 1970, it was also discovered that another organic molecule, tetrathiafulvalene, (TTF) [13], could be halogenated to yield conducting salts [14]. However, the first 'Organic Metal' was only obtained in 1973, when these two molecules were combined in a 1:1 donor–acceptor compound [15]. In TTF.TCNQ, a partial charge transfer between separately stacked donor (TTF) and acceptor (TCNQ) molecules leads to two types of co-linear, one-dimensionally delocalised systems and the formation of partially-filled bands. A tremendous number of studies have been carried out on TTF.TCNQ and related compounds and several reviews have appeared [16–18].

In spite of this tremendous effort, up to 1980 none of the TTF.TCNQ-like organic metals retained their metallic characteristics down to low temperatures. As in the case of the KCP compounds, they all underwent a metal-to-insulator transition when the temperature was lowered.

In 1980, Bechgaard *et al.* used a molecule derived from TTF, tetramethyltetraselenafulvalene (TMTSF; Figure 3) to prepare radical salts of the type (TMTSF)$_2$X, in which X was an inorganic cation such as PF$_6^-$, ClO$_4^-$, etc.

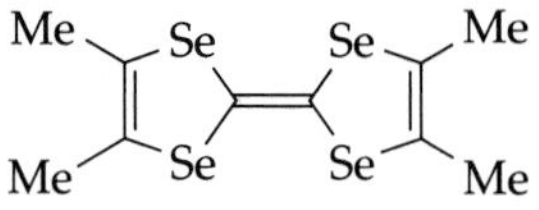

TMTSF 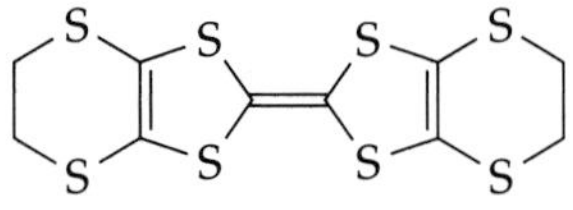   BEDT-TTF

Figure 3 The TMTSF and BEDT-TTF molecules

At last, the holy grail was found: (TMTSF)$_2$PF$_6$ was the first molecular compound to become superconducting under pressure [19] and (TMTSF)$_2$ClO$_4$ was the first 'Molecular Superconductor' at ambient pressure [20]. Then, in 1984, another chemical modification of TTF, the bis(ethylenedithio)-tetrathiafulvalene (BEDT-TTF), also yielded ambient pressure molecular superconductors of the type (BEDT-TTF)$_2$X, in which X was I$_3^-$ [21] or an inorganic cation such as [Cu(NCS)$_2$]$^-$ [22] or [Cu{N(CN)$_2$}][Y] (Y = Cl, Br) [23, 24]. To date, the κ-phase of (BEDT-TTF)$_2$[Cu{N(CN)$_2$}][Cl] exhibits the highest critical temperature T$_c$ (12.8 K, 0.3 kbar) of any molecular superconductor [24].

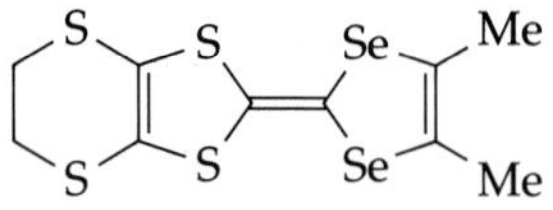

DMET 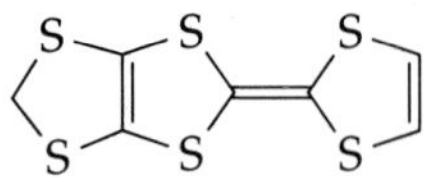   MDT-TTF

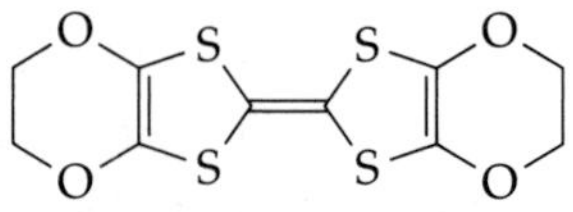

BEDO-TTF

Figure 4 The DMET, MDT-TTF and BEDO-TTF molecules

Three additional modifications of TTF have also been used for the preparation of molecular superconductors, namely DMET (Figure 4) [25],

MDT-TTF [26] and BEDO-TTF [27]. To date, thirty two organic superconductors are known [17, 28].

At this point an important question arises, "What is the key feature that makes some molecular systems conducting and others not?" This will be discussed in the next section.

1.2 USE OF GUIDELINES FOR SELECTING CANDIDATE SYSTEMS

1.2.1 Laying Down the Guidelines

A number of structural and electronic criteria required for the formation of conducting, molecular, partially-oxidised (KCP-like) or partial charge-transfer (TTF.TCNQ-like), one-dimensional systems have been suggested (*a posteriori*, as usual) on the basis of tight-binding band structure considerations and analysis of available experimental data [5, 7, 29, 30].

The following guidelines [5] are quite straightforward, but at least they were useful for the synthetic chemist:

a. 1-D Stacking: Either inorganic or organic molecules should stack in one direction to provide a framework for possible band structure formation. Therefore, the repeat units should preferably have an overall planar geometry.

b. Overlap: The stacking molecules should have an uneven number of electrons and have an unfilled orbital with a large extension perpendicular to the plane of the molecule, thus allowing good overlap between sites. Moreover, the molecules should closely stack in order to increase the overlap, i.e. the bandwidth, through metal-metal bonding as in KCP or through π-orbital overlap as in TTF.TCNQ. Therefore, no appreciable steric obstacles (bulky groups) should hinder close packing.

c. Non-Integral Oxidation State: The partial filling of the conduction band, i.e. the metallic state, should be ensured either by partial oxidation as in KCP or by partial charge transfer as in TTF.TCNQ.

d. Regular Stacking: The molecules should be uniformly spaced to avoid splitting of the conduction band (Peierls-like distortion). Moreover, in the case of donor–acceptor compounds, the stacks of donor and acceptor molecules should be segregated. This criterion, however, cannot be used as an effective guideline as there does not seem to be any means of controlling the mode of stacking.

While a number of other effects, such as the Coulombic repulsion between electrons on the same molecule or electrons on neighbouring molecules, the exchange interactions between neighbouring spins, the polarisability, size and symmetric or asymmetric nature of the cations, the crystallographic disorder, vibronic or libronic couplings, etc. have also been considered [30], they do not lead to easily applicable guidelines for the synthetic chemist.

By contrast, in the case of the charge-transfer compounds, the guideline based on matching the redox potentials of the donor and the acceptor [30] has been widely used:

e. Redox Potentials Criterion: Coupling a strong donor with a good acceptor will lead to a total charge transfer, while no charge transfer at all will occur between a weak donor and a reluctant acceptor. Therefore, partial charge transfer can be best obtained when using moderate donors and moderate acceptors. The donor or acceptor strength, i.e. the ionisation potential of the donor molecule and the electron affinity of the acceptor molecule, although not directly measurable, can be estimated in solution from the redox potential values of the donor (oxidation) and the acceptor (reduction) molecules. Matching these redox potentials translates to Equation (1):

$$E_{1/2}(D \rightarrow D^+ + e^-) - E_{1/2}(A + e^- \rightarrow A^-) = 0.1\text{--}0.4 \text{ V} \qquad (1)$$

which means that the difference between the redox potentials of the donor and acceptor molecules should preferably lie within the recommended 0.1–0.4 V range.

1.2.2 Application to Molecular Inorganic Conductors

These guidelines may be, and have been applied in exploring the use of molecular inorganic materials, especially transition metal complex systems, as a potential source of conductive compounds.

1.2.2.1 *Linear-chain Iridium Complexes*

A number of partially-oxidised dihalodicarbonyliridate(I) complexes, $C_x[Ir(CO)_2X_2]$ (with for example C = H$^+$, K$^+$, Na$^+$ and TTF$^+$; $0 < x < 1$, X = Cl, Br), have been prepared and studied [31]. They exhibit metal-like conductivity in the range 0.1–5 S cm^{-1}. Their structural arrangement is similar to that of KCP and consequently the conduction mechanism seems to arise from electron delocalisation along overlapped iridium d_{z^2}-orbitals (Ir–Ir distance ≈ 2.86 Å) and from the formation of a partially-filled band induced by partial oxidation (Ir oxidation state ≈ 1.4). However, non-reproducibility of the stoichiometry, low stability and poor crystallinity have hampered much of the characterisation of these compounds.

The case of the halotricarbonyliridium complexes, $[IrX(CO)_3]$ (with X = Cl, Br and I), is even more puzzling. The conductivity of the chlorinated compound is respectable (up to 0.2 S cm^{-1}), but there is a still unresolved controversy about the stoichiometry [31].

1.2.2.2 Macrocyclic Metal Complexes

A number of macrocyclic metal complexes have been selected as potential donor molecules for preparing one-dimensional conductive systems. This choice probably resulted from the observation of the overall planarity of these complexes, and in some cases the extended delocalised π system.

Halogenated linear-chain systems have been obtained from nickel or palladium complexes of glyoximate ligands [32]. Likewise, nickel, palladium and copper complexes of tetraazaannulene ligands, phthalocyanines and porphyrins have been subjected to partial oxidation with iodine [32]. All these compounds with the exception of [Ni(phthalocyanine)][I], were found to be semiconducting.

Semiconducting donor–acceptor compounds using a nickel tetraaza-annulene complex as a donor molecule have been also reported [33].

1.2.2.3 Metal Bis(dithiolene) Complexes

Many metal complexes of the *cis*-1,2-disubstituted ethylene-1,2-dithiol ligands, $[M(S_2C_2R_2)_2]^{n-}$ (with for example M = Ni, Pd, Pt and Cu; R = H, C and CF$_3$; Figure 5) have been used as possible acceptor molecules for preparing π donor–acceptor compounds with a large number of organic or inorganic donor molecules [34].

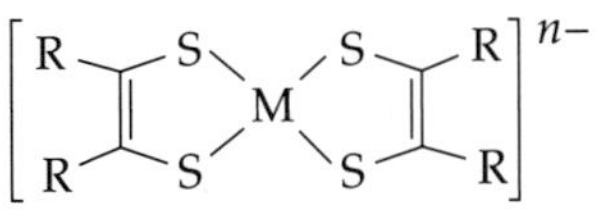

$$\left[\begin{array}{c} R \diagdown \diagup S \diagdown \diagup S \diagdown \diagup R \\   M   \\ R \diagup \diagdown S \diagup \diagdown S \diagdown R \end{array} \right]^{n-}$$

Figure 5 The metal bis(dithiolene) complexes

These metal bis(dithiolene) complexes were selected because of their overall planar geometry and tendency to 1-D stacking, their extended π-electron system, their reversible redox behaviour and the availability of stable oxidation states with low net charges, thus perhaps facilitating close intermolecular packing and electron transfer between complex units

[35–37]. In that sense, they were expected to meet criteria (a) and (b) (see Section 1.2.1). With respect to criteria (c) and (e), the half-wave potentials corresponding to the first redox step i.e. the electrochemical couple $[M(S_2C_2R_2)_2]^{-1}/[M(S_2C_2R_2)_2]^0$, were in some cases in the appropriate range (≈ 0.2 V versus SCE) such that a partial charge transfer with appropriate donor molecules might be anticipated.

The metal bis(dithiolene) complexes were also attractive for two additional reasons. First, they combined in one molecule two key features, an extended π-electron system as in TTF on one hand, and a transition metal, thus introducing possible metal-metal interactions as in KCP, on the other. Second, the metal bis(dithiolene) complexes bore some resemblance to the TTF molecule as one system can be derived from the other by substituting the central C=C bond for a metal atom. In fact, following on from the concept of isolobality developed by Hoffmann [38], a metal d^8 (M^{2+}) ion is isolobal to an ethylene (C_2^{4+}) fragment.

In spite of these many blessings, very few systems derived from metal bis(dithiolene) complexes exhibit high conductivity and metallic behaviour. The highest conductivities are found in the series [perylene]$_2$[M(S$_2$C$_2$(CN)$_2$)$_2$] with M = Ni, Cu, Pt, Pd and Au. The platinum, palladium and gold derivatives exhibit a metal-like conducting behaviour at high temperature, but undergo a metal-to-insulator transition at low temperatures [39]. A similar behaviour was observed in the [TTF]$_{1.2}$[M{S$_2$C$_2$(CH$_3$)$_2$}$_2$] compound [29, 40]. But, in the above compounds, only the organic donor component is responsible for the high conductivity! The [H$_3$O]$_{0.33}$[Li]$_{0.82}$[Pt(S$_2$C$_2$(CN)$_2$)$_2$] platinum salt was the first, and the only one to date, complex of the metal bis(dithiolene) series in which metal-like behaviour was really due to the intermolecular interaction of the metal complex anion, but unfortunately, this compound also undergoes a metal-to-insulator transition at about 270 K [41].

The essential clue, i.e. dimensionality, through which the gap between the non-superconducting TTF-based compounds and the superconducting BEDT-TTF-based compounds was bridged, had yet to be found in the case of the metal bis(dithiolene) compounds. Interestingly, it happened to be the same clue.

1.2.3 One-dimensionality versus Two-dimensionality

The time has come now to answer the question, 'What is the key feature that makes some molecular systems conducting and others not?' Most of the non-superconducting molecular systems studied so far were 1-D (Figure 6).

In the cases of the $(TMTSF)_2X$ and $(BEDT\text{-}TTF)_2X$ superconductors, several factors such as the size and the symmetry of the X anions, the resulting unit cell volume, some short cation-anion distances and some anion orderings, have been proposed and discussed [17]. However, the most important feature rested on the observation of an extended 'infinite 2-D sheet network' of Se···Se or S···S interstack interactions [17, 21, 42] (Figure 6).

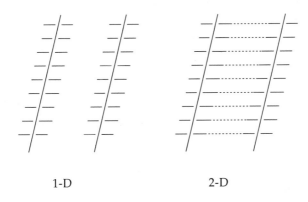

1-D   2-D

Figure 6 1-D versus 2-D systems

In fact, not all $(TMTSF)_2X$ and $(BEDT\text{-}TTF)_2X$ molecules exhibiting these Se···Se or S···S interactions are superconductors, but all superconductors in both series do exhibit this feature.

This observation confirmed the hypothesis that increasing the dimensionality (1-D to quasi-1-D as in $(TMTSF)_2X$, or even to 2-D as in $(BEDT\text{-}TTF)_2X$ could be a way to avoid the ubiquitous metal-to-insulator transition encountered in cooling most previously studied 1-D molecular metals [9].

1.2.4 The [M(dmit)$_2$] Complexes

In both the $(TMTSF)_2X$ and $(BEDT\text{-}TTF)_2X$ compounds, the short interstack interactions involved selenium or sulphur atoms on the periphery of the molecules. A simple strategy for enhancing interstack interactions emerged [43], which involved the incorporation of additional selenium or sulphur atoms on the periphery of molecular candidate systems.

Applying this new guideline to the metal bis(dithiolene) complexes led to the consideration of the [M(dmit)$_2$] complexes (dmit^{2-} = 1,3-dithiol-2-thione-4,5-dithiolato; Figure 7). The [M(dmit)$_2$] complexes are derived from the metal bis(dithiolene) complexes by extending the central bicyclic MS_4C_4 fragment with two fused CS_3 rings on both ends of the molecule. These [M(dmit)$_2$] complexes have ten peripheral sulphur atoms at their disposal to establish interstack interactions.

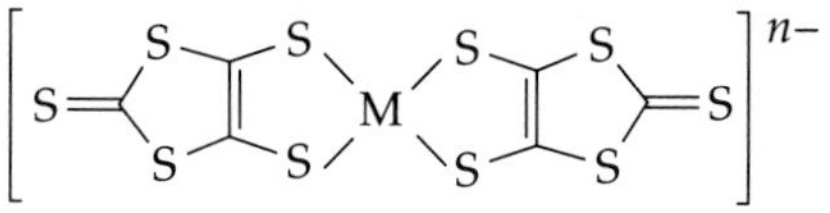

Figure 7 The [M(dmit)$_2$] complexes

This almost naïve strategy has led to the characterisation of four superconducting phases which are the only molecular inorganic superconductors known to date which are derived from transition metal complexes.

1.3   THE SYNTHESIS OF CONDUCTORS AND SUPERCONDUCTORS BASED ON [M(DMIT)$_2$] COMPLEXES

Most preparations within this field involve as a starting material one of the tetraalkylammonium salts of [M(dmit)$_2$]$^{n-}$ (n = 2, 1). These precursor complexes can be used for the preparation of either non-integral oxidation state complexes such as the superconducting [Me$_4$N]$_{0.5}$[Ni(dmit)$_2$], or donor–acceptor compounds such as the superconducting [TTF][M(dmit)$_2$]$_2$ (M = Ni, Pd). In passing, it should be noted that in these molecular inorganic superconductors the cationic:anionic species ratio is 1:2, i.e. the inverse of that observed in the organic superconductors such as (TMTSF)$_2$X and (BEDT-TTF)$_2$X.

1.3.1   The dmit Ligand

In 1975, Steimecke et al. [44] reported that reducing carbon disulphide with sodium in dimethylformamide yielded a mixture of the sodium salt of the dmit ligand ($C_3S_5Na_2$) and the sodium trithiocarbonate salt (CS_3Na_2) (Figure 8). The rather unstable dmit ligand could best be isolated and stabilised as the zinc complex. To do this, the reaction mixture was treated with a zinc salt in aqueous ammonia, then with tetraalkyl-ammonium bromide R_4NBr (R = Et or Bu) in methanol to give, exclusively, a precipitate of $[R_4N]_2[Zn(dmit)_2]$ since the trithiocarbonate derivative was soluble. This general procedure has been reproduced and improved by several authors [45, 46], including ourselves [47].

A further way to protect the dmit ligand consisted of treating $[R_4N]_2[Zn(dmit)_2]$ with benzoyl chloride which gave the stable 4,5-bis(benzoylthio)-1,3-dithiol-2-thione (dmit(COPh)$_2$) [45]. This thioester could be conveniently stored as precursor for further work: when needed the dmit^{2-} anion could be regenerated from dmit(COPh)$_2$ by treatment with sodium methoxide [45].

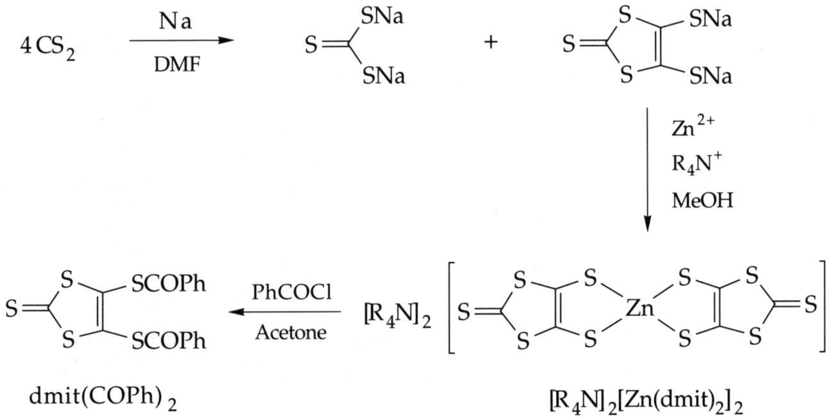

Figure 8 Synthesis of the dmit ligand and related starting materials

In some cases, it may even be convenient to isolate the highly reactive Na$_2$dmit salt by precipitation from diethyl ether using standard Schlenk techniques [48] (Figure 9). This method allows not only very clean further synthesis but also synthesis in which high pH values of the reaction mixture would prohibit the precipitation of the desired complex; for example, when mono-, bi- or trialkylammonium salts are used as counter-cations.

Alternatively, the dmit^{2-} anion can be prepared by electrochemical reduction of carbon disulphide [49, 50].

It should be noted that the dmit ligand and dmit-based compounds may also be used as starting materials for a convenient preparation of organic donor molecules such as BEDT-TTF [51].

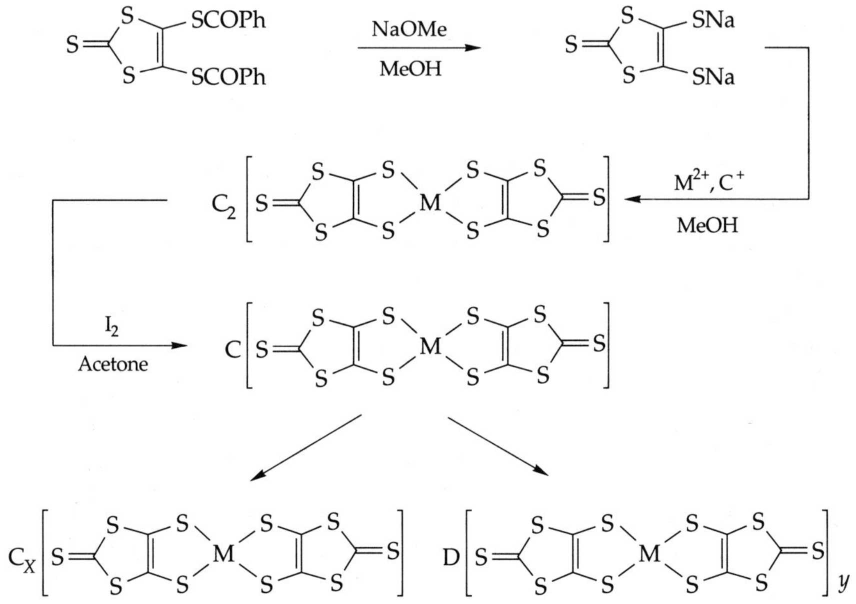

Figure 9 Synthesis of [M(dmit)$_2$] compounds

1.3.2   The [C]$_n$[M(dmit)$_2$] Precursor Complexes

A number of [C]$_n$[M(dmit)$_2$] complexes (C = H$_4$N$^+$, R$_4$N$^+$, R$_4$As$^+$, R$_4$P$^+$ and Me$_3$SO$^+$; n = 2, 1, 0; M = Ni, Pd, Pt, Fe, Cu, Au and Rh) have been prepared [52]. For example, the dianion salt [R$_4$N]$_2$[M(dmit)$_2$] can be readily obtained by treating a solution of Na$_2$dmit in methanol with the appropriate M^{2+} metal salt and the appropriate tetraalkylammonium bromide (Figure 9). The corresponding monoanionic salt [R$_4$N][M(dmit)$_2$] is obtained from the dianionic salt by iodine oxidation. These procedures have been fully described by several authors [45, 53, 54].

1.3.3   The [C]$_x$[M(dmit)$_2$] Non-integral Oxidation State Complexes

These complexes can be obtained by further chemical oxidation of the corresponding [C]$_n$[M(dmit)$_2$] precursor complex (Figure 9). However, in many cases direct aerial oxidation, or oxidation with an oxidizing agent such as iodine or bromine, leads to a mixture of partially-oxidised species with different stoichiometries, the separation of which may prove difficult; an example of this is given by the bromine oxidation of

[Bu$_4$N][Ni(dmit)$_2$] [47]. Also, direct chemical oxidation mostly yields poor quality crystals.

In fact, the main problem encountered in the synthesis of any type of molecular conductor is its insolubility in all common solvents. As it cannot be recrystallised, specific crystal growth techniques have to be used in order to obtain good quality crystals, convenient for structural and physical studies. In a few propitious circumstances, well-defined and crystallised phases can be obtained, such as for example [Ph$_4$As]$_{0.25}$[Ni(dmit)$_2$] [55], by slow interdiffusion of solutions of a precursor complex and the appropriate oxidizing agent.

The best, and most widely used method of preparation of the [C]$_x$[M(dmit)$_2$] complexes proceeds through electrochemical techniques. Moreover, the formation of conducting [C]$_x$[M(dmit)$_2$] complexes can be detected in a first step by the observation of several characteristic features in the cyclic voltammogram of the corresponding [C]$_n$[M(dmit)$_2$] precursor complexes. These features (fast increase of the oxidation peak, presence of several redissolution peaks in the reverse scan, increase of the intensity of the reduction backward peak; Figure 10) indicate a modification of the nature of the electrode during the oxidation step arising from the build-up of a conductive oxidised species [56].

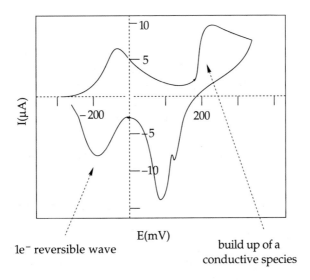

1e$^-$ reversible wave

build up of a
conductive species

Figure 10 Cyclic voltammogram of [Bu$_4$N][Ni(dmit)$_2$] in CH$_3$CN/TBAP (0.1M) on a solid Pt electrode (from [56])

It is also expected that when using electrochemical oxidation, non-conducting and conducting species may be discriminated and that the most conducting species will preferentially grow on the electrode. However, it often happens that several different conducting phases are collected on the electrode. In this case, the phase selectivity may depend on the nature, size and treatment of the electrodes, the nature and purity of the solvent, the concentration of the solution, the current density, etc. [20, 54, 57, 58], so that electrochemical crystal-growth looks like an art rather than a science.

Typically, solutions of the appropriate $[C][M(dmit)_2]$ precursor complex, and sometimes of a salt of the C cation (to increase solubility and conductivity of the solutions) in an organic solvent (acetonitrile is mostly used) are placed in both compartments, separated by a frit, of a H- or U-shaped cell (Figure 11).

The electrodes (mostly platinum) are then inserted in both compartments, and, in order to control the crystal growth rate, galvanostatic electrolysis involving current intensities in the 0.1–5 µA range (current density $\approx$ 0.4–20 µA cm^{-2}) is preferred to potentiostatic electrolysis. The crystals are grown on the anode according to the reaction shown in Equation (2):

$$n\,[C][M(dmit)_2] \longrightarrow \{[C]_x[M(dmit)_2]\}_n + n(1-x)C^+ + n(1-x)e^- \quad (2)$$

When the $[C]_n[M(dmit)_2]$ precursor complex with a given cation C (Li, Na, K or Cs, for example) is not available, $[C]_x[M(dmit)_2]$ salts of those cations can still be obtained by electrocrystallisation of acetonitrile solutions of $[Bu_4N]_n[M(dmit)_2]$ (n = 2 or 1) in the presence of a large excess of the appropriate C cation [59, 60].

A number of non-integral oxidation state complexes, $[C]_x[M(dmit)_2]$, (approximately 30) have been produced using chemical or electrochemical oxidation techniques [52]. Most of them exhibit semiconducting behaviour, or at best, metal-like behaviour at high temperatures, followed by the standard metal-to-insulator transition.

Only one superconducting phase out of this series of complexes has been characterised, namely $[Me_4N]_{0.5}[Ni(dmit)_2]$. Black plates of this compound were obtained by the electrochemical method from an acetonitrile solution of $[Me_4N][Ni(dmit)_2]$ and $[Me_4N][ClO_4]$ [61].

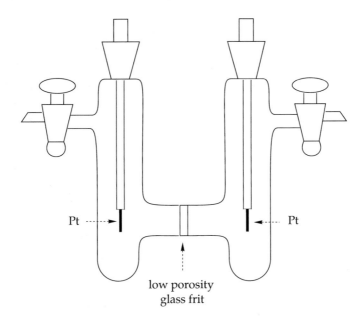

Figure 11 Electrosynthesis cell

1.3.4   The [D][M(dmit)$_2$]$_y$ Donor–acceptor Compounds

The half-wave redox potential of the electrochemical couple [M(dmit)$_2$]$^{-1}$/[M(dmit)$_2$]^0 can be estimated around 0.2 V (versus SCE) [53, 54, 62]. Following on from guideline (e) (see Section 1.2.1), the [M(dmit)$_2$] complexes can be considered as moderate acceptors and used in combination with appropriate donor molecules for preparing donor–acceptor complexes in which partial charge-transfer may be anticipated.

A number of [D][M(dmit)$_2$]$_y$ donor–acceptor compounds (approximately 25) have been obtained with various D donor molecules such as TTF, TMTSF, BEDT-TTF and related molecules, and various stoichiometry, y being mostly equal to 2 [52, 62–69].

These [D][M(dmit)$_2$]$_y$ compounds can be either chemically or electro-chemically prepared. In the first case, the [M(dmit)$_2$]^0 neutral complex being available only with difficulty, direct reaction between the neutral donor and acceptor components was not possible. Thus, the metathesis reaction between the appropriate D$^+$ and the [M(dmit)$_2$]$^{-1}$ salts was used.   Good quality crystals were generally obtained by slow interdiffusion of solutions of both salts (Figure 12).

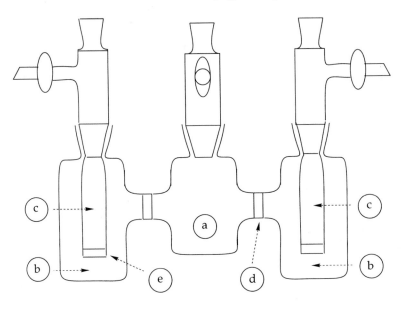

Figure 12 Diffusion cell. (a) solvent compartment; (b) saturated solutions of starting materials; (c) reservoirs containing powder of starting materials; (d) low-porosity glass frit; (e) high-porosity glass frits

The $[D][M(dmit)_2]_y$ compounds can also be prepared using electrocrystallisation techniques similar to those employed in the case of the $C_x[M(dmit)_2]$ complexes, according, in the present case, to the reactions shown in Equations (3) and (4):

$$n\,D \;\longrightarrow\; n\,D^+ \;+\; n\,e^- \tag{3}$$

$$n\,D^+ \;+\; ny[M(dmit)_2]^{-1} \;\longrightarrow\; \{D[M(dmit)_2]_y\}_n \tag{4}$$

Among the approximately twenty five $[D][M(dmit)_2]_y$ compounds, only three superconducting phases have been characterised, namely $[TTF][Ni(dmit)_2]_2$ and two phases, noted α and α' of $[TTF][Pd(dmit)_2]_2$.

1.3.4.1 *[TTF][Ni(dmit)₂]₂*

It was reported for the first time in 1981 that a 1:2 adduct between TTF and $[Ni(dmit)_2]$ could be obtained by a direct metathetical reaction between a TTF salt, $(TTF)_3(BF_4)_2$ and $[Bu_4N]_2[Ni(dmit)_2]$ or $[Bu_4N][Ni(dmit)_2]$ [63].

Good quality single crystals suitable for conductivity measurements and X-ray diffraction studies were subsequently obtained by slow interdiffusion of saturated solutions of $(TTF)_3(BF_4)_2$ and $[Bu_4N][Ni(dmit)_2]$ [64]. The experiments were carried out in a three-compartments H-tube inspired by those described by several authors [70], with a central solvent chamber and porous glass frits between compartment (see Figure 12). The concentrations of the solutions were kept close to saturation by means of additional containers filled with an excess of starting reagents, placed in the appropriate compartment and communicating with it through a glass frit. The cell was set in a thermostated dark chamber for 15 days at a temperature of 40 °C. A mixture of fine needles (majority product) and a few platelets was obtained: the needle-shaped phase was identified as $[TTF][Ni(dmit)_2]_2$ [64] and the platelets as the neutral $[Ni(dmit)_2]$ [47].

Crystals of $[TTF][Ni(dmit)_2]_2$ may also be obtained by galvanostatic electrolysis of an acetonitrile solution containing $[Bu_4N][Ni(dmit)_2]$ and a large excess of neutral TTF [66]. It is interesting to note that this electrochemical synthesis of $[TTF][Ni(dmit)_2]_2$ would have been predicted as impossible on the basis of the redox potential values of the starting components. In fact, it is expected that oxidation of a mixture of TTF and $[Bu_4N][Ni(dmit)_2]$ should yield the TTF^+ cation which could react with the $[Ni(dmit)_2]^-$ to give $[TTF][Ni(dmit)_2]_2$. Unfortunately, the first oxidation potential of isolated TTF is higher than that of $[Bu_4N][Ni(dmit)_2]$ (≈ 0.30 and 0.20 V, respectively). Therefore, one should directly obtain the partially-oxidised $[Bu_4N]_{0.29}[Ni(dmit)_2]$ [47] complex instead of the desired $[TTF][Ni(dmit)_2]_2$ compound. Cyclic voltammetry experiments have shown that, in the mixture, no oxidation process involving either TTF or $[Bu_4N][Ni(dmit)_2]$ takes place separately, which could explain the feasibility of the electrochemical synthesis of $[TTF][Ni(dmit)_2]_2$ [62].

1.3.4.2 [TTF][Pd(dmit)₂]₂

Metathesis of $(TTF)_3(BF_4)_2$ and $[Bu_4N][Pd(dmit)_2]$ in solution yields a mixture of two different kinds of crystals, needle-shaped and platelet-shaped [71]. The reaction rate seems to be a determining factor: rapid mixing of the component salts gives a majority of platelets, whereas slow interdiffusion gives a majority of needles. These two types of crystals were later characterised as the α- and α'-phases (needles) and the δphase (platelets) of [TTF][Pd(dmit)$_2$]$_2$, each phase having a different structure and physical properties [72].

1.4   QUESTIONS RELATED TO THE [M(DMIT)$_2$] SYSTEMS

From the beginning of the studies on [M(dmit)$_2$] conducting and superconducting systems to the present time, a number of questions have arisen concerning these compounds. Some of these questions are related to the guidelines that were followed for selecting these systems and the relevance of these guidelines in the case of molecular inorganic systems. Apparently, it was right to follow these guidelines, because they did lead to the discovery of molecular inorganic conductors and superconductors. However, do these systems really meet these criteria? Other questions, some of which are still today not fully answered, concern the origin and mechanisms of superconductivity in the [M(dmit)$_2$] systems. In the following section we turn our attention to those questions which can easily be answered.

1.4.1   May the [M(dmit)$_2$] Chains Alone be Conducting and Superconducting ?

[TTF][Ni(dmit)$_2$]$_2$ was the first [M(dmit)$_2$] system reported to be highly conducting [63], to retain its metal-like characteristics down to low temperatures [64], and to become superconducting under pressure [73]. One of the first legitimate questions was the following: "Which stack is responsible for the conductivity, that of TTF or of [Ni(dmit)$_2$]?"

As said above, in previously-studied metal bis(dithiolene) based all conductive systems, with the sole exception of [H$_3$O]$_{0.33}$[Li]$_{0.82}$ [Pt(S$_2$C$_2$(CN)$_2$)$_2$], only the organic donor component is responsible for high conductivity. Given this 'case-law', and given the many examples of TTF-derived conductors and superconductors, it was tempting to ascribe the conductive rôle in [TTF][Ni(dmit)$_2$]$_2$ to the TTF chain.

Later on however, several non-integral oxidation state complexes such as [Bu$_4$N]$_{0.33}$[Pd(dmit)$_2$] and [Bu$_4$N]$_{0.5}$[Pd(dmit)$_2$] [74], [Me$_4$N]$_{0.5}$ [Pd(dmit)$_2$] [75], [Et$_2$Me$_2$N]$_{0.5}$[Ni(dmit)$_2$] [76], or [HMe$_3$N]$_{0.5}$[Ni(dmit)$_2$] [77] were obtained which exhibited high conductivity and metal-like behaviour. One of these complexes, [Me$_4$N]$_{0.5}$[Ni(dmit)$_2$], even becomes superconducting under pressure [78]. In all these complexes the conducting behaviour was clearly not due to the R$_4$N$^+$ cations but to the [M(dmit)$_2$] chains.

It remains that the conductivity of the $[D][M(dmit)_2]_y$ donor–acceptor compounds may be due to both the donor chains and the $[M(dmit)_2]$ chains. For example, thermopower experiments have shown that in $[TTF][Ni(dmit)_2]_2$ electrons (arising from the $[Ni(dmit)_2]$ chains) are the predominant carriers at high temperature, whereas holes (arising from the TTF and the $[Ni(dmit)_2]$ chains) are the predominant carriers at low temperatures [79]. Moreover, solid state ^1H and ^{13}C NMR studies on this compound have shown that both TTF and $[Ni(dmit)_2]$ chains remain metallic down to low temperatures [80, 81].

1.4.2   What is the Oxidation State in these Systems ?

In the so called mixed-valence or non-integral oxidation state compounds as well as in the donor–acceptor compounds, the oxidation state is usually inferred from the stoichiometry.   This straightforward deductive reasoning may be deceiving in some cases.

For example, the $(TTF)_3(BF_4)_2$ compound [82], the formula of which could be written $TTF(BF_4)_{0.67}$, has a 'non-stoichiometric look' and one might infer from its formula that the TTF radical cation bears a formal partial charge of +0.67.   However, a detailed X-ray structural determination of this compound and a careful analysis of the stacking distances and bond lengths have shown that two definitely different TTF entities are found, TTF^+ cations and TTF^0 neutral molecules [83]. Therefore, $TTF(BF_4)_{0.67}$ is not a mixed valence compound but an ionic salt, and consequently an insulator, the formula of which could be more clearly written $(TTF^+)_2(TTF^0)(BF_4)_2$.

Likewise, the 1:1 stoichiometry of the organic metal TTF.TCNQ does not give any hint of the partial charge transfer between TTF and TCNQ which had to be experimentally determined, by neutron scattering techniques [84], as equal to 0.59 electrons and can be inferred from bond length comparisons in various TTF.TCNQ-like compounds [85, 86].

1.4.2.1   *Non-integral Oxidation State Complexes*

In the case of the $[C]_x[M(dmit)_2]$ complexes the stoichiometry, i.e. the value of x, can be roughly determined by elemental analysis, although the possible presence of solvent molecules of crystallisation might warp this determination.   Cyclic voltammetry on solid samples of $[C]_x[M(dmit)_2]$ using a carbon paste electrode allows direct determination of the value of x by measuring the ratio of the area of the $[C]_x[M(dmit)_2] \rightarrow [C][M(dmit)_2]$

reduction peak (C1) (formally $1 - x$ electron transfer) to the area of the $[C][M(dmit)_2] \rightarrow [C]_2[M(dmit)_2]$ reduction peak (C2) area (one-electron transfer internal reference; Figure 13) [87]. However, these measurements do not tell us whether all $[M(dmit)_2]$ species are in the same non-integral oxidation state.

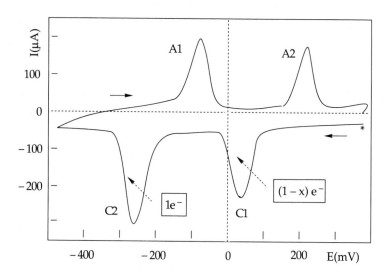

Figure 13 Cyclic voltammogram of solid $[Bu_4N]_{0.29}[Ni(dmit)_2]$ using a carbon paste electrode; * = starting point (from [87])

Even X-ray structural determinations do not always give a clear answer to this question. In contrast to the situation encountered in the TTF.TCNQ-like compounds [85], bond length comparisons in the $[M(dmit)_2]^{n+}$ units of various compounds with different oxidation states ($n = 0, x, 1, 2$) show no clear trends within the limits of uncertainties in the measurements. This is partly due to insufficient accuracy in the various structure determinations. There are two main reasons for this lack of accuracy: (i) crystals are usually of poor crystallographic quality; (ii) the anisotropy of the structure is reflected in the reciprocal space, leading to a large number of very weak reflections in the diffraction pattern. Therefore, an averaging over idealised molecular geometry (*mmm* symmetry) is necessary [47]. However, and in spite of possible variations in the stacking distances observed in some cases, all the $[M(dmit)_2]$ species in the $[C]_x[M(dmit)_2]$ complexes can be considered as nearly equivalent (as far as the molecular geometry is considered) and therefore can be considered as being in nearly the same non-integral oxidation state which can be, in this case, rather safely derived from the value of x.

For example, the structure of the superconducting $[Me_4N]_{0.5}[Ni(dmit)_2]$ complex consists of sheets of 'dimerised' $[Ni(dmit)_2]$ pairs stacked along (*110*), separated by slabs of Me_4N^+ cations [61]. All $[Ni(dmit)_2]$ units being crystallographically equivalent, the formal charge on each $[Ni(dmit)_2]$ unit in $[Me_4N]_{0.5}[Ni(dmit)_2]$ can be reasonably estimated at -0.5.

1.4.2.2 Donor–acceptor Compounds

In the $[TTF][M(dmit)_2]_2$ (M = Ni, Pd) superconducting compounds also, the charge transfer can be only indirectly estimated (Table 1).

Table 1   Central C=C bond length in various TTF-derived compounds

Compound	Charge	$d(C=C)$ (Å)	Reference
TTF	0	1.349	[92]
TTF.TCNQ	0.6	1.372	[18]
$[TTF][Ni(dmit)_2]_2$	0.8	1.35	[66]
α'-$[TTF][Pd(dmit)_2]_2$	0.76	1.39	[72]
$(TTF)_3(BF_4)_2$	1	1.393	[83]
$(TTF)ClO_4$	1	1.404	[90]

In the case of the α'-$[TTF][Pd(dmit)_2]_2$ phase, a charge transfer close to unity (≈ 0.76) has been suggested on the basis of diffuse X-ray scattering investigations [88] and band structure calculations [89]. This value is in agreement with the central C=C bond length of TTF in this compound, 1.39 Å, obtained from an accurately resolved structure ($R = 0.03$) [72], which typically corresponds to the values 1.393 and 1.404 Å, found for the TTF^+ cation in $(TTF)_3(BF_4)_2$ [83] and $(TTF)ClO_4$ [90], respectively.

In the case of $[TTF][Ni(dmit)_2]_2$, a similar joint analysis of diffuse X-ray scattering results [91] and band structure calculations [89] also suggests a charge-transfer close to one (0.8). This estimation is in agreement with thermopower measurements which show that holes are the major carriers at low temperature [79] but is in contradiction with the central C=C bond length of TTF in this compound, 1.33 Å [66], which is even shorter than in neutral TTF, 1.34 Å [92]. However, this discrepancy may be explained by the rather poor accuracy of the structure resolution of [TTF][Ni(dmit)$_2$]$_2$ (R = 0.097 [66]).

In conclusion, in both α'-[TTF][Pd(dmit)$_2$]$_2$ and [TTF][Ni(dmit)$_2$]$_2$, all [M(dmit)$_2$] species being identical within a regular stack, their formal charge can be reasonably estimated at a value close to -0.4.

1.4.3   Are the [M(dmit)$_2$] Systems Polymorphic ?

In the case of the organic (BEDT-TTF)$_2$X systems it is known that as many as five or more different crystallographic phases, with differing electrical properties, may form in one growth cycle [17, 93]. The same trend is observed in the [M(dmit)$_2$] systems, although more pronounced when the metal is palladium.

Only one crystal form has been obtained for the [Me$_4$N]$_{0.5}$[Ni(dmit)$_2$] [61] and [TTF][Ni(dmit)$_2$]$_2$ [64] superconducting compounds. However, this is not a fixed rule for all the [Ni(dmit)$_2$]-based compounds: for example, two phases of [Me$_2$Et$_2$N]$_{0.5}$[Ni(dmit)$_2$] have been characterised [76].

In the case of the [Pd(dmit)$_2$]-based compounds, several crystal phases are often obtained in the same growth experiment. Sample preparation of [TTF][Pd(dmit)$_2$]$_2$ by slow interdiffusion of saturated solutions of (TTF)$_3$(BF$_4$)$_2$ and [Bu$_4$N][Pd(dmit)$_2$] yields mainly black shiny needles, noted the α-phase [66]. In some experiments, needles of another phase, noted the α'-phase, have been isolated. Moreover, a few platelets, noted the δ-phase, could also be sorted out from the predominant needle-shaped crystals. As proven by X-ray diffraction structure determination, the three phases have the same [TTF][Pd(dmit)$_2$]$_2$ stoichiometry (Figure 14).

The α-[TTF][Pd(dmit)$_2$]$_2$ phase is isostructural with the analogous [TTF][Ni(dmit)$_2$]$_2$ compound. Its structure consists of segregated regular stacks of TTF and [Pd(dmit)$_2$] molecules [66, 72]. At room temperature, the structure of the α'-[TTF][Pd(dmit)$_2$]$_2$ phase is strictly identical to that of the α-phase.

The only way to distinguish between these two phases is to study their temperature-dependent behaviour: when cooling, the crystals of the α'-phase retain their initial monoclinic symmetry down to 100 K, whereas the crystals of the α-phase undergo a structural transition from monoclinic to triclinic symmetry [72]. This transition leads to the triclinic low-temperature phase, called the β-phase, and induces the stabilisation of the triclinic symmetry which is retained upon warming back to room temperature where a high-temperature crystalline modification of the β-phase, called the γ-phase is finally obtained. This irreversible

$\alpha \to \beta \leftrightarrow \gamma$ structural transition is reflected in the electrical behaviour (see Section 1.5.2).

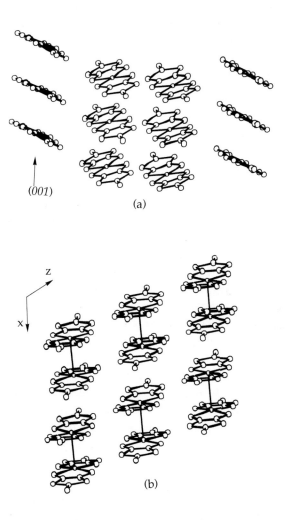

Figure 14 The stacking motifs for (a) [TTF][Ni(dmit)$_2$]$_2$, α- and α'-[TTF][Pd(dmit)$_2$]$_2$ and (b) δ-[TTF][Pd(dmit)$_2$]$_2$ (from [66, 72])

The structure of the platelet-shaped crystals of the δ-[TTF][Pd(dmit)$_2$]$_2$ phase is entirely different from that of the α- and α'-phases. It consists of stacks of real [Pd(dmit)$_2$]$_2$, with an intra-dimer Pd–Pd bond length of 3.11 Å and an inter-dimer Pd···Pd distance of 5.31 Å [72]. These stacks are arranged in layers alternating with layers containing unstacked TTF molecules.

This tendency of the palladium compounds to polymorphism makes it more difficult to study their physical properties and the analysis of their structure-properties relationships. The situation may be even more complicated when several phases with different stoichiometries are obtained in the same run. For example, at least two phases of $[Bu_4N]_x[Pd(dmit)_2]$ with $x = 0.33$ and 0.5 have been isolated [74].

1.4.4   Is Regular Stacking a Prerequisite to Metal-like Conductivity or Superconductivity ?

This question refers to guideline (e) (see Section 1.2.1) according to which, in order to obtain a molecular metallic, and eventually superconducting compound, one should select molecules that will stack one-dimensionally with a constant spacing. This was believed to avoid the splitting of the conduction band, but, ironically, it was afterwards recognised that such a Peierls-like distortion [9] is inherent in true 1-D systems. In order to avoid the ubiquitous metal-to-insulator transition encountered on cooling the 1-D systems, it was then suggested that the dimensionality be increased [43], but the overall stacking arrangement was maintained in the structure of the first-discovered molecular organic superconductors [19–21]. In the $(TMTSF)_2X$ series, slight alternations in stacking distances could be considered as meaningless [17, 43].

In $[TTF][Ni(dmit)_2]_2$ and α- and α'-$[TTF][Pd(dmit)_2]_2$, the TTF and $[M(dmit)_2]$ units are uniformly spaced within their respective stacks (3.65 and 3.55 Å, respectively for M = Ni [66]; 3.52 and 3.44 Å, respectively, for M = Pd [72]). In $[Me_4N]_{0.5}[Ni(dmit)_2]$ a slight 'dimerisation' is observed within the $[Ni(dmit)_2]$ stacks with alternating distances of 3.53 and 3.58 Å [61].

Conductivity studies of δ-$[TTF][Pd(dmit)_2]_2$ under pressure have not yet been carried out, and therefore we do not know whether this phase might become superconducting under pressure. However, it is interesting to note that a relatively strong dimerisation is observed in the $[Pd(dmit)_2]$ stacks of this phase with alternating Pd···Pd distances of 3.11 and 5.31 Å [72]. Nevertheless, this phase exhibits metal-like properties down to 120 K, showing that a regular stacking is not absolutely necessary to obtain a molecular metal (see Figure 14). The same conclusion could have been drawn from the study of metallic dimerised $[Bu_4N]_x[Pd(dmit)_2]$ ($x = 0.33$ and 0.5) compounds [74].

After all, is a stacking arrangement actually necessary for obtaining a molecular superconductor? Certainly not if one considers the structures of the κ-(BEDT-TTF)$_2$I$_3$ [93], κ-[BEDT-TTF]$_2$[Cu(NCS)$_2$] [22] or κ-[BEDT-TTF]$_2$[Cu{N(CN)$_2$}Cl] [24] organic superconductors, which exhibit the highest observed critical temperatures. In these κ-phases, 'face-to-face dimers' of BEDT-TTF molecules are arranged in 2-D sheets (Figure 15).

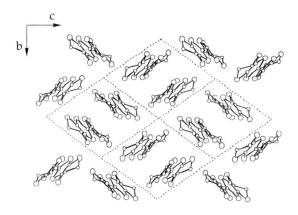

Figure 15 The structure of κ-[(BEDT-TTF)]$_2$[Cu(NCS)$_2$] projected along the *a*-axis. The 'dimeric' units are encompassed by dotted lines (from [22])

In each 'dimer', the planes of the BEDT-TTF molecules are nearly parallel. The orientation of adjacent 'dimers' is almost perpendicular to each other. At this point, it was not an overstatement to say that the 'stacking criterion' should be audited.

1.4.5 Is Metal-like Behaviour down to Low Temperatures a Prerequisite to Superconductivity ?

After good quality crystals of [TTF][Ni(dmit)$_2$]$_2$ were obtained, it was soon discovered that, when cooling this compound, it does not undergo a metal-to-insulator transition, but retains its metal-like characteristics down to low temperatures [64]. This behaviour, and its resemblance to that of the (TMTSF)$_2$X superconductors, urged us to check whether [TTF][Ni(dmit)$_2$]$_2$ might become superconducting at lower temperatures or higher pressures. It does [73]!

However, neither of the α- and α'-[TTF][Pd(dmit)$_2$]$_2$ phases, still metallic at high temperatures, remain metallic down to low temperatures

[72], and yet both phases undergo a transition under pressure to a superconducting state [94].

Likewise, a metallic behaviour is observed when cooling [Me$_4$N]$_{0.5}$[Ni(dmit)$_2$] down to 100 K and below this temperature a transition to an insulating state occurs [61], but this compound is also superconducting under pressure [78].

The effect of pressure is not even absolutely necessary to suppress a metal-to-insulator transition in molecular superconductors. For example, κ-[BEDT-TTF]$_2$[Cu(NCS)$_2$] undergoes a metal-to-insulator transition at high temperatures, followed by an insulator-to-metal transition and, finally, by a metal-to-superconductor transition; this behaviour is believed to be related to some disorder in the crystals [22].

In conclusion, metal-like behaviour down to low temperatures at ambient pressure is not a prerequisite to superconductivity. Worse than that, a transition from, not a metallic state, but a *semiconducting* state, to a superconducting state has been observed under pressure in the case of [TTF][Ni(dmit)$_2$]$_2$ [95, 96].

1.5   DIMENSIONALITY AND THE ORIGIN OF SUPER-CONDUCTIVITY IN [M(DMIT)$_2$]-BASED SUPERCONDUCTORS

In the previous section, we screened a number of questions relating to the conducting and superconducting [M(dmit)$_2$]-based systems. We know that in these compounds the [M(dmit)$_2$] chains may, alone, be conducting and superconducting and that the charge on the [M(dmit)$_2$] units is fractional. We also found out that some characteristic features observed in purely 1-D conductors, such as the stacking and the regularity of the stacking or a metal-like conductivity down to low temperatures, were not prerequisites to superconductivity. Does it mean that the [M(dmit)$_2$]-based superconducting systems are not 1-D and that the related conduction mechanism is different?

1.5.1   Structure and Dimensionality of [M(dmit)$_2$]-based Superconductors

As mentioned above (see Section 1.4.3 and Figure 14), [TTF][Ni(dmit)$_2$]$_2$ and α- and α'-[TTF][Pd(dmit)$_2$]$_2$ are isostructural and their structure can be described as segregated stacks along the *(010)* direction of the TTF and [M(dmit)$_2$] molecules. The spacing between the stack sites, i.e. between the metal atoms, is constant at 3.73 Å in the nickel compound [64, 66] and 3.60 Å in the palladium compound [72]. From this viewpoint, these three phases could be considered as strictly 1-D with uniform stacking.

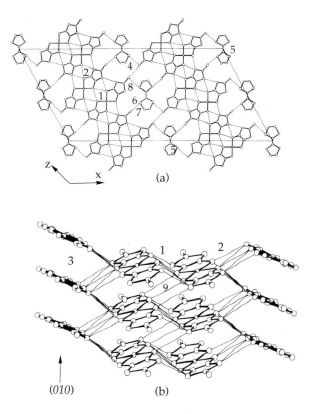

Figure 16 3-D S···S short contacts (thin lines) in the structure of [TTF][Ni(dmit)$_2$]$_2$. (a) projection onto the {010} plane; (b) parallel view along the (010) direction of the [Ni(dmit)$_2$] slabs. The different types of interactions numbered from 1 to 9 are reported in Table 2 (from [89])

On the other hand, an examination of the interatomic distances between molecules in adjacent stacks (Figure 16) reveals a number of sulphur-sulphur contacts which are appreciably shorter than the van der Waals' separation of 3.70 Å [97]. Such S···S contacts (interactions of types 1, 2, 4, and 9 in Figure 16) as short as 3.39 Å in α- and α'-[TTF][Pd(dmit)$_2$]$_2$ and 3.45 Å in [TTF][Ni(dmit)$_2$]$_2$ are observed between [M(dmit)$_2$] units in adjacent stacks (arranged along (001)) involving molecules at different levels along (001). Consequently, a two-dimensional array of closely-spaced [M(dmit)$_2$] molecules is formed in the {100} plane. This high degree of molecular connectivity in this plane suggests the existence of substantial electronic coupling along both the *b* and *c*-direction. It is interesting to note that the rather large interplanar separation within the TTF and the [M(dmit)$_2$] stacks results in *intrastack* S···S distances larger than the van der Waals' separation (interactions of type 5 and 3, respectively, in Figure 16).

Short S···S contacts are also found between the terminal sulphur atom of the [M(dmit)$_2$] units and sulphur atoms of the TTF molecules (interactions of types 6, 7 and 8 in Figure 16), possibly extending the range of electronic interactions into the third direction. Thus, on the basis of these structural observations, we would anticipate appreciable deviation from the usual '1-D metal' description, and indeed [TTF][Ni(dmit)$_2$]$_2$ and α- and α'-[TTF][Pd(dmit)$_2$]$_2$ appear to have a *quasi-three-dimensional* network of intermolecular S···S interactions.

In [Me$_4$N]$_{0.5}$[Ni(dmit)$_2$] such S···S interactions are not possible between sulphur atoms of TTF and [Ni(dmit)$_2$] molecules. Nevertheless, the structure of this compound (Figure 17) can be described as *two-dimensional*. Indeed, its structure does consist of sheets of 'dimerised' [Ni(dmit)$_2$] pairs stacked along the *(110)* direction, separated by slabs of Me$_4$N$^+$ cations, but intermolecular S···S contacts as short as 3.49 Å are observed along the *b*-direction in the [Ni(dmit)$_2$] sheets [61].

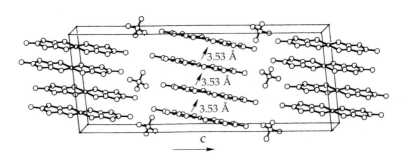

Figure 17 Structure of [Me$_4$N]$_{0.5}$[Ni(dmit)$_2$] (from [61])

The observation of a high degree of molecular connectivity as reflected by short S···S distances may not necessarily mean that these contacts correspond to effective electronic interactions (orbital symmetry should obviously be also taken into account), and that high conductivity is observed in several directions of the crystals.

In some cases, the size and morphology (platelets) of the crystals is such that the anisotropy in the conductivities may be determined by using standard two- or four-contact resistance measurements, or by the

Montgomery method [98]. It is interesting to note that a low anisotropy in the conductivities has been observed for the semiconducting $[Bu_4N]_{0.29}[Ni(dmit)_2]$ [47] and $[Et_4N]_{0.5}[Ni(dmit)_2]$ [99] compounds in which intermolecular S···S contacts have also been observed. In $[Me_4N]_{0.5}[Ni(dmit)_2]$ the anisotropy in the conductivities is small in the crystal plane, but the resistivity along the c-direction is 10^3 larger than along the in-plane directions [61]. This seems to be quite consistent with the two-dimensional structure described above.

The small size and needle-like morphology of $[TTF][Ni(dmit)_2]_2$, α- and α'-$[TTF][Pd(dmit)_2]_2$ does not allow the anisotropy in the conductivities to be assessed in these compounds, neither by linear four-contact resistance measurements or by the Montgomery method, nor, to date, by other techniques such as microwave conductivity measurements.

The magnetoresistance of $[TTF][Ni(dmit)_2]_2$ (i.e. the conductivity variation induced when applying a magnetic field) has been measured as a function of the orientation of the crystal with respect to the direction of the field [100]. These measurements reveal quite a low anisotropy compared to that observed in $(TMTSF)_2X$ compounds [101]. On the other hand, the ^1H relaxation data at ambient pressure indicate that, as far as the TTF chains are concerned, their electronic properties are typically 1-D [80].

The latter result is in agreement with the theoretical calculation of the intermolecular overlap integrals of the LUMOs of $[TTF][Ni(dmit)_2]_2$ which indicates that despite the close side-by-side S···S contacts, the transverse interstack integrals are much lower than the intrastack integrals [61, 102]. This has been confirmed by a more detailed calculation of the β_{ij} interaction energies involving both the LUMOs of $[Ni(dmit)_2]$ and the HOMOs of TTF and $[Ni(dmit)_2]$ [89]. The β_{ij} values gathered in Table 2 clearly show that the intrastack interactions for TTF (type 5) and $[Ni(dmit)_2]$ (type 3) are comparable to those obtained for most organic metals [103], but much larger than all the interstack interactions.

In conclusion, neither experimental nor theoretical results allow an unambiguous decision concerning the dimensionality of $[M(dmit)_2]$ systems — either strictly 1-D or quasi 1-D with small interstack interactions. As will be seen later, such apparently negligible 2-D or 3-D interactions may have critical effects on the properties of these systems.

Table 2   Interaction energies (β_{ij}) for [TTF][Ni(dmit)$_2$]$_2$ (from [89])

Interaction	Type [*]	Interaction Energy (eV)
LUMO$_{(Ni)}$-LUMO$_{(Ni)}$ [**]	1	0.0362
	2	0.0072
	3	0.2785
	4	−0.0063
	9	−0.0015
HOMO$_{(TTF)}$-HOMO$_{(TTF)}$	5	0.2964
HOMO$_{(TTF)}$-LUMO$_{(Ni)}$	6	0.0176
	7	0.0092
	8	−0.0038
HOMO$_{(Ni)}$-HOMO$_{(Ni)}$	1	−0.1078
	2	0.0229
	3	0.3402
	4	−0.0044
	9	0.0206
HOMO$_{(TTF)}$-HOMO$_{(Ni)}$	6	−0.0279
	7	−0.0146
	8	−0.0026

[*] see Figure 16 for definition; [**] Ni stands for [Ni(dmit)$_2$]

1.5.2   Electrical Properties of [M(dmit)$_2$]-based Superconductors

Four-probe resistance measurements were carried out on the four [M(dmit)$_2$]-based superconductors.

1.5.2.1 Ambient Pressure Conductivity

The ambient pressure, room temperature electrical conductivity of [TTF][Ni(dmit)$_2$]$_2$, measured along the needle axis is quite high: 300 S cm^{-1} (the needle axis is parallel to the *(010)* direction) [64, 66]. This compound exhibits a metal-like conductivity behaviour down to 4 K and the ratio of the conductivities at 4 and 300 K, $\sigma_{4\,K}/\sigma_{300\,K}$, is $\approx$ 500 (Figure 18).

Over the whole 300–4 K temperature range, the observed data can be fitted by a $\sigma \propto T^{-1.65}$ power law [66]. This behaviour is completely reversible on warming the sample back to 300 K. A weak anomaly is observed in the resistivity curve at 10 K, and finally maximum conductivity is reached below 4 K at about 1.5-3 K, depending on the sample [96, 100]. Radiofrequency penetration depth [104] as well as SQUID magnetisation measurements [73], gave no indication of a superconducting state above 50 mK.

The room-temperature conductivity of α- or α'-[TTF][Pd(dmit)$_2$]$_2$, measured along the *(010)* needle axis is 750 S cm^{-1}, even higher than the value observed for the nickel analogue compound [66]. As noted above (see Section 1.4.3), the only way to distinguish between the α- and α'- phases of [TTF][Pd(dmit)$_2$]$_2$, which are structurally identical at room temperature, is to study their temperature-dependent behaviour.

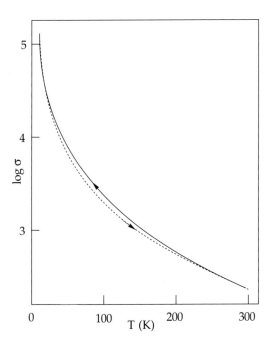

Figure 18   Ambient pressure conductivity (σ) of [TTF][Ni(dmit)$_2$]$_2$ as a function of temperature (from [64])

On cooling, α-[TTF][Pd(dmit)$_2$]$_2$ undergoes a metal-to-insulator transition at ≈ 220 K (Figure 19); below this temperature, and on subsequent warming and cooling cycles, the sample behaves as a semiconductor and the overall behaviour shows a noticeable temperature hysteresis. By contrast, α'-[TTF][Pd(dmit)$_2$]$_2$ also undergoes a metal-to-insulator transition, but does not exhibit any hysteresis, and this transition is completely reversible.

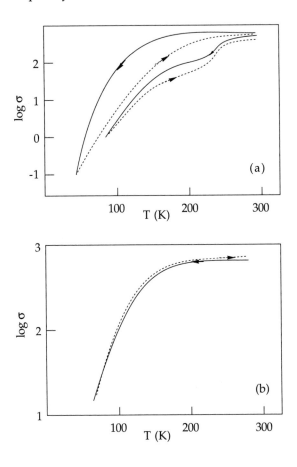

Figure 19 Temperature-dependent conductivity of (a) α-[TTF][Pd(dmit)$_2$]$_2$ and (b) α'-[TTF][Pd(dmit)$_2$]$_2$ (from [72])

The room-temperature conductivity of [Me$_4$N]$_{0.5}$[Ni(dmit)$_2$], measured along two directions of the {001} plane is ≈ 50 S cm^{-1} [61], lower than that of the [TTF][M(dmit)$_2$]$_2$ phases. The temperature-dependent conductivity behaviour is sample dependent (Figure 20) [61, 78].

Most crystals exhibit a metal-like behaviour down to ≈ 100 K. Below this temperature, they undergo a sharp metal-to-insulator transition with a large hysteresis on subsequent warming, but this hysteresis tends to fade upon repeated cooling and warming temperature cycles [61, 78]. A few samples do not exhibit this hysteresis-accompanied transition, but all crystals show a thermoactivated behaviour below $\approx$ 10 K.

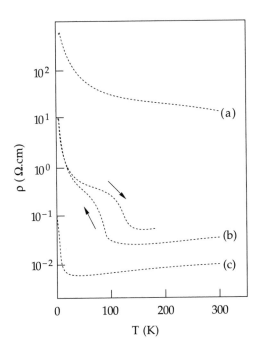

Figure 20 Temperature-dependent resistivity of [Me$_4$N]$_{0.5}$[Ni(dmit)$_2$]. (a) resistivity perpendicular to the *ab*-conducting plane; (b) resistivity in the conducting plane of most crystals; (c) resistivity in the conducting plane of several crystals (from [78])

1.5.2.2 *Effect of Pressure*

The room-temperature conductivity of both [TTF][Ni(dmit)$_2$]$_2$ and α'-[TTF][Pd(dmit)$_2$]$_2$ increases with pressure [96]. A larger compressibility of the nickel compound lattice compared to that of the palladium compound is observed in agreement with the shrinking of the unit cell of α'-[TTF][Pd(dmit)$_2$]$_2$ [72, 94].

Under pressure, [TTF][Ni(dmit)$_2$]$_2$ undergoes a complete transition to a superconducting state (Figure 21) [73].

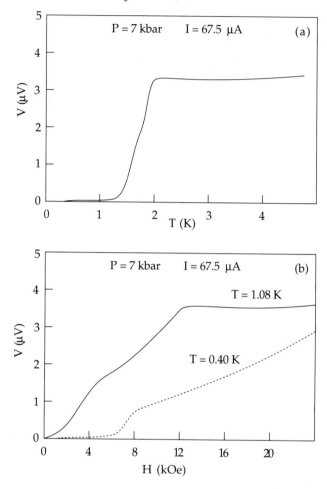

Figure 21 (a) Superconducting transition of [TTF][Ni(dmit)$_2$]$_2$; (b) Magnetic field dependence of the resistance of [TTF][Ni(dmit)$_2$]$_2$ (from [73])

For example, at 7 kbar the superconducting critical temperature, T_c, is 1.62 K. At 7 kbar and 400 mK, the critical current density which restores the normal metallic state is larger than 12 A cm^{-2}, which precludes any filamentary-type superconductivity and means that the superconducting state is established in the bulk. When applying a magnetic field parallel to the c-axis of the crystals, the critical field which brings the sample back to its normal metallic state at 7 kbar is larger than 1.3 and 2.5 Tesla at 1.08 and 0.4 K, respectively.

The effect of pressure on the superconducting transition temperature of [TTF][Ni(dmit)$_2$]$_2$ has been determined by low-field EPR (Figure 22) [104] and resistivity measurements [95, 96]: T_c is found to increase monotonically with increasing pressure.

This behaviour, opposite to that observed in both the (TMTSF)$_2$X and (BEDT-TTF)$_2$X families of organic superconductors [105, 106], is quite unusual for molecular superconductors and has been observed for the first time in such molecular compounds in [TTF][Ni(dmit)$_2$]$_2$.

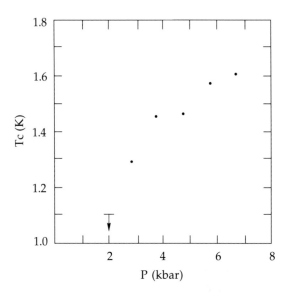

Figure 22 Pressure-dependency of the superconducting critical temperature, T_c, of [TTF][Ni(dmit)$_2$]$_2$ as determined by low-field EPR (from [104])

In the case of α'-[TTF][Pd(dmit)$_2$]$_2$, superconducting transitions were observed at remarkably high temperatures, but at pressures above 16 kbar. In fact, the superconducting transition is really complete at even higher pressure (5.93 K at 24.2 kbar; Figure 23) [94, 107].

Superconducting properties of α'-[TTF][Pd(dmit)$_2$]$_2$ were confirmed by applying a magnetic field parallel to the *a*-axis and determining the critical fields [94]. By contrast with [TTF][Ni(dmit)$_2$]$_2$, but similarly to what was observed in most molecular superconductors, the value of T_c of α'-[TTF][Pd(dmit)$_2$]$_2$ decreases with increasing pressure above 20 kbar.

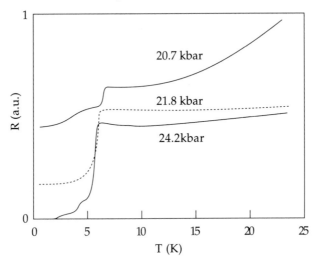

Figure 23 Superconducting transition of α'-[TTF][Pd(dmit)$_2$]$_2$ (from [94])

In the case of α-[TTF][Pd(dmit)$_2$]$_2$, superconducting transitions have also been observed at high pressures, but at lower temperatures (1.7 K at 22 kbar, for example) than for the α'-[TTF][Pd(dmit)$_2$]$_2$ phase [94]. However, for a change, the value of T_c of the α-phase increases with increasing pressure.

Hydrostatic pressure suppresses the anomalies observed for [Me$_4$N]$_{0.5}$[Ni(dmit)$_2$] at ambient pressure (see Section 1.5.2.1) and superconducting transitions are observed at relatively high temperatures and moderate pressures: 5 K at 7 kbar (Figure 24) [78].

The magnetic field dependence of the conductivity also confirms the superconducting state of this compound under pressure [78]. As for [TTF][Ni(dmit)$_2$]$_2$, the value of T_c of [Me$_4$N]$_{0.5}$[Ni(dmit)$_2$] increases with increasing pressure.

Some anomalies in the temperature-pressure dependent conductivity behaviour and the pressure effect on T_c, sometimes positive, sometimes negative, observed in these [M(dmit)$_2$]-based superconductors, prompted physicists to determine their complete temperature/pressure phase diagram.

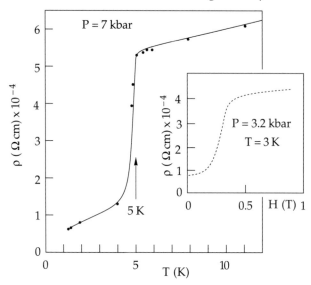

Figure 24 Superconducting transition of $[Me_4N]_{0.5}[Ni(dmit)_2]$ and its magnetic field effect (from [78])

1.5.3 Phase Diagrams of α'-[TTF][Pd(dmit)$_2$]$_2$ and [TTF][Ni(dmit)$_2$]$_2$

The purpose of a temperature/pressure phase diagram is to determine the temperature-pressure domains in which a system is in its various possible states, insulating, metallic, or surperconducting.

Usually, such a determination proceeds as follows: the conductivity measurements are carried out in a 'pressure bomb' filled with a fluid (a mixture of alkanes, for example) used as pressure-transmitting and sample-thermalising medium (experimental details can be found in [94, 96]). Once a given hydrostatic pressure is established, the sample resistance is measured as a function of the temperature. The observation of a minimum in the resistivity-temperature curve gives notice of a gradual condensation of the carriers. The temperature at which this minimum is observed is noted T_{min}. In some cases, it is possible to spot an inflection point in the resistivity-temperature curve at a temperature, noted T_{inf}, below T_{min}. It is generally accepted that the condensation of the carriers occurs gradually between T_{min} and T_{inf}, and that T_{inf} actually is the temperature below which the metal-to-insulator transition takes place. As mentioned above, the temperature of a possible superconducting transition (zero resistance) is noted T_c. These measurements, and the determination of T_{min}, T_{inf} and T_c, are repeated for different pressures. The resulting temperature/pressure phase diagram is obtained by plotting the values of T_{min}, T_{inf} and T_c as functions of the pressure. The curves joining these points are boundaries between the different domains of existence of the insulating, metallic, or superconducting states of the studied system.

Given the extremely large number of measurements needed for obtaining such a phase diagram, only those of α'-[TTF][Pd(dmit)$_2$]$_2$ and [TTF][Ni(dmit)$_2$]$_2$, have been determined to date.

1.5.3.1 Phase Diagram of α'-[TTF][Pd(dmit)₂]₂

The phase diagram of α'-[TTF][Pd(dmit)$_2$]$_2$ consists of three domains (Figure 25).

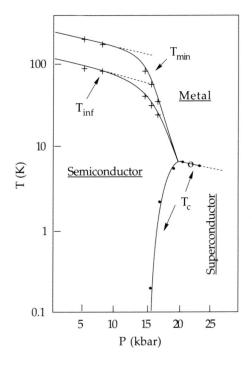

Figure 25 Temperature/pressure phase diagram of α'-[TTF][Pd(dmit)₂]₂
(from [94])

At high temperatures, the compound is in a metallic state. At low temperatures and high pressures it is superconducting. When decreasing the temperature at high pressures, the compound undergoes a metal-to-superconductor transition, the temperature of which, T_c, decreases with increasing pressure. At low temperatures and low pressures the compound is in an insulating state. When decreasing the temperature at low pressures, the compound undergoes a metal-to-insulator transition below T_{inf}.

It is interesting to note that the same exponential pressure dependence, $d(\log T_i)/dP = -3.9 \, 10^{-2} \, kbar^{-1}$ for $T_i = T_{min}$, T_{inf} and T_c, governs the carrier condensation between 2 and 9 kbar as well as the superconducting regime above 20 kbar. This means that the same basic physical process leading to the insulating and superconducting states is at work at low and high pressures, respectively [94]. From this result, and from the analysis of other features in the temperature/pressure-dependent conductivity behaviour, it has been inferred that the insulating state of α'-[TTF][Pd(dmit)$_2$]$_2$ cannot be related to 1-D instabilities such as Spin Density Wave (SDW) [108] or Spin Peierls (SP) [109] ground states which had been observed in organic superconductors such as (TMTSF)$_2$PF$_6$ and (TMTTF)$_2$PF$_6$, respectively [105]. In fact, the insulating state in α'-[TTF][Pd(dmit)$_2$]$_2$ is a Charge Density Wave (CDW) state [110, 111], as experimentally confirmed by diffuse X-ray scattering (see Section 1.5.4.2).

1.5.3.2 *Phase Diagram of [TTF][Ni(dmit)$_2$]$_2$*

The diagram of the nickel compound (Figure 26) is more complicated than that of α'-[TTF][Pd(dmit)$_2$]$_2$. In the nickel compound the superconducting critical temperature increases with pressure, and as said before, this behaviour had never been previously observed for any molecular superconductor.

There are now four domains instead of three as in the palladium compound: a metallic state, a low-pressure insulating state, a superconducting state, and an additional high pressure insulating state. Indeed, when decreasing the temperature at high pressures, the compound undergoes as a first step a metal-to-insulator transition, then an insulator-to-superconductor transition [96].

Other striking features are observed at low pressures. It must be recalled that, at ambient pressure, [TTF][Ni(dmit)$_2$]$_2$ remains metallic down to 1.5 K. When going from ambient pressure to pressures of only a few hundred bar, a metal-to-insulator transition is observed and the characteristic critical temperatures, T_{min} and T_{inf}, increase dramatically within that very narrow pressure range. Moreover, between 1 kbar and ≈ 4 kbar, T_{min} and T_{inf} exhibit peculiar oscillations. Finally, at very low temperature and in a narrow pressure range around 5.3 kbar, the superconducting ground state is 're-entrant' into the low-pressure insulating state [95, 96].

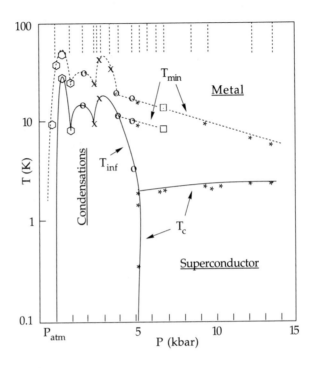

Figure 26 Temperature/pressure phase diagram of [TTF][Ni(dmit)$_2$]$_2$ (from [96])

1.5.4. Charge-density-wave Condensations in α'-[TTF][Pd(dmit)$_2$]$_2$ and [TTF][Ni(dmit)$_2$]$_2$

As mentioned above, some unexpected features in the temperature-pressure-dependent electrical behaviour of α'-[TTF][Pd(dmit)$_2$]$_2$ allow us to suppose that the insulating state in this compound was not an SP or SDW ground state. Diffuse X-ray scattering experiments provide evidence that the insulating state is a CDW state in α'-[TTF][Pd(dmit)$_2$]$_2$, as well as in [TTF][Ni(dmit)$_2$]$_2$.

1.5.4.1 SP, SDW and CDW States

It may now be useful to recall briefly the different possible mechanisms which drive a metal-to-insulator transition or a superconducting transition. A more detailed description of these mechanisms can be found in several books and review articles [108–112].

The well-known BCS theory describes the superconducting state as resulting from the progressive coupling of two electrons with opposite spins and wave vectors in a so-called Cooper pair [112]. The BCS coupling is due to electron-phonon interactions. Another coupling may arise from antiferromagnetic fluctuations [113].

The Spin Density Wave instability is related to magnetic exchange interactions. Below a given temperature, T_s, the material shows an itinerant antiferromagnetic state, which means that two electrons with opposite spins will couple in pairs. This coupling induces a spatial modulation of the spin density which results in a gap opening at the Fermi level and, possibly, in a metal-to-insulator transition [108].

Moreover, Coulombic interactions tend to localise the electrons, and in the resulting Mott insulator the magnetic moment localised on each site will be antiferromagnetically coupled with the nearest neighbours. Finally, a Spin-Peierls transition may occur through interchain spin-phonon coupling [109].

The Charge Density Wave instability usually results from an important electron-phonon interaction which induces a periodical distortion of the crystal lattice, and simultaneously a spatial modulation of the charge density [110].

In the 1-D or quasi 1-D molecular conductors, a spontaneous CDW instability due to the formation of electron-hole pairs may induce a periodical lattice distortion, and thus drive an activated metal-to-insulator Peierls transition [9, 111]. This CDW-induced lattice distortion may be detected by diffuse X-ray scattering studies.

1.5.4.2 Diffuse X-ray Scattering Study of α'-[TTF][Pd(dmit)$_2$]$_2$

At room temperature, two sets of diffuse lines are observed at the reduced wave-vector $q_1 = 0.5\ b^*$ and $q_2 = 0.31\ b^*$. These 1-D fluctuations become correlated along the c-axis.

A study of the temperature-dependent intensities of the corresponding superlattice spots shows that these two kinds of scattering condense into satellite reflections at $T_1 \approx 150$ K and $T_2 \approx 105$ K, respectively [88]. It should be noted that the latter temperature corresponds to the critical temperature of the metal-to-insulator transition, T_{inf}, observed in the resistivity curve at ambient pressure. Moreover, additional weaker lines are observed at low temperatures at the reduced wave vector $2q_2 = \pm 0.62\ b^*$ and $q_1 \pm q_2 = 0.81\ b^*$ and $0.196\ b^*$.

The main result of this X-ray diffuse scattering investigation was to provide the first experimental evidence of CDW instabilities in α'-[TTF][Pd(dmit)$_2$]$_2$. It should be noted that the q_1 and q_2 instabilities involve only the [Pd(dmit)$_2$] stacks.

Thus, this compound is the first example of a molecular metal exhibiting competition between superconductivity and CDW (and not SDW as in the quasi 1-D organic superconductors such as (TMTSF)$_2$PF$_6$ [105]). Such a competition had previously been observed only in 2*H*-polytypes of the group (V) metal dichalcogenides [114] and [NbSe$_3$] [110].

1.5.4.3 *Diffuse X-ray Scattering Study of [TTF][Ni(dmit)$_2$]$_2$*

A parallel ambient pressure diffuse X-ray scattering study of [TTF][Ni(dmit)$_2$]$_2$ reveals one main 1-D structural instability at the reduced wave-factor $q_1 = 0.40\ b^*$ and two quasi 1-D diffuse scatterings of weaker intensity at $q_2 = 0.22\ b^*$ and $q_3 = 0.18\ b^*$ [88, 91]. This, again, provided experimental evidence for a low-pressure CDW insulating state in this compound. The q_1 instability leads to a structural transition at ≈ 40 K. This result is totally surprising and in apparent contradiction to the fact that [TTF][Ni(dmit)$_2$]$_2$ remains metallic down to much lower temperatures (1.5 K [96]; see Section 1.5.2.1). This contradiction will be further discussed and resolved later (see Section 1.6.2). The q_2 and q_3 scatterings have also been associated with CDW instabilities [96].

Diffuse X-ray scattering studies of α-[TTF][Pd(dmit)$_2$]$_2$ have not yet been carried out, but from preliminary resistivity results one could predict that this phase also exhibits CDW instabilities. A CDW instability has also been observed in another, non superconducting, compound of this series, [Cs]$_{0.5}$[Pd(dmit)$_2$] [91]. Therefore, it is most puzzling that no diffuse lines have yet been observed in the case of [Me$_4$N]$_{0.5}$[Ni(dmit)$_2$] [115].

1.5.4.4 [13]*C NMR Study of [TTF][Ni(dmit)$_2$]$_2$*

The paramagnetic Knight shifts in [TTF][Ni(dmit)$_2$]$_2$ enriched by ^{13}C isotope in the [Ni(dmit)$_2$] moiety only, have been determined by broad band and high resolution ^{13}C NMR [81]. The Knight shifts of the outer carbon atoms decrease by a factor of 1.7 when decreasing the temperature from 295 K down to 160 K, whereas those of the inner carbon atoms remain approximately constant. It should be noted, in passing, that this is the first observation of different temperature behaviour of the local Knight shifts for the same molecule in a molecular conductor.

These differences in the temperature-dependent behaviour of the Knight shifts for the outer and inner carbon atoms have been analyzed in terms of a band structure calculation (that we will discuss later; see Section 1.6) and have been associated with the q_1 CDW instability known from the above mentioned X-ray diffuse scattering studies. Moreover, these data clearly show that this CDW instability involves only the [Ni(dmit)$_2$] stacks [81].

1.5.5 Intermediate Conclusions

We now realise that the title of Section 1.5 was somewhat over-ambitious. A number of, mainly experimental, but also theoretical results have been obtained from various studies on the [M(dmit)$_2$]-based superconductors. These results have been analysed and several contradictory interpretations could be suggested concerning the dimensionality and the various conducting states of these systems.

A good illustration of this may be given by the EPR study of α'-[TTF][Pd(dmit)$_2$]$_2$. Powder EPR spectra of this compound show a continuous decrease in the linewidth from 350 to 150 K and also of the intensity of a broad conduction electron signal [94, 116]. It was thus suggested that the spin degrees of freedom were not frozen around the T_{min} critical temperature. Moreover, below 150 K a narrower signal appears with a Curie-Weiss spin susceptibility behaviour. This was, logically, interpreted as spin localisation arising from the conduction electrons. Later on, diffuse X-ray scattering studies invalidated this interpretation by providing evidence of CDW condensations. Consequently, the appearance of the narrow signal below 150 K should be due to charge localisation on one chain and not to spin localisation [116].

Therefore at this point the experimental data gathered on the [M(dmit)$_2$]-based superconductors, and their interpretation, cannot give a clear picture of the mechanisms which determine the various electronic properties of these compounds. For example, one basic contradiction remains to be removed: usually CDW induces a Peierls metal-to-insulator transition in a low-dimensional molecular conductor and metal-like conductivity, or superconductivity, is suppressed. How then is it possible that, in [TTF][Ni(dmit)$_2$]$_2$, the onset of CDW states does not induce any metal-to-insulator transition at ambient pressure?

1.6   BAND STRUCTURE CALCULATIONS AND GENERAL DISCUSSION

The above-described puzzling contradictions, the differences in the phase diagrams of α'-[TTF][Pd(dmit)$_2$]$_2$ and [TTF][Ni(dmit)$_2$]$_2$, and other peculiar features of the phase diagram of the nickel compound, such as the fact that the metallic state is maintained at ambient pressure down to very low temperature, the dramatic increase of the metal-to-insulator transition temperature, etc., has prompted theoretical physicists to calculate the band structure of these systems.

1.6.1   Band Structure Calculations

Two types of tight binding band calculations have been performed on [M(dmit)$_2$] systems, using the X-ray structural data obtained at room temperature. The first ones by Kobayashi *et al.* [102] included just the LUMO bands of [M(dmit)$_2$]. The most recent ones by Canadell *et al.* [89] included the LUMO bands of [M(dmit)$_2$] as well as the HOMO bands of TTF and [M(dmit)$_2$].

1.6.1.1   [TTF][Ni(dmit)$_2$]$_2$

Figure 27 shows the band structure of a [Ni(dmit)$_2$] slab of [TTF][Ni(dmit)$_2$]$_2$ calculated using the latter methodology [89]. The most significant, and surprising result is that the LUMO-based bands overlap appreciably with the HOMO-based ones; this is due to the weak HOMO-LUMO energy splitting (0.04 eV) [89]. Consequently, both the HOMO and LUMO series of bands are partially filled.

1.6.1.2 α'-[TTF][Pd(dmit)$_2$]$_2$

Figure 28 shows the band structure of a [Pd(dmit)$_2$] slab of α'-[TTF][Pd(dmit)$_2$]$_2$ calculated using the same method [89]. This band structure is very similar to that of [TTF][Ni(dmit)$_2$]$_2$. There are however two meaningful differences. A larger energy dispersion is observed in the band structure of α'-[TTF][Pd(dmit)$_2$]$_2$ compared to that of [TTF][Ni(dmit)$_2$]$_2$; this results from the shortening of the *b*-axis of the unit cell when Pd is substituted for Ni [66]. The position of the Fermi level with respect to the lowest HOMO band is different in both compounds.

In the palladium compound the Fermi level is lower than the lowest HOMO band while in the nickel compound the Fermi level intersects the top of the lowest HOMO band. Consequently, the palladium compound

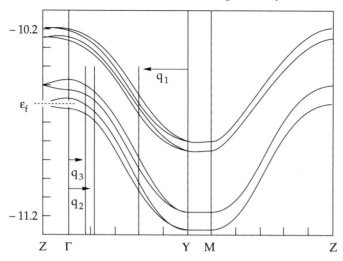

Figure 27 Band structure of the [Ni(dmit)$_2$] slabs in [TTF][Ni(dmit)$_2$]$_2$. Γ, Y, Z and M refer to the wave vectors (0, 0), (b*/2, 0), (0, c*/2) and (b*/2, c*/2), respectively (from [89]). Half the 1-D critical wave vectors (Q$_i$ = q$_i$/2) measured by diffuse X-ray scattering are indicated. ε_f is the Fermi level for a charge transfer of ≈ 0.8 (from [89, 117])

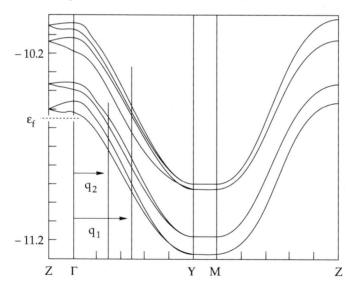

Figure 28 Band structure of the [Pd(dmit)$_2$] slabs in α'-[TTF][Pd(dmit)$_2$]$_2$. Half the 1-D critical wave vectors (Q$_i$ = q$_i$/2) measured by diffuse X-ray scattering are indicated. ε_f is the Fermi level for a charge transfer of ≈ 0.76 (from [89, 117])

has a more pronounced 1-D character, whereas in the nickel compound a 2-D hole pocket centred at the Γ point (corresponding to the wave vector (0, 0)) will induce a 2-D character of the Fermi surface [96, 118].

1.6.1.3 [Me₄N]₀.₅[Ni(dmit)₂]

1.6.1.3 [Me₄N]₀.₅[Ni(dmit)₂]

Figure 29 shows the band structure of a [Ni(dmit)₂] slab of [Me₄N]₀.₅[Ni(dmit)₂] calculated using the method including only the LUMO bands of [Ni(dmit)₂] [102]. In fact, it has been shown that in spite of the 'dimerisation' observed in the [Ni(dmit)₂] stacks of this compound, the need for inclusion of both HOMO- and LUMO-orbitals is less crucial [89, 119].

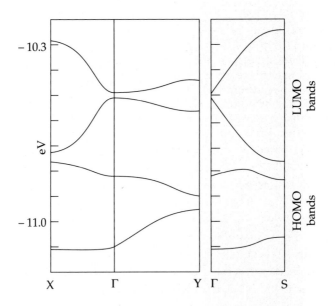

Figure 29 Band structure of the [Ni(dmit)₂]₂ slabs in [Me₄N]₀.₅[Ni(dmit)₂]. Γ, X, Y and S refer to the wave vectors (0, 0), $(a_0{}^*/2, 0)$, $(0, b_0{}^*/2)$ and $(-a_0{}^*/2, b_0{}^*/2)$, respectively, where a_0 and b_0 are the primitive vectors defined as $a_0 = (a + b)/2$ and $b_0 = -b$ (after [89, 102])

1.6.2 General Discussion

The results of the band structure calculation mentioned above allow us now to go deeper in the interpretation of the physical data, especially when taking into account the *multi-sheets Fermi surface* character of these

[M(dmit)$_2$] systems, which might be the essential difference with the organic molecular superconductors.

In α'-[TTF][Pd(dmit)$_2$]$_2$, structure factor calculations have shown that both diffuse X-ray scattering lines observed at $q_1 = 0.5\ b^*$ and $q_2 = 0.31\ b^*$ involve the [Pd(dmit)$_2$] stacks [88]. Both instabilities can thus be ascribed to CDW instabilities within each set of LUMO and HOMO bands, respectively (see Figure 28), of the [Pd(dmit)$_2$] acceptor molecules [89, 117–118]. These instabilities drive two successive broad phase transitions at $T_1 = 150$ K and $T_2 = 105$ K, temperatures at which a metal-to-insulator transition occurs as observed by resistivity measurements at ambient pressure [94]. Moreover, the appearance at low temperature of weaker $2q_2$ scatterings [88] is consistent with a CDW formation on the donor TTF chains, leading thus to the destruction of all the carriers [89].

We have previously stressed that the CDW instability observed at $q_1 = 0.4\ b^*$ in [TTF][Ni(dmit)$_2$]$_2$ [88] leads to a structural transition at ≈ 40 K which surprisingly does not significantly affect the transport properties [96]. This q_1 instability can be associated with the bunch of LUMO bands (see Figure 27) of the [Ni(dmit)$_2$] acceptor molecules [91]. This assignment is in agreement with the thermopower data showing that holes become the dominant carriers below 70 K [79], and with the ^{13}C NMR data [81]. The two other weaker diffuse scatterings observed at $q_2 = 0.22\ b^*$ and $q_3 = 0.18\ b^*$ [91] could be associated with the bunch of the highest HOMO bands of the [Ni(dmit)$_2$] molecules. Thus, the weak anomaly observed at 10 K in the resistivity versus temperature curve and the increase in resistivity below 1.5 K (see Section 1.5.2.1) could result from the q_2 and q_3 instabilities [96].

[TTF][Ni(dmit)$_2$]$_2$ staying (semi)metallic below 10 K, the remaining carriers at low temperatures could come from the TTF band. This would be in agreement with ^{1}H NMR data which did not reveal any gap opening in the electronic excitation of the TTF stacks down to 1.5 K [80]. On the other hand, the band structure calculations have shown that the Fermi level intersects the top of the lowest [Ni(dmit)$_2$] HOMO band (see Figure 27). This results in a 2-D hole pocket centred at the Γ point which induces a partially 2-D character of the Fermi surface. Consequently, the remaining carriers at low temperatures could also come from this partially filled HOMO band. This would be in agreement for the departure from 1-D behaviour of the magnetoresistance at 4.2 K [100].

The information inferred from the band structure calculation may also explain the differences in the phase diagrams of α'-[TTF][Pd(dmit)$_2$]$_2$ and [TTF][Ni(dmit)$_2$]$_2$ and the unusual features of that of the nickel compound.

In α'-[TTF][Pd(dmit)$_2$]$_2$, the Fermi level is lower than the lowest HOMO band and the palladium compound exhibits a more pronounced 1-D character than the nickel compound for which the Fermi level intersects the lowest HOMO band.   Therefore, the insulating state in α'-[TTF][Pd(dmit)$_2$]$_2$ is a CDW state and, for the first time in this compound, there is strong competition between a CDW state and a high pressure superconducting state.

When applying pressure to [TTF][Ni(dmit)$_2$]$_2$, the Fermi level will shift below the lowest HOMO band and this will result in a progressive suppression of the 2-D hole pocket centred at the Γ point. The dramatic continuous increase of T_{min} and T_{inf} from ambient pressure to a few hundred bars may result from this progressive suppression of the 2-D character of the Fermi surface. Moreover, correlation effects may be at the origin of the oscillating character of T_{min} and T_{inf} at low pressures [96]. Above 4 kbar, the continuous decrease of T_{min} and T_{inf} — as observed in the palladium compound — suggests that [TTF][Ni(dmit)$_2$]$_2$ is entering the 1-D regime. Hence, in opposition with most organic conductors, low pressures induce 1-D electronic properties in the nickel compound. At higher pressures, transverse couplings probably suppress this 1-D instability.

Other unexpected features of the phase diagram of [TTF][Ni(dmit)$_2$]$_2$, such as the coexistence or weak competition between CDW and superconducting states at high pressures, the increase of T_c with increasing pressure, and the re-entrance of the superconducting state into the low-pressure insulating state, may find their explanation in the effect of the CDW instabilities at various pressures, provided that these instabilities remain under pressure. These interpretations have been discussed in detail elsewhere [96] and would be beyond the scope of the present work. Moreover, they need additional measurements, such as isothermal excursions of the phase diagrams with He gas pressure, in order to be better argued.

1.7   CONCLUSION

A number of *firsts* have been collected when studying the [M(dmit)$_2$] superconductors:

(i)   These systems form the first and unique series of molecular inorganic superconductors derived from transition metal complexes in which the inorganic component is responsible for superconductivity.

(ii)   [TTF][Ni(dmit)$_2$]$_2$ and [Me$_4$N]$_{0.5}$[Ni(dmit)$_2$] are the first molecular conductors, among the organic or inorganic ones, in which the superconducting critical temperature increases with increasing pressure.

(iii) α'-[TTF][Pd(dmit)$_2$]$_2$ is the first example of a molecular superconductor in which a CDW state (and not a SDW state) is in strong competition with a superconducting state.

(iv)   [TTF][Ni(dmit)$_2$]$_2$ is the first molecular compound in which a CDW instability does not induce a metal-to-insulator transition.

(v)   [TTF][Ni(dmit)$_2$]$_2$ is the first compound in which CDW states coexist or are in weak competition with a superconducting state.

(vi)   The complexity of the temperature-pressure phase diagram of [TTF][Ni(dmit)$_2$]$_2$ is unique.

(vii)   The band-electronic structure of the [M(dmit)$_2$] superconductors is characterised by a unique *multi-sheets Fermi surface*.

The rôle of the dimensionality, though not as conspicuous as firstly thought, is nevertheless important as shown by the band structure calculations.

In spite of the absence of any direct metal-metal interaction as in KCP, the rôle of the metal atom, though subtle, is nevertheless important. When going from Ni to Pd, important differences in the results of band structure calculations are obtained, concerning for example the crucial position of the Fermi level with respect to the lowest HOMO band and its effect on the dimensionality of the system.

From the results reviewed and discussed in the present work it is clear that small changes in the chemistry may lead to novel compounds exhibiting different behaviours. By taking advantage of the large number of chemical modifications offered by molecular inorganic chemistry, we may anticipate the preparation of new compounds having novel or improved solid state properties. Indeed, next to the four superconductors which have been the central subject of this work, other [M(dmit)$_2$] compounds exhibit unusual properties. The last example is the [HMe$_3$N]$_{0.5}$[Ni(dmit)$_2$] compound in which a sharp metal-to-insulator transition is followed at lower temperatures by an equally sharp insulator-to-metal transition [77, 120].

The physical properties of the [M(dmit)$_2$] compounds seem to us so novel that we may state that a new area in the physics of superconductors has been opened. Our happiness would be fulfilled if one last frustration could be removed: none of the [M(dmit)$_2$] compounds are *ambient-pressure* superconductors. However, it has been said that "*Hoping is not necessary*

to undertake, neither is succeeding to persevere" [121] (but it helps!). We
shall persevere.

1.8   ACKNOWLEDGEMENTS

A number of *miners* have dug in the [M(dmit)$_2$] *mine* in the last ten years
and we cannot pay a tribute to all of them by name. We acknowledge the
contributions of the authors indicated in the citations. However, we wish
to single out J-P Legros and L Brossard for distinction for their major
involvement in our work and in the writing of this chapter.

1.9   REFERENCES

1. W Knop, *Justus Liebig's Ann Chem*, **43**, 111 (1842); W Knop and
    G Schnedermann, *J Prakt Chem*, **37**, 461 (1846).
2. L A Levy, *J Chem Soc*, 1081 (1912).
3. K Krogmann and H-D Hausen, *Z Anorg Allg Chem*, **358**, 67 (1968).
4. H R Zeller, *Phys Rev Lett*, **28**, 1452 (1972); H R Zeller and A Beck,
    J Phys Chem Solids, **35**, 77 (1973).
5. J S Miller and A J Epstein, *Prog Inorg Chem*, **20**, 46 (1976).
6. J M Williams and A J Schultz, in: *Molecular Metals*, W E Hatfield,
    Ed, Plenum Press, New York, p 337 (1979); J M Williams, A J
    Schultz, A E Underhill and K Carneiro, in: *Extended Linear Chain
    Compounds*, J S Miller, Ed, Plenum Press, New York, **1**, 73 (1982);
    J M Williams, *Adv Inorg Chem Radiochem*, **26**, 235 (1983).
7. J R Ferraro and J M Williams, in: *Introduction to Synthetic
    Electrical Conductors*, Academic Press, New York, p 139 (1987).
8. A E Underhill, D M Watkins, J M Williams and K Carneiro, in:
    Extended Linear Chain Compounds, J S Miller, Ed, Plenum Press,
    New York, **1**, 120 (1982).
9. R E Peierls, in: *Quantum Theory of Solids*, Oxford University Press,
    London, p 108 (1955).
10. W A Little, *Phys Rev A*, **134**, 1416 (1964); D Davis, H Gutfreund and
    W A Little, *Phys Rev B*, **13**, 4766 (1976).
11. L R Melby, R J Harder, W R Hertler, W Mahler, R E Benson and
    W E Mochel, *J Am Chem Soc*, **84**, 3374 (1962).
12. B A Scott, S J LaPlaca, J B Torrance, B B Silverman and B Welber,
    Ann N Y Acad Sci, **313**, 369 (1978).
13. F Wudl, G M Smith and E J Hufnagel, *J Chem Soc, Chem Commun*,
    1453 (1970).
14. J S Miller, *Ann N Y Acad Sci*, **313**, 25 (1978); L C Isett and E A Perez-
    Albuerne, *ibid*, **313**, 395 (1978).

15. L B Coleman, M J Cohen, D J Sandman, F G Yamagishi, A F Garito and A J Ferraris, *Solid State Commun*, **12**, 1125 (1973); J P Ferraris, D O Cowan, V Valatka and J H Perlstein, *J Am Chem Soc*, **95**, 948 (1973).

16. F Wudl, *Acc Chem Res*, **17**, 227 (1984); J M Williams, M A Beno, H H Wang, P C W Leung, T J Emge, U Geiser and K D Carlson, *Acc Chem Res*, **18**, 261 (1985).

17. J R Ferraro and J M Williams, in: *Introduction to Synthetic Electrical Conductors*, Academic Press, New York, p 1 (1987).

18. T J Kistenmacher, T E Phillips and D O Cowan, *Acta Crystallogr Sect B*, **30**, 763 (1974).

19. D Jérome, A Mazaud, M Ribault and K Bechgaard, *J Phys Lett (Orsay, France)*, **41**, 95 (1980).

20. K Bechgaard, K Carneiro, F B Rasmussen, H Olsen, G Rindorf, C S Jacobsen, H Pedersen and J E Scott, *J Am Chem Soc*, **103**, 2440 (1981).

21. S S P, Parkin, E M Engler, R R Schumaker, R Lagier, V Y Lee, J C Scott and R L Greene, *Phys Rev Lett*, **50**, 270 (1983); E B Yagubskii, I F Shchegolev, V N Laukhin, P A Konovich, M V Kartsovnik, A V Zvarykina and L I Buravov, *JETP Lett*, **39**, 12 (1984); G W Crabtree, K D Carlson, L N Hall, P T Copps, H H Wang, T J Emge, M A Beno and J M Williams, *Phys Rev B*, **30**, 2958 (1984); J M Williams, T J Emge, H H Wang, M A Beno, P T Copps, L N Hall, K D Carlson and G W Crabtree, *Inorg Chem*, **23**, 2558 (1984).

22. H Urayama, H Yamochi, G Saito, K Nozawa, T Sugano, M Kinoshita, S Sato, K Oshima, A Kawamoto and J Tanaka, *Chem Lett*, 55 (1988); H Urayama, H Yamochi, G Saito, S Sato, A Kawamoto, J Tanaka, T Mori, Y Maruyama and H Inokuchi, *ibid*, 463 (1988); K Nozawa, K Sugano, H Urayama, H Yamochi, G Saito, M Kinoshita, *ibid*, 617 (1988).

23. A M Kini, U Geiser, H H Wang, K D Carlson, J M Williams, W K Kwok, K G Vandervoort, J E Thomson, D L Stupka, D Jung and M-H Whangbo, *Inorg Chem*, **29**, 2555 (1990).

24. H H Wang, K D Carlson, U Geiser, A M Kini, A J Schultz, J M Williams, L K Montgomery, W K Kwok, U Welp, K G Vandervoort, S J Boryschuk, A V Strieby Crouch, J M Kommers, D M Watkins, J E Schirber, D L Overmyer, D Jung, J J Novoa and M-H Whangbo, *Synth Met*, **41–43**, 1983 (1991).

25. K Kikuchi, M Kikuchi, T Namiki, K Saito, I Ikemoto, K Murata, T Ishiguro and K Kobayashi, *Chem Lett*, 931 (1987); K Kikuchi, K Murata, Y Honda, T Namiki, K Saito, H Anzai, K Kobayashi and I Ikemoto, *J Phys Soc Jpn*, **56**, 4241 (1987).

26. G C Papavassiliou, G A Mousdis, J S Zambounis, A Terzis, A Hountas, B Hilti, C W Mayer and J Pfeiffer, *Synth Met*, **27**, B379 (1988).

27. M A Beno, H H Wang, K D Carlson, A M Kini, G M Frankenbach, J R Ferraro, N F Larson, G D McCabe, J Thompson, C Purnama, M Vashon, J M Williams, D Jung and M-H Whangbo, *Mol Cryst Liq Cryst*, **181**, 145 (1990).

28. *International Conference on Science and Technology of Synthetic Metals (ICSM'90)*, Tübingen, Germany, Sept 2–7, 1990, Proceedings published in: *Synth Met*, **41–43** (1991).

29. L V Interrante, J W Bray, H R Hart, Jr, J S Kasper, P A Piacente and G D Watkins, *Ann N Y Acad Sci*, **313**, 407 (1978).

30. J B Torrance, *Acc Chem Res*, **12**, 79–86 (1979); *Molecular Metals*, W E Hatfield, Ed, Plenum Press, New York, p 7 (1979).

31. A H Reis, Jr, in: *Extended Linear Chain Compounds*, J S Miller, Ed, Plenum Press, New York, **1**, 157 (1982).

32. T J Marks and D W Kalina, in: *Extended Linear Chain Compounds*, J S Miller, Ed, Plenum Press, New York, **1**, 197 (1982).

33. P Cassoux, L V Interrante and J Kasper, *C R Acad Sci (Paris), Série C*, **291**, 25–28 (1980); P Cassoux, L V Interrante and J Kasper, *Mol Cryst Liq Cryst*, **81**, 293 (1982).

34. L Alcacer and H Novais, in: *Extended Linear Chain Compounds*, J S Miller, Ed, Plenum Press, New York, **3**, 319 (1983).

35. J A McCleverty, *Prog Inorg Chem*, **10**, 49 (1968).

36. G N Schrauzer, *Trans Met Chem*, **4**, 299 (1968); *Acc Chem Res*, **2**, 72 (1969).

37. E Hoyer, W Dietzsch and W Schroth, *Z Chem*, **11**, 41 (1971).

38. S Alvarez, R Vicente and R Hoffmann, *J Am Chem Soc*, **107**, 6253 (1985).

39. L J Alcacer and A H Maki, *J Phys Chem*, **78**, 215 (1974); *ibid*, **80**, 1912–1916 (1976); L J Alcacer, H M Novais and F P Pedroso, in: *Molecular Metals*, W E Hatfield, Ed, Plenum Press, New York, p 415 (1979).

40. J W Bray, H R Hart, Jr, L V Interrante, I S Jacobs, J S Kasper, P A Piacente and G D Watkins, *Phys Rev B*, **16**, 1359 (1977).

41. M M Ahmad, D J Turner, A E Underhill, C S Jacobsen, K Mortensen and K Carneiro, *Phys Rev B*, **29**, 4796 (1984).

42. J M Williams, M A Beno, E K Appelman, J M Capriotti, F Wudl, E Aharon-Shalom and D Nalewajek, *Mol Cryst Liq Cryst*, **79**, 319 (1982).

43. F Wudl, *J Am Chem Soc*, **103**, 7064–7069 (1981); *Pure Appl Chem*, **54**, 1051 (1982).

44. G Steimecke, R Kirmse and E Hoyer, *Z Chem*, **15**, 28 (1975).

45. G Steimecke, *PhD Thesis*, Leipzig (1977); G Steimecke, H J Sieler, R Kirmse and E Hoyer, *Phosphorus and Sulphur*, **7**, 49 (1979).

46. K Hartke, T Kissel, J Quante and R Matusch, *Chem Ber*, **113**, 1898 (1980).

47. L Valade, J-P Legros, M Bousseau, P Cassoux, M Garbauskas and L V Interrante, *J Chem Soc, Dalton Trans*, 783 (1985).
48. K S Varma, A Bury, N J Harris and A E Underhill, *Synthesis*, 837 (1987).
49. S Wawzonek and S M Heilmann, *J Org Chem*, **39**, 511 (1974).
50. G Bontempelli, F Magno, G A Mazzochin and R Seeber, *J Electroanal Chem Interfacial Electrochem*, **63**, 231 (1975); M F Hurley and J Q Chambers, *J Org Chem*, **46**, 775 (1981).
51. M Mizuno, A F Garito and M Cava, *J Chem Soc, Chem Commun*, 18 (1978); E M Engler, V Y Lee, R R Schumaker, S S S Parkin, R L Greene and J C Scott, *Mol Cryst Liq Cryst*, **107**, 19 (1984).
52. P Cassoux, L Valade, H Kobayashi, A Kobayashi, R A Clark and A E Underhill, *Coord Chem Rev*, **110**, 115 (1991).
53. L Valade, *Thèse de Doctorat de 3ème Cycle*, Toulouse (1983); M Bousseau, *Thèse de Doctorat de 3ème Cycle*, Toulouse (1984).
54. L Valade, *Thèse de Doctorat d'Etat*, Toulouse (1987).
55. L Valade, J-P Legros, P Cassoux and F Kubel, *Mol Cryst Liq Cryst*, **140**, 335 (1986).
56. L Valade, P Cassoux, A Gleizes and L V Interrante, *J Phys (Paris)*, **44** (C3), 1203 (1983).
57. D A Stephens, A E Rehan, S J Compton, R A Barkhau and J M Williams, *Inorg Synth*, **24**, 135 (1986).
58. H H Wang, L K Montgomery, C A Husting, B A Vogt, J M Williams, S M Budz, M J Lowry, K D Carlson, W-K Kwok and V Mikheyev, *Chem Mat*, **1**, 484 (1989).
59. R A Clark, *PhD Thesis*, University of Wales (1989); R A Clark and A E Underhill, *J Chem Soc, Chem Commun*, 228 (1989).
60. R A Clark and A E Underhill, *Synth Met*, **27**, B515 (1988).
61. H Kim, A Kobayashi, Y Sasaki, R Kato and H Kobayashi, *Chem Lett*, 1799 (1987).
62. L Valade, J-P Legros, D de Montauzon, P Cassoux and L V Interrante, *Israel J Chem*, **27**, 353 (1986).
63. J Ribas and P Cassoux, *C R Acad Sci (Paris), Série II*, **293**, 287 (1981).
64. M Bousseau, L Valade, M-F Bruniquel, P Cassoux, M Garbauskas, L V Interrante and J Kasper, *Nouv J Chim*, **8**, 3 (1984).
65. L Valade and P Cassoux, *C R Acad Sci (Paris), Série II*, **301**, 999 (1985).
66. M Bousseau, L Valade, J-P Legros, P Cassoux, M Garbauskas and L V Interrante, *J Am Chem Soc*, **108**, 1908 (1986).
67. R Kato, H Kobayashi, A Kobayashi and Y Sasaki, *Chem Lett*, 131 (1985); H Kobayashi, R Kato, A Kobayashi and Y Sasaki, *Chem Lett*, 191 (1985); H Kobayashi, R Kato, A Kobayashi and Y Sasaki, *Chem Lett*, 535 (1985); A Kobayashi, R Kato, H Kobayashi, T Mori and H Inokuchi, *Physica B & C (Amsterdam)*,

143, 562 (1986); A Kobayashi, Y Sasaki, R Kato and H Kobayashi, *Chem Lett*, 387 (1986); R Kato, H Kobayashi, A Kobayashi, T Naito, M Tamura, H Tajima and H Kuroda, *Chem Lett*, 1839 (1989).

68. G C Papavassiliou, *Mol Cryst Liq Cryst*, **86**, 159 (1982).
69. I Johansen, K Bechgaard, C Rindorf, N Thorup, C S Jacobsen and K Mortensen, *Synth Met*, **15** , 333–343 (1986); K Bechgaard, *Stud Org Chem (Amsterdam) (New Methodol Funct Interesting Compd)*, **25**, 391 (1986).
70. M L Kaplan, *J Cryst Growth*, **33**, 161 (1976); J Anzai, *ibid*, **33**, 185 (1976); J R Andersen, E M Engler and K Bechgaard, *Ann N Y Acad Sci*, **313**, 293 (1978).
71. P Cassoux, L Valade, J-P Legros, L Interrante and C Roucau, *Physica B & C (Amsterdam)*, **143**, 313 (1986).
72. J-P Legros and L Valade, *Solid State Commun*, **68**, 599 (1988).
73. L Brossard, M Ribault, M Bousseau, L Valade and P Cassoux, *C R Acad Sci (Paris), Série II*, **302**, 205 (1986); L Brossard, M Ribault, L Valade and P Cassoux, *Physica B & C (Amsterdam)*, **143**, 378 (1986).
74. P Legros, L Valade and P Cassoux, *Synth Met*, **27**, B347 (1988).
75. A Kobayashi, H Kim, Y Sasaki, K Murata, R Kato and H Kobayashi, *J Chem Soc Faraday Trans*, **86**, 361 (1990).
76. R Kato, H Kobayashi, H Kim, A Kobayashi, Y Sasaki, T Mori and H Inokuchi, *Chem Lett*, 865 (1988); R Kato, H Kobayashi, H Kim, A Kobayashi, Y Sasaki, T Mori and H Inokuchi, *Synth Met*, **27**, B359 (1988).
77. C Tejel, B Pomarède, J-P Legros, L Valade, P Cassoux and J-P Ulmet, *Chem Mat*, **1**, 578 (1989).
78. A Kobayashi, H Kim, Y Sasaki, R Kato, H Kobayashi, S Moriyama, Y Nishio, K Kajita and W Sasaki, *Chem Lett*, 1819 (1987); A Kobayashi, H Kim, Y Sasaki, S Moriyama, Y Nishio, K Kajita, W Sasaki, R Kato and H Kobayashi, *Synth Met*, **27**, B339 (1988); K Kajita, Y Nishio, S Moriyama, R Kato, H Kobayashi and W Sasaki, *Solid State Commun*, **65**, 361 (1988).
79. W Kang, *Thèse de Doctorat d'Etat*, Univ Paris Sud, Orsay (1989); W Kang, D Jérome, L Valade and P Cassoux, *Synth Met*, **41–43**, 2343 (1991).
80. C Bourbonnais, P Wzietek, D Jérome, F Creuzet, L Valade and P Cassoux, *Europhys Lett*, **6**, 177 (1988).
81. A Vainrub, D Jérome, M-F Bruniquel and P Cassoux, *Europhys Lett*, **12**, 267 (1990); A Vainrub, E Canadell, D Jérome, P Bernier, T Nunes, M-F Bruniquel and P Cassoux, *J Phys (Paris)*, **51**, 2465 (1990).

82. F Wudl, *J Am Chem Soc*, **97**, 1962 (1975); F Wudl and M L Kaplan, *Inorg Synth*, **19**, 27 (1979).

83. J-P Legros, M Bousseau, L Valade and P Cassoux, *Mol Cryst Liq Cryst*, **100**, 181 (1983).

84. R Comes, S M Shapiro, G Shirane, A F Garito and A J Heeger, *Phys Rev Lett*, **35**, 1518 (1975).

85. S Flandrois and D Chasseau, *Acta Cryst B*, **33**, 2744 (1977).

86. T C Umland, S Allie, T Kuhlmann and P Coppens, *J Phys Chem*, **92**, 6456 (1988).

87. D Medus, D de Montauzon, L Valade and P Cassoux, *unpublished results*.

88. S Ravy, J-P Pouget, L Valade and J-P Legros, *Europhys Lett*, **9**, 391 (1989).

89. E Canadell, E I Rachidi, S Ravy, J-P Pouget, L Brossard and J-P Legros, *J Phys (Paris)*, **50**, 2967 (1989).

90. K Yakushi, S Nishimura, T Sugano, H Kuroda and I Ikemoto, *Acta Cryst Sect B*, 36, 358 (1980).

91. S Ravy, E Canadell and J-P Pouget, in: *The Physics and Chemistry of Organic Superconductors*, G Saito and S Kagoshima, Eds, Springer Proceedings in Physics, Springer Verlag, Berlin, **51**, 252 (1990).

92. W F Cooper, N C Kenny, J W Edmonds, A Nagel, F Wudl and P Coppens, *J Chem Soc, Chem Commun*, 889 (1971).

93. A Kobayashi, R Kato, H Kobayashi, S Moriyama, Y Nishio, K Kajita and W Sasaki, *Chem Lett*, 459 (1987).

94. L Brossard, M Ribault, L Valade and P Cassoux, *J Phys (Paris)*, **50**, 1521 (1989).

95. L Brossard, M Ribault, L Valade and P Cassoux, *C R Acad Sci (Paris) Série II*, **309**, 1120 (1989).

96. L Brossard, M Ribault, L Valade and P Cassoux, *Phys Rev B*, **42**, 3935 (1990).

97. L Pauling, in: *The Nature of the Chemical Bond*, Cornell University Press, Ithaca, N Y (1960).

98. H C Montgomery, *J Appl Phys*, **42**, 2971 (1971).

99. R Kato, T Mori, A Kobayashi, Y Sasaki and H Kobayashi, *Chem Lett*, 1 (1984).

100. J-P Ulmet, P Auban, A Khmou, L Valade and P Cassoux, *Phys Lett A*, **120**, 217 (1985).

101. J-P Ulmet, P Auban and S Askenazy, *Solid State Commun*, **52**, 547 (1984).

102. A Kobayashi, H Kim Y Sasaki, R Kato and H Kobayashi, *Solid State Commun*, **62**, 57 (1987).

103. J M Williams, H H Wang, T J Emge, U Geiser, M A Beno, P C W Leung, K D Carlson, R J Thorn, A J Schultz and M-H Whangbo, *Prog Inorg Chem*, **35**, 51 (1987).

104. J E Schirber, D L Overmyer, J M Williams, H H Wang, L Valade and P Cassoux, *Phys Lett A*, **120**, 87 (1987).

105. D Jerome and H J Schulz, *Adv Phys*, **31**, 299 (1982).

106. J E Schirber, L J Azevedo, J F Kwak, E L Venturini, P C W Leung, M A Beno, H H Wang and J M Williams, *Phys Rev*, **33**, 1987 (1986).

107. L Brossard, H Hurdequint, M Ribault, L Valade, J-P Legros and P Cassoux, *Synth Met*, **27**, B157 (1988).

108. A W Overhauser, *Phys Rev Lett*, **4**, 462 (1960); W M Lomer, *Proc Phys Soc (London)*, **80**, 489 (1962).

109. J W Bray, L V Interrante, I S Jacobs and J C Bonner, in: *Extended Linear Chain Compounds*, J S Miller, Ed, Plenum Press, New York, **3**, 353 (1983).

110. P Monceau, in: *Electronic Properties of Quasi 1-D Compounds*, Reidel, Dordrecht (1985).

111. J-P Pouget, S K Khanna, F Denoyer, R Comes, A F Garrito and A J Heeger, *Phys Rev Lett*, **37**, 437 (1976); G A Toombs, *Phys Rep C*, **40**, 181 (1978).

112. J Bardeen, L N Cooper and J R Schrieffer, *Phys Rev*, **106**, 162 (1957); *ibid*, **108**, 1175 (1957).

113. C Bourbonnais, *Fermi School Series*, North Holland (1989).

114. T F Smith, R N Shelton and R E Schwall, *J Phys (Paris)*, **5**, 1713 (1975).

115. S Ravy, *personal communication*.

116. C Coulon, *personal communication*.

117. S Ravy, E Canadell, J-P Pouget, P Cassoux and A E Underhill, *Synth Met*, **41–43**, 2191 (1991).

118. L Brossard, E Canadell, S Ravy, J-P Pouget, J-P Legros and L Valade, *Fizika*, **21**, 15 (1989).

119. E Canadell, S Ravy, J-P Pouget and L Brossard, *Solid State Commun*, **75**, 633 (1990).

120. J-P Ulmet, M Mazzaschi, C Tejel, P Cassoux and L Brossard, *Solid State Commun*, **74**, 91 (1990).

121. William of Orange (1650–1702).

2 Molecular Inorganic Magnetic Materials

Olivier Kahn, Yu Pei and Yves Journaux

Inorganic Materials. Edited by Duncan W Bruce and Dermot O'Hare
© 1992 John Wiley & Sons Ltd

2.1   INTRODUCTION

One of the main challenges in the field of molecular materials concerns the design of molecular-based compounds exhibiting a spontaneous magnetisation below a critical temperature. The first compounds of this kind have been described in the last four years; in spite of many efforts, these compounds are still very rare. Quite a long maturation was necessary before obtaining the first achievements along this line. The pioneering ideas were put forward at the beginning of the sixties.

The goals of this contribution are to introduce the basic knowledge absolutely necessary to enter the field of molecular-based magnetic materials, then to present the state-of-the-art in 1991. In writing this chapter, we have assumed that quite a few readers are not at all familiar with molecular magnetism; that is why the first section summarises the key definitions and equations in this discipline, and reminds the reader of the magnetic behaviour of molecules containing a unique magnetic centre. The following two sections are devoted to the phenomenon of interaction between magnetic centres, first in binuclear, then in one-dimensional compounds. The last section deals with the various strategies which have already been explored to design molecular-based ferromagnets, and emphasises the latest achievements.

2.2 SOME GENERALITIES IN MOLECULAR MAGNETISM

2.2.1 Definitions and Units

In this short first Section, we establish some key equations in molecular magnetism. To start with, we consider a sample containing one mole of a molecular compound within an homogeneous magnetic field H. The sample acquires a molar magnetisation M related to H through:

$$\frac{\partial M}{\partial H} = \chi \tag{1}$$

where χ is the molar magnetic susceptibility. M, also called molar magnetic moment, is a vector, H an axial vector and χ a second rank tensor. It is always possible to choose the reference axes in order for χ to be diagonal with χ_u (u = x, y, z) principal values. If the sample is magnetically isotropic, χ becomes a scalar.

Generally, in a large magnetic field range, χ is independent of H, so that one can write:

$$M = \chi H \tag{2}$$

The problem of units deserves a few words. The SI is the legal system [1], but legality is not science. This system, indeed, is particularly inappropriate in molecular magnetism and, like most researchers involved in this field, we very much prefer to use the unrationalised c.g.s. e.m.u. system. The unit of magnetic field is then the Gauss. The volume magnetic susceptibility is a dimensionless quantity traditionally expressed in e.m.u./cm^3, so that the dimension of e.m.u. is formally the cm^3. Therefore, the molar magnetic susceptibility is expressed in cm^3 mol^{-1}. Very often the quantity of interest is χT, the product of the molar magnetic susceptibility and the temperature. χT is expressed in cm^3 K mol^{-1}. As for the molar magnetisation M, it is expressed in cm^3 Gauss mol^{-1}. Alternatively, M may be expressed in Bohr magnetons per mole, with the symbol β mol^{-1}, the relationship between the two units being:

$$1\ \beta\ mol^{-1} = 5585\ cm^3\ Gauss\ mol^{-1} \tag{3}$$

It should be stressed here that all energies will be expressed in cm^{-1}.

2.2.2 Fundamental Equations

The molar paramagnetic susceptibility characterises the way an applied magnetic field H interacts with the angular momenta associated with the thermally populated states of a molecule. In classical mechanics, when a sample is perturbed by an external magnetic field its magnetisation is related to its energy variation through Equation (4).

$$M = - \frac{\partial E}{\partial H}$$
(4)

This relationship may be easily translated into the language of quantum mechanics. For a molecule with an energy spectrum E_n ($n = 1, 2, ...$) in the presence of a magnetic field H, the macroscopic molar magnetisation M is obtained by summing the microscopic magnetisations weighted according to the Boltzmann distribution law, which leads to Equation (5).

$$M = \frac{\sum_n - \left(\frac{\partial E}{\partial H}\right) \exp\left(\frac{-E_n}{kT}\right)}{\sum_n \exp\left(\frac{-E_n}{kT}\right)}$$
(5)

where N is the Avogadro number, T the temperature, and k the Boltzmann constant equal to 0.6950388 cm^{-1} K^{-1}.

Equation (5) may be considered as the fundamental expression in molecular magnetism. It does not depend on any approximation. The molar magnetic susceptibility is deduced from Equation (5), or, if H/kT is not too large, from Equation (2). Equation (5) may be re-expressed as:

$$M = NkT \left(\frac{\partial LnZ}{\partial H}\right)$$
(6)

where Z is the partition function defined as:

$$Z = \sum_n \exp\left(\frac{-E_n}{kT}\right)$$
(7)

If the relationship given by Equation (5) is quite general, it is often difficult to apply. Indeed, it requires a knowledge of $E_n = f(H)$ variations for all thermally populated states in order to calculate the $\partial E_n / \partial H$ derivatives. In 1932, Van Vleck proposed a simplification based on a few approximations [2]. The first of these approximations is that it is legitimate to expand the energies E_n according to the increasing powers of H:

$$E_n = E_n^{(0)} + E_n^{(1)}H + E_n^{(2)}H^2 + ...$$
(8)

$E_n^{(0)}$ is the energy of the level n in zero field. $E_n^{(1)}$ and $E_n^{(2)}$ are called the first- and second-order Zeeman coefficients, respectively. The second approximation is that H/kT is small. If so, Van Vleck showed that the magnetic susceptibility may be expressed by Equation (9).

$$\chi = \frac{N \sum\limits_n \left(\dfrac{(E_n^{(1)})^2}{kT} - 2E_n^{(2)} \right) \exp\left(\dfrac{-E_n^{(0)}}{kT} \right)}{\sum\limits_n \exp\left(\dfrac{-E_n^{(0)}}{kT} \right)} \tag{9}$$

To apply this formula we only need to know the $E_n^{(0)}$, $E_n^{(1)}$ and $E_n^{(2)}$ quantities. It is not necessary anymore to calculate the partial derivatives $\partial E_n / \partial H$. When all energies E_n are linear in H, the second-order Zeeman coefficients $E_n^{(2)}$ vanish and Equation (9) becomes:

$$\chi = \frac{N \sum\limits_n (E_n^{(1)})^2 \exp\left(\dfrac{-E_n^{(0)}}{kT} \right)}{\sum\limits_n \exp\left(\dfrac{-E_n^{(0)}}{kT} \right)} \tag{10}$$

2.2.3   Molecules Containing a Unique Magnetic Centre

The simplest situation is that of molecules for which the $^{2S+1}\Gamma$ ground state has no first-order angular momentum and is very separated in energy from the first excited states, so that any kind of coupling between ground and excited states may be neglected. The 2S+1 spin degeneracy is then retained in the absence of an external magnetic field. It is convenient to choose the energy of this $^{2S+1}\Gamma$ state as the energy origin. When applying the field, the energies of the 2S+1 Zeeman components are given by:

$$E_n = M_S g \beta H \tag{11}$$

with M_S varying by integer values from $-S$ to $+S$. The energies E_n being linear in H, it is possible to use the simplified form of Van Vleck's formula (Equation (10)) provided that H/kT is small, with:

$$E_n^{(0)} = 0 \tag{12}$$

$$E_n^{(1)} = M_s g \beta, \quad M_s = -S, -S + 1, \dots S - 1, S \tag{13}$$

which leads to:

$$\chi = \frac{Ng^2\beta^2}{3kT}S(S + 1) \qquad (14)$$

The molar magnetic susceptibility varies with C/T, the constant C depending on the spin multiplicity of the ground state; this is the *Curie law*. The most convenient way, experimentally, to establish that a compound obeys this Curie law is to obtain a horizontal straight line for the χT versus T plot. In the 'c.g.s e.m.u' unit system, $N\beta^2/3k$ is equal to 0.12505 (very close to 1/8).

The Curie law (Equation (14)) is valid only when H/kT is small. The molar magnetisation is then linear in H. When H/kT becomes large, M must be calculated from Equation (5) or (6). This calculation is rather tedious; it leads to:

$$M = Ng\beta S\, B_S(y) \qquad (15)$$

with:

$$y = \frac{g\beta SH}{kT} \qquad (16)$$

where $B_S(y)$ is the Brillouin function defined by:

$$B_S(y) = \frac{2S + 1}{2S} \coth\left(\frac{2S + 1}{2S}y\right) - \frac{1}{2S}\coth\left(\frac{1}{2S}y\right) \qquad (17)$$

When H/kT and hence y is small, $B_S(y)$ may be replaced by:

$$B_S(y) = \frac{y(S + 1)}{3S} + \text{term in } y^3 + \dots \qquad (18)$$

so that, in this approximation, $\chi = M/H$ is effectively given by the Curie law (Equation (14)). When, on the contrary, H/kT becomes very large, $B_S(y)$ tends to unity and M tends to the saturation value M_S:

$$M_S = Ng\beta S \qquad (19)$$

If the saturation magnetisation is expressed in Bohr magnetons per mole (β mol^{-1}), its value is simply given by gS. Equation (15) is valid for an assembly of non-interacting magnetic molecules. If these molecules interact within the crystal lattice in a ferromagnetic fashion (see below), the increase of M versus H is faster than the Brillouin function; the saturation magnetisation, on the other hand, remains the same. When a ferromagnetic ordering is achieved, the saturation magnetisation is, in principle, reached even in zero field.

When the spin multiplicity of the ground state is larger than 2, the coupling of this state with the excited states through the spin-orbit coupling may provoke a splitting of its Zeeman components in a zero applied magnetic field. This phenomenon, called *zero-field splitting*, leads to an anisotropy of the magnetic properties and to a deviation of the average magnetic susceptibility with respect to the Curie law. In the low-temperature range, χT decreases instead of being constant as the temperature is lowered.

Phenomenologically, the zero-field splitting within a $^{2S+1}\Gamma$ state without first-order angular momentum is expressed by the Hamiltonian:

$$H_{ZFS} = \mathbf{S} \cdot \mathbf{D} \cdot \mathbf{S} \qquad (20)$$

where $\mathbf{D}$ is a symmetric and traceless tensor. The total spin Hamiltonian taking into account the Zeeman perturbation is then in matrix notation:

$$H = \beta \mathbf{S} \cdot \mathbf{g} \cdot \mathbf{H} + \mathbf{S} \cdot \mathbf{D} \cdot \mathbf{S} \qquad (21)$$

where $\mathbf{g}$ = Zeeman-tensor.

Until now, we have supposed that the magnetic molecules were perfectly isolated within the crystal lattice. If not, the Curie law has to be replaced by:

$$\chi = \frac{Ng^2 \, \beta^2 S(S+1)}{3kT - zJS(S+1)} \qquad (22)$$

where J is the interaction parameter between two nearest neighbour magnetic molecules and z the number of nearest neighbours around a given magnetic molecule. Accordingly, as J is positive or negative, the intermolecular interaction is said to be ferro- or antiferromagnetic. In the former case, the spins tend to align in a parallel fashion; in the latter case, they tend to align in an antiparallel fashion. χ may be rewritten as:

$$\chi = \frac{C}{(T - \Theta)} \qquad (23)$$

which is known as the Curie-Weiss law. C is the Curie constant and Θ the Weiss temperature (or Weiss constant) defined by:

$$\Theta = \frac{zJS(S+1)}{3k} \qquad (24)$$

A plot of χ^{-1} = f(T) for a system obeying the Curie-Weiss law gives a straight line, the slope of which is C^{-1}. The intercept with the T axis yields both the sign and the value of Θ. Θ positive indicates ferromagnetic intermolecular interactions and Θ negative anti-ferromagnetic intermolecular interactions. It is important to point out here that a deviation with respect to the Curie law may have other origins than intermolecular interactions. Actually, the zero-field splitting has quite similar effects on the average magnetic susceptibility as $\Theta < 0$.

2.3   INTERACTION IN BINUCLEAR COMPOUNDS

There is no doubts that the most important developments in molecular magnetism in the last two decades have essentially concerned the compounds in which several magnetic centres interact. These magnetic momentum carriers may be transition metal ions, rare earths or organic radicals.

2.3.1   Isotropic Interaction in Binuclear Compounds

The most investigated binuclear compounds are by far those involving Cu(II) ions [3–11]. The interaction then occurs between two local doublet states. Let us consider a compound of this nature, in which two Cu(II) ions, noted A and B, in the same molecular entity are bridged by a diamagnetic ligand capable of transmitting the electronic effects between A and B. If the two metal ions interact through the bridge, the local spins $S_A = S_B = 1/2$ are not good quantum numbers. The good spin quantum numbers are S = 0 and 1. Let E(S = 0) and E(S = 1) be the energies of the two pair states. Due to electrostatic reasons, E(S = 0) and E(S = 1) are in general not equal, but separated by a J energy gap defined as :

$$J = E(S = 0) - E(S = 1) \qquad (25)$$

J is often referred to as the isotropic interaction parameter. When the state S = 0 is the ground state, the interaction is said to be antiferromagnetic; J is then negative. When the state S = 1 is the ground state, the interaction is said to be ferromagnetic and J is positive. This latter situation is much less frequent than the former one. The overwhelming majority of the Cu(II) binuclear compounds present an antiferromagnetic interaction.

In the absence of any other phenomenon, like those studied in Section 2.3.2, the Zeeman perturbation does not affect the singlet state and splits the components of the triplet state (Figure 1), which leads to:

$$\chi = \frac{2Ng^2\beta^2}{kT\left[3 + \exp\left(\dfrac{-J}{kT}\right)\right]} \tag{26}$$

Equation (26) was derived for the first time by Bleaney and Bowers [3]. For $J < 0$, the magnetic susceptibility rises to a maximum, and then tends to zero when the temperature approaches absolute zero. For $J > 0$, upon cooling, χ increases faster than in C/T. The best way to reveal a ferromagnetic interaction is to plot χT versus T. For $J = 0$, χT is constant and equal to $Ng^2\beta^2/2k$. For $J > 0$, χT is close to $Ng^2\beta^2/2k$ when $kT \gg J$. Upon cooling, χT increases, due to the depopulation of the diamagnetic excited state in favour of the triplet ground state, and tends to a plateau with $\chi T = 2Ng^2\beta^2/3kT$ corresponding to the temperature range where the excited singlet state is fully depopulated. The ratio $(\chi T)_{LT}/(\chi T)_{HT}$ between the low- and high-temperature limits is equal to 4/3. Of course, for $J < 0$, χT continuously decreases upon cooling.

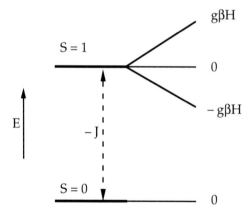

Figure 1 Effect of the magnetic field on the low lying states of a Cu(II) pair

The isotropic interaction phenomenon is purely electrostatic in nature, and its mechanism has been discussed in many papers [12, 13]. For our purpose, it is enough to remind the reader that J results from a competition between ferromagnetic J_F and antiferromagnetic J_{AF} contributions:

$$J = J_F + J_{AF} \qquad (27)$$

with: $J_F = 2k$ and $J_{AF} = 4\beta S$ (or $-4\beta^2/U$)

where k is the two-electron exchange integral, S the overlap integral and β the one-electron transfer integral:

$$k = <a(1)b(2)\left|\frac{1}{r_{12}}\right|a(2)b(1)>$$

$$S = <a(1)\,|\,b(1)> \qquad (28)$$

$$\beta = <a(1)\,|\,h(1)\,|\,b(1)>$$

and a and b are the magnetic orbitals centred on A and B respectively. $J_{AF} = 4\beta S$ corresponds to the natural orbital approach; the antiferromagnetism is then governed by the overlap integral between the magnetic orbitals. $J_{AF} = -4\beta^2/U$ corresponds to the orthogonalised orbital approach; the antiferromagnetism then arises from the stabilisation of the low-lying singlet state through coupling with the excited singlet state in which the two electrons occupy the same magnetic orbital a (or b).

This isotropic interaction is often phenomenologically described by a coupling between the local spins $\mathbf{S}_A$ and $\mathbf{S}_B$ as :

$$H = -J\,\mathbf{S}_A.\mathbf{S}_B \qquad (29)$$

and the energies E(S) of the low-lying states are given by :

$$E(S) = -JS(S+1)/2 \qquad (30)$$

Equation (29) is defined as the Heisenberg-Dirac-Van Vleck (HDVV) Hamiltonian. The HDVV Hamiltonian is valid for any pair of interacting magnetic centres with local spins S_A and S_B, provided that the local states have no first-order angular momentum. The pair states are defined by S varying by integer value from $|S_A - S_B|$ to $S_A + S_B$ with the relative energies given by Equation (30). Whatever S_A and S_B may be, the spin S varies monotonically versus the energy of the pair states. The system is said to present a regular spin state structure [14]. This regularity leads to a simple correspondence between the nature of the interaction and the shape of the χT versus T plot. The high-temperature limit of χT, for $kT >> |J|$, is the sum of what is expected for each of the magnetic centres, i.e.

$$(\chi T)_{HT} = \left(\frac{N\beta^2}{3k}\right)\left[g_A^2 S_A(S_A + 1) + g_B^2 S_B(S_B + 1)\right] \tag{31}$$

where g_A and g_B denote the principal values of the local g-tensors assumed to be isotropic. In the absence of interaction ($J = 0$), χT remains constant in the whole temperature range with the value given in Equation (31); this is the Curie law. If the interaction is antiferromagnetic ($J < 0$), the ground state has the smallest spin $|S_A - S_B|$ and the most excited state the highest spin $S_A + S_B$. Between these two limits, S increases when going up in energy. χT continuously decreases upon cooling and tends toward the low-temperature limit:

$$(\chi T)_{LT} = \left(\frac{N\beta^2 g_S^2}{3kT}\right)\left[(S_A - S_B)^2 + |S_A - S_B|\right] \tag{32}$$

where g_S ($S = |S_A - S_B|$) is related to the local g-factors as [15–17]:

$$g_S = (1 + c)g_A/2 + (1 - c)g_B/2 \tag{33}$$

with:

$$c = \frac{S_A(S_A+1) - S_B(S_B+1)}{S(S+1)} \tag{34}$$

Only in the particular case of $S_A = S_B$, the ground state is diamagnetic and the χ versus T plot exhibits the characteristic maximum. Otherwise, although χT decreases upon cooling, χ continuously increases. If the interaction is ferromagnetic, the spectrum of the low-lying states is reversed. χT continuously increases upon cooling and tends toward the low-temperature limit:

$$(\chi T)_{LT} = \left(\frac{N g_S^2 \beta^2}{3kT}\right)\left[(S_A + S_B)(S_A + S_B + 1)\right] \tag{35}$$

where g_S ($S = S_A + S_B$) is again related to the local g-factors as in Equation (33).

2.3.2 Dipolar, Anisotropic and Antisymmetric Interactions

The interaction between two ions within a binuclear entity, in addition to the splittings between the pair states, leads to splittings within these pair states with $S > 1/2$. This effect has two origins, namely the dipolar and anisotropic interactions.

The dipolar interaction arises from the influence of the magnetic field created by one of the magnetic ions on the magnetic moment due to the other magnetic ion. The anisotropic interaction results from the combined effect of the local spin-orbit coupling and the interaction between the magnetic centres. Both phenomena can be described by the same Hamiltonian $S_A \cdot D \cdot S_B$ where D is a traceless tensor. D is actually the sum of dipolar D^{dip} and anisotropic D^{ani} contributions.

When the binuclear entity is not symmetrical anymore, but rather of low symmetry, another phenomenon may be operative, namely the antisymmetric interaction, which contributes to the zero-field splittings within the pair states and, in addition, couples these pair states. The origin of the antisymmetric interaction is the same as that of the anisotropic interaction, i.e. the synergistic effect of the spin-orbit coupling and of the interaction between the magnetic centres. It is phenomenologically described by $\vec{d}\ S_A \wedge S_B$ where $\vec{d}$ is a vector. This effect vanishes not only when the binuclear unit is centrosymmetric but also when its molecular symmetry is C_{nv} with $n \geq 2$ and the n-fold axis joining the interacting centres, or higher. Therefore, for most of the compounds, the antisymmetric interaction is actually zero.

To our knowledge the presence of antisymmetric interaction in antiferromagnetically coupled pairs has not yet been demonstrated. On the other hand, in extended lattices of low symmetry, the antisymmetric interaction is at the origin of the weak ferromagnetism. This phenomenon may be visualised as follows: the antisymmetric interaction tends to orientate the neighbouring spins perpendicularly to each other while the isotropic interaction tends to orientate them in either parallel or antiparallel directions.

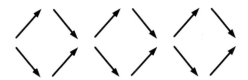

Figure 2   Schematic representation of spin canting in a two-dimensional array of spins

It results in a canting between these spins in the state which otherwise would be diamagnetic, and the onset of a magnetic ordering with a weak spontaneous magnetisation as shown in Figure 2 as the example of a two-dimensional array of spins.

It should be noted, however, that in most of the cases the onset of a spontaneous magnetisation results from a three-dimensional transition. In no case can it result from a one-dimensional transition.

2.4 MAGNETIC CHAIN COMPOUNDS

The one-dimensional magnetic compounds occupy an intermediate situation between the magnetic clusters and the three-dimensional extended lattices. Several review papers on the subject have already appeared [18, 19]. In this Section, we do not intend to review or even to survey this field. Rather, we will focus on some specific aspects which in our mind are more directly related to the molecular magnetic materials.

2.4.1 Chains of Equally Spaced Ions

The simplest case of a magnetic chain is provided by an array of equally spaced Cu(II) ions with $S_{Cu} = 1/2$ local spins, as schematised below:

$$— A_i \xrightarrow{\quad J \quad} A_{i+1} \xrightarrow{\quad J \quad} A_{i+2} —$$

Figure 3 Schematic representation of a one-dimensional array of identical spins

The spin Hamiltonian in zero-field adapted to describe the isotropic interaction between nearest neighbour ions is:

$$H = -J \sum_{i=1}^{n-1} S_{A_i} \cdot S_{A_{i+1}} \tag{36}$$

where the summation runs over the n sites of the chain. There is no analytical method to determine the energies of the low-lying states and the magnetic susceptibility when n tends to infinite. Nonetheless, the problem can be solved numerically by considering ring chains of increasing size and extrapolating for n becoming infinity. This method was applied for the first time by Bonner and Fisher in 1964 who calculated explicitly the magnetic susceptibility of ring chains up to $n = 11$ and proposed an extrapolation for the infinite ring [20]. As far as the extrapolation is concerned, a difficulty appears in the case where J is negative, related to the parity of n. If the interaction between nearest neighbours is antiferromagnetic, the ground state for n even is diamagnetic and the magnetic susceptibility χ (per Cu site) tends to zero when T approaches zero. On the other hand, the ground state for n odd is a doublet $S = 1/2$ and χ diverges when T approaches zero. The problem is then: what is the low-temperature limit of χ when n becomes infinite? Bonner and Fisher have shown that this limit is actually finite and given by:

$$(\chi)_{LT} = \frac{0.14692Ng^2\beta^2}{|J|} \tag{37}$$

This result may be understood as follows: when J is infinite, the energy levels form a continuum from one of the $S = 0$ levels up to the unique $S = n/2$ level. At zero K, only the bottom of this continuum is thermally populated, but since there is no gap between the $S = 0$ ground level and the levels immediately above, χ does not tend to zero.

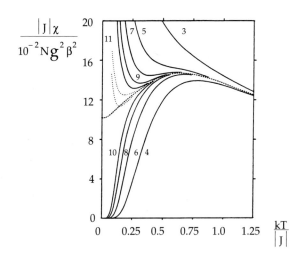

Figure 4 $|J|\chi/Ng^2\beta^2$ versus $kT/|J|$ curves for finite rings of n Cu(II) ions and extrapolation for $n \to \infty$

The results of Bonner and Fisher for J < 0 are represented in Figure 4 in the form of the $|J|\chi/Ng^2\beta^2$ versus $kT/|J|$ plot. χ passes through a rounded maximum at a temperature T_{max} defined by :

$$\frac{kT_{max}}{|J|} = 0.641 \tag{38}$$

For J > 0, not only χ but also χT continuously increase when the temperature decreases and diverge when T approaches zero. What is quite important to realise is that, whatever the nature (ferro- or antiferromagnetic) of the interaction, a one-dimensional magnetic compound does not order magnetically above absolute zero. This result was established for the first time by Ising in 1925. In a real solid, however, the chains are never perfectly isolated from each other; interchain interactions superimpose to dominant intrachain interactions, which leads to a three-dimensional magnetic ordering at a finite temperature. This important question will be discussed in detail in the next section.

An example of a ferromagnetically coupled Cu(II) chain is provided by hexylammonium copper(II)trichloride, $[CuCl_3(C_6H_{11}NH_3)]$, often abbreviated as CHAC [21]. The structure of this compound is represented in Figure 5. It consists of symmetrically doubly-bridged linear chains parallel to the c-axis of the orthorhombic lattice. Cu(II) is in square pyramidal surroundings with a non-bridging chlorine atom at the apex. The hexylammonium cations hydrogen bond the chains together into sheets in the bc-plane through the apical chlorine atoms and efficiently separate these sheets from each other along the a-direction. The CuClCu bridging angles are equal to 85.3° and 86°, respectively, which favours a ferromagnetic interaction along the chain. A weak interchain coupling in the bc-sheets may also be anticipated.

The interaction along the a-axis may be expected to be very small, and rather of a dipolar nature. The χT versus T plot for a powder sample of CHAC increases rapidly when T decreases. At 2.2 K, χT is approximately 30 times as large as at room temperature. From these data, an intrachain interaction parameter of 100 cm^{-1} has been extracted. At 2.18 K, χT shows a sharp maximum corresponding to the onset of a long-range magnetic order.

Magnetisation studies have revealed a metamagnetic behaviour, i.e. a change in the nature of the ground state when applying a magnetic field stronger than a critical value H_c. In the present case, H_c is close to 100 G. For $H < H_c$, the ground state below 2.18 K is antiferromagnetic, with a small canting of the spins, leading to a weak residual moment along the a-direction. From the magnitude of this residual moment, the angle between the spins is estimated as 17°. For $H > H_c$, the ground state is ferromagnetic, with a parallel alignment of all spins. The small value of H_c is due to the fact that the antiferromagnetic interchain coupling is very weak. Actually, this coupling is found along the a-axis with a ratio

$|J_a/J_c|$ between the interaction parameters along the a- and c- (chain) axes less than 10^{-4}. As for the interaction along the b-axis, it is ferromagnetic with J_b/J_c of the order of 10^{-3}.

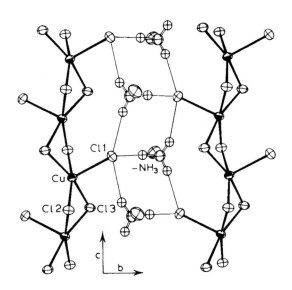

Figure 5 Crystal structure of $[CuCl_3(C_6H_{11}NH_3)]$ (CHAC)

To sum up, the compound CHAC may be described as follows: along c, we have ferromagnetic chains ($J_c = 100$ cm^{-1}); these chains interact ferromagnetically along the b-axis ($J_b \approx 10^{-1}$ cm^{-1}) to give ferromagnetic bc-planes. These planes, finally, are very weakly antiferromagnetically coupled along the a-axis ($J_a < -10^{-2}$), so that the three-dimensional ground state in zero-field is antiferromagnetic (ignoring the small canting). An applied field of 100 G is sufficient to overcome J_a.

The same method as that introduced by Bonner and Fisher may be utilised to determine the energy level spectrum and the magnetic susceptibility of a chain compound with local spins larger than $1/2$, provided that the Hamiltonian (Equation (**36**)) remains valid. However, the larger the value of the local spin, the smaller the number n of sites that is possible to take into account in the ring chain calculation. For instance, with $S_A = 5/2$, the calculation so far has been limited to 5 sites so that the accuracy of the extrapolation for infinite n is questionable. A conceptual difficulty appears for the chains with integer local spins. In 1983, Haldane suggested that the energy spectrum for such chains presents a gap between a singlet ground state and the first excited states so that, even in the absence of local anisotropy, the low temperature limit of χ would be zero [22]. Recent experimental data on Ni(II) chains seem to confirm Haldane's conjecture [23, 24].

2.4.2 Regular Ferrimagnetic Chains; Theory

The newest aspect in the field of magnetic chain compounds concerns the design and the investigation of systems in which two kinds of magnetic centres A and B regularly alternate. These compounds are called either ordered bimetallic chains, or ferrimagnetic chains; they may be arranged schematically as shown in Figure 6. The former name implicitly supposes that the magnetic centres are metal ions, which is not always the case; the latter supposes that the interaction parameter J is negative. The first compound of this type was reported in 1981. Since then, several families of ferrimagnetic chains have been described, and the main theoretical concepts adapted to this new class of materials have been discussed. It can be argued that the design of ferrimagnetic chains represents one of the most significant contributions of synthetic molecular chemistry in the magnetic materials area. For this reason we have devoted a rather large part of this chapter to both the theoretical and the experimental results in this area. We will initially focus on the theoretical aspects, then we will present some experimental results.

$$
\underline{} A_i \xrightarrow{\quad J \quad} B_{i+1} \xrightarrow{\quad J \quad} A_{i+2} \xrightarrow{\quad J \quad} B_{i+3} \underline{}
$$

Figure 6 Schematic representation of a one-dimensional array of alternating spins

2.4.2.1 *Qualitative Approach*

To introduce the basic concepts [25], we consider an ordered and regular ring chain $(AB)_n$ where n may become infinite. A and B carry the local spins S_A and S_B, respectively, with $S_A \neq S_B$. The spin Hamiltonian in zero-field appropriate to describe the isotropic interaction between nearest neighbours is:

$$
H = -J \sum_{i=1}^{2n} S_i \cdot S_{i+1} \tag{39}
$$

with: $\qquad S_{2i-1} = S_A, \quad S_{2i} = S_B \text{ and } S_{2n+i} = S_i$ $\qquad$ (40)

We suppose first that J is negative. The states of lowest E_g and highest E_e energies are represented in Figure 7.

(a)   (b)

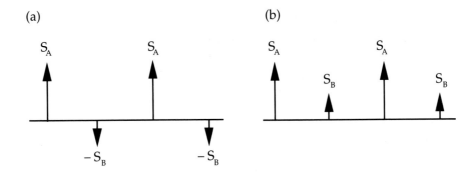

Figure 7 Schematic representations of (a) the ground state, and (b) the most excited state for a ferrimagnetic chain

The spins are $S_g = n(|S_A - S_B|)$ and $S_e = n(S_A + S_B)$. Between the two limits E_g and E_e, states with spins less than S_g do exist. In particular, there are states with $S = 0$ if the total number of unpaired electrons $2n(S_A + S_B)$ is even (Figure 8) or with $S = 1/2$ if this number is odd.

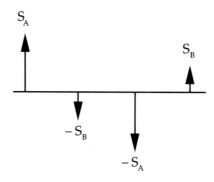

Figure 8 Schematic representation of a state with $S = 0$ for a ferrimagnetic chain

Let us study the temperature dependence of χT, χ being the molar magnetic susceptibility per AB unit. The Zeeman perturbation to add to Equation **(39)** is:

$$H = \sum_{i=1}^{n} \beta\left(g_{A_{2i-1}}S_{A_{2i-1}} + g_{B_{2i}}S_{B_{2i}}\right) \cdot H \tag{41}$$

We assume that the local tensors g_A and g_B are isotropic with g_A and g_B principal values, respectively. The high-temperature limit $(\chi T)_{HT}$ of χT corresponds to the sum of the local contributions, i.e.:

$$(\chi T)_{HT} = \left(\frac{N\beta^2}{3k}\right)\left[g_A^2 S_A(S_A+1) + g_B^2 S_B(S_B+1)\right] \tag{42}$$

The low-temperature limit $(\chi T)_{LT}$ of χT is reached when only the ground state is thermally populated. This limit is:

$$(\chi T)_{LT} = \left(\frac{Ng^2\beta^2}{3k}\right)\left[n(S_A - S_B)^2 + |S_A - S_B|\right] \tag{43}$$

where the Zeeman factor g for the ground state is a linear combination of g_A and g_B. For a peculiar size n_0 of the ring chain, $(\chi T)_{HT}$ and $(\chi T)_{LT}$ are equal. In the approximation where g_A and g_B are equal, n_0 is given by:

$$n_0 = \frac{(S_A^2 + S_B^2 + 2S_<)}{(S_A - S_B)^2} \tag{44}$$

where $S_<$ is the smaller of S_A and S_B. For $n > n_0$, $(\chi T)_{LT}$ is larger than $(\chi T)_{HT}$, and when n tends to infinity, $(\chi T)_{LT}$ diverges. Upon cooling from the high temperatures, the state of highest spin S_e depopulates first, and χT decreases. Therefore, this leads to the following result: for an AB ferrimagnetic chain, χT first decreases upon cooling, reaches a minimum for a finite temperature, then diverges when T approaches zero. This behaviour is valid for any couple of S_A and $S_B \neq S_A$ local spins, provided that there is no accidental compensation of the local magnetic momenta. Another way to express the same result is to say that at high temperature χT tends to the paramagnetic limit. The minimum of χT corresponds to a short-range order state where the spins S_A and S_B of adjacent magnetic centres are antiparallel, but without correlation between neighbouring AB units. When T decreases, the correlation length within the chain increases. The divergence of χT at low temperature may be associated with the onset of a magnetic ordering at 0 K. Below the temperature of the minimum of χT, the magnetic behaviour is qualitatively equivalent to what happens for a chain of n spins $|S_A - S_B|$ ferromagnetically coupled.

If the intrachain interaction is ferromagnetic, i.e. J > O, the order of the spin levels is reversed and no extremum of the χT versus T plot can be predicted *a priori*.

2.4.2.2 *Quantitative Approach*

Several quantitative approaches of the magnetic susceptibility of ferrimagnetic chains have been proposed so far. The first one consists of carrying out the calculation on $(AB)_n$ ring chains of increasing size with quantum numbers S_A and S_B and to extrapolate to n $\to$ ∞. The method is then an extension of what has been performed for the first time by Bonner and Fisher in the case of uniform chains of local spins $1/2$ (Section 2.4.1).

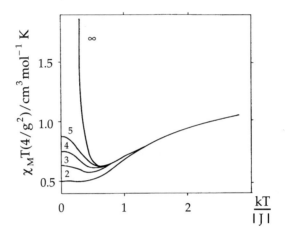

Figure 9 $\chi_M T$ versus $kT/|J|$ curves for finite rings $(AB)_n$ and extrapolation for $n \to \infty$

Even by taking into account the full D_N symmetry of the problem, this method is severely limited by the storage capacity of the computer as well as by the computing time. The first utilisation of this approach was for the case $S_A = 1/2$ and $S_B = 1$ [25–27]. Drillon *et al.* performed the calculation up to $n = 5$ and extrapolated for $n \to \infty$. The results for $g_A = g_B$ are shown in Figure 9. The minimum of χT is obtained for $kT/|J| = 0.570$. The divergence of χT when T tends to zero follows a law in $T^{-0.80}$, which coincides with the law found for a ferromagnetic chain of $1/2$ local spins. The $(AB)_n$ ring chain technique has been extended to several other cases, in particular for $S_A = 1/2$ and S_B taking all other values up to $5/2$. Except for $S_B = 1$, the calculation was performed only up to $n = 3$.

Another approach, of particular interest when $S_A - S_B$ is large, consists of treating S_A as a classical spin and S_B as a quantum spin. This approach was first introduced by Seiden in the case where $S_A = 5/2$ and $S_B = 1/2$ with $g_A = g_B$ [28, 29], then generalised by Georges *et al.* not only to any S_A,

S_B couple with possibly $g_A \neq g_B$, but also to the case of alternating ferrimagnetic chains with two interaction parameters $J_{AB}(1 + \alpha)$ and $J_{AB}(1 - \alpha)$ [30, 31].

2.4.3 Regular Ferrimagnetic Chains; Some Examples

The first structurally characterised ferrimagnetic chain is $[MnCu(dto)_2(H_2O)_3.4.5H_2O]$ (dto = dithiooxalato) [29, 32]. This compound has been investigated by Gleizes and Verdaguer. The structure of two adjacent chains as well as the detail of the structure of one of the chains are shown in Figure 10.

(a)

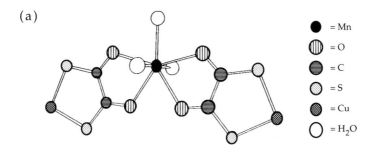

● = Mn
◐ = O
⊜ = C
◉ = S
▨ = Cu
○ = H_2O

(b)

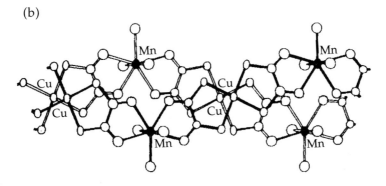

Figure 10 (a) Crystal structure and (b) packing for $[MnCu(dto)_2(H_2O)_3.4.5H_2O]$

Cu(II) is tetracoordinated with a planar environment and Mn(II) is heptacoordinated in a rather unusual fashion. The metal ions are bridged by the dithiooxalato ligand, the oxygen atoms being bound to manganese and the sulfur atoms to copper. The magnetic data closely follow the theoretical predictions. Upon cooling, χT decreases, reaches a weakly pronounced minimum around 130 K and then increases as T is lowered further down to 7.5 K. Finally, below 7.5 K, χT decreases rapidly. In the range 7.5–300 K, these data are well fitted by the classical-quantum model, with $J = -30.3$ cm^{-1}. The maximum of χT at 7.5 K is clearly due to antiferromagnetic interchain interactions. Magnetisation studies at 1.3 and 4.2 K show a saturation corresponding to a spin $S_A - S_B = 2$ per MnCu unit. The relatively large value of J in [MnCu(dto)$_2$(H$_2$O)$_3$.4.5H$_2$O] confirms the efficiency of the dithiooxalato bridge to propagate a strong antiferromagnetic interaction between metal centres separated by more than 5 Å.

Another example of a Mn(II)–Cu(II) bimetallic chain is provided by [MnCu(pba)(H$_2$O)$_3$.2H$_2$O], (pba = 1,3-propylenebis(oxamato)) [33]. The structure of the chain is shown in Figure 11.

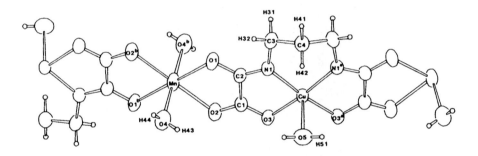

Figure 11   Crystal structure for [MnCu(pba)(H$_2$O)$_3$.2H$_2$O]

Mn(II) is in distorted octahedral surroundings and Cu(II) in square pyramidal surroundings. The metal ions are aligned along the b-axis of the orthorhombic structure, whereas in the previous example they formed zig zag chains. Within the chains, two nearest neighbour ions are bridged by an oxamato group with a Mn···Cu separation of 5.41 Å. As expected, the χT versus T plot exhibits a rounded minimum about 115 K and below this temperature a very fast increase upon cooling. A maximum of χT is observed, but at much lower temperature than for the previous example, namely 2.3 K instead of 7.5 K. The one-dimensional character is more pronounced. The fitting of the magnetic data above 4.2 K leads to $J = -23.4$ cm^{-1}. The maximum of χT at 2.3 K is related to the onset of a long-range antiferromagnetic ordering. This point will be treated in the next Section.

Our last example of a ferrimagnetic chain concerns a compound in which the spin carriers are alternately Mn(II) ions with local spins 5/2 and nitronyl nitroxide radicals with local spins 1/2. The structure of these organic radicals is shown in Figure 12. R may be an alkyl or an aromatic group.

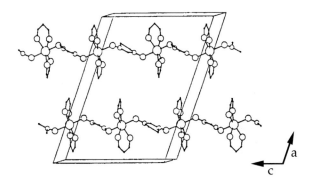

Figure 12 Structure of the nitronyl nitroxide radical

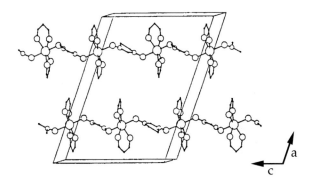

Figure 13 Unit cell for [Mn(hfa)$_2$NITPr^i]. The chains develop parallel to the c-axis

These radicals can bind to two different metal atoms with their two equivalent oxygen atoms, provided that the metal atoms are strong Lewis acids [34–37]. The crystal structure of [Mn(hfa)$_2$NITPi^i] (hfa = hexafluoroacetylacetonato, NITPr^i = isopropyl nitronyl nitroxide) is shown in Figure 13. This structure consists of chains of Mn(hfa)$_2$ units bridged by NITPi^i radicals. The Mn(II) ions are hexacoordinated by four oxygen atoms of two hfa groups and two oxygen atoms of two radicals in trans positions. The χT versus T plot continuously increases upon cooling from room temperature down to 7.6 K. At this temperature, χT reaches a sharp maximum, due to three-dimensional effects, with a value of the order of 350 cm^3 K mol^{-1}. We will discuss later the three-dimensional ordering in compounds of this kind. Here, we restrict ourselves to the one-dimensional behaviour. [Mn(hfa)$_2$NITPr^i] is a very good example of a ferrimagnetic chain compound with a large interaction parameter. The absence of a minimum in the χT variation is due to the fact that |J| is larger than $kT_{max}/2.98$, T_{max} being here the highest temperature investigated, namely the room temperature. Actually, the fitting of the magnetic data leads to a J value of the order of -250 cm^{-1}.

2.4.4 Alternating Ferrimagnetic Chains

Since the possibilities of synthesis in molecular chemistry are almost limitless, it was foreseeable that shortly after the regular ferrimagnetic chains, the first alternating ferrimagnetic chains would be reported [30]. This happened in 1988 along with a theoretical model appropriate for describing the magnetic properties of such rather exotic systems.

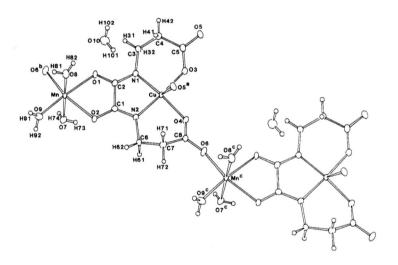

Figure 14 Crystal structure for [MnCu(obp)(H$_2$O)$_3$.H$_2$O]

The first reported example of an alternating ferrimagnetic chain is [MnCu(obp)(H$_2$O)$_3$.H$_2$O] (obp = oxamidobis(propionato)). The chain structure shown in Figure 14 consists of octahedral Mn(II) and planar Cu(II) ions alternately bridged by oxamido and carboxylato groups. The Mn$\cdots$Cu separations are equal to 5.452 and 6.066 Å, respectively. The variation of χT versus T presents a rounded minimum about 40 K, characteristic of the one-dimensional ferrimagnetic behaviour, and a sharp maximum at 2.9 K, related to the onset of a three-dimensional antiferromagnetic ordering at 2.3 K. The magnetic data above 10 K were interpreted with $J_{AB}(1 + \alpha) = -32(1)$ cm^{-1} and $J_{AB}(1 - \alpha) = -7(1)$ cm^{-1}.

There is no doubt that the former interaction parameter has to be associated with the oxamido bridge and the latter with the carboxylato bridge.

2.5 MAGNETIC LONG-RANGE ORDERING IN MOLECULAR COMPOUNDS; DESIGN OF MOLECULAR-BASED FERROMAGNETS

2.5.1 Three-dimensional Magnetic Ordering and Molecular-based Ferromagnets

In most of the molecular compounds, the molecular entities are rather well isolated from each others in the crystal lattice. However, at low temperature, the intermolecular interactions cannot be ignored. For any assembly of molecular species with a magnetic ground state, a three-dimensional magnetic ordering is expected at a finite temperature T_C. When the intermolecular contacts are only of the van der Waals' type, T_C may be very low, sometimes in the 10^{-2} or even 10^{-3} Kelvin range. On the other hand, when hydrogen bonds provide some interaction pathways, T_C may reach a few Kelvin. At temperatures much higher than T_C the orientation of the molecular spins is independent of that of their neighbours. The length along which two adjacent molecular spins are correlated, or correlation length, is zero. When the temperature is lowered and approaches the critical temperature, this correlation length increases. Just above T_C the correlation length becomes very large but still finite. At T_C it becomes infinite. If the orientation between the nearest neighbour molecules is ferromagnetic, the molecular spins are oriented in a parallel fashion. If the interaction is antiferromagnetic, they are oriented in a way that two adjacent molecular spins are antiparallel. In a few cases, when the symmetry of the crystal structure is low enough, the molecular spins may be not rigorously antiparallel but canted, which gives rise to weak ferromagnetism (Section 2.3.2).

Transitions from a paramagnetic state to an ordered magnetic state are second order phase transitions with typical λ-like discontinuities in the heat capacity curves, and characteristic susceptibility and magnetisation behaviours. Those aspects are treated in details in numerous books dealing with magnetic materials [38, 39]. As far as we are concerned, we will focus on the behaviour of assemblies of molecules exhibiting a three-dimensional ferromagnetic behaviour. For many years, propositions or speculations have been put forward to favour ferromagnetic interactions between magnetic molecules, and so to design molecular-based ferromagnets, but only since 1986 have compounds of this kind really been characterised.

Let us provide some information concerning the magnetic behaviour of a molecular-based ferromagnet. We consider an assembly of magnetic molecules interacting ferromagnetically, and we assume that these molecules are all of the same type. At temperatures much above T_c, the intermolecular interactions do not influence the magnetic properties. If each molecule contains a unique magnetic centre, the magnetic susceptibility χ varies according to a Curie law provided that there is no first-order orbital momentum. If each molecule contains several magnetic centres, the intramolecular interactions in general dominate. They may be either ferro- or antiferromagnetic, but in this latter case, they must lead to a non-diamagnetic ground state. Such a situation is achieved, for instance, for an odd number of equivalent local spins, or for a pair of non-equivalent local spins. In the absence of intermolecular interaction, in the temperature range where only the molecular ground state is thermally populated, the molar magnetic susceptibility χ should follow a Curie law with a Curie constant depending on the spin multiplicity of the ground state. Actually, we assume that significant intermolecular ferromagnetic effects are operative. If so, when the temperature decreases and approaches T_c, χ increases faster than anticipated for a Curie law, which is easily detected when plotting χT versus T. χT increases more and more rapidly upon cooling. This increase of χT is related to the increase of the correlation length. At temperatures not too close to T_c, the intermolecular interactions can be fairly well accounted for by the mean field approximation. In principle, at T_c, χ and χT should diverge and a spontaneous magnetisation appear. Upon cooling further, the magnetisation increases up to a saturation value $M_S = Ng\beta S$, where S is the spin associated with the molecular ground state and g the Zeeman factor.

In practice, most often, one detects no magnetisation in zero field. This is due to two factors, the domain formation and the demagnetisation field. A sample of a ferromagnetic material is generally broken in domains. Each domain has a net magnetisation in zero field, directed in a given direction. Within the sample the magnetic moments of the domains are randomly oriented so that the resulting magnetisation is zero. An external magnetic field provokes a displacement of the domain walls along with the formation of a new domain structure. The magnetic moments of the domains are not randomly oriented anymore but tend to align along the field.   A magnetisation M is then observed.   If the sample was a monodomain, the M = f(H) variation at T ≤ T_c would be as shown in Figure 15, with an infinite zero-field susceptibility $(\partial E_n / \partial H)_{H=0}$.   Actually, owing to the domain formation, this slope of the M versus H plot for H = 0, although very large, is not infinite.

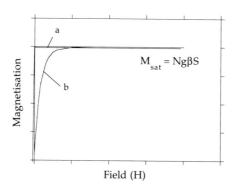

$$M_{sat} = Ng\beta S$$

Field (H)

Figure 15 M = f(H) plots for a ferromagnetic material below T_c; (a) ideal behaviour, (b) actual behaviour

It must be stressed here that the increase of M versus H is much faster for a ferromagnetic material than for a paramagnetic system, so that no confusion can be made.   The other factor limiting the zero field susceptibility, the demagnetizing field, is due to uncompensated moments at the surface of the sample, opposite to the direction of the magnetisation of this sample.   This demagnetizing field can be, in principle, estimated knowing the geometry of the sample.

So far, in this Section, we have ignored the anisotropic effects due to spin-orbit coupling.   These effects are anticipated to be negligible for purely organic materials, the magnetic properties of which are associated with s- or p-electrons.   The EPR spectroscopy of organic radicals clearly shows that the orbital contribution is extremely weak.   The resonance is always observed for g-values very close to 2.0023.   On the other hand, the magnetic anisotropy may be important for metal ion-containing compounds.   It leads to a preferred spin orientation, called the easy magnetisation axis.   For T ≤ T_c, the zero-field susceptibility $(\partial E_n / \partial H)_{H=0}$ is larger, and the saturation magnetisation M_S is reached for a weaker field when this field is applied along this easy magnetisation axis.

There is another aspect which is quite typical of the ferromagnetic materials, including the molecular-based ferromagnets, namely the remanence and the hysteresis. When the magnetic field is switched on below T_c, then switched off, the field-induced magnetisation does not disappear totally, in contrast with what happens for a paramagnetic system. This remnant magnetisation, in some cases, may be equal to the field-induced magnetisation. All the information is retained. To suppress the remnant magnetisation, it is necessary a apply a coercive field in the opposite direction. Remnant magnetisation and coercive field define the hysteresis loop conferring a memory effect to the material. The mechanism of the hysteresis phenomenon for a ferromagnet is beyond the scope of this book. We restrict ourselves to saying that it takes much more time for the magnetic moments of the ferromagnetic domains to reorient randomly when the field is switched off than for those of the paramagnetic molecules.

2.5.2   Orbital Degeneracy and Ferromagnetic Interaction

In Section 2.3, we have seen that the antiferromagnetic interaction in an AB pair may be interpreted as resulting from a coupling between the singlet state arising from the ground configuration and one of the singlet states arising from the charge transfer configuration. This is the orthogonalised magnetic orbital view. In such an approach, it is supposed that the two magnetic orbitals a and b, centred on A and B, respectively, are well separated in energy from both the doubly occupied orbitals of lower energy and the vacant orbitals of higher energy. Quite a different situation might be anticipated if actually the two interacting local states were orbitally degenerate or quasi-degenerate. Let us assume that there are two degenerate orbitals a_1 and a_2 around A, and b_1 and b_2 around B, with one unpaired electron around each centre. In the ground configuration the two electrons may have parallel or antiparallel spins. The state of lowest energy arising from the charge transfer configuration A^+B^- (or $A^- B^+$) is certainly a triplet state; the electrons occupy different orbitals to respect Hund's rule. The coupling between the triplet states arising from ground and excited configurations stabilises the former and favours a ferromagnetic interaction, as shown in Figure 16.

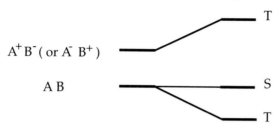

Figure 16 Schematic representation of the stabilisation of the low-lying triplet state (T) by coupling with the excited triplet state arising from the charge-transfer configuration

These arguments have been theoretically tested on a simple system, the benzene dimer dianion with parallel rings and D_{6h} symmetry [40, 41]. The energies of the 10 singlet and 6 triplet states coming from the various occupations of the π-orbitals have been calculated using a full configuration interaction, and, for any separation between the rings, the ground state is found to be a triplet. For a separation of 3.2 Å corresponding to the equilibrium geometry, the energy gap between the ground triplet and the first excited singlet is about 1100 cm^{-1}. If so, this excited singlet is completely depopulated, even at ambient temperature. Actually, the first excited state is not the singlet of lowest energy but another triplet.

If, instead of a pair of benzene anions, one considers an infinite stack, the same arguments suggest that the system will behave as a ferromagnetic chain. In principle, the requirements for applying these ideas are rather simple; one needs a stack of radicals of which the symmetry is high enough to give an orbital degeneracy. This is realised if the radical has a C_n symmetry axis with $n \geq 3$. All the synthetic efforts along this line, however, has been unsuccessful so far.

Much the same ideas had been put forward by McConnell as early as 1967 [42], but in a slightly different context. Always, in the perspective of designing organic ferromagnets, McConnell considered alternating stacks of donor cations D^+ and acceptor anions A^- where either A or D, but not both, has a symmetry high enough to allow an orbital degeneracy. We present here an extension of this McConnell idea proposed by Breslow [43] where it is assumed that D and A are so efficient donor and acceptor, respectively, that $D^{2+}A^{2-}$ is lower in energy than D^0A^0. In other words, the forward charge transfer is favoured with respect to the backward charge transfer. We suppose first that D is highly symmetrical with two orbitally degenerate highest occupied molecular orbitals (HOMO) and A of lower symmetry with a non-degenerate lowest unoccupied molecular orbital (LUMO). The ground states of D^+ and D^{2+} are then an orbitally degenerate spin doublet and an orbitally non-degenerate spin triplet, respectively. Those of A^- and A^{2-} are an orbitally non-degenerate spin doublet and a closed-shell spin singlet, respectively, as shown in Figure 17.

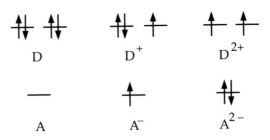

Figure 17 Schematic representation of the configurations formed when D has orbitally degenerate HOMOs

If, on the contrary, A is highly symmetrical with two degenerate LUMOs and D of lower symmetry with a non-degenerate HOMO, the situation is as shown in Figure 18.

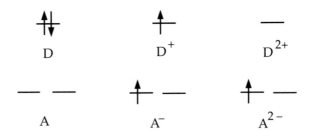

Figure 18 Schematic representation of the configuations formed when A has orbitally degenerate LUMOs

In both cases, due to Hund's rule, the lowest state arising from $D^{2+}A^{2-}$ is a triplet and the coupling between D^+A^- and $D^{2+}A^{2-}$ stabilises the triplet of lower energy. At the scale of an infinite stack, this would lead to a one-dimensional ferromagnetic behaviour. In order to test this strategy, Breslow *et al.* have synthesised a great many new aromatic molecules of high symmetry [44–46]. However, they never observed any ferromagnetic interaction and *a fortiori* any bulk ferromagnetic behaviour. Actually, all the efforts to test *McConnell's mechanism* have until now failed, a situation which pushed the theoreticians to investigate the status of this mechanism in a more thorough manner. Indeed, the arguments above are greatly over-simplified. It is true that the lowest state arising from $D^{2+}A^{2-}$ is a spin triplet, but this state is not the only one. In addition to this triplet state, there are three singlet states, two of them being possibly degenerate. This situation is reminiscent of what happens in the dioxygen molecule. The states arising from the $(\pi_g)^2$ ground configuration are $^3\Sigma_g^-$, which is the lowest, $^1\Delta_g$ and $^1\Sigma_g^+$. The energy separations between triplet and singlet states arising from the $D^{2+}A^{2-}$ configuration are proportional to the one-site exchange integral $k^0 = <a_1(1)a_2(2)|1/r_{12}|a_1(2)a_2(1)>$, a_1 and a_2 being the two degenerate orbitals for D or A. The coupling between the singlet states coming from $D^{2+}A^{2-}$ and D^+A^- stabilises the low-lying singlet state, in the same way as the coupling between the two triplet states stabilises the low-lying triplet state. It has been demonstrated that if k^0 is smaller than the energy gap U between $D^{2+}A^{2-}$ and D^+A^-, then the low-lying singlet state is more stabilised than the low-lying triplet, and the overall interaction is antiferromagnetic. $k^0 < U$ seems to be the most probable situation. If so, *McConnell-Breslow's mechanism* does not operate [47].

2.5.3 Ferromagnetic Transition in Decamethylferrocenium Tetracyanoethenide

Miller *et al.* in 1987 characterised a ferromagnetic transition in the organometallic donor-acceptor salt decamethylferrocenium tetracyano-ethenide [48]. This compound is obtained by a simple electron transfer reaction between an electron donor, decamethylferrocene $Fe(Cp^*)_2$ ($Cp^* = \eta$-C_5Me_5), and an electron acceptor, tetracyanoethylene TCNE (Figure 19). The crystal structure of an acetonitrile solvate $[Fe(Cp^*)_2][TCNE].MeCN$ consists of chains of alternating $[Fe(Cp^*)_2]^+$ and $[TCNE]^-$ units with an Fe···Fe separation along the chain of 10.42 Å. $[TCNE]^-$ lies between essentially parallel Cp^* rings, but the TCNE and Cp^* adjacent planes are not rigorously parallel. They make a dihedral angle of 2.8°. The shortest interchain separation is equal to 8.23 Å. It is associated with pairs of chains in which the cation of one of the chains has the anion of the other chain as nearest neighbour (Figure 20).

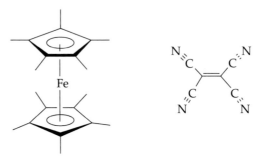

Figure 19 The structure of Fe(Cp*)$_2$ and TCNE

It should be noticed that this MeCN solvate is not the compound for which the physical properties have been investigated. This solvate easily loses its MeCN molecule with a transformation of the lattice from monoclinic to orthorhombic. The structure of the orthorhombic phase is strongly disordered so that it has not been possible to refine it properly. All the magnetic studies, however, were performed with this desolvated orthorhombic phase.

Let us present now the main physical properties of [Fe(Cp*)$_2$][TCNE]. The χT versus T plot increases when T is lowered, and follows the behaviour expected for a Heisenberg ferromagnetic chain compound. The intrachain interaction parameter is found as $J = 26$ cm^{-1}. The temperature dependence of the magnetisation within the magnetic field of the earth for a polycrystalline sample shows a break at $T_C = 4.8$ K, then saturates at a value close to 2×10^3 cm^3 G mol^{-1} (Figure 21) when cooling below T_C.

The magnetisation at 4.2 K for a single crystal saturates at a value of 1.6×10^4 cm^3 G mol^{-1} for magnetic fields parallel to the chain axis of a few tens of Gauss. This value, M_S, of the saturation magnetisation is interpreted as resulting from the ferromagnetic alignment of the spins S = 1/2 of the donor D$^+$ and acceptor A$^-$ units with parallel g-factors $g_D = 4$ and $g_A = 2$.

$$M_S = N(g_D + g_A)\beta/2 \tag{45}$$

The magnetisation versus field loop at 2 K shows a pronounced hysteresis effect with a coercive field of the order of 10^3 G (Figure 23).

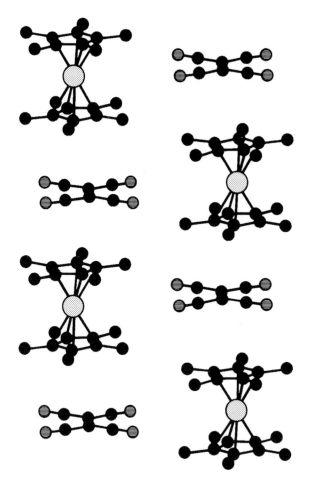

Figure 20 Crystal structure for [Fe(Cp*)$_2$][TCNE].MeCN

All the experimental data are consistent with the following representation of [Fe(Cp*)$_2$]]TCNE]. The spins 1/2 associated with the donor and acceptor units are strongly coupled along the chains in a ferromagnetic fashion (J = 26 cm^{-1}); the ferromagnetic chains are weakly coupled at the scale of the crystal lattice, again in a ferromagnetic fashion. This shows that the compound exhibits bulk ferromagnetic properties with a spontaneous magnetisation below T$_C$ = 4.8 K. The problem at hand now is to understand the mechanism of the intra- and intermolecular ferromagnetic interactions.

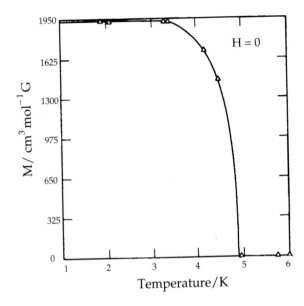

Figure 21 Zero-applied field temperature dependence of the magnetisation for [Fe(Cp*)₂][TCNE]

Miller, Epstein *et al.* have attributed the ferromagnetic behaviour of [Fe(Cp*)₂][TCNE] to McConnell's mechanism [48–50]. The 3d metal orbitals of iron(II) in the crystal field due to the Cp* rings are split into $a_1(d_z^2)$, $e_2(d_x^2 {}_- {}_y^2$ and $d_{xy})$ and $e_1(d_{yz}$ and $d_{zx})$ with the relative energies as $a_1 \leq e_2 << e_1$. a_1 is very close to, or accidentally degenerate with, e_2 (Figure 22).

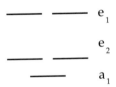

Figure 22 Energy level diagram for the d-orbitals in decamethylferrocene

For neutral Fe(Cp*)₂ the ground configuration is $(a_1)^2(e_2)^4$, and for the monocation [Fe(Cp*)₂]⁺ it is $(a_1)^2(e_2)^3$ with a local spin doublet state. For neutral TCNE, the first antibonding orbital π^* is empty and for the monoanion $[TCNE]^-$, it is singly occupied so that the local ground state is also a spin doublet. The interaction between the unpaired electrons of D^+ and A^- leads to singlet and triplet states. Let us assume now that the dication-dianion $D^{2+}A^{2-}$ is low enough in energy to interact with D^+A^-. The configuration of lower energy is $(a_1)^2(e_2)^2$ for D^{2+}, and $(\pi^*)^2$ for A^{2-}, which leads to a triplet state. The interaction between the two triplets, arising from D^+A^- and $D^{2+}A^{2-}$, respectively, stabilises the former, which corresponds to a ferromagnetic interaction as shown in Figure 24.

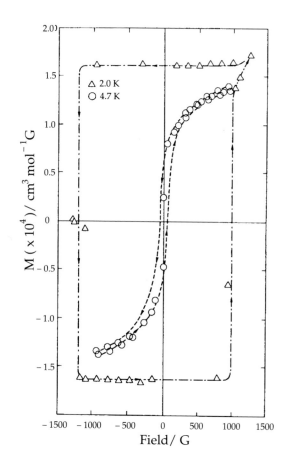

Figure 23 Hysteresis curve for $[Fe(Cp^*)_2][TCNE]$

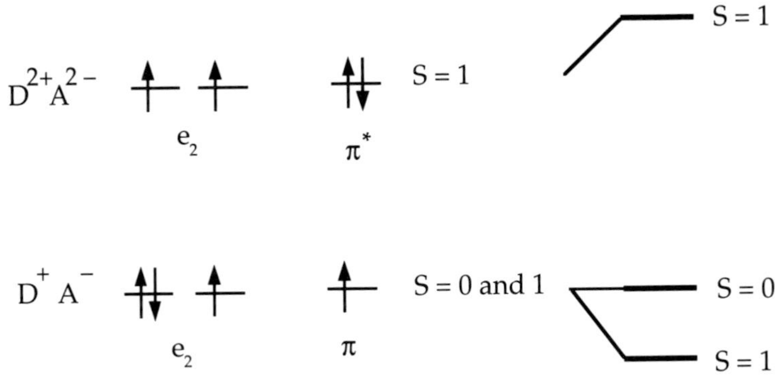

$$D^{2+}A^{2-} \quad \underset{e_2}{\uparrow \quad \uparrow} \quad \underset{\pi^*}{\uparrow\downarrow} \quad S = 1 \qquad \underline{\qquad} \; S = 1$$

$$D^{+}A^{-} \quad \underset{e_2}{\uparrow\downarrow \quad \uparrow} \quad \underset{\pi}{\uparrow} \quad S = 0 \text{ and } 1 \qquad \underline{\qquad} \; S = 0 \\ \underline{\qquad} \; S = 1$$

Figure 24 Schematic diagram showing the interaction of the two triplet states arising from D^+A^- and $D^{2+}A^{2-}$

This mechanism is assumed to work both for intra- and interchain interactions. Miller, Epstein *et al.* have extended this model to other D^+A^- stack compounds. At the first view, this interpretation based on McConnell's mechanism seems to be quite satisfying, especially as it allows us to rationalise a series of experimental data. However, some severe objections may be raised. First of all, we have pointed out that McConnell's mechanism could not be of the general validity assumed during two decades. Furthermore, nothing proves that $D^{2+}A^{2-}$ is lower in energy that D^0A^0. Finally, and this might well be the most serious objection, the stabilisation of the low-lying triplet state due to the D^+A^- /$D^{2+}A^{2-}$ interaction is proportional to the square of a matrix element of the form $<e_2|H|\pi^*>$. If we neglect the small dihedral angle between the Cp* and TCNE planes, then the e_2 orbitals of $Fe(Cp^*)_2$ and π^*-orbital of TCNE are strictly orthogonal, so that the two integrals $<e_2|H|\pi^*>$ are zero. Even by taking into account this angle, the value of these integrals remains extremely weak, and does not seem to explain a stabilisation of the low-lying triplet by *ca* 26 cm^{-1}. An alternative and more simple interpretation would be to say that in $[Fe(Cp^*)_2][TCNE]$ the magnetic orbitals e_2 for D^+ and π^* for A^- are orthogonal (if we neglect the dihedral angle between Cp* and TCNE planes) or quasi-orthogonal, which favours a ferromagnetic interaction. But, here again, an objection may be raised. The stabilisation of the triplet state with respect to the singlet state in the case of strict orthogonality of the magnetic orbitals is governed by the two-electron exchange integrals $<e_2(1)\pi^*(2)|1/r_{12}|e_2(2)\pi^*(1)>$. Such integrals are non-vanishing only when there are some zones of the space where the overlap densities $e_2(1)\pi^*(1)$ have significant values. This is apparently not the case. The two e_2 orbitals are very weakly delocalised toward the Cp* rings, and, more importantly toward the TCNE units. In turn, the π^*-orbital is vanishing in the close environment of the iron atom.

In 1990, Hoffmann *et al.* synthesised [Mn(Cp*)$_2$][TCNQ], and found that this novel charge-transfer compound exhibits a ferromagnetic transition at T_c = 6.2K [51]. These authors also invoked McConnell's mechanism. The characterisation of ferromagnetic transitions in [Fe(Cp*)$_2$][TCNE] and [Mn(Cp*)$_2$][TCNQ] is a results of the utmost importance in molecular magnetism. The interpretation of these results, however, remains an open problem.

2.5.4   Ferromagnetic Ordering of Ferrimagnetic Chains

2.5.4.1   [MnCu(pbaOH)(H$_2$O)$_3$] and [MnCu(pbaOH)(H$_2$O)$_2$]

Another approach to the design of molecular-based compounds exhibiting a ferromagnetic transition consists of assembling ferrimagnetic chains within the crystal lattice in a ferromagnetic fashion. Such an approach was successfully applied for the first time in the case of the compound [MnCu(pbaOH)(H$_2$O)$_3$] (pbaOH = 2-hydroxy-1,3-propylenebisoxamato) [52]. The structure of this compound consists of Mn(II)Cu(II) chains almost identical to those encountered in [MnCu(pba)(H$_2$O)$_3$.2H$_2$O] (Section 2.4.2 and Figure 11). The two compounds hereafter are abbreviated as [MnCu(pbaOH)] and [MnCu(pba)], respectively. The crystal lattices, however, are different. For both compounds, the structure is orthorhombic, the chains running along the *b*-axis. Along the *c*-axis, the chains are related by a unit-cell translation with Mn···Mn and Cu···Cu shortest interchain separations along this direction (Figure 23). The structural difference between the two compounds concerns the relative positions of the chains along the *a*-axis. The shortest metal···metal separations between neighbouring chains in the *a*-direction are Cu···Cu and Mn···Mn in [MnCu(pba)], and Mn···Cu in [MnCu(pbaOH)]. Such a difference may be described as follows: in the pbaOH compound, as compared to the pba one, every other chain is displaced by almost half of a repeat unit along the *b*-axis as schematised in Figure 25.

96   *Olivier Kahn, Yu Pei and Yves Journaux*

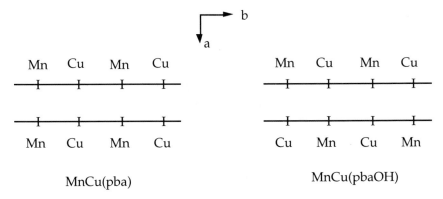

Figure 25 Schematic diagram showing the relative positions of the Cu^{2+} and Mn^{2+} ions in [MnCu(pba)] and [MnCu(pbaOH)]

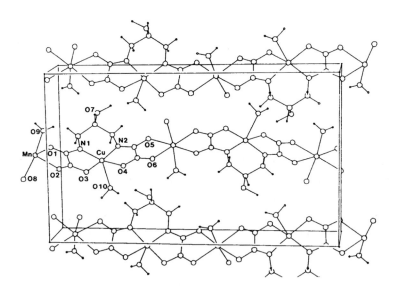

Figure 26 Relative positions of the chains along the *a*-axis in [MnCu(pbaOH)(H$_2$O)$_3$]

The magnetic properties of MnCu(pbaOH) down to 30 K are identical, within the experimental uncertainties, to those of [MnCu(pba)]. The χT versus T plot exhibits a minimum around 115 K, characteristic of the one-dimensional ferrimagnetic behaviour. On the other hand, when T is lowered below 30 K, the increase of χT is much more abrupt for the pbaOH compound than for the pba one. Below 5 K, the two compounds behave quite differently. We have already seen that [MnCu(pba)] exhibits a maximum of χT at 2.3 K and a maximum of χ at 2.2 K, due to the onset of a three-dimensional antiferromagnetic ordering. In contrast, χT for [MnCu(pbaOH)] reaches values as high as 100 cm^3 K mol^{-1} and becomes strongly field-dependent.

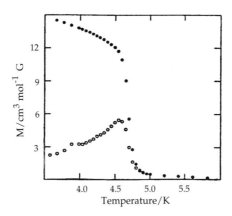

Figure 27 Magnetisation M versus T curves for [MnCu(pbaOH)(H$_2$O)$_3$]

The variation of the molar magnetisation M versus T within a field of 3×10^{-2} G for a polycrystalline sample confirms that a ferromagnetic transition occurs (Figure 27). The field-cooled magnetisation (FCM) obtained on cooling within the field shows the typical features of a ferromagnetic transition, i.e. a rapid increase of M when T decreases below 5 K, then a break in the curve at T_C = 4.6 K and finally the beginning of the saturation. If the field is switched off below T_C, a remnant magnetisation is observed. The zero-field-cooled magnetisation (ZFCM) is obtained by cooling down below T_C in zero field and then applying the field and heating. At any temperature below T_C, the ZFCM is smaller than the FCM, due to the fact that in this temperature range the applied field is too weak to move the domain walls. The ZFCM displays a maximum at T_C, as expected for a polycrystalline ferromagnet. [MnCu(pbaOH)] also exhibits a hysteresis loop M = f(H) characteristic of a soft ferromagnet.

Magnetic anisotropy measurements and single-crystal EPR data have shown that the preferred spin orientation for both [MnCu(pba)] and [MnCu(pbaOH)] is along the c-axis. The dramatic difference for the magnetic behaviours has been related to the difference of the crystal packings. Assuming that the interactions between nearest neighbour magnetic ions are antiferromagnetic both along the chain (*b*-axis) and along the *a*-axis, the relative positions of the chains along *a* lead to a cancellation of the resulting spin for [MnCu(pba)]. In contrast, the displacement of every other chain in the *ab*-plane of [MnCu(pbaOH)] by roughly half of a repeat unit leads to a parallel alignment of the $S_{Mn} = 5/2$ local spins as shown in Figure 28.

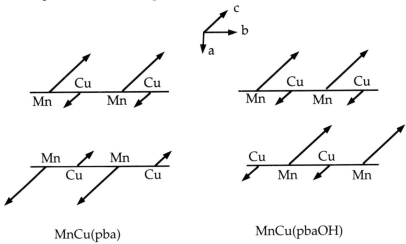

MnCu(pba)   MnCu(pbaOH)

Figure 28 Schematic representation of the preferred spin alignment in [MnCu(pba)] and [MnCu(pbaOH)]

It has been suggested that the interchain interaction along the *c*-axis is dominated by the dipolar contribution, and favours the parallel alignment of the spins.

Quite recently, it has been possible to manipulate [MnCu(pbaOH)(H$_2$O)$_3$] in order to shift T_C toward higher temperatures. T_C is governed by both intrachain J_{intra} and interchain J_{inter} interaction parameters. J_{intra} is essentially determined by the nature of the bridges, therefore by the nature of the Cu(II) precursor [Cu(pbaOH)]$^{2-}$. As for J_{inter}, it might be related to the interchain distances. Along this line, it appeared that the chains could be closer to each others if the weakly coordinated water molecule occupying the apical position in the copper coordination sphere was removed. Heating [MnCu(pbaOH)(H$_2$O)$_3$] at 100 °C under vacuum affords [MnCu(pbaOH)(H$_2$O)$_2$]. Spectroscopic data indicate that the water molecule which has been removed actually belonged to the Cu(II) chromophore. The electronic absorption spectrum of

[MnCu(pbaOH)(H$_2$O)$_3$] shows a broad band at 600 nm characteristic of Cu(II) in square pyramidal surroundings, and several Mn(II) spin-forbidden bands activated by an exchange mechanism, one of them (^{6}A$_1$ → ^{4}A$_1$ + ^{4}E) located at 411 nm being very narrow and intense [53] (Figure 29). For the partially dehydrated compound [MnCu(pbaOH)(H$_2$O)$_2$], the Cu(II) d–d band is shifted to 550 nm, as expected when passing from a square pyramidal to a square planar environment. In contrast, the Mn(II) bands are unchanged.

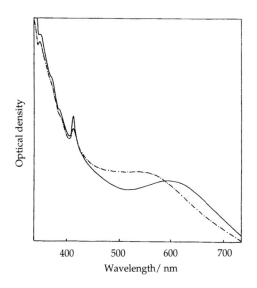

Figure 29 Absorption spectrum for (——) [MnCu(pbaOH)(H$_2$O)$_3$], (– · –) MnCu(pbaOH)(H$_2$O)$_2$]

In the 300–50 K temperature range, the χT versus T curve for [MnCu(pbaOH)(H$_2$O)$_2$] is essentially identical to that of [MnCu(pbaOH)(H$_2$O)$_3$]. In particular, the minimum characteristic of the one-dimensional ferrimagnetism is observed at the same temperature. Upon cooling down further, χT for [MnCu(pbaOH)(H$_2$O)$_2$] increases even faster than for [MnCu(pbaOH)(H$_2$O)$_3$]. The magnetisation versus temperature curves reveal a magnetic transition at 30 K (Figure 30).

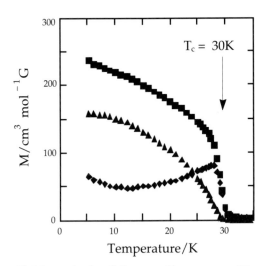

Figure 30 Magnetisation (M) versus Temperature (T) curves for
[MnCu(pbaOH)(H$_2$O)$_2$], (♦) ZFCM, (■) FCM, (▲) RM

2.5.4.2 *[MnCu(obbz).H$_2$O]*

Bimetallic Mn(II)–Cu(II) compounds may be prepared using the Cu(II)
precursors of the type shown in Figure 31 which are able to bind the Mn(II)
ion through both the oxamido group and one of the carboxylato groups.

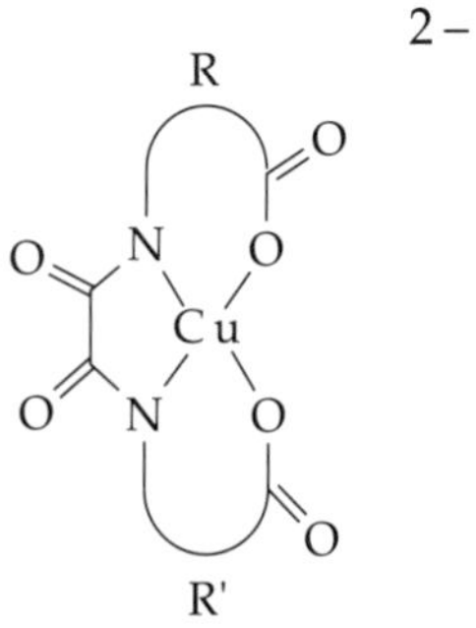

Figure 31 General form of oxamidobis(carboxylato)Cu(II) dianions

When R = R' = phenyl, the precursor noted [Cu(obbz)]$^{2-}$ (obbz =
oxamidobis(N,N'-benzoato)) is shown in Figure 32.

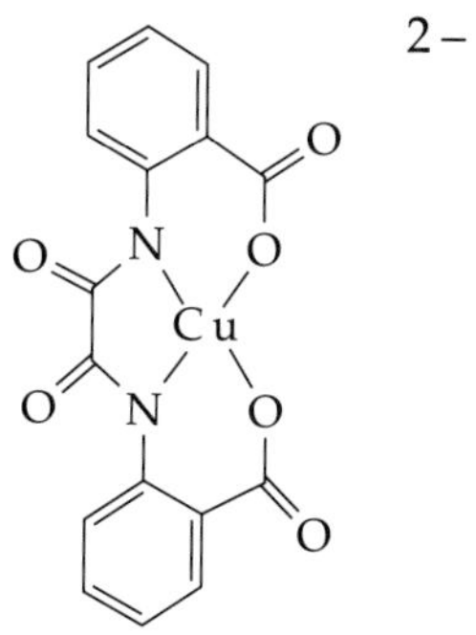

2 –

Figure 32 Structure of oxamidobis(N,N'-benzoato)Cu(II)

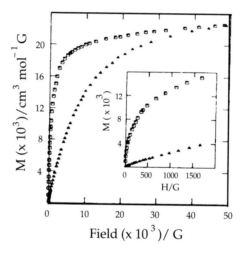

Figure 33 Magnetisation versus H curves for (■) [MnCu(obbz).H$_2$O] and (▲)
[MnCu(obbz).5H$_2$O]

The reaction of [Cu(obbz)]$^{2-}$ with Mn(II) affords two phases of formula [MnCu(obbz).5H$_2$O] and [MnCu(obbz).H$_2$O], respectively [54]. The former phase orders antiferromagnetically at 2.3 K, but the latter shows a ferromagnetic transition at 14 K, as indicated by the magnetisation versus field curves of Figure 33. The magnetisation versus magnetic field curve displays a rapid saturation with a saturation magnetisation of 4 β mol^{-1}. The crystal structure of MnCu(obbz).H$_2$O is not known but that of a compound of formula [MnCu(obbz)(H$_2$O)$_3$.DMF], with DMF = dimethylformamide, has been determined [55] (Figure 34). It consists of alternating bimetallic chains with alternation of both the spin carriers (Mn and Cu), and the exchange pathways (oxamido and carboxylato).

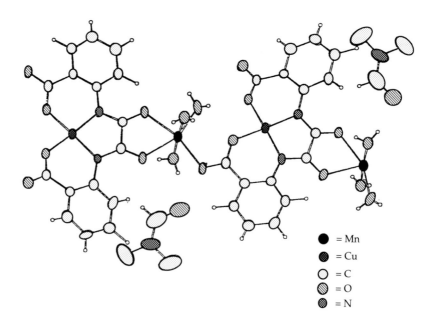

= Mn
= Cu
= C
= O
= N

Figure 34 Crystal structure of [MnCu(obbz)(H$_2$O)$_3$.DMF]

2.5.4.3 *[MnCu(obze)(H$_2$O)$_4$.2H$_2$O]*

The reaction of [Cu(obze)]$^{2-}$ (Figure 35) with Mn(II) affords a compound the structure of which consists of discrete units [MnCu(obze)(H$_2$O)$_4$] in which Mn(II) and Cu(II) ions are bridged by an oxamido group, and non-coordinated water molecules (Figure 36).

The magnetic behaviour of this compound is that of anti-ferromagnetically coupled Mn(II)–Cu(II) pairs with $\chi_M T$ decreasing continuously upon cooling. This behaviour quite surprisingly is completely modified when treating this compound under vacuum at room temperature. Four water molecules are very easily lost. The $\chi_M T$ versus T curve for the resulting material of formula [MnCu(obze)(H$_2$O)$_2$] shows a rounded minimum at 85 K and a rapid increase when cooling down further below this temperature, which is characteristic of a ferrimagnetic behaviour (Figure 37). The magnetisation versus temperature curves reveal a ferromagnetic transition at 4.6 K.

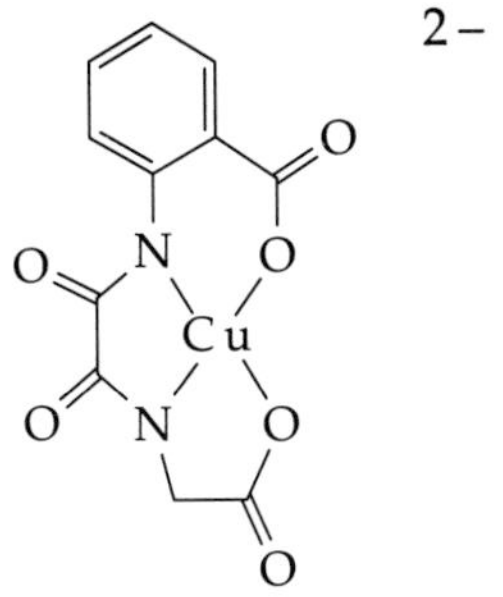

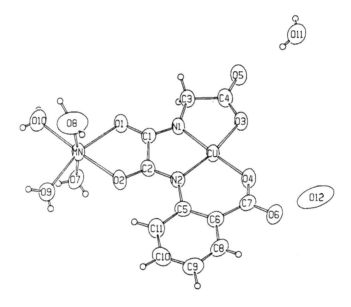

2−

Figure 35 Structure of [Cu(obze)]$^{2-}$

Figure 36 Crystal structure for [MnCu(obze)(H$_2$O)$_4$.2H$_2$O]

The structure of [MnCu(obze)(H$_2$O)$_2$] is not known but the spectroscopic data indicate that Mn(II) remains in octahedral surroundings, which requires the formation of Mn–O(carboxylato) bonds. Instead of isolated oxamido-bridged Mn(II)–Cu(II) pairs, we have now a polymeric structure. This kind of polymerisation reaction taking place in the solid phase at ambient temperature is quite unusual and might open very interesting perspectives in the field of the molecular materials [56].

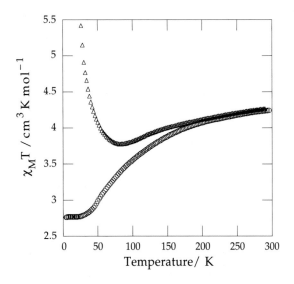

Figure 37 $\chi_M T$ versus T plots for (o) MnCu(obze)(H$_2$O)$_4$.2H$_2$O] and (△) [MnCu(obze)(H$_2$O)$_2$]

In the present case, it provides some exchange pathway between the negative spin density of an oxamido-bridged Mn(II)–Cu(II) unit and the positive spin density of the adjacent unit. The strategy used to design this novel molecular-based ferromagnet is shown in Figure 38. For the sake of simplicity, we have represented a one-dimensional array of local spins while the actual structure would be rather two- or three-dimensional.

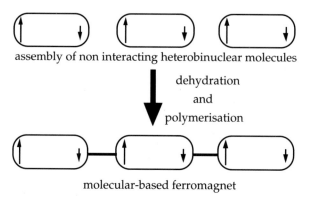

assembly of non interacting heterobinuclear molecules

dehydration
and
polymerisation

molecular-based ferromagnet

Figure 38 One of the current strategies used to design novel molecular-based ferromagnets

This strategy is reminiscent of an idea put forward by McConnell as early as 1963 in the context of the organic chemistry [57]. McConnell suggested that a magnetic molecular entity may possess regions of non-compensating positive and negative spin densities. If so, the interaction between a positive spin density of a unit and a negative spin density of the adjacent unit may lead to ferromagnetic intermolecular interactions.

2.5.4.4 *Metal Nitroxide Compounds*

Rather similar results have been reported by Gatteschi, Rey *et al.,* concerning the metal-nitronyl nitroxide ferrimagnetic chain compounds already mentioned in Section 2.4.4 [34–37]. The compound [Mn(hfa)$_2$NITPr^i] exhibits a three-dimensional ferromagnetic transition at T$_C$ = 7.61 K. This compound crystallises in the monoclinic system, the ferrimagnetic chains running along the *c*-axis. The packing is shown in Figure 13. It should be noticed that along one of the directions perpendicular to the chain axis, the shortest interchain separation involves the Mn(II) ion of a chain and the organic radical of the adjacent chain, which is reminiscent of the situation encountered in [MnCu(pbaOH)].

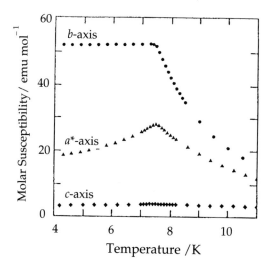

Figure 39 Experimental molar susceptibility versus temperature for [Mn(hfa)$_2$NITPr^i] along three orthogonal axes

Single crystal magnetic susceptibilities are shown in Figure 39. The preferred spin orientation is along the *b*-axis. The temperature independence of the magnetic susceptibility along this direction below T_c is due to demagnetisation effects and is typical of a bulk ferromagnet. From single crystal EPR data, it is postulated that the interchain interactions are dominated by dipolar effects. Along this line, it can be noticed that the analogous compound in which Mn(II) is replaced by Ni(II) orders at a slightly lower temperature, namely 5 K; the smaller local spin of Ni(II) as compared to Mn(II) makes the dipolar interactions between ferrimagnetic chains less effective.

Three-dimensional ferromagnetic transitions above 20 K have also been observed for compounds of formula $[Mn(F_5benz)_2]_2$ where F_5benz is pentafluorobenzoato and R an alkyl group [58]. The structure of these compounds is not yet known. The χT versus T plot for the R = Et derivative continuously increases when T is lowered down to 20.5 K and reaches a very high value close to 800 cm^3 K mol^{-1} at this temperature. The field-cooled magnetisation curve shows the typical break at T_c = 20.5 K, and the M = f(H) curves below T_c show hysteresis loops characteristic of a soft ferromagnet.

2.5.5   Other Strategies to Design Inorganic Molecular-based Ferromagnets

2.5.5.1   *Orthogonality of the Magnetic Orbitals*

Many studies have already been devoted to this acpect of molecular magnetism [59–61]. We will restrict ourselves to a reminder of the key concept. We consider two spin doublets A and B with unpaired electrons occupying a and b magnetic orbitals, respectively. The singlet-triplet energy gap J resulting from the interaction is given in the Heitler-London formalism by:

$$J = 2k + 4\beta S \qquad (46)$$

In most of the cases, $4\beta S$ dominates 2k and the singlet molecular state is the lowest. However, if the overlap integral S is zero, J reduces to the ferromagnetic contribution 2k which may be important if the overlap density $\rho(i) = a(i)b(i)$ is important in some regions of the space; the coupling is then ferromagnetic.

An interesting example of quasi-orthogonality of the magnetic orbitals is provided by the tetranuclear cluster $\{Cr[Ni(ox)(Me_6-[14]ane-N_4)]_3\}^{3+}$, the skeleton of which is shown in Figure 40 [62]. Both Cr(III) and Ni(II) ions are in octahedral surroundings. Cr(III) has three unpaired electrons occupying t_{2g} orbitals, and each Ni(II) has two unpaired electrons occupying e_g orbitals. In the C_{2v} symmetry of each Cr(III)–Ni(II) linkage, the t_{2g} and e_g orbitals are quasi-orthogonal and all the local spins are expected to be parallel, providing an S = 9/2 ground state. The $\chi_M T$ versus T plot shown in Figure 41 confirms this. Upon cooling down, $\chi_M T$ increases continuously and reaches a plateau below 6 K with a $\chi_M T$ value (11.3 cm^3 K mol^{-1}) corresponding to what is expected for an S = 9/2 state. The J_{CrNi} interaction parameter (H = $- J S_{Cr} \cdots S_{Ni}$) is found equal to 5.3 cm^{-1}.

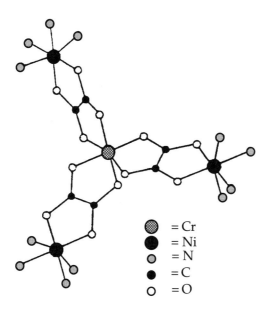

= Cr
= Ni
= N
= C
= O

Figure 40 Schematic drawing of $\{Cr[(ox)Ni(Me_6\text{-}[14]ane\text{-}N_4)]_3\}^{3+}$

Using this concept of orthogonality, Okawa *et al.* recently synthesised a material of formula [NBu$_4$][CuCr(ox)$_3$] exhibiting a ferromagnetic transition at T_c = 7 K with a remnant magnetisation and a hysteresis loop below this critical temperature [63].

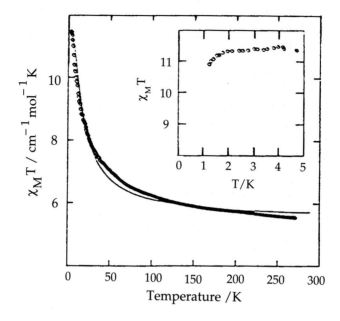

Figure 41 $\chi_M T$ versus T plot for $\{Cr[Ni(ox)(Me_6\text{-}[14]ane\text{-}N_4)]_3\}(ClO_4)_3$

2.5.5.2 Double Exchange

The concept of double exchange was introduced during the sixties by solid state physicists [64, 65] and has been recently reformulated in the context of molecular inorganic chemistry [66, 67]. The basic idea may be introduced by considering first an AB magnetic pair with the local spins $S_A = 1$ and $S_B = 1/2$, and the orbital scheme in Figure 42.

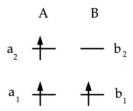

Figure 42 Orbital scheme for a $S_A = 1$, $S_B = 1/2$ magnetic pair

The interaction gives rise to molecular $S = 1/2$ and $S = 3/2$ states, the former being the lowest in energy if, as it is most likely, the interaction is antiferromagnetic. The doublet-quartet energy gap is given by:

$$3J/2 = 3\beta_1 S_1 + (3/2)(k_{11} + k_{12}) \tag{47}$$

with $|2\beta_1 S_1| > (k_{11} + k_{22})$, the notations being obvious from Equation (46). We assume now that one of the electrons is fully delocalised between the a_2 and b_2 orbitals, so that in addition to the $S = 1/2$ and $3/2$ states associated with the $S_A = 1$ and $S_B = 1/2$ local spins, we must consider the new $S = 1/2$ and $3/2$ states associated with the $S_A = 1/2$ and $S_B = 1$ local spins. This situation may be encountered in mixed-valence compounds. Owing to the electron delocalisation, the two $S = 1/2$ states on the one hand and the two $S = 3/2$ states on the other hand couple, which leads to the energy level diagram shown in Figure 43.

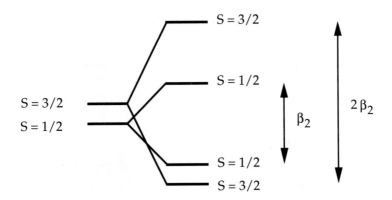

Figure 43 Low-lying states in a fully delocalised $S = 3/2$ mixed-valence dinuclear species with local doublet and triplet states

The ground state is the quartet and the energy involved in this double exchange phenomenon is of the order of β_2, i.e. much larger than the energy involved in the classical exchange phenomenon. Not only the ground state has the highest spin, but this state is strongly stabilised with respect to the first excited state.

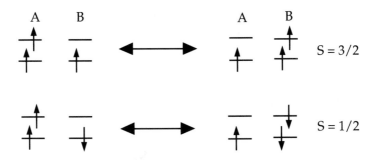

Figure 44 Electron delocalisation in a mixed-valence dinuclear species with local doublet and triplet states

This phenomenon arises from the fact that the delocalised electron can move from one site to the other without spin flip in the quartet state, but not in the doublet state, as shown in Figure 44. Some mixed-valence Fe(III)-Fe(II) binuclear complexes with $S_A = 5/2$ and $S_B = 2$, in which the metal sites are crystallographically equivalent, have recently been reported to exhibit an $S = 9/2$ ground state, due to this double exchange mechanism [68–69].

To our knowledge, there is no report yet of molecular magnetic materials in which the double exchange would be operative. This strategy, however, seems quite appealing because it leads to a strong stabilisation of the ferromagnetic state.

2.6 CONCLUSION

In this contribution, we have spoken about ferro-, ferri- and antiferromagnetism, about double exchange and spin canting, etc. All these words, and all the concepts they define, arise from solid state physics. The question at hand might then be: what is the specificity of the molecular magnetism with respect to the magnetism investigated by the physicists for so many decades? As a conclusion, we would like to suggest some elements of answer.

The first idea coming to mind is that molecular chemistry has a flexibility which is obviously unknown in solid state chemistry. It is possible to control a synthesis in quite a subtle fashion, to adjust the relative positions of the molecular units within the crystal lattice in order to tune the magnetic properties [3]. The fact that the first one-dimensional ferrimagnetic compounds have been synthesised by molecular chemists well illustrates this situation. Molecular chemistry can afford exotic new materials involving unknown combinations of interactions between the magnetic centres, thus can lead to new problems.

Some of the interaction mechanisms encountered in the molecular magnetic materials apparently have no equivalent in the ionocovalent compound chemistry. So, the magnetic orbitals may be much more strongly delocalised, which favours the coupling between spin carriers far apart from each others. The spin polarisation effect seems to be crucial in many cases. It might play the key role in the ferromagnetism of $[Fe(\eta\text{-}C_5Me_5)_2]^+[TCNE]^-$ and $[Mn(\eta\text{-}C_5Me_5)_2]^+[TCNQ]^-$.

In other respects, the molecular magnetic materials could exhibit specific optical properties. Several of them are weakly coloured and their absorption spectrum is strongly correlated with their magnetic properties. Of course, one of the very challenging issues in molecular magnetism would be to obtain soluble ferromagnets. This type of processability would open quite interesting perspectives. Preliminary results suggest that this step could soon be reached.

The difficulties in this research area, however, should not be masked. One of them arises from the fact that till now chemists have been more skilled at designing one- than two- or three-dimensional edifices. This is an obvious limitation since the bulk ferromagnetism is a three-dimensional property. Maybe the main endeavours in the near future should be directed along this line.

2.7   REFERENCES

1. T I Quickenden and R C Marshall, *J Chem Ed*, **49**, 114 (1972).
2. Van Vleck: *The Theory of Electric and Magnetic Susceptibilities*, Oxford University Press, Oxford (1932).
3. B Bleaney and K D Bowers, *Proc Roy Soc (London)*, **A214**, 451 (1952).
4. B C Guha, *Proc Roy Soc (London)*, **A206**, 353 (1951).
5. M Kato, H B Jonassen and J C Fanning, *Chem Rev*, **64**, 99 (1964).
6. R L Martin, in: *New Pathways in Inorganic Chemistry*, E A V Ebsworth, A Maddock and A G Sharpe, Eds, Cambridge University Press, Cambridge (1968).
7. A P Ginsberg, *Inorg Chim Acta Rev*, **5**, 45 (1971).
8. R J Doedens, *Prog Inorg Chem*, **19**, 173 (1975).
9. O Kahn, *Angew Chem Int Ed Engl*, **24**, 834 (1985).
10. M Kato and Y Muto, *Coord Chem Rev*, **92**, 45 (1988).
11. C J O'Connor, *Prog Inorg Chem*, **29**, 203 (1982).

112   *Olivier Kahn, Yu Pei and Yves Journaux*

12. R D Willett, D Gatteschi and O Kahn, Eds *Magneto-Structural Correlations in Exchange Coupled Systems*, NATO ASI Series, Plenum Press, Reidel (1985).
13. O Kahn and M F Charlot, in: *Valence Bond Theory and Chemical Structure*, D J Klein and N Trinajstic Eds, p 489, Elsevier, Amsterdam (1990).
14. Y Pei, Y Journaux and O Kahn, *Inorg Chem*, **27**, 399 (1988).
15. C C Chao, *J Mag Reson*, **10**, 1 (1973).
16. E Buluggiu, *J Phys Chem Solids*, **41**, 1175 (1980).
17. R P Scaringe, D Hodgson and W E Hatfield, *Mol Phys*, **35**, 701 (1978).
18. R C Carlin: *Magnetochemistry*, Springer-Verlag, Berlin Heidelberg (1986).
19. L J de Jongh and A R Miedema, *Adv Phys*, **23**, 1 (1974).
20. J C Bonner and M E Fisher, *Phys Rev*, **A135**, A640 (1964).
21. R D Willett, C P Landee, R M Gaura, D D Swank, H A Groenendijk and A J van Duynevelt, *J Magn Mat*, **15–18**, 1055 (1980).
22. F D M Haldane, *Phys Rev Lett*, **50**, 1153 (1983).
23. J P Renard, S Clement and M Verdaguer, *Proc Indian Acad Sci (Chem Sci)*, **98**, 131 (1987).
24. J P Renard, M Verdaguer, L P Regnault, W A C Erkelens, J Rossat-Mignod, J Ribas, W G Stirling and C Vettier, *J Appl Phys*, **63**, 3538 (1988).
25. M Verdaguer, M Julve, A Michalowicz and O Kahn, *Inorg Chem*, **22**, 2624 (1983).
26. M Drillon, J C Gianduzzo and R Georges, *Phys Lett*, **96A**, 413 (1983).
27. M Drillon, E Coronado, D Beltran and R Georges, *J Appl Phys*, **57**, 3353 (1985).
28. J Seiden, *J Phys (Paris) Lett*, **44**, L947 (1983).
29. M Verdaguer, A Gleizes, J P Renard and J Seiden, *Phys Rev*, **B29**, 5144 (1984).
30. Y Pei, O Kahn, J Sletten, J P Renard, R Georges, J C Gianduzzo, J Curely and X Qiang, *Inorg Chem*, **27**, 47 (1988).
31. R Georges, J Curely, J C Gianduzzo, X Qiang, O Kahn and Y Pei, *Physica*, **B153**, 77 (1988).
32. A Gleizes and M Verdaguer, *J Am Chem Soc*, **106**, 3727 (1984).
33. Y Pei, M Verdaguer, O Kahn, J Sletten and J P Renard, *Inorg Chem*, **26**, 138 (1987).
34. A Caneschi, D Gatteschi, P Rey and R Sessoli, *Inorg Chem*, **27**, 1756 (1988).
35. A Caneschi, D Gatteschi, J P Renard, P Rey and R Sessoli, *Inorg Chem*, **28**, 1976 (1989).
36. A Caneschi, D Gatteschi, J Laugier, P Rey, R Sessoli and C Zanchini, *J Am Chem Soc*, **110**, 2795 (1988).

37. A Caneschi, D Gatteschi, R Sessoli and P Rey, *Acc Chem Res*, **22**, 392 (1989).
38. A H Morrish, *The Physical Principle of Magnetism*, R E Krieger Publ Co, New York (1980).
39. D H Martin: *Magnetism in Solids*, M I T press, Cambridge (1967).
40. J B Torrance, P S Bagus, I Johannsen, A I Nazzal, S S P Parkin and P Batail, *J Appl Phys*, **63**, 2962 (1988).
41. P S Bagus and J B Torrance, *Phys Rev*, **B39**, 7301 (1989).
42. H M McConnell, *Proc R A Welch Found Conf*, **11**, 144 (1967).
43. R Breslow, B Jaun, R Q Kluttz and C Z Xia, *Tetrahedron*, **38**, 863 (1982).
44. R Breslow, P Maslak and J S Thomaides, *J Am Chem Soc*, **106**, 6453 (1984).
45. R Breslow, *Mol Cryst & Liq Cryst*, **125**, 261 (1985).
46. T J LePage and R Breslow, *J Am Chem Soc*, **109**, 6412 (1987).
47. C Kollmar and O Kahn, *J Am Chem Soc*, **113**, 7987 (1991).
48. J S Miller, J C Calabrese, H Rommelmann, S R Chittipeddi, J H Zhang, W M Reiff and A J Epstein, *J Am Chem Soc*, **109**, 769 (1987).
49. J S Miller and A J Epstein, *J Am Chem Soc*, **109**, 3850 (1987).
50. J S Miller, A J Epstein and W M Reiff, *Science*, **240**, 40, (1988); *Chem Rev*, **88**, 201 (1988); *Acc Chem Res*, **21**, 114 (1988).
51. W E Broderick, J A Thompson, E P Day and B M Hoffman, *Science*, **249**, 401 (1990).
52. O Kahn, Y Pei, M Verdaguer, J P Renard and J Sletten, *J Am Chem Soc*, **110**, 782 (1988).
53. K Nakatani, O Kahn, C Mathonière, Y Pei and C Zakine, *New J Chem*, **14**, 861 (1990).
54. K Nakatani, J Y Carriat, Y Journaux, O Kahn, F Lloret, J P Renard, Y Pei, J Sletten and M Verdaguer, *J Am Chem Soc*, **111**, 5739 (1989).
55. F Lloret, M Julve, J Sletten, Y Journaux, K Nakatani, J C Colin and O Kahn, to be published
56. Y Pei, O Kahn, K Nakatani, E Codjovi, C Mathonière and J Sletten, *J Am Chem Soc*, **113**, 6558 (1991).
57. H M Mc Connell, *J Chem Phys*, **39**, 1910 (1963)
58. A Caneschi, D Gatteschi, J P Renard, J P Rey and R Sessoli, *J Am Chem Soc*, **111**, 785 (1989).
59. O Kahn, J Galy, Y Journaux, J Jaud and I Morgenstern-Badarau, *J Am Chem Soc*, **104**, 2165 (1982).
60. P De Loth, P Karafiloglou, J P Daudey and O Kahn, *J Am Chem Soc*, **110**, 5676 (1988).
61. O Kahn, R Prins, J Reedijk and J S Thompson, *Inorg Chem*, **26**, 3557 (1987).
62. Y Pei, Y Journaux and O Kahn, *Inorg Chem*, **28**, 100 (1989).

63.   Z J Zhong, N Matsumoto, H Okawa and S Kida, *Chem Letters*, 87 (1990).
64.   C Zener, *Phys Rev*, **82**, 403 (1951).
65.   P W Anderson and H Hasawaga, *Phys Rev*, **100**, 675 (1955).
66.   J J Girerd and G Blondin, *Chem Rev*, **90**, 1359 (1990).
67.   E Münck, V Papaefthimiou, K K Surerus and J J Girerd, in: *Metal Clusters in Proteins*, L Que, Ed, ACS Symposium Series no. 372 (1988).
68.   S Drüeke, P Chaudhuri, K Pohl, K Wieghardt, X Q Ding, E Bill, A Sawaryn, A X Trautwein, H Winkler and S J Gurman, *J Chem Soc Chem Comm*, 59 (1989).
69.   B S Snyder, G S Patterson, A J Abrahamson and R H Holm, *J Am Chem Soc*, **111**, 5214 (1989).

3 Metal-containing Materials for Nonlinear Optics

Seth R Marder

Inorganic Materials. Edited by Duncan W Bruce and Dermot O'Hare
© 1992 John Wiley & Sons Ltd

3.1   OVERVIEW

This chapter introduces the reader to the rôle of metal containing molecular materials in the field of nonlinear optics. Since no prior knowledge of nonlinear optics has been assumed, the first part of the chapter is a short introduction to this field. It considers first how the passage of light perturbs the electron density distribution in a material (i.e. polarises it) in a linear manner. Next, we examine the consequences of this polarisation upon the behaviour of the light. Building on this foundation, we then describe, in an analogous manner, how the interaction of light with materials can give rise to nonlinear polarisation. We then review the literature for metal-containing molecular materials exhibiting nonlinear optical behaviour.

3.2   BASIC CONCEPTS OF NONLINEAR OPTICS

3.2.1   The Big Picture

As light travels through a material a variety of nonlinear optical (NLO) effects may occur [1–8]. The interaction of light with such a material will cause the material's properties to change, such that another photon that arrives will 'see' a different material. As light goes through a material, its electric field interacts with charges in the material. The time dependent electron density distribution in the material resulting from this interaction can affect the propagation of other light waves, if for example, two or more light sources are used. These interactions can cause the original optical beam to have its frequency, phase, polarisation or path changed significantly. The ability to manipulate light in this manner has many important technological ramifications in optical signal processing, generation of variable frequency laser light, tunable filters and optical data storage. In order to control light, materials chemists must design and synthesise an optimal medium within which the modulation or combination of photons (wave mixing) can take place, and in which both the magnitude and response time of these optical processes can be controlled. For the sake of this paper we will refer to these materials as NLO materials, although it should be noted that all materials will exhibit nonlinear optical effects at some level. Most research has focused on so-called second-order and third-order NLO effects. The structural features that optimise these types of effects are often different. The discipline of nonlinear optics can be opaque to chemists, in part because it tends to be presented as a series of intimidating equations, that does not provide an intuitive grasp of the processes. Therefore, graphical representations of these processes are used as much as possible, starting with the interaction of light with a molecule or atom.

3.2.2 Polarisability: A Microscopic View

What causes the electron density of an optical material to couple to and be polarised by the electromagnetic field of a light wave? To understand this process, we need to consider quantitatively what is happening at the molecular level. How does light perturb or couple to the electrons in a molecule? Light has an electric field, E, that interacts with the charges in a material producing a force ($F = q \cdot E$, where q is the charge) [9]. Figure 1 schematically shows how the electron density of the atom is displaced (polarised), subject to the time-dependent force induced by the electric field of the light, if the response is instantaneous. The displacement of the centre of electron density away from the nucleus results in a charge separation, an *induced dipole* with moment μ, (Figure 1b). For small fields the displacement of charge from the equilibrium position is proportional to the strength of the applied field (Figure 1c). Since μ and E are vector quantities with both direction and magnitude, μ is given by:

$$\text{Polarisation} = \mu(\omega) = \alpha_{ij}(\omega)E(\omega) \tag{1}$$

where $\alpha_{ij}(\omega)$ is the *linear polarisability tensor at frequency* ω of the molecule or atom since it defines the linear variation of polarisation (induced dipole moment) with the electric field. If the field oscillates with some frequency, (as in electromagnetic radiation or light) then the induced polarisation will have the same frequency and phase if the response is instantaneous (Figure 1a).

In this (classical) model of linear polarisability, the electrons are bound to the atoms by a *harmonic potential* (Figure 2), i.e. the restoring

118

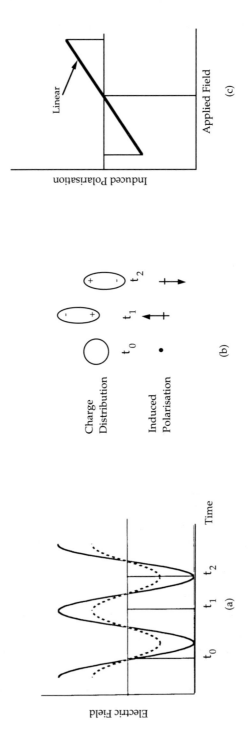

Figure 1 (a) Plots of the electric field of the applied light wave (——) and the induced polarisation wave (- -) as a function of time, for a linear material; (b) cartoon depicting the polarisation of the material as a function of time; (c) plot of *induced* polarisation versus applied field

force for the electron is linearly proportional to its displacement from the nucleus:

$$\mathbf{F} = -\kappa x \qquad (2)$$

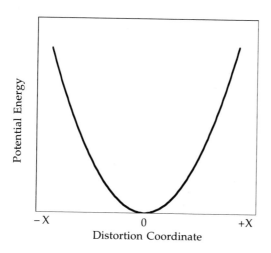

Figure 2 Plot of potential energy versus distortion coordinate for a material with a harmonic potential

The electrons see a potential energy surface:

$$V = \left(\frac{1}{2}\right)\kappa x^2 \qquad (3)$$

This means that there is a symmetrical distribution of electron density around the atom with an equal ease of *charge* displacement in both the $+x$ and $-x$ directions. In bulk materials, the linear polarisation per unit volume (compare with Equation (1) for atoms or molecules) is given by:

$$\mathbf{P}(\omega) = \chi_{ij}(\omega)\mathbf{E}(\omega) \qquad (4)$$

where $\chi_{ij}(\omega)$ is the *linear susceptibility* of an ensemble of molecules. To relate $\chi_{ij}(\omega)$ to the atomic or molecular polarisation described above, the assumption is frequently made that the atoms or molecules that make up the optical material are independently polarised by the light with no interatomic or intermolecular coupling (see below). Within this approximation, $\chi_{ij}(\omega)$ is related to the sum of all the individual polarisabilities, $\alpha_{ij}(\omega)$.

When the electronic charge in the material is displaced by the electric field (E) of the light and polarisation takes place, the *total electric field* (the 'displaced' field, **D**) *within the material* becomes:

$$D = E + 4\pi P = (1 + 4\pi\chi)E \tag{5}$$

where $4\pi\chi E$ is the internal electric field created by the induced displacement (polarisation) of charges. The induced polarisation may cause the spatial orientation of the internal electric field to differ from that of the applied electric field. That is, like $\alpha_{ij}(\omega)$, $\chi_{ij}(\omega)$ is a tensor quantity that describes the anisotropy of the internal electric field.

Two common bulk parameters that characterise the susceptibility of a material are the *dielectric constant*, $\varepsilon(\omega)$ and the *refractive index*, $n(\omega)$. The dielectric constant in a given direction is defined as the ratio of the displaced internal field to the applied field ($\varepsilon = D/E$) in that direction. Therefore from Equation (5),

$$\varepsilon(\omega) = 1 + 4\pi\chi(\omega) \tag{6}$$

We have been concerned with the effect of light on the medium (ε and χ). Since the study of NL optics addresses how the optical material changes the propagation characteristics of light, we must now ask what happens to the light as it goes through the medium.

3.2.3 Linear Polarisation of Matter and Linear Optical Effects

As shown in Figure 1, the light wave moves electronic charge back and forth. This motion of charge will re-emit radiation at the frequency of oscillation. For linear polarisation, the radiation has the same frequency as the incident light. However, the polarisation changes the velocity of the light wave.

We know from everyday experience that when light travels from one medium to another its path can change. Thus, a straight stick entering water at an angle appears to bend as it goes below the surface. This apparent bending is due to the difference between the velocity of light in air and in water. The ratio of the velocity of light in a vacuum, c, to the velocity of light in a material, v, is called the *index of refraction* (n)

$$n = \frac{c}{v} \tag{7}$$

At optical frequencies, and far from molecular resonances, the dielectric constant is equal to the square of the refractive index:

$$\varepsilon_\infty(\omega) = n^2(\omega) \tag{8}$$

Consequently, we can relate the refractive index to the bulk linear (first-order) susceptibility:

$$n^2(\omega) = 1 + 4\pi\chi(\omega) \tag{9}$$

Since $\chi(\omega)$ is related to the individual atomic or molecular polarisabilities, this equation relates a property of light (its velocity) to a property of the electron density distribution (the polarisability). This is an example of how the optical properties of a material depend on the electron density distribution that is dictated by chemical structure: i.e. if the molecular structure is altered, then the optical properties are changed.

3.2.4 Nonlinear Polarisability: A Microscopic View

Until now it has been assumed that the polarisation of a molecule or material is proportional to the applied electric field. However, when a material is subjected to very high intensity electric fields, the material can become sufficiently polarised that its polarisability can change.

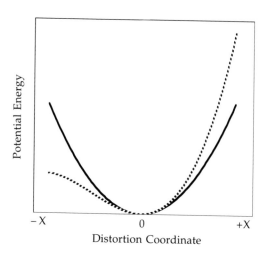

Figure 3 Plot of potential energy versus distortion coordinate for a material with a harmonic potential (—) and a material with an additional cubic anharmonic (- -) term

122

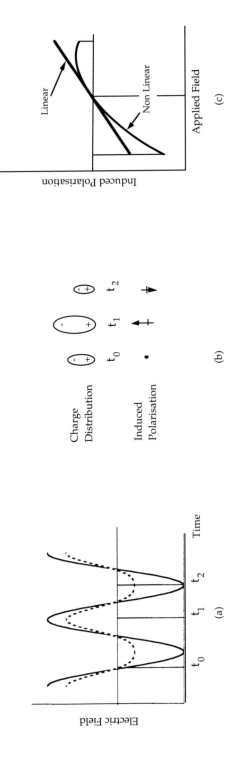

Figure 4 (a) Plots of the electric field of the applied light wave (——) and the induced polarisation wave (- -), as a function of time, for a second-order nonlinear material; (b) cartoon depicting the polarisation of the material as a function of time; (c) plots of induced polarisation versus applied field for both linear and second-order nonlinear materials

Thus, the induced polarisation is a nonlinear function of the field strength. It has been previously stated that polarisability depends on the frequency of the applied field. One way to modify the polarisability is to change the restoring force constant, κ (Equation (2)). The potential surface can also be modified if an anharmonic term (a cubic term in Figure 3) is added.

The restoring force on the electron is no longer proportional to its displacement, it is now *nonlinear* (Figure 4). A first approximation (in one dimension) for the restoring force is:

$$F = -kx - \left(\frac{1}{2}\right)k'x^2 \qquad (10)$$

Here, the polarisation depends on the direction of displacement (Figure 4). For the covalent C=O bond in acetone, for example, one expects that the electron cloud would be more easily polarised towards the oxygen atom. The electric field may be applied along the x, y or z directions of the molecule, and since application of the field along each of these directions can polarise the molecule in the x, y or z direction the molecular polarisability is a tensor.

Just as linear polarisation leads to linear optical effects, such as refraction and birefringence, nonlinear polarisation leads to nonlinear effects that are usually more subtle. Figure 4 shows that application of a symmetric field (i.e. the electric field associated with the light wave) to the anharmonic potential leads to an asymmetric polarisation response. This polarisation wave has diminished maxima in one direction and accentuated maxima in the opposite direction. This asymmetric polarisation can be Fourier decomposed into a DC polarisation component and components at the fundamental and second harmonic frequencies.

This Fourier analysis of the resultant second-order polarisation is shown in Figure 5. Since only the time averaged asymmetrically induced polarisation leads to second-order NLO effects, only molecules and materials lacking a centre of symmetry possess them.

A common approximation is to expand the polarisability as a Taylor series:

$$\mu = \mu_0 + E\left(\frac{\partial \mu_i}{\partial E_j}\right)_{E_0} + \left(\frac{1}{2}\right)E{\cdot}E\left(\frac{\partial \mu_i}{\partial E_j E_k}\right)_{E_0} + \left(\frac{1}{6}\right)E{\cdot}E{\cdot}E\left(\frac{\partial \mu_i}{\partial E_j E_k E_l}\right)_{E_0} + \dots \qquad (11)$$

$$\mu = \mu_0 + \left(\alpha_{ij}\right)E + \left(\frac{\beta_{ijk}}{2}\right)E{\cdot}E + \left(\frac{\gamma_{ijkl}}{6}\right)E{\cdot}E{\cdot}E + \dots \qquad (12)$$

124   *Seth R Marder*

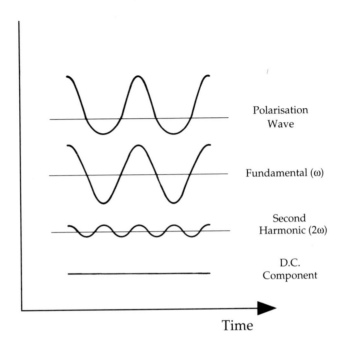

Figure 5 Fourier analysis of an asymmetric polarisation wave showing that it is comprised of components at the fundamental frequency, second harmonic frequency and zero frequency (DC)

The terms beyond αE are not linear in E and are therefore referred to as the *nonlinear polarisation* and give rise to *nonlinear optical effects*. Note that nonlinear polarisation becomes more important with increasing field strength, since it scales with higher powers of the field. Under normal conditions, $\alpha_{ij}E > (\beta_{ijk}/2)E \cdot E > (\gamma_{ijkl}/6)E \cdot E \cdot E$, thus, there were few observations of NLO effects before the invention of the laser with its associated large electric fields.

Just as α is the *linear polarisability*, the higher order terms β *and* γ (Equation (**12**)) are called the *first and second hyperpolarisabilities* respectively. The observed bulk polarisation density is given by an expression analogous to (**12**):

$$P = P_0 + \chi^{(1)} \cdot E + \chi^{(2)} \cdot \cdot EE + \chi^{(3)} \cdot \cdot \cdot E + \ldots \tag{13}$$

where the $\chi^{(i)}$ susceptibility coefficients are tensors of order i+1 (e.g. $\chi^{(2)}_{ijk}$). P_0 is the intrinsic static dipole moment density of the sample.

3.2.5 Second-order Nonlinear Polarisation of Matter and Second-order NLO Effects

3.2.5.1 *Frequency Doubling and Sum-Frequency Generation*

An electric field can polarise the electronic distribution in a material, but what does this have to do with observable nonlinear optical properties? The electronic charge displacement (polarisation) induced by an oscillating electric field (e.g. light) can be viewed as a classical oscillating dipole that itself emits radiation at the oscillation frequency. For linear first-order polarisation, the radiation has the same frequency as the incident light. What is the frequency of the re-emitted light for a nonlinear optical material? Recalling that the electric field of a plane light wave can be expressed as:

$$E = E_0 \cos(\omega t) \tag{14}$$

equation (14) can be rewritten as:

$$P = P_0 + \chi^{(1)} E_0 \cos(\omega t) + \chi^{(2)} E_0^2 \cos^2(\omega t) + \chi^{(3)} E_0^3 \cos^3(\omega t) + \ldots \tag{15}$$

Since $\cos^2(\omega t)$ equals $\{1/2 + 1/2 \cos(2\omega t)\}$, the first three terms of Equation (15) become:

$$P = \left\{ P_0 + \left(\frac{1}{2}\right)\chi^{(2)}E_0^2 \right\} + \chi^{(1)}E_0 \cos(\omega t) + \left(\frac{1}{2}\right)\chi^{(2)}E_0^2 \cos(2\omega t) + \ldots \tag{16}$$

Physically, Equation (16) states that the polarisation consists of a second-order DC field contribution to the static polarisation (first term), a frequency component ω corresponding to the incident light frequency (second term) and a new *frequency doubled* component, 2ω (third term), as in Figure 5. Thus, if an intense light beam passes through a second-order NLO material, light at twice the input frequency will be produced as well as a static electric field. The first process is called second harmonic generation (SHG) and the second is called optical rectification. SHG is a form of three wave mixing, since two photons with frequency ω have combined to generate a single photon with frequency 2ω. Since the oscillating dipole re-emits at all of its polarisation frequencies, one observes light at both ω and 2ω. This analysis can be extended to third-

and higher-order terms. By analogy, third-order processes involve four wave mixing.

As written, Equation (15) is a simplified picture in which a single field, $E(\omega, t)$ acts on the material. The general picture of second-order NLO effects involves the interaction of two distinct waves with electric fields E_1 and E_2 with the electrons of the NLO material. Suppose, for example, that two laser beams with different frequencies are used. With two interacting waves of amplitudes E_1 and E_2, the second-order term of Equation (16) becomes:

$$\chi^{(2)}E_1\cos(\omega_1 t)E_2\cos(\omega_2 t) \tag{17}$$

From trigonometry we know that Equation (17) is equivalent to:

$$\left(\frac{1}{2}\right)\chi^{(2)}E_1 E_2 \cos\{(\omega_1 + \omega_2)t\} + \left(\frac{1}{2}\right)\chi^{(2)}E_1 E_2 \cos\{(\omega_1 - \omega_2)t\} \tag{18}$$

This equation shows that when two light beams of frequencies ω_1 and ω_2 interact in an NLO material, polarisation occurs at sum $(\omega_1 + \omega_2)$ and difference $(\omega_1 - \omega_2)$ frequencies. This electronic polarisation will therefore re-emit radiation at these frequencies, with contributions that depend on the magnitude of the NLO coefficient, $\chi^{(2)}$. The combination of frequencies is called sum (or difference) frequency generation (SFG). SHG is a special case of SFG, where the two frequencies are equal. The sum is the second harmonic and the difference is the DC component.

3.2.5.2 *Changing the Propagation Characteristics of Light: The Pockels Effect*

As noted above, refractive indices of a material for different frequencies are usually not the same. Furthermore, it is possible to change the amplitude, phase or path of light at a given frequency by using a static DC electric field to polarise the material and modify the refractive indices. Consider the special case $\omega_2 = 0$ (Equation (17)) in which a DC electric field is applied to the material. The optical frequency polarisation (P_{opt}) arising from the second-order susceptibility is:

$$\chi^{(2)}E_1 E_2 \cos(\omega_1 t) \tag{19}$$

where E_2 is the magnitude of the electric field caused by voltage applied to the nonlinear material. Remember that the refractive index is related to the linear susceptibility (Equation (9)) that is given by the second term of Equation (16):

$$\chi^{(1)}E_1\cos(\omega_1 t) \tag{20}$$

so the total optical frequency polarisation is:

$$\mathbf{P_{opt}} = \chi^{(1)}E_1\cos(\omega_1 t) + \chi^{(2)}E_1E_2\cos(\omega_1 t) \tag{21}$$

$$\mathbf{P_{opt}} = \left[\chi^{(1)} + \chi^{(2)}E_2\right]E_1\cos(\omega_1 t) \tag{22}$$

The applied field in effect changes the linear susceptibility and thus the refractive index of the material. This is known as the linear electrooptic (LEO) or Pockels effect, and is used to modulate light by changing the applied voltage. At the atomic level, the applied voltage is anisotropically distorting the electron density within the material. Thus, application of a voltage to the material causes the optical beam to 'see' a material with a different polarisability and a different anisotropy of the polarisability than in the absence of the voltage. As a result, a beam of light can (1) have its polarisation state (i.e. ellipticity) changed by an amount related to the strength and orientation of the applied voltage, and (2) travel at a different speed and possibly in a different direction.

Quantitatively, the change in the refractive index as a function of the applied electric field is approximated by the general expression:

$$\frac{1}{n_{ij}^2} = \frac{1}{n_{ij}^2} + r_{ijk}E_k + s_{ijkl}E_kE_l \tag{23}$$

where n_{ij} are the induced refractive indices, n_{ij} are the refractive indices in the absence of the electric field, r_{ijk} are the linear or Pockels coefficients and s_{ijkl} are the quadratic or Kerr coefficients (see below). The optical indicatrix (that characterises the anisotropy of the refractive index) therefore changes as the electric field within the sample changes. The 'r' coefficients form a tensor (just as do the coefficients of α). The first subscript refers to the resultant polarisation of the material along a defined axis and the following subscripts refer to the orientations of the applied electric fields. Since the Pockels effect involves two fields mixing to give rise to a third, then r_{ijk} is a third rank tensor.

The Pockels effect has many important technological applications. Light travelling through an electrooptic material can be phase or polarisation modulated by refractive index changes induced by an applied electric field. Devices exploiting this effect include optical switches, modulators, and wavelength filters.

3.2.6   Third-order Nonlinear Polarisation of Matter and Third-order NLO Effects

Second-order optical nonlinearities result from the introduction of a cubic term in the potential function for the electron and third-order optical nonlinearities result from the introduction of a quartic term (Figure 6).

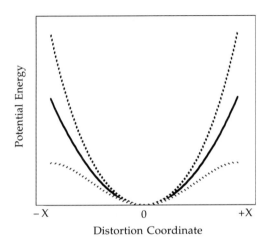

Figure 6 Plot of potential energy versus distortion coordinate for a material with a harmonic potential ($-$) and a material with positive ($- -$) and negative ($\cdot$ $\cdot$) quartic anharmonic term

   Two important points relate to the symmetry of this contribution to the potential. First, while negative and positive β both give rise to the same potential and therefore the same physical effects (the only difference being the orientation of the coordinate system), a negative γ leads to a different electronic potential than does a positive γ. Secondly, the quartic contribution to the potential has mirror symmetry with respect to the distortion coordinate; as a result both centrosymmetric and noncentrosymmetric materials will have third-order optical nonlinearities. If we reconsider Equation (**16**) for the expansion of polarisation of a molecule as a function of electric field and assume that the even-order terms are zero (i.e. that the molecule is centrosymmetric) we see that

$$\mu = \mu_0 + \alpha E_0 \cos(\omega t) + \left(\frac{\gamma}{6}\right) E_0^3 \cos^3(\omega t) + \dots \tag{24}$$

If a single field, $E(\omega,t)$, is acting on the material, we know from trigonometry that:

$$\left(\frac{\gamma}{6}\right) E_0^3 \cos^3(\omega t) = \left(\frac{\gamma}{6}\right) E_0^3 \left\{\left(\frac{3}{4}\right)\cos(\omega t) + \left(\frac{1}{4}\right)\cos(3\omega t)\right\} \quad (25)$$

thus,

$$\mu = \mu_0 + \alpha E_0 \cos(\omega t) + \left(\frac{\gamma}{6}\right) E_0^3 \left(\frac{3}{4}\right) \cos(\omega t) + \left(\frac{\gamma}{6}\right) E_0^3 \left(\frac{1}{4}\right) \cos(3\omega t) \quad (26)$$

or:

$$\mu = \mu_0 + \left\{\alpha + \left(\frac{\gamma}{6}\right) E_0^2 \left(\frac{3}{4}\right)\right\} E_0 \cos(\omega t) + \left(\frac{\gamma}{6}\right) E_0^3 \left(\frac{1}{4}\right) \cos(3\omega t) \quad (27)$$

Thus, the interaction of light with third-order NLO molecules will create a polarisation component at its third harmonic. In addition, there is a component at the fundamental, and we note that the $\{\alpha + \gamma/6$ $E_0^2(3/4)]\}$ term of Equation (27) is similar to the term leading to the linear electrooptic effect. Likewise, the induced polarisation for a bulk material would lead to third harmonic generation through $\chi^{(3)}$, the material susceptibility analogous to γ. Similarly, it can be shown that the application of an intense voltage will also induce a refractive index change in a third-order NLO material. These effects are known as the optical and the DC Kerr effects, respectively. For materials, the sign of $\chi^{(3)}$ will determine if the third-order contribution to the refractive index is positive or negative in sign. Materials with positive $\chi^{(3)}$ have the property of self focusing a laser beam and those with negative $\chi^{(3)}$ will be self defocusing.

Another interesting manifestation of third-order NLO effects is degenerate four wave mixing. Two coherent beams of light interacting within a material will create an interference pattern (Figure 7c) that will lead to a spatially periodic variation in light intensity across the material. As we have noted before, the induced change in refractive index of a third-order nonlinear optical material is proportional to the intensity of the applied field. Thus, if two beams (1 and 2 in Figure 7) interact with a third-order NLO material, the result will be a refractive index grating. When beam 3, which is counterpropagating with beam 2, is incident on this grating a fourth beam, called the phase conjugate of beam 1, is diffracted from the grating. This process is called four wave mixing: two writing beams and a probe beam result in a fourth phase conjugate beam.

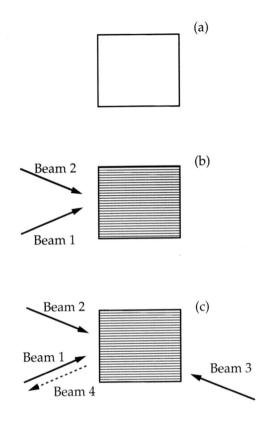

Figure 7 (a) A phase conjugating material in the absence of an applied field; (b) Beams 1 and 2 create a refractive index grating; (c) Beam 3, which is counterpropagating with Beam 2, interacts with the grating creating Beam 4 that is the *phase onjugate* of Beam 1

A potential use of this phenomenon is in phase conjugate optics. Phase conjugate optics takes advantage of a special feature of the diffracted beam: its path exactly retraces the path of one of the writing beams. As a result, a pair of diverging beams impinging on a phase conjugate mirror will converge after 'reflection'. In contrast, a pair of diverging beams reflected from an ordinary mirror will continue to diverge (Figure 8). Thus, distorted optical wavefronts can be reconstructed using phase conjugate optical systems (Figure 9).

(a)   (b)

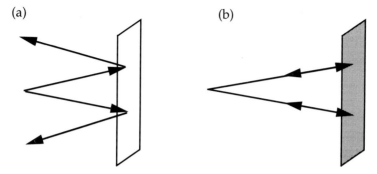

Figure 8 (a) A diverging set of beams reflected off a normal mirror continue to diverge. (b) A diverging set of beams reflected off a phase conjugate mirror exactly retrace their original path and are therefore recombined at their point of origin.

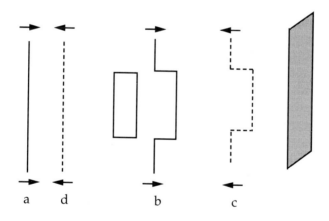

a   d   b   c

Figure 9 (a) A planar wave (b) passes through a distorting material that introduces an aberration and (c) the light interacts with a phase conjugate mirror creating the phase conjugate wavefront. (d) Phase conjugate wave passes through the distorting material on the reverse path cancelling the original aberration , thus producing an undistorted wavefront

3.2.7 Engineering Linear and Nonlinear Polarisability

As chemists we have some intuitive understanding of what factors affect polarisability. For example, the organic chemist knows that the electrons in polyacetylene are more polarisable than those in butane. Likewise the inorganic chemist knows that semiconductors are more polarisable than insulators. These simple realisations suggest that the extent of electron delocalisation is related to the linear polarisability. In organic molecules the extent of delocalisation is affected by the number of orbitals in the electronic system of interest, their hybridisation and the degree of coupling between the orbitals. The linear polarisability derived from perturbation theory is:

$$\alpha \approx \sum_n \left(\frac{\mu_{gn}^2}{E_{ng}^2} \right) \tag{28}$$

where μ_{gn} is the transition dipole moment between the ground-state (g) and an excited-state (n) and E_{ng} is an energy denominator determined by the energy gaps between electronic states and the energy of perturbing field (related to its frequency). We see that molecules/materials with strong, low-energy, absorption bands tend to be highly polarisable. It is not surprising therefore, that cyanine dyes and semiconductors with their large transition dipole moments and small highest occupied molecular orbital to lowest unoccupied molecular orbital (HOMO-LUMO) gaps are highly polarisable.

For second-order nonlinear polarisation, the problem becomes more complex. The anharmonic polarisation shows the largest deviation from the linear polarisation at large distortion (Figure 6). Therefore, if the material is not polarisable (i.e. if the electrons can only be perturbed a small distance from their equilibrium positions), then the anharmonicity will not be manifested. For large second-order nonlinearities therefore, a material must have a large linear polarisability and a large anharmonicity. In organic donor-acceptor molecules, it is easier to polarise the electrons toward the acceptor than toward the donor. Thus, these systems have an asymmetric anharmonic term, and not surprisingly, organic donor-acceptor molecules have some of the largest known values of β.

Theoretical models for individual tensorial components of β can also be derived from perturbation theory that leads to a sum over states formulation, where:

$$\beta \approx 3 \left\{ \sum_{nn'} \left(\frac{\mu_{gn}\mu_{nn'}\mu_{n'g}}{E_{ng}E_{n'g}} \right) - \mu_{gg} \sum_n \left(\frac{\mu_{gn}\mu_{n'g}}{E_{ng}^2} \right) \right\} \tag{29}$$

where μ_{gn} and E_{ng} are defined as before and n' is another excited-state [10–16]. For donor-acceptor substituted benzenes such as 4-nitroaniline, it was found that a two-state model, in which a single charge transfer excited-state dominated the above expression, reasonably accounted for the observed β values [12, 15]. In this approximation, the expression for the dominating components of β reduces to:

$$\beta \approx 3(\mu_{nn} - \mu_{gg})\left(\frac{\mu_{gn}^2}{E_{ng}^2}\right) \tag{30}$$

It is not sufficient for a molecule to have a large β. In order to observe second-order NLO effects the bulk material must also be noncentrosymmetric. Since ~75% of all achiral molecules crystallise in centrosymmetric space groups and thus have zero $\chi^{(2)}$, proper alignment of the chromophore in the bulk material is a major impediment to achieving the goal of engineering materials with large $\chi^{(2)}$ [17, 18]. The electronic driving force for crystallisation in a centrosymmetric space group may be the cancellation of destabilizing dipole-dipole interactions between adjacent molecules in the crystal lattice. Strategies employed to overcome this major obstacle include: the use of chiral molecules [19], the incorporation of functional groups that encourage asymmetric intermolecular hydrogen bonding [19], the synthesis of molecules with very small ground-state dipole moments but larger excited-state dipole moments [20], and the use of ionic chromophores [21–23]. The incorporation of the chromophore into organic [24–27] and inorganic hosts [28] has also been implemented with some success. In addition, it is possible, using large electric fields, to orient the NLO active molecules in a glassy polymer matrix [29–31].

The structure/property relationships that govern third-order NLO polarisation are not well understood. Like second-order effects, they seem to scale with the linear polarisability. As a result, most research to date has been on highly polarisable molecules and materials such as polyacetylene, polythiophene and various semiconductors [3]. In order to optimise the third-order NLO response, a quartic, anharmonic term must be introduced into the electronic potential of the material. However, an understanding of the relationship between chemical structure and quartic anharmonicity must still be developed.

The synthesis of materials for device applications has very different requirements. Here the most important questions are: What does the device do and what factors will affect its performance? The magnitude of the desired optical nonlinearity will only be one of many criteria that will ultimately dictate the material of choice. In many instances the magnitude of the nonlinearity will not be the most important parameter.

Depending on the device applications, other considerations such as optical transparency, processability, one- and two-photon optical stability, thermal stability, orientational stability, and speed of nonlinear response will all be important. Our current understanding of NLO materials suggests that these variables are frequently interrelated and that there is often no ideal NLO material. The material of preference for a given application will typically be one that is the best compromise among a variety of variables.

3.3   NONLINEAR OPTICAL PROPERTIES OF ORGANOMETALLIC AND COORDINATION COMPOUNDS

3.3.1   Motivation for Studying Organometallic and Coordination Compounds for Nonlinear Optics

Organometallic and coordination compounds provide new opportunities for engineering nonlinear optical hyperpolarisabilities and susceptibilities. One can change the transition metal element, its oxidation state and hence the number of d-electrons, examine the differences between diamagnetic and paramagnetic complexes and the effect of novel bonding geometries and coordination patterns. Some attractive features of organometallic and coordination compounds are:

These compounds can have metal to ligand or ligand to metal charge transfer bands in the UV to visible region of the spectrum. These optical absorption bands are often associated with large second-order optical nonlinearities.

Chromophores containing metals are among the most *intensely* coloured materials known. The intensity of the optical absorption band is related to its transition dipole moment. Large optical nonlinearities are associated with such strongly allowed transitions [32].

Organometallic and coordination compounds are often strong oxidizing or reducing agents, since metal centres may be electron rich or poor depending on their oxidation state and ligand environment. Thus, the metal centre may be an extremely strong donor or acceptor, in comparison to conventional donors and acceptors incorporated in organic molecules for NLO.

Transition metals can also stabilise unusual or unstable organic fragments such as carbenes, carbynes or cyclobutadienes; therefore new classes of materials may be investigated [33]. Metals can be used to tune the electronic properties of organic fragments (see below). Because of these properties, these compounds have been the subject of active study as NLO materials.

3.3.2 Second-order Materials

3.3.2.1 Introduction

The two-state model (Equation (**30**)) suggests that β will be large in molecules that have: (1) large changes in dipole moment upon excitation, (2) large transition dipole moments, and (3) a small energy gap between the excited- and ground-states. Molecules with donor-acceptor interactions (resulting from charge transfer between electron donating and withdrawing groups) are therefore promising candidates to fulfil the above requirements [10–16]. Theoretical and experimental studies have demonstrated significant enhancement of hyperpolarisability by extending the conjugation length between the donor and the acceptor (by increasing the number of double bonds, for example). This enhancement to a great extent results from decreasing the energy gap and increasing the transition dipole moment between the relevant states. In addition, the electronic coupling between the donor and the acceptor can be increased, which sometimes leads to a large difference between the excited-state and ground-state dipole moments. Since second-order NLO effects result from asymmetric polarisation, they can only be induced in molecules lacking a centre of symmetry (there will be no dipole moment and therefore no net change in dipole moment between the excited- and ground-states in centrosymmetric materials). If asymmetric molecules are oriented noncentrosymmetrically within a material then the molecular nonlinearities will result in observable NLO effects. However, if the orientation of the asymmetric molecules is centrosymmetric then the molecular nonlinearities will cancel and no effects will be observed.

The original studies of metal organic materials relied on the Kurtz powder technique [34]. In this technique a powdered sample is illuminated with laser light and, if the material is SHG active, light at the second harmonic frequency is collected and its intensity is compared to some reference. Relatively large powder SHG efficiencies of several ferrocene complexes and square planar platinum and palladium aryl complexes provided the impetus for systematic studies of structure-property relationships for metal organic complexes. The effects on the hyperpolarisability arising from variation of the metal centre, bonding patterns, conjugation, and ligands on the metal for various classes of organometallic and coordination complexes are of interest.

Unfortunately, results obtained from the Kurtz powder test provide little insight into molecular structure-property relationships since they are strongly influenced by crystallographic, phase matching, and dispersive factors [34]. In addition, since molecular structure modification is often accompanied by crystallographic changes, powder testing cannot, in general, be used to systematically probe molecular structure-property relationships. Solution-phase DC electric-field-induced second-harmonic (EFISH) generation [35–39] is a more appropriate method for hyperpolarisability studies. The EFISH experiment allows for extraction of a vectorial projection of the hyperpolarisability tensor (β) along the molecular dipole (μ) direction. When experiments are carried out with radiation of sufficiently long wavelength to minimise dispersive enhancements, EFISH provides direct information on the intrinsic optical nonlinearity of a molecule. When the second harmonic is not far from the energy of a molecular transition, corrections for dispersion are often made using the two state approximation. However, corrections based on this approximation assume that the ground-state and a charge-transfer state dominate the perturbation theory expression for β. For many organic compounds, such as *p*-nitroaniline, this approximation appears to be reasonable, however, it is still an open question whether it is applicable to extended donor-acceptor substituted polyenes. Likewise, in metal organic systems, there is no reason, *a priori*, that two states will necessarily dominate the sum-on-states expression for β. Therefore, care must be exercised when correcting for dispersion using this approximation. Since the EFISH experiments provide more insight into the structure-property relationships for β, Kurtz powder test studies will be reviewed only briefly. Accordingly, emphasis will be placed on molecular classes where molecular hyperpolarisabilities have been measured.

3.3.2.2 *η-Arene Metal Carbonyl Complexes*

The first SHG study reported for transition metal organometallic complexes appeared in 1986 and focused on arene metal carbonyl complexes and on metal carbonyl pyridine complexes (discussed below) [40]. Of the 60 complexes studied only four had SHG efficiencies roughly equivalent to urea. Given the propensity for organic and organometallic compounds to crystallise in centrosymmetric space groups (leading to vanishing $\chi^{(2)}$), the results should neither be surprising nor particularly discouraging. It was later shown that inclusion complexes of metal arene carbonyl and related complexes with thiourea and tris(*o*-thymotide) hosts often crystallised in noncentrosymmetric space groups. For example, the thiourea complex of benzene chromium tricarbonyl gave an SHG efficiency of 2.3 times that of urea but the uncomplexed material gave no signal [25, 26].

Table 1 Summary of linear and nonlinear optical data for complexes of the form $[Cr(\eta^6\text{-}C_6H_5X)(CO)_3]$

Compound	X	λ_{CT} (nm)	μ ($\times 10^{-18}$) (esu) ± 0.1	β ($\times 10^{-30}$) (esu) $\pm 10\%$
N,N-Dimethylaniline		293	1.6	1.1
Benzaldehyde		320	2.8	0.8
(1)	H	310	4.4	-0.8
(2)	OMe	310	4.7	-0.9
(3)	NH_2	313	5.5	-0.6
(4)	NMe_2	318	5.5	-0.4
(5)	COOMe	318	4.0	-0.7

A series of η^6-arene chromium tricarbonyl complexes was examined by EFISH (Table 1) [41]. In these complexes, the arene acts as a donor to the metal centre via a d-π interaction, and the carbonyl as ground-state acceptor by virtue of strong d-π^* backbonding interactions. The complexes are strongly polarised in the ground-state with the metal positively charged and the carbonyl oxygen atoms negatively charged. Since the arene ring acts as a net donor in this system it is expected that donors on the arene ring will increase the dipole moments and acceptors on the arene ring will diminish them. However, since the ring substituent is presumably directed perpendicular to the ring-metal axis these perturbations are expected to be small. The experimentally measured dipole moments on a series of arene chromiumtricarbonyl complexes agreed with this expectation. Thus, μ: (5) < (1) < (2) < (3).

As noted before, the EFISH experiment provides information on the vectorial projection of β on μ. Thus, for the η^6-arene chromium tricarbonyl complexes, it is somewhat difficult to evaluate the observed trends in the data since substituent modifications result in changes: (1) of the orientation of the dipole moment axis with respect to a molecular coordinate, and(2) in components of the β tensor that are not collinear with the dipole moment axis. It is not surprising that the measured β

values of substituted η^6-arene chromium tricarbonyls are found to be quite insensitive to the donor–acceptor substituents on the benzene ring since, because the metal is perpendicularly bonded to the arene ring, the coupling between ring substituent and the metal centre will not be as strong as between donor and acceptor in 1,4-disubstituted benzenes. The insensitivity of the CT band in these π-complexes to both donor and acceptor ring substitution is consistent with relatively weak metal–substituent interactions.

The experimental values of β in this system are all negative. Within the context of a two state model, this implies that the excited-state dipole moment in these complexes is lower than the ground-state dipole moment, or is opposite in sign. η-Benzene chromium tricarbonyl has a band assigned as an MLCT excitation [42] at about 310 nm. In the ground-state, the dipole moment is directed from the metal to the carbonyls, however the change in dipole moment between the ground- and excited-state is in the opposite direction, from the metal to the arene ring [41]. These results suggest that the metal-to-benzene CT transition is in large part responsible for the observed nonlinearities in this system and that higher lying metal to CO π^*-transitions are less important.

3.3.2.3 Metal Pyridine Carbonyl Complexes

Initial studies on metal pyridine and bipyridine complexes focused on powder SHG measurements [43, 44]. It was found that several pyridine and bipyridine complexes had SHG efficiencies roughly comparable to urea (SHG at 532 nm). For example, [ReCl(CO)$_3$(2,2'-bipyridine)] had a powder SHG of 1.6–3 times that of urea, and the structurally characterised [Re(CF$_3$SO$_3$)(CO)$_3$(2,2'-bipyridine)] had a powder SHG efficiency of 1.7–2 times that of urea [43]. The rationale for investigating complexes of this general class stemmed from the fact that metal pyridine complexes often exhibit strong metal-to-ligand charge-transfer (MLCT) excitations [32].

The values of β for a series of tungsten pyridine metal σ-complexes determined from EFISH studies (Table 2) showed this class of compounds to be much more nonlinear than the previously discussed arene chromium tricarbonyl complexes [41, 45]. Furthermore, hyperpolarisabilities of the pyridine complexes were highly sensitive to the 4-pyridine substituents, in contrast to the relative insensitivity of the η-arene complexes to substituents. This is most likely a manifestation of the strong backbonding between filled metal d-orbitals and the pyridine π^*-orbitals, improving the coupling between the donor and the acceptor. Also, in these complexes, it is likely that μ and the major component of the β tensor are roughly collinear.

Examination of Table 2 reveals two important features. First, the signs of β for the metal pyridine complexes are all negative. Second, whereas the dipole moment of the complexes increases with increasing donor ability of the 4-substituent, the absolute magnitude of β decreases. These observations were explained in the following manner. The pyridine acts as a strong ground-state donor while the metal centre is a potent ground-state acceptor, as a result of substitution with the five carbonyl ligands that are strong π-acceptors. Therefore, the ground-state dipole moments of these complexes are likely to be oriented along the molecular four-fold axes with excess charge on the carbonyl oxygen atoms. Substitution of the pyridine in the 4-position enhances its donor ability and is therefore expected to increase the ground-state dipole moment.

The two-state approximation for β derived from perturbation theory [12, 15] predicts that a reduction of dipole moment upon excitation leads to a negative sign for the hyperpolarisability. The MLCT transition results in a transfer of electron density from the metal centre to the ring, leading to reduced excited-state dipole moments relative to their ground-state values. Thus, observed negative β values suggest that MLCT excitations play an important role in determining the quadratic nonlinearity of metal pyridine complexes.

The observation that the optical transition moves to higher energy in more polar solvents (negative solvatochromic behaviour) for these compounds [32] lends further credence to the notion that the excited-state dipole moment in the complexes is lower than that of the ground-state. Since the change in dipole moment that appears to be responsible for the observed nonlinearity results from an MLCT transition, substitution of the pyridine with donors, that inhibits this back donation from the metal centre, should lower the magnitude of β, consistent with the observed results. The magnitude of these β values (Table 2) is comparable to that of conventional nonlinear organics such as *p*-nitroaniline ($\mu = 6.2 \times 10^{-18}$ esu and $\beta = 9.2 \times 10^{-30}$ esu) [38, 39].

Table 2 Summary of linear and nonlinear optical data for complexes of the form $[W(NC_5H_4X)(CO)_5]$

Compound	X	Solvent	λ_{CT} (nm)	μ (x 10^{-18}) (esu)	β (x 10^{-30}) (esu)
Pyridine		Neat	256	2.3	1.0
4-Acetylpyridine		CHCl$_3$	276	2.7	<0.2
4-Aminopyridine		CHCl$_3$	260	4.3	1.2
(6)	H	Toluene	332	6.0	−4.4
(7)	Phenyl	CHCl$_3$	330–340	4.6	−4.5
(8)	Butyl	p-Dioxane	328	7.3	−3.4
(9)	NH$_2$	DMSO	290	8.0 (± 1)	−2.1 (± 3)
(10)	COMe	CHCl$_3$	420–440	4.5	−9.3
(11)	CHO	CHCl$_3$	420–440	4.6	−2.0

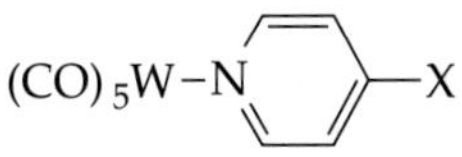

$$(CO)_5W-N\diagdown\diagup\diagup\diagdown-X \qquad \begin{array}{l} X = H, CHO, COCH_3 \\ C_6H_5, NH_2 \end{array}$$

Figure 10 Structures of the metal pyridine complexes given in Table 2

3.3.2.4 Square-Planar Metal Complexes

Interest in square-planar metal aromatic complexes arose from the knowledge that the fragments, $[MX\{P(Et)_3\}_2]$, where M = Ni, Pd and Pt; X = I, Br and Cl, are good electron donors. Powder SHG testing revealed that this class of compounds substituted with various aromatic acceptors can exhibit relatively large powder SHG signals as high as 14 times that of urea [46, 47]. Furthermore, it was observed that many compounds crystallised in noncentrosymmetric space groups. More recently, donor–acceptor substituted bis(acetylide) complexes of the form trans-$[Pt\{P(Me)_2(Ph)\}_2(C\equiv C-X)(C\equiv C-Y)]$, where X is a π-donor and Y is a π-acceptor, have been synthesised. Some of these compounds crystallise in noncentrosymmetric space groups and exhibit powder SHG efficiencies on the order of urea [48]. In these systems, the rôle of the metal centre as a linker between organic donors and acceptors was of interest.

Results from EFISH measurements on several square-planar platinum and palladium benzene derivatives are given in Table 3 [41]. The magnitude of the β values has been explained in terms of both the resonance and inductive donating ability of the metal centre. The inductive contribution is quite strong and is sensitive to the nature of the *trans*-ligand. Spectroscopic studies [49] have also revealed a *cis*-influence that is comparable in strength but opposite in sign to the *trans*-effect.

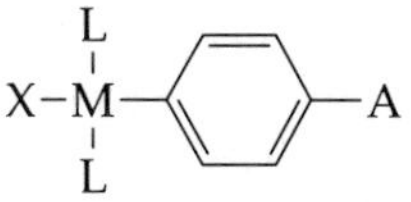

$$X = Br, I$$
$$M = Pd, Pt$$
$$A = CHO, NO_2$$
$$L = P(Et)_3$$

Figure 11 Structures of the square-planar metal complexes given in Table 3

For the compounds in Table 3, the plane of the aromatic ring is orthogonal to the plane of the metal and two phosphorous atoms. Therefore, the resonance donation is mediated by an interaction of the largely d_{xy} HOMO, (where the z-axis is perpendicular to the ligand plane) with the π-system of the aromatic ligand. Thus, in this system and likewise in the metal pyridine complexes discussed in the previous section, the para substituent should be in conjugation with the metal centre and is expected to have a significant effect on hyperpolarisability. Both dipole moments and β values are found to increase with acceptor strength, giving the higher value for a nitro derivative relative to a formyl derivative ((16) > (12)). Since the metal centre acts as both a ground- and excited-state donor, the dipole moment is expected to increase in the excited-state. Consistent with this expectation, all the compounds in Table 3 have positive β.

The importance of the inductive donating strength of the metal centres is evidenced by the observation that compound (15) is more nonlinear than the palladium analogue (13). This reflects the substantial difference in inductive donating strength of the two metal centres [50]. The *trans*-influence is also evident with the bromo derivative significantly more nonlinear than the iodo derivative ((16) versus (15)). The strong *cis*-influence is seen by replacing the triethylphosphine ligands with the less electron rich triphenylphosphine ((13) versus (14)).

Table 3 Summary of nonlinear optical data for complexes of the form
$trans$-$[M(L)_2X(\sigma\text{-}C_6H_5A)]$

Compound	A	M	L	X	μ ($\times 10^{-18}$) (esu) ± 0.2	β ($\times 10^{-30}$) (esu) $\pm 15\%$
(12)	CHO	Pt	PEt$_3$	Br	2.5	2.1
(13)	NO$_2$	Pd	PEt$_3$	I	3.6	0.5
(14)	NO$_2$	Pd	PPh$_3$	I	5.5	1.5
(15)	NO$_2$	Pt	PEt$_3$	I	3.0	1.7
(16)	NO$_2$	Pt	PEt$_3$	Br	3.4	3.8

3.3.2.5 Second-order NLO Properties of Metallocenes

The observation that the ferrocene complexes (Z)-[1-ferrocenyl-2-(4-nitrophenyl)ethylene] [51] and [Fe(η-C$_5$H$_5$)(η-C$_5$H$_4$)-CH=CH-(4)-C$_5$H$_4$N(CH$_3$)]$^+$ [I]$^-$ [52, 53] have SHG efficiencies 62 and 220 times that of urea respectively demonstrate that organometallic compounds, in particular metallocenes, could exhibit large $\chi^{(2)}$. The latter value is the largest efficiency yet observed for an organometallic compound [52, 53]. Furthermore, the magnitude of the powder SHG signal depends on the counterion, consistent with results for organic salts [21–23]. Related ferrocene and ruthenocene complexes have been prepared and exhibit small SHG efficiencies as seen in Table 4 [54]. Asymmetry and optical activity were introduced by substituting the η-cyclopentadienyl ring that contains the acceptor moiety with an additional methyl group. It was hoped that by analogy to approaches used in organic systems, this would lead to a more desirable alignment of the molecular dipoles in the crystal lattice. In each case the (η-C$_5$H$_3$CH$_3$)CH=CHR compound exhibited a larger efficiency than the analogous (η-C$_5$H$_4$)CH=CHR compound.

More recently, it has been shown that bimetallic complexes [Mo(NO)L(Cl)(NHC$_6$H$_4$-4-N=N-C$_6$H$_4$-4'-Fc] {Fc = Fe(η-C$_5$H$_5$)(η-C$_5$H$_4$) and L = {HB(3,5-Me$_2$-(C$_3$N$_2$H)$_3$}, [W(NO)L(Cl)(NHC$_6$H$_3$-3-Me-4-N=N-C$_6$H$_4$-4'-Fc] and [Mo(NO)L(Cl)(NHC$_6$H$_3$-3-Me-4-N=N-C$_6$H$_4$-4'-Fc] gave SHG efficiencies at 1.9 µm of 59, 53, and 123 times urea respectively [55]. Of the nineteen compounds containing molybdenum or tungsten and the pyrazolylborate ligand {HB(3,5-Me$_2$-C$_3$N$_2$H)$_3$}, only the complexes that also contained the Fc fragment exhibited large SHG efficiencies. It was proposed that the Mo or W centre acted as an acceptor in these molecules [55], however, it is difficult to speculate even qualitatively on the rôle of the molybdenum or tungsten centres until EFISH measurements have been performed. Donor–acceptor acetylenes with the Fc fragment as the donor have also been prepared and exhibit small powder efficiencies [56]. Recently, polymeric materials with pendent Fc–acceptor moieties have been prepared. It was suggested that these compounds may be of interest for poled polymer applications, but this has yet to be demonstrated [57].

Table 4 Powder SHG efficiencies of compounds of the form
$Fe(\eta\text{-}C_5H_5)(\eta\text{-}C_5H_3R')CH=CHR]$

R	Compound	Isomer	SHG [a,b] R'= H	R' = CH₃ [c]
$p\text{-}C_6H_4(NO_2)$	(17)	Z	62.0	—
$p\text{-}C_6H_4(NO_2)$	(18)	E	0.0	8.0
$p\text{-}C_6H_4CN$	(19)	Z	0.95	—
$p\text{-}C_6H_4CN$	(20)	E	0.87	1.2
$p\text{-}C_6H_4CHO$	(21)	Z	0.0	—
$p\text{-}C_6H_4CHO$	(22)	E	0.72	2.5
$(E)\text{-}CH=CH\text{-}p\text{-}C_6H_4(NO_2)$	(23)	Z	0.0	—
$(E)\text{-}CH=CH\text{-}p\text{-}C_6H_4(NO_2)$	(24)	E	0.04	6.4
$\overline{C=CH\text{-}CH=C(NO_2)}O$	(25)	E	0.03	17.5

[a] All measurements were performed at 1.907 μm; [b] The SHG signal is the magnitude of the signal at 0.953 μm for the compound of interest relative to a urea reference standard measured under the same conditions; [c] (S)-[Fe(η-C₅H₅)(η-C₅H₃CH=CHR(Me)] (82% enantiomeric excess of S-isomer) [54]

These observations provided the motivation to study the molecular hyperpolarisabilities of several ferrocene and ruthenocene derivatives. Several structural variations, including different metal centres, *cis* and *trans* isomers, extension of conjugation and symmetric electron donating substituents in the form of η-pentamethylcyclopentadienyl rings (Cp*), were examined and the results summarised in Table 5 [41, 58].

Table 5   Summary of linear and nonlinear optical data for complexes of the form $[M(\eta\text{-}C_5X_5)(\eta\text{-}C_5H_4)\text{-}(CH=CH)_n\text{-}C_6H_4]$ [a]

Compound	M	X	n	Isomer	Y	λ_{CT} (nm)	μ^b	β^c
(26)	Fe	H	1	E	NO_2	356/496	4.5	31.0
(27)	Fe	H	1	Z	NO_2	325/480	4.0	13.0
(28)	Fe	Me	1	E	NO_2	366/533	4.4	40.0
(29)	Fe	H	1	E	CN	324/466	4.6	10.0
(30)	Fe	H	1	Z	CN	308/460	3.9	4.0
(31)	Fe	H	1	E	CHO	338/474	3.9	12.0
(32)	Ru	H	1	E	NO_2	350/390	5.3	12.0
(33)	Ru	Me	1	E	NO_2	370/424	5.1	24.0
(34)	Fe	H	2	E,E	NO_2	382/500	4.5	66.0

[a] All measurements were performed at 1.907 µm; [b] ($\times 10^{-18}$ esu); [c] ($\times 10^{-30}$ esu)

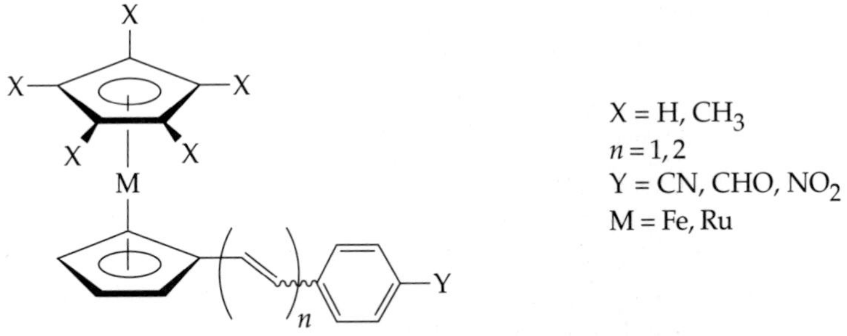

$X = H, CH_3$
$n = 1, 2$
$Y = CN, CHO, NO_2$
$M = Fe, Ru$

Figure 12   Structures of the metallocene derivatives discussed in Table 5

The UV-visible spectra of acceptor substituted ferrocenylethylenes typically have two strong bands that are attributed to charge-transfer transitions.   For example, $[Fe(\eta\text{-}C_5H_5)(\eta\text{-}C_5H_4)\text{-}CH=CH\text{-}(4)\text{-}C_5H_4N(Me)_2]^+$ [I]$^-$ has two strong bands at $\lambda_{max} = 380$ nm ($\varepsilon = 29{,}000$ M^{-1} cm^{-1}) and at $\lambda_{max} = 550$ nm [53]. Extended Hückel molecular orbital (EHMO) calculations on (26) suggest that the lowest energy transition in these systems was due to an MLCT band and the highest energy transition to effectively a ligand π to π^* transition, *with some metal character.*

Electron density is substantially redistributed in both transitions and therefore *both* probably contribute to β [58].

Derivatives that have a well defined charge transfer direction along the 4-nitrophenylvinyl group show respectable nonlinearities in comparison with nitrostilbene (β = 9.1 x 10^{-30} esu), 4,4'-methoxynitrostilbene (β = 29 x 10^{-30} esu), and 4,4'-dimethylamino-nitrostilbene (β = 75 x 10^{-30} esu) [38, 39]. Since these compounds have long wavelength absorption bands, the measured nonlinearities are somewhat dispersively enhanced. The *cis* compound (27) is less nonlinear than the *trans* compound (26) as expected, considering the lower second hyperpolarisability of *cis*-stilbene and the shorter conjugation length because of the bent charge-transfer pathway, and the reduced donor–acceptor coupling due to the nonplanar geometry [51]. Pentamethyl substitution at the opposite ring significantly increases both the dipole moment and the nonlinearity (28). This agrees with the notion that the added electron density on the ligand raises the energy of the metal orbital and thus increases the donating strength of the metal d-electrons. The HOMO is clearly destabilised as evidenced by the large spectral red shift. The effect of the conjugation length is dramatic, with compound (34) exhibiting a significantly higher β value. The ruthenium compounds are found to be less nonlinear than their iron counterparts, ((32) versus (26) and (33) versus (28)) consistent with ruthenium being a weaker donor as evidenced by the higher oxidation potential of ruthenium versus iron metallocene complexes. Recently, ZINDO calculations were performed on these and other organometallic systems discussed above [59]. In general, there was excellent agreement between the experimental values and those predicted by theory. Thus, it is expected that computational methods will play a significant rôle in the future design of organometallic compounds for NLO applications.

3.3.2.6 *Donor–Acceptor Substituted Silanes*

A series of donor–acceptor substituted silanes of the form [(Me)$_2$N(C$_6$H$_4$)-{Si(Me)$_2$}$_n$(C$_6$H$_4$)-CH=C(CN)$_2$], where n = 1, 2, and 6, have been synthesised [60]. The authors ascribe their interest in these compounds to improved transparency relative to donor–acceptor polyenes and to improved coupling between the σ-π system relative to saturated hydrocarbon linkages and to the σ-delocalisation through the Si–Si single bonds. Results of the EFISH study are shown in Table 6 [60].

Table 6 Summary of EFISH data measured in $CHCl_3$ on compounds of the form $[N(Me)_2(C_6H_4)\{Si(Me)_2\}_n(C_6H_4)CH=C(CN)_2]$ with the 1.34 μm fundamental

Compound	n	$\lambda_{max}(\varepsilon)$ (nm)	μ (x10^{-18}) (esu)	β (x 10^{-30}) (esu)	β_{add} (x 10^{-30}) (esu)	β_{CT} (x 10^{-30}) (esu)
(35)	1	320 (27,200)	6 ± 0.5	16 ± 4	14 ± 4	0
(36)	2	334 (25,680)	7 ± 1	22 ± 5	16 ± 4	6 ± 4
(37)	6	276 (36,600) 320, 385 (CT)	6.8 ± 0.6	38 ± 4	16 ± 4	22 ± 4

Table 6 shows that there is a substantial increase in the hyperpolarisability with increasing number of $SiMe_2$ groups. This was attributed to the increased σ delocalisation along the silicon backbone. The measured β was analyzed as a sum of $\beta_{add} + \beta_{CT}$, where β_{add} is the vectorial sum of the contribution of each side group, and β_{CT} is the difference between β and β_{add}. Using this analysis, the authors show that in σ-delocalised systems long range charge transfer can play an important rôle. It was also shown that relative to *p*-nitroaniline (λ_{max} 365 nm, dioxane) $[N(Me)_2(C_6H_4)-\{Si(Me)_2\}_6(C_6H_4)-CH=C(CN)]_2$ has a dispersion corrected β value roughly 2–3 times larger, thus it was suggested that the compound exhibits a large nonlinearity for its transparency window. However, the comparison did not account for the fact that the donor/acceptor silane effectively has two chromophores and is therefore roughly twice the size of *p*-nitroaniline, thus, the volume normalised values are equivalent, within experimental error.

3.3.2.7 *Other Studies*

Several other classes of compounds have been examined by the Kurtz powder method including diiron bridging alkenylidyne complexes [61]. Also, coordination polymers containing the tetradentate SALEN ligand (SALEN = N,N'-bis(salicylideneaminato)ethylene) and a bifunctional ligand capable of bonding to the metal in both a dative and covalent manner have been examined [62]. In addition, polymers of the form $[(RO)_3M\equiv N]_n$, (M = W; R = $C(Me)_3$, and M = Mo; R = $C(Me)_3$, $C(Me)_2(CF_3)$, or $C(Me)_2Et$ and M = Mo), were synthesised and examined by the Kurtz powder method [63]. As of yet none of these materials exhibit large powder SHG efficiencies (> urea) nor have any been examined by EFISH.

Langmuir-Blodgett films of $[Ru(\eta\text{-}C_5H_5)\{P(Ph)_3\}_2L]^+$ $[PF_6]^-$ complexes, where L is a substituted nitrile have been examined for nonlinear optical applications. The unsubstituted nitrile, exhibited only modest hyperpolarisabilities as measured by EFISH and the authors suggest the red shift observed upon complexation to the cationic ruthenium centre would enhance the acceptor strength of the nitrile and therefore improve the hyperpolarisability of the organic fragment [64]. Pentaamine ruthenium complexes of the form $[(NH_3)_5RuN\equiv C\text{-}C(R)\text{-}C\equiv NRu(NH_3)_5]^{5+}$ $[PF_6^-]_5$, where R = H or Bu^t, were considered as chromophores for Langmuir-Blodgett films to frequency double the fundamental of a Nd:YAG laser in a waveguide [65]. The motivation for this study was the potential to achieve phase matched SHG in a waveguide by using anomalous dispersion. The above ruthenium complexes were attractive because they met the requirement of having a strong charge transfer band between the fundamental and the second harmonic frequencies. Unfortunately, the above complexes exhibited significant decomposition upon irradiation with the Nd:YAG fundamental, rendering them unsuitable for this particular application.

3.3.2.8 Conclusions

These preliminary studies indicate that donor-bridge-acceptor molecules with organometallic end groups are useful for engineering molecular hyperpolarisability, resulting in species with nonlinearities comparable to those found in organics. Many of the same chemical factors that influence the magnitude of hyperpolarisabilities of organic materials, i.e. nature of the π-electron bridge and donor–acceptor strength, are operative in the metal containing systems as well. In addition, interesting behaviour not commonly seen in organic compounds is observed with organometallic species. Absorption bands in the visible region of the electronic spectrum may limit the utility of many metal organic compounds for harmonic generation of ultraviolet or visible light. However, they may have some utility for telecommunication applications that utilise 0.8, 1.3 and 1.5 μm radiation.

3.3.3 Third-order Materials

3.3.3.1 Introduction

Third-order NLO effects include frequency tripling, optical phase conjugation (potentially useful for restoring distorted optical images), and optical limiting [1–7]. In contrast to the relatively well developed models that have guided the synthesis of second-order NLO molecules, the models for molecular third-order optical nonlinearities are less well developed. Frequently molecules with strong, low energy optical absorption bands have large second hyperpolarisabilities, γ [1–7]. Such strong transitions are often seen in materials with highly delocalised electrons, like conjugated polymers and semiconductors. Using metal systems, it may be possible to develop completely new design criteria for materials with large third-order nonlinearities that do not invoke the use of large delocalised electronic systems. To date, investigations of metal-containing materials with third-order optical nonlinearities have been relatively limited. Most research has focused on resonant or near resonant nonlinearities, that is, those which occur in a spectral region where the molecule absorbs light.

Many third-order NLO materials absorb in the visible and the near IR. These absorptions introduce additional difficulties for assessing their electronic hyperpolarisability [66]. One powerful technique, third harmonic generation (THG), suffers from the problem that the third harmonic (TH) is often in or near an absorbing region of the spectrum (even when using 1.907 μm light from a H_2 Raman shifted Nd:YAG laser, TH at 636 nm). This does not preclude obtaining a value for γ, however, interpretation of this value is often not straightforward since the measured γ will often be a complex quantity and/or dispersively enhanced. Thus, extreme caution must be exercised when comparing hyperpolarisabilities for materials measured at different frequencies. Using degenerate four wave mixing (DFWM) to determine γ circumvents some of the problems associated with absorption since one need not be concerned with three photon resonances. In the THG experiment, only the electronic hyperpolarisability is fast enough to contribute effectively to the observed nonlinearity. In contrast, in the DFWM experiment, if the laser pulses are not sufficiently short (< 10 ps), then vibrational and orientational mechanisms can contribute to the observed nonlinearity. These contributions can be large and although they may be useful for certain applications, they will complicate the interpretation of results for developing structure/property relationships of the purely electronic part of γ.

3.3.3.2 Metal-containing Organic Polymers

A series of papers has appeared on the nonlinear optical properties of the polymers shown in Table 7 [67–70], originally synthesised by Takahashi *et al.* [71]. These liquid crystalline polymers have interesting magnetic and electronic properties [71]. The presence of low energy MLCT transitions and the possibility of extended delocalisation through the polymer chain provided the rationale for studying these materials for NLO applications. In the original report, γ for $[Pd\{P(Bu^n)_3\}_2\text{-}C\equiv C\text{-}C_6H_4\text{-}C\equiv C\text{-}]_n$ as measured by THG was 1.47×10^{-33} esu in benzene [67]. The authors note, however, that this value is resonance enhanced since the third harmonic (355 nm) was near λ_{max} (344 nm) of the MLCT band. For comparison, γ for benzene is 3.8×10^{-36} esu.

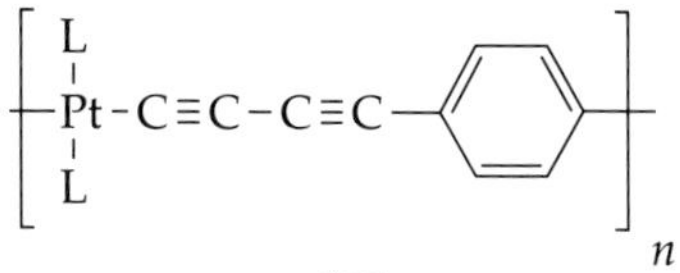

(**38**)

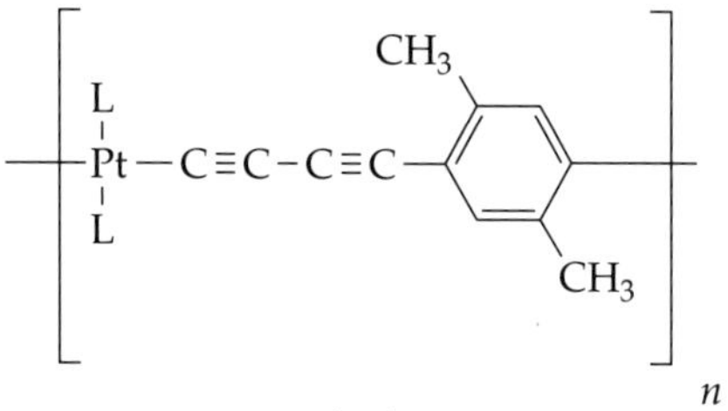

(**39**)

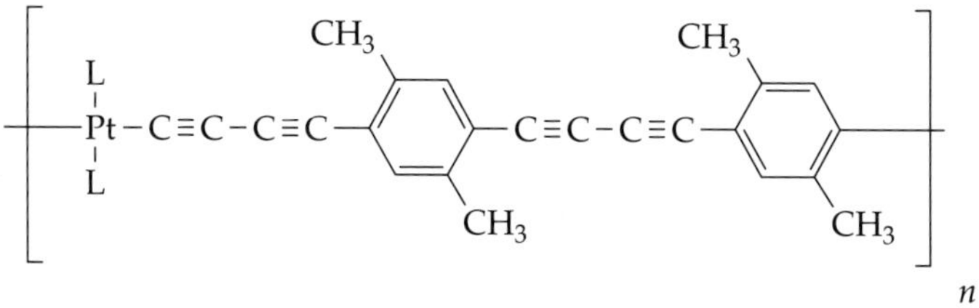

(**40**)

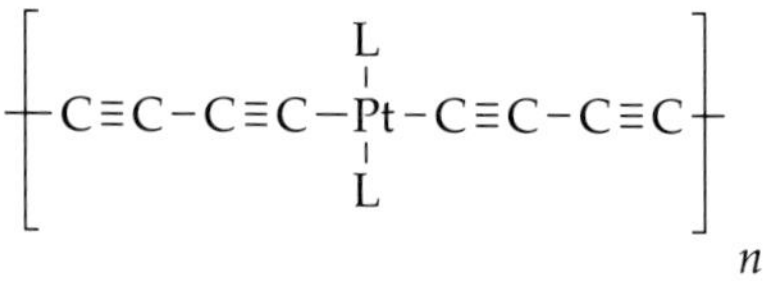

(**41**)

Figure 13 Structures of some platinum polymers discussed in Table 7 (L = $P(Bu^n)_3$)

In a later study, the polymers were examined by a four wave mixing experiment in which two laser beams at ω_1 and ω_2 were mixed to create a third beam at ω_3, where $\omega_3 = 2\omega_1 - \omega_2$. In this manner, the authors extracted the real and imaginary hyperpolarisabilities for a series of materials of the form $[Pt\{P(Bu^n)_3\})_2\text{-}C{\equiv}C\text{-}X\text{-}C{\equiv}C\text{-}]_n$, where X is phenyl, *p*-xylyl or nothing. The results are summarised in Table 7 [68].

Table 7 Summary of nonlinear optical data on rigid platinum polymers[a]

Compound	Formula	Solvent	M_n	Hyperpolarisabilities	
				γ'	γ''
(38)	$[Pt(L)_2DEB]_n$	THF	61,110	890	1300
(38)	$[Pt(L)_2DEB]_n$	C_6H_6	80,800	328	949
(39)	$[Pt(L)_2DEX]_n$	C_6H_6	19,600	471	464
(40)	$[Pt(L)_2(DEX)_2]_n$	C_6H_6	60,300	1552	344
(41)	$[Pt(L)_2C{\equiv}C\text{-}C{\equiv}C\text{-}]_n$	THF	63,000	743	1296
(42)	$[Pt(L)_2(C{\equiv}C\text{-}C{\equiv}C\text{-}H)_2]$	THF	—	20	108
(43)	*trans*-$[Pt(L)_2(DEB)_2]$	C_6H_6	—	245	221
(44)	DEB	THF	—	12	0

[a] γ' and γ'' are real and imaginary components of the hyperpolarisabilities (x 10^{-36} esu); [b] M_n = number average molecular weight, L = $P(Bu^n)_3$, DEB = diethynylbenzene, DEX = diethynylxylene

The contribution of the imaginary component of the hyperpolarisability (γ'') that is thought to result from two-photon absorption is roughly as large as the real component (γ'). The presence of two-photon absorption has been determined in separate experiments. The dependence of the hyperpolarisability upon solvents ((38) in THF versus C_6H_6) suggests that the two solvents have different interactions with the π system and affect the relative energetics of the excited-states. This behaviour is well known in the spectroscopy of polyenes [72]. There is little enhancement of either γ' or γ'' upon polymerisation ((43) versus (38) in C_6H_6) of the DEB bridge. However, when the length of the bridge between the metals was lengthened there was a modest increase in γ' whereas γ'' remained roughly unchanged. Since γ' for conjugated organic molecules is known to increase dramatically with conjugation length (in the short conjugation length limit) this behaviour is quite understandable. Likewise, the larger value of γ' for (38) (in THF) as compared with (41) is consistent with normal dependence of γ' on conjugation length. It is interesting to observe that both γ' and γ'' for monomer (42) are dramatically lower than those for the related polymer (41). The authors suggest that the increase observed here, as opposed to the case of (**43**) versus (**38**), may result from the lack of aromatic groups in chains of the former series, that may adopt conformations in which the coupling between the metal centres is diminished. Molecular orbital calculations on these systems would provide insight into the bonding and would therefore be helpful in understanding the origin of the NLO effect in these molecules.

3.3.3.3 Phthalocyanines

Metals may also be used to tune the optical properties of an organic fragment. A metal may dramatically perturb the ground-state structure of a material, as when a late metal forms a π-complex with an acetylene, or it may tune the excited-state properties of the complex by effecting lifetimes, fluorescence quantum yields, or facilitating intersystem crossing, for example. Phthalocyanines are known to have intense low energy optical absorptions [73], and are therefore promising candidates for third-order NLO applications. Metallation of the ring often has only a small effect on the linear absorption spectrum but serves to tune the excited-state photophysical properties of these materials, in a subtle manner. Several groups have examined various phthalocyanines and metallated phthalocyanines for third-order NLO applications [74–78]. Many studies focus on using the resonant γ of these molecules for efficient optical limiting or switching. The details of the excited-state photophysics are therefore critical to the overall response of the chromophore.

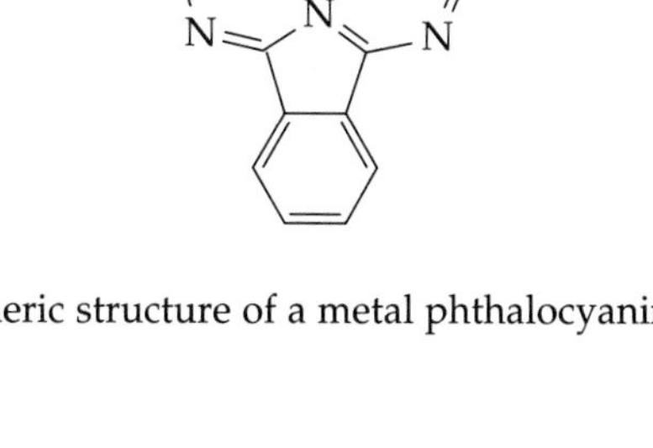

Figure 14 Generic structure of a metal phthalocyanine complex

Phthalocyanines are well known saturable absorbers [74], materials whose ground-state populations can be depleted (by excitation) resulting in diminished absorption at the frequency of excitation. This bleaching typically occurs in materials with excited-states that have reasonably long lifetimes, such that relaxation back to the ground-state is slow relative to formation of the excited-state under intense illumination. Based on this process, soluble derivatives of a silicon naphthalocyanine have shown optical bistability, both in solution and in polymethylmethacrylate. Thus, a plot of output power as a function of input power in a silicon naphthalocyanine-polymethylmethacrylate film exhibited hysteresis. The behaviour of optically bistable materials is likened to that of ferroelectric materials [75]. In optically bistable materials, on the increasing side of the hysteresis loop, output power slowly increases until the field strength effectively causes a 'phase transition' in the material (in this case, bleaching) leading to a high output power level [79]. The material, now in a different state, has a different internal electric field. If there is some cooperativity of the internal fields, then on decreasing the applied field from the peak value, the total internal field in the material will be different from that on the increasing side of the loop. As a result, the material resists switching back to the original state and the decreasing side of the hysteresis loop will return to the original state via a different path.

Phthalocyanines have also been investigated as optical power limiters [76]. In optical limiting, light of low incident intensity is transmitted through a sample as a linear function of the input intensity, whereas at input intensities above a threshold level, the intensity of the transmitted light approaches a limiting value. At very high input intensities (but below the damage threshold intensity for the medium or below the level where the *excited-state* bleaches) the output intensity saturates at a value determined by the nonlinearity of the medium. This process can be used to protect detectors (including eyes) from very bright light.

Materials that undergo two-photon absorption or two sequential one-photon absorptions are promising candidates for optical power limiters. In the latter case, materials that exhibit efficient excited-state absorption from either the singlet or the triplet manifolds are of interest. For a material to be useful as an optical limiter in a given spectral region by the previous mechanism, it must fulfil several criteria: the material must have relatively high transparency at low fluence, strong excited-state absorption, and a relatively long excited–state lifetime, such that excited absorption can effectively compete with relaxation to the ground-state. Several metallophthalocyanines meet the above requirements and have therefore been examined as the active elements in optical limiters. They have a transparency window from roughly 400–600 nm and also exhibit strong excited-state absorption in this region. Nanosecond optical limiting experiments on these materials reveal that several compounds exhibit significant limiting even within an order of magnitude of the response needed for eye protection in the visible. Results for selected metallophthalocyanines are given in Table 8 [76].

Table 8 Nonlinear transmission data for phthalocyanines[a]

Compound	Formula	Solvent	$T(0)$[b]	$T(I)$[c]	$T(I)/T(0)$
(45)	[Al(Cl)Pc]	EtOH	0.75	0.57	0.76
(46)	[Si(OR)$_2$Pc]	Toluene	0.755	0.50	0.66
(47)	[Ge(OR)$_2$Pc]	Toluene	0.75	0.46	0.62
(48)	[Sn(OR)$_2$Pc]	Toluene	0.75	0.42	0.56
(49)	[{Si(OR)Pc}$_2$O]	Toluene	0.745	0.66	0.89
(50)	[Si(OR)$_2$NPc]	Toluene	0.75	0.47	0.63
(51)	[Zn{N(Bu^t)}$_4$NPc]	DMF	0.75	0.60	0.80
(52)	[Al{N(Bu^t)}$_4$NPc(Cl)]	Toluene	0.76	0.75	0.99

[a] NPc = naphthalocyanine, Pc = phthalocyanine, R = Si(n-hexyl)$_3$; [b] $T(0)$ = low intensity transmission; [c] $T(I)$ = transmission at 0.032 J/cm^2

Although substantial work has been devoted to understanding mechanisms of limiting in the metallophthalocyanines, studies directed towards understanding structure/property relationships of metallophthalocyanines are few in number. As can be seen in the table several dyes are efficient limiters. First, in comparison to [Al(Cl)Pc], [M(OR)$_2$Pc] (M = Si, Ge or Sn), exhibit stronger excited-state absorption and thus are more efficient limiters. Within the latter series, the limiting efficiency is in the order Sn > Ge > Si, indicating that the excited-state absorption increases as the atomic number Z of the metalloid increases [76]. One possible explanation for this increased efficiency is that the triplet absorption is stronger than the singlet absorption and that increasing Z enhances the rate of intersystem crossing. A definitive explanation, however, awaits a more complete photophysical study of these materials.

DFWM experiments at 1.064 μm have also been used to characterise the hyperpolarisability of *tetrakis*(cumylphenoxy)phthalocyanines of Co, Ni, Cu, Zn and Pt [77, 78]. For the first row metals a decrease in γ was observed with increasing number of d electrons. In addition, there was a qualitative correlation between a large hyperpolarisability and the existence of a weak band in the near infrared [77]. Further theoretical and photophysical studies are required for a definitive interpretation of these results.

3.3.3.4 *Metallocenes*

The interest in studying metallocenes stems from the desire to understand the contrast between organometallic compounds and organic compounds. Thus, the wide variety of metallocene analogues to organic compounds may allow specific comparisons between organic and organometallic chromophores to be made. Of particular interest are questions such as what is the role of d–d transitions and MLCT transitions. What is the effect of conjugation through the metal of a metallocene in comparison to through a carbon-carbon double bond or a benzene ring?

There have been several studies on the third-order nonlinear optical properties of various metallocenes. One study used optical power limiting measurements to determine n_2 (the nonlinear refractive index) of several metallocene derivatives, including ferrocene and ruthenocene [80–82]. Using this technique, γ for ferrocene was measured to be roughly twice that of nitrobenzene. Further polarisation experiments suggest that this nonlinearity is electronic, not rotational in origin. More recently, the near resonance enhanced optical nonlinearity of $[Yb(\eta-C_5H_5)_3]$ was examined by optical limiting experiments. Irradiation of this compound in the near IR (1064 nm), close to a sharp, but relatively weak band (λ_{max}, 1030 nm) assigned to f–f transitions, resulted in efficient optical limiting [83]. However, the compound also undergoes efficient two-photon absorption that limits its utility for device applications.

Third harmonic generation studies have also demonstrated that metallocene complexes can exhibit large third-order NLO susceptibilities. For example, at 1907 nm fundamental radiation it was shown that $\chi^{(3)}$ of 1,4-bis(ferrocenyl)butadiyne was roughly 2.5 times that of 1,4-diphenylbutadiyne, though the former value is somewhat dispersively enhanced due to its low absorption edge [56].

More recently, a series of substituted ferrocene complexes shown in Table 9 have been examined by DFWM at 602 nm using subpicosecond pulses [84]. Despite the very short pulses, the authors were cautious that there may be some orientational contribution to the observed nonlinearity, although it was expected to be small. In addition, all of the compounds had weak but non-zero absorption at 632 nm (ε = 0.6–40 M^{-1} cm^{-1}) therefore the measured values of γ are somewhat dispersively enhanced. The results of the DFWM experiments are given in Table 9.

Table 9 Linear polarisabilities and hyperpolarisabilities for some ferrocenyl derivatives

Compound	Formula[a]	α (x 10^{-23} esu)	γ (x 10^{-36} esu)
(53)	[HFc-CHCH-C_6H_5]	3.96	85.5 ± 19.8
(54)	[CHOFc-CHCH-C_6H_5]	4.38	305.0 ± 36
(55)	[C_6H_5CHCHFc-CHCH-C_6H_5]	5.26	270.0 ± 26
(56)	[Fc-CHCH-C_6H_4CHCHFc]	7.04	504.0 ± 52
(57)	[CHOFc-CHCH-C_6H_4CHCHFc]	7.59	925.0 ± 86
(58)	[CHO(Fc-CHCH-C_6H_4CHCH)$_n$][b]	4.66	1550.0 ± 270
(59)	[FcH_2]	1.90	16.1 ± 1.8
(60)	[CH_2CH-C_6H_5]	1.44	16.9 ± 0.8
(61)	[C_6H_5CHCHC_6H_5]	2.90	41.6 ± 4.9

[a] Fc = {Fe(η-C_5H_4)(η-C_5H_4)}; [b] n = 6–8

In organic molecules, such as polyenes, γ is known to increase dramatically with increasing number of conjugated double bonds until it saturates at a large value [85]. Although styrene (60) and ferrocene (59) have similar values, compounds (55) and (56) differ by roughly a factor of two. Similarly if one compares the values of (57) and (58) (that is the hyperpolarisability per repeat unit) there is only a modest increase. Both results suggest that the electronic coupling through the metal centre is less than that through a double bond or a phenyl group.

Substitution of a cyclopentadienyl ring with a formyl group results in a substantial enhancement of γ, ((54) versus (53) and (57) versus (56)). This enhancement cannot simply be attributed to an increase in the conjugation length since (54) has a larger value than (55). The formyl substituent may add some charge transfer character to the low energy absorption transition in the molecules. This results in a red shift of this transition and increases the absorption at 602 nm. Thus, the formyl derivatives show increased dispersive enhancement relative to the other compounds in the study.

Alternatively, introducing the formyl acceptor might introduce a significant β that enhances the observed value of γ, by a cascading process [86].

Concentration-dependent studies demonstrated that all of the hyperpolarisabilities of the compounds in this study were both real and positive. The authors argue that these results are not consistent with the nonlinearity resulting from bleaching the low lying ferrocene d-d transitions. Thus it is believed that the nonlinearity results from a combination of d–π* and π–π* transitions [84]. The measured hyperpolarisability of ferrocene is a factor of four greater than the theoretically predicted value. This discrepancy may either reflect the effect of dispersion in the experimental value or the inadequacy of the quantum mechanical calculation used to predict the hyperpolarisability [87].

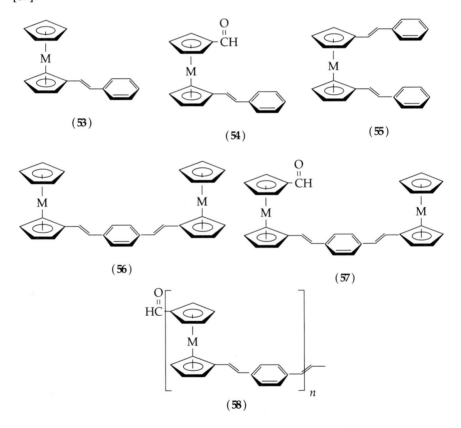

Figure 15 Structures of some of the metallocene complexes given in Table 9

3.3.3.5 Metal Dithiolenes

Transition metal dithiolenes are known to have strong π-π^* transitions in the region 700–1400 nm, depending on the transition metal [88]. Like phthalocyanines, dithiolenes are known to be saturable absorbers. This property, coupled with their photochemical stability has led to their use as laser Q-switches. As a result of these properties, the bis{1,2-diphenyl-1,2-ethenedithiolato-(2)-S–S'}nickel and bis{1,2-methyl-1,2-ethene-dithiolato-(2)-S–S'}nickel, two-photon absorption coefficients were determined, and measurements of the critical power for self focusing (related to the nonlinear refractive index, n_2) were carried out using 10 ns pulses [83].

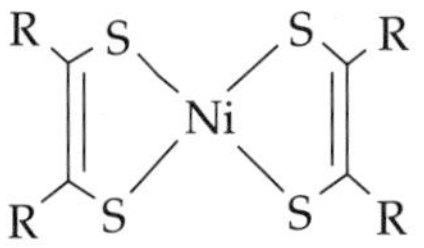

$$R = C_6H_5, CH_3$$

Figure 16 Structures of bis{1,2-diphenyl-1,2-ethenedithiolato-(2)-S–S'}nickel and bis{1,2-methyl-1,2-ethenedithiolato-(2)-S–S'}nickel

As noted earlier, even if a material has a large optical nonlinearity, if losses due to either one- or two- photon absorption are substantial, the material will be unsuitable for all-optical switching applications. Therefore a figure of merit, $2\lambda\beta/n_2$ (where λ is the wavelength of the laser, β is the two-photon absorption coefficient and n_2 is as defined as above) was suggested. Here a lower value is desirable, and it has been suggested that a ratio less than unity is needed for switching in waveguide devices. Comparison of the n_2 values using linearly and circularly polarised light suggested that, in the nanosecond regime, the observed nonlinearities were dominated by molecular rotation. Therefore, DFWM experiments were performed with 100 ps pulses to measure directly the electronic contribution to n_2. The measurement gave roughly similar values of n_2 for the diphenyldithiolato complex and the dimethyldithiolato complex (0.3×10^{-11} esu and 1.2×10^{-11} esu respectively), however, the value of β for the diphenyl compound (2.92 cm/GW) was much greater than that for the methyldithiolato complex (< 0.01 cm/GW). The figure of merit for the diphenyldithiolato complex, 18.0, is unacceptable for all-optical switching in waveguides. However, β for the dimethyldithiolato complex, < 0.2, suggests that this class of materials can possess sufficiently good figures of merit for device applications [83].   Clearly, further detailed structure/property relationship studies are needed to fully understand and utilise the potential of this promising system.

3.3.3.6 Poly(silanes) and Poly(germanes)

Organic materials with π-conjugation have been shown to have large electronic hyperpolarisabilities arising from strong electric field induced mixing of ground- and excited-states. The strongly coupled p-orbitals on adjacent carbons form the required band structure to facilitate such mixing. Poly(silanes) and poly(germanes) that contain silicon or germanium atoms in the backbone are thought to have extensive σ-delocalisation [89]. These materials have a variety of unusual electronic and optical properties, including low energy transitions in the UV–visible region ascribed to σ–σ^* transitions.   Given these observations, the NLO properties of poly(silanes) and poly(germanes) were investigated [90–95].

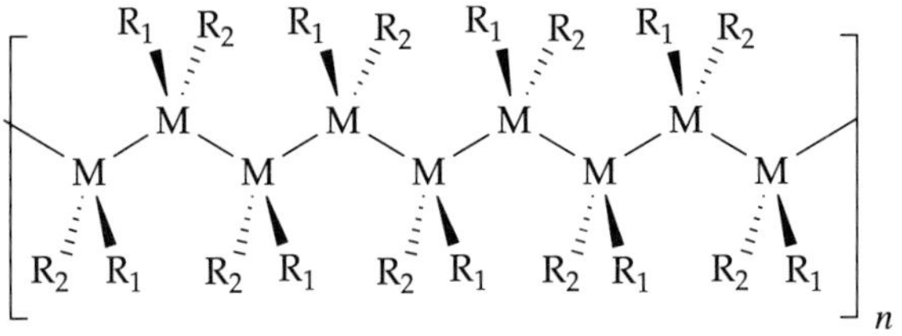

Figure 17 Schematic structures of poly(silanes) and poly(germanes)

Relative to standard conjugated organic polymers such as poly(phenylene), poly(phenylenevinylene), poly(thiophene), poly-(acetylene) and various substituted poly(acetylenes), that typically have absorption bands at 450 nm or lower in energy, the poly(silanes) are quite transparent (λ_{max} 300–375 nm) in the visible [95]. Initial THG studies of $[Si(Me)Ph]_n$ at 1.064 μm gave a $\chi^{(3)}$ of 1.5 x 10^{-12} esu, however, this material has considerable absorption at the TH and it was argued that much of the observed nonlinearity was due to a three-photon resonance [90]. In a later study, working in a region of improved transparency, THG studies at 1.907 μm reveal that $[Si(n\text{-}hexyl)_2]_n$ has a $\chi^{(3)}$ of 1.3 x 10^{-12} esu and the corresponding germanium polymer $[Ge(n\text{-}hexyl)_2]_n$ has a $\chi^{(3)}$ of 1.4 x 10^{-12} esu. In comparison, poly(phenylenevinylene) (with 1.85 μm fundamental radiation) has a $\chi^{(3)}$ of 7.8×10^{-12} esu. These results clearly demonstrate that σ-delocalised polymers can exhibit significant third-order optical nonlinearities. The electronic absorption spectra of poly(silanes) are temperature dependent. For example, at 53 °C λ_{max} for $[Si(n-hexyl)_2]_n$ is at 318 nm whereas at 23 °C λ_{max} is at 372 nm. This change results from a transition from a somewhat rigid, all *trans* conformation of the silicon backbone to more conformationally mobile and disordered structures above a transition temperature (42 °C for $[Si(n-hexyl)_2]_n$) [89, 95]. This additional conformational flexibility interferes with the σ-delocalisation as evidenced by the hypsochromic shift observed in the absorption spectrum. In accord, the $\chi^{(3)}$ of $[Si(n-hexyl)_2]_n$ at 53 °C, 0.9×10^{-12} esu, is lower than the value quoted at 23 °C.

In summary, both poly(silanes) and poly(germanes) exhibit $\chi^{(3)}$ roughly comparable to some conjugated organic polymers. For polymers with like substituents, the poly(silanes) and poly(germanes) have similar $\chi^{(3)}$. Although there are diminutions in values of $\chi^{(3)}$ for the conformationally mobile phases of the poly(silanes) in comparison to the all *trans*-phase, the differences are relatively small, indicating that susceptibilities of the poly(silanes) are not particularly sensitive to conformation. The transparency, processability and susceptibilities of these materials make them worthy of further consideration for NLO device applications.

3.3.3.7 Conclusions

The preliminary studies of metal-containing materials for third-order NLO applications are encouraging. In particular, work on phthalocyanines [74–76] and dithiolenes [83] may lead to devices based on coordination complexes. As noted earlier, the design requirements for second-order NLO materials are relatively well understood [16], and it may be possible to synthesise organic molecules that are optimised for second-order NLO device applications. In contrast, the prognosis for all-optical devices based on third-order NLO materials is uncertain, at best. Many think that conjugated organic materials with sufficiently large optical nonlinearities and sufficiently low loss coefficients cannot be made. The inorganic and organometallic chemist is therefore presented with an opportunity to enhance optical nonlinearities by exploiting the unique characteristics of metal-containing materials. To succeed in this challenge, it will be necessary to go beyond the models currently employed for the design of organic materials, and to explore fundamentally new mechanisms for enhancing third-order optical nonlinearities.

3.4   ACKNOWLEDGEMENTS

I would like to thank Bruce Tiemann, Joseph Perry, Bob Denning, Lap-Tak Cheng, Paul Cahill, Christopher Gorman, Daniel Alvarez, Todd Marder and Wilson Tam, for helpful discussions and/or careful reading of this manuscript.

3.5   REFERENCES

1.   D J Williams, *Angew Chem Int Ed Engl*, **23**, 690 (1984).
2.   D J Williams, Ed, *Nonlinear Optical Properties of Organic and Polymeric Materials*, ACS Symp Ser, **233**, American Chemical Society, Washington, DC (1983).
3.   S R Marder, J E Sohn and G D Stucky, Eds, *Materials for Nonlinear Optics: Chemical Perspectives*, ACS Symp Ser, **455**, American Chemical Society, Washington, DC (1991).
4.   D S Chemla and J Zyss, Eds, *Nonlinear Optical Properties of Organic Molecules and Crystals*, Academic Press, Orlando, Vol 1 and 2 (1987).
5.   A J Heeger, J Orenstein and D R Ulrich, Eds, *Nonlinear Optical Properties of Polymers*, Materials Research Society Proceedings, Vol 109, Materials Research Society, Pittsburgh (1988).
6.   J Messier, F Kajzar, P Prasad and D R Ulrich, Eds, *Nonlinear Optical Effects in Organic Polymers*, NATO ASI Series E, **162**, Kluwer Academic Publishers, Boston (1988).
7.   R A Hann and D Bloor, Eds, *Organic Materials for Nonlinear Optics*, Royal Soc Chem Special Pub, no 69, Royal Society of Chemistry, Burlington House, London (1989).
8.   I P Kaminow, *An Introduction to Electrooptic Devices*, Academic Press, New York (1974).
9.   R P Feynman, R B Leighton and M Sands, *The Feynman Lectures on Physics*, Vol 1, Addison Wesley Publishing Company, Reading (1963).
10.   J Ward, *Rev Mod Phys*, **37**, 1 (1965).
11.   B J Orr and J F Ward, *Mol Phys*, **20**, 513 (1971).
12.   J L Oudar and D S Chemla, *J Chem Phys*, **66**, 2664 (1977).
13.   B F Levine and C G Bethea, *J Chem Phys*, **66**, 1070 (1977).
14.   S J Lalama and A F Garito, *Phys Rev A*, **20**, 1179 (1979).
15.   J L Oudar, *J Chem Phys*, **67**, 446 (1977).
16.   S R Marder, D N Beratan and L T Cheng, *Science*, **252**, 103 (1991).
17.   R W Twieg and J F Nicoud, in: *Nonlinear Optical Properties of Organic Molecules and Crystals*, D S Chemla and J Zyss, Eds, Academic Press, Orlando, **1**, 242 (1987).

18. J S Zyss and J L Oudar, *Phys Rev A,* **26**, 2028 (1982).
19. R W Twieg and K Jain, in: *Nonlinear Optical Properties of Organic and Polymeric Materials, ACS Symp Ser, 233,* D J Williams, Ed, American Chemical Society, Washington, DC, p 57 (1983).
20. J S Zyss, D S Chemla and J F Nicoud, *J Chem Phys,* **74**, 4800 (1981).
21. G R Meredith, in: *Nonlinear Optical Properties of Organic and Polymeric Materials, ACS Symp Ser,* **233**, 30 (1983).
22. S R Marder, J W Perry and W P Schaefer, *Science,* **245**, 626 (1989).
23. S R Marder, J W Perry, B G Tiemann, R E Marsh and W P Schaefer, *Chem Mater,* **2**, 685 (1990).
24. S Tomaru, S Zembutsu, M Kawachi and M Kobayashi, *J Chem Soc, Chem Comm,* 1207 (1984).
25. D F Eaton, A G Anderson, W Tam and Y Yang, *J Amer Chem Soc,* **109**, 1886 (1987).
26. W Tam, D F Eaton, J C Calabrese, I D Williams, Y Wang and A G Anderson, *Chem Mater,* **1**, 128 (1989).
27. I Weissbuch, M Lahav, L Leiserowitz, G R Meredith and H Vanherzeele, *Chem Mater,* **1**, 114 (1989).
28. S D Cox, T E Gier, J D Bierlein and G D Stucky, *J Am Chem Soc,* **110**, 2986 (1989).
29. K D Singer, J E Sohn and S J Lalama, *Appl Phys Lett,* **49**, 248 (1986).
30. K D Singer, W R Holland, G L Wolk, H E Katz, M L Schilling and P A Cahill, *Proc SPIE,* **1147**, 233 (1989).
31. K D Singer, M G Kuzyk, W R Holland, J E Sohn, S J Lalama, R B Comizoli, H E Katz and M L Schilling, *Appl Phys Lett,* **53**, 1800 (1988).
32. G L Geoffroy and M S Wrighton, *Organometallic Photochemistry,* Academic Press, New York (1979).
33. J P Collman and L S Hegedus, *Principles and Applications of Organotransition Metal Chemistry,* Second edition, University Science Books, Mill Valley (1987).
34. S K Kurtz and T T Perry, *J Appl Phys,* **39**, 3798 (1968).
35. J L Oudar and H Le Person, *Opt Commun,* **15**, 258 (1975).
36. B F Levine and C G Bethea, *Appl Phys Lett,* **24**, 445 (1974).
37. K D Singer and A F Garito, *J Chem Phys,* **75**, 3572 (1981).
38. L-T Cheng, W Tam, G R Meredith, G Rikken and E W Meijer, *Proc SPIE,* **1147**, 61 (1989).
39. L-T Cheng, W Tam, S H Stevenson, G R Meredith, G Rikken and S R Marder, *J Phys Chem,* **95**, 10631 (1991).
40. C C Frazier, M A Harvey, M P Cockerham, E A Chauchard and C H Lee, *J Phys Chem,* **90**, 5703 (1986).
41. L-T Cheng, W Tam, G R Meredith and S R Marder, *Mol Cryst Liq Cryst,* **189**, 137 (1990).
42. D G Carroll and S P McGlynn, *Inorg Chem,* **7**, 1285 (1968).

43. J C Calabrese and W Tam, *Chem Phys Lett*, **133**, 244 (1987).
44. A G Anderson, J C Calabrese, W Tam and I D Williams, *Chem Phys Lett*, **134**, 392 (1987).
45. L-T Cheng, W Tam and D F Eaton, *Organometallics*, **9**, 2857 (1990).
46. W Tam and J C Calabrese, *Chem Phys Lett*, **144**, 79 (1988).
47. W Tam, Y Wang, J C Calabrese and R A Clement, *Proc SPIE*, **971**, 107 (1988).
48. T B Marder, G Lesley, Z Yuan, H B Fyfe, P Chow, G Stringer, I R Jobe, N J Taylor, I D Williams and S K Kurtz, *Materials for Nonlinear Optics: Chemical Perspectives, ACS Symp Ser*, **455**, 605 (1991).
49. D P Arnold and M A Bennett, *Inorg Chem*, **23**, 2117 (1984).
50. G W Parshall, *J Am Chem Soc*, **96**, 2360 (1974).
51. M L H Green, S R Marder, M E Thompson, J A Bandy, D Bloor, P V Kolinsky, and R J Jones, *Nature*, **330**, 360 (1987).
52. S R Marder, J W Perry, W P Schaefer, B G Tiemann, P C Groves and K J Perry, *Proc SPIE*, **1147**, 108 (1989).
53. S R Marder, J W Perry, W P Schaefer and B G Tiemann, *Organometallics*, **10**, 1896 (1991).
54. J A Bandy, H E Bunting, M L H Green, S R Marder, M E Thompson, D Bloor, P V Kolinsky and R J Jones, in: *Organic Materials for Nonlinear Optics; Royal Society of Chemistry Special Publication, No 69*, R A Hann and D Bloor, Eds, Royal Society of Chemistry, Burlington House, London, p 219 (1989).
55. B J Coe, C J Jones, J A McCleverty, D Bloor, P V Kolinsky and R J Jones, *J Chem Soc Chem Commun*, 1485 (1989).
56. J W Perry, A E Stiegman, S R Marder, D R Coulter, D N Beratan, D E Brinza, F L Klavetter and R H Grubbs, *Proc SPIE*, **971**, 17 (1988).
57. M E Wright and S A Svejda, in: *Materials for Nonlinear Optics: Chemical: Perspectives, ACS Symp Ser* Vol 455, S R Marder, J E Sohn and G D Stucky, Eds, American Chemical Society, Washington, DC, p 602 (1991).
58. J C Calabrese, L-T Cheng, J C Green, S R Marder and W Tam, *J Am Chem Soc*, **113**, 7227 (1991).
59. D R Kanis, M A Ratner and T J Marks, *J Am Chem Soc*, **112**, 8203 (1990).
60. G Mignani, A Krämer, G Puccetti, I Ledoux, G Soula, J Zyss and T Meyueix, *Organometallics*, **9**, 2640 (1990).
61. J A Bandy, H E Bunting, M H Garcia, M L H Green, S R Marder, M E Thompson, D Bloor, P V Kolinsky and R J Jones, in *Organic Materials for Non-linear Optics; Royal Society of Chemistry Special Publication*, No 69; R A Hann and D Bloor, Eds, Royal Society of Chemistry, Burlington House, London, p 225 (1989).

62. W Chaing, M E Thompson, D Van Engen and J W Perry, in: *Organic Materials for Non-linear Optics Vol 2; Royal Society of Chemistry Special Publication*, R A Hann and D Bloor, Eds, Royal Society of Chemistry:, Burlington House, London, p 210 (1991).
63. T P Pollagi, T C Stoner, R F Dallinger, T M Gilbert and M D Hopkins, *J Am Chem Soc*, **113**, 703 (1991).
64. T Richardson, G G Roberts, M E C Polywka and S G Davies, *Thin Solid Films*, **179**, 405 (1989).
65. P A Cahill, *Mat Res Soc Proc*, **109**, 319 (1988).
66. J W Perry in *Materials for Nonlinear Optics: Chemical Perspectives, ACS Symp Ser*, **455**, S R Marder, J E Sohn and G D Stucky, Eds, American Chemical Society, Washington, DC, p 67 (1991).
67. C C Frazier, S Guha, W P Chen, M P Cockerham, P L Porter, E A Chauchard and C H Lee, *Polymer*, **28**, 553 (1987).
68. C C Frazier, S Guha, P L Porter, M P Cockerham and E A Chauchard, *Proc SPIE*, **971**, 186 (1988).
69. C C Frazier, E A Chauchard, M P Cockerham and P L Porter, *Mater Res Soc Sym Proc*, **109**, 323 (1988).
70. S Guha, C C Frazier, K Kang and S E Finberg, *Optics Lett*, **14**, 952 (1989).
71. S Takahashi, M Kariya, T Yatake, K Sonogashira and N Hagihara, *Macromolecules*, **11**, 1063 (1978).
72. P Karrer and C H Eugster, *Helv Chem Acta*, **34**, 1805 (1951).
73. *The Phthalocyanines 2nd edition*, F H Moser and A L Thomas, Eds, CRC Press, Boca Raton (1983), Vols 1 and 2.
74. J W Wu, J R Heflin, R A Norwood, K Y Wong, O, Zamani-Khamiri, A F Garito, P Kalyanaraman and J Sounik, *J Opt Soc Am*, **B6**, 707 (1989) and references therein.
75. A F Garito, J W Wu, G F Lipscomb and R Lytel, *Mater Res Soc Sym Proc*, **173**, 467 (1990).
76. J W Perry, L R Khundkar, D R Coulter, D Alvarez, S R Marder, T H Wei, M J Sence, E W Van Stryland and D J Hagan, in: *Organic Molecules for Nonlinear Optics and Photonics, NATO ASI Series E*, **194**, J Messier, F Kajzar and P Prasad, Eds, Kluwer Academic Publishers, Boston, p 369 (1991).
77. J S Shirk, J R Lindle, F J Bartoli, Z H Kafafi and A W Snow, in: *Materials for Nonlinear Optics: Chemical Perspectives, ACS Symp Ser*, **455**, S R Marder, J E Sohn and G D Stucky, Eds, American Chemical Society, Washington, DC, p 626 (1991).
78. J S Shirk, J R Lindle, F J Bartoli, C A Hoffman, Z H Kafafi and A W Snow, *Appl Phys Lett*, **55**, 1287 (1989).
79. H M Gibbs, *Optical Bistability-Controlling Light with Light*, Academic Press, New York (1985).

80. C S Winter, S N Oliver, and J D Rush in, *Nonlinear Optical Effects in Organic Polymers, NATO ASI Series E*, **162**, J Messier, F Kajzar, P Prasad and D R Ulrich, Eds, Kluwer Academic Publishers, Boston, p 247 (1988).

81. C S Winter, S N Oliver and J D Rush, in: *Organic Materials for Non-linear Optics; Royal Society of Chemistry Special Publication*, No 69; R A Hann and D Bloor, Eds, Royal Society of Chemistry, Burlington House, London, p 232 (1989).

82. C S Winter, S N Oliver and J D Rush, *Opt Commun*, **69**, 45 (1988).

83. C S Winter, S N Oliver, J D Rush, R J Manning, C Hill and A Underhill, in: *Materials for Nonlinear Optics: Chemical Perspectives, ACS Symp Ser*, **455**, S R Marder, J E Sohn and G D Stucky, Eds, American Chemical Society, Washington, DC, p 616 (1991).

84. S Ghosal, M Samoc, P N Prasad and J J Tufariello, *J Phys Chem*, **94**, 2847 (1990).

85. D N Beratan, J N Onuchic and J W Perry, *J Phys Chem*, **91**, 2696 (1987).

86. B Buchalter and G R Meredith, *Appl Optics*, **21**, 3221 (1982).

87. J Waite and M G Papadopoulos, *Z Naturforsch*, **42a**, 749 (1987).

88. J McCleverty, *Prog Inorg Chem*, **10**, 49 (1968).

89. R D Miller and J Michl, *Chem Rev*, **89**, 1359 (1989).

90. F Kazjar, J Messier and C Rosilio, *J Appl Phys*, **60**, 3040 (1986).

91. J-C Baumert, G Bjorklund, D H Jundt, M C Jurich, H Looser, R D Miller, J Rabolt, J Sooriyakumaran, J D Swalen and R J Twieg, *Appl Phys Lett*, **53**, 1147 (1988).

92. D J McGraw, A E Siegman, G M Wallgraff and R D Miller, *Appl Phys Lett*, **84**, 1713 (1989).

93. L Yang, Q Z Wang, P P Ho, R Dorsenvill, R R Alfona, W K Zou and N L Yang, *Appl Phys Lett*, **53**, 1245 (1988).

94. F M Schellenberg, R L Byer and R D Miller *Chem Phys Lett*, **166**, 331 (1990).

95. R D Miller, F M Schellenberg, J-C Baumert, H Looser, P Shukla, W Torruellas, G Bjorklund, S Kano, Y Takahashi and F M Schellenberg, in: *Materials for Nonlinear Optics: Chemical Perspectives, ACS Symp Ser*, **455**, S R Marder, J E Sohn and G D Stucky, Eds, American Chemical Society, Washington, DC, p 636 (1991).

4   Inorganic Intercalation Compounds

Dermot O'Hare

Inorganic Materials. Edited by Duncan W Bruce and Dermot O'Hare
© 1992 John Wiley & Sons Ltd

4.1   INTRODUCTION

4.1.1   General Considerations

The phenomenon of intercalation appears to have been discovered in China *circa* 600-700 A.D. [1] when they produced porcelain from the intercalation of alkali metal ions into commonly found naturally occurring minerals such as quartz, feldspar and kaolin [2]. In fact, the origins of the modern word 'kaolin' used to describe the clay used for ceramics and refractories derives from the Chinese mountain 'Kao-Ling'. The first intercalation compound to be reported in the scientific literature was described by C. Schafhäutl in 1840 [3] when he reported his observations on attempting to dissolve graphite in sulphuric acid. However, the contemporary resurgence of the subject dates from 1926 when Karl Fredenhagen and Gustav Cadenbach described the uptake of potassium vapour into graphite [4]. Since their report, intercalation reactions have fascinated inorganic, organic and organometallic chemists, and to date over 5000 scientific papers have been published describing the synthesis, reactivity, and physical characterisation of inorganic intercalation compounds.

Literally, the term *intercalation*[§] refers to the act of inserting into a calendar some extra interval of time. Today the usage of the term intercalation by chemists is to describe the (reversible) insertion of mobile guest species (atoms, molecules or ions) into a crystalline host lattice that contains an interconnected system of empty lattice sites ($\square$) of appropriate size. The reaction can be generalised by Equation (1).

$$x_{\text{Guest}} + \square_x[\text{Host}] \rightleftharpoons \{\text{Guest}\}_x[\text{Host}] \qquad (1)$$

As indicated by Equation (1), intercalation reactions are usually reversible, and they may also be characterised as topochemical processes, since the structural integrity of the host lattice is formally conserved in the course of the forward and reverse reactions. Typically these reactions occur near room temperature, but this is in sharp contrast to most conventional solid state synthetic procedures which often require temperatures in excess of 600 °C. Remarkably, a wide range of host lattices have been found to undergo these low temperature reactions, including: framework (3-D), layer (2-D), and linear chain (1-D) lattices, as shown in Figure 1.

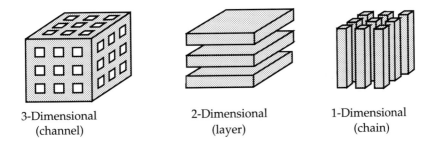

| 3-Dimensional
(channel) | 2-Dimensional
(layer) | 1-Dimensional
(chain) |

Figure 1 Schematic representation of basic host lattices with different structural dimensionality (from Schöllhorn [5])

[§] **Intercalation** [ad. L. *intercàlatiòn-em*, n. of action f. *intercalàre* to intercalate. Cf. F. *intercàlation* (15th c.)] **1.** The insertion of an additional day, days, or month into the ordinary or normal year; **2.** transf. The insertion of any addition between the members of an existing or recognised series; interposition or interlection (of something additional or foreign).

The study of intercalation reactions and their products draws widely from many aspects of organic, inorganic and physical chemistry. For example, from the design and synthesis of a potential organometallic guest to the detailed investigation of the structure and electronic properties of the products. Today a battery of modern physical techniques are available to help elucidate the structure, bonding, and guest dynamics, such as X-ray and neutron diffraction, EXAFS, infrared spectroscopy, solid state NMR, electrochemistry, electronic spectroscopy, magnetic measurements, electronic transport measurements and surface spectroscopy. A feature of intercalation reactions which makes the study of these materials particularly intriguing is that the guest and host may experience a spectrum of perturbations in their geometrical, chemical, and electronic environment depending on the individual characteristics of either the chosen host or guest. More importantly, these environments may be subtly controlled and tailored to meet specific requirements, such as catalytic activity, electrochromic displays, battery technology or as lubricants. It is perhaps this element of tuning, or incremental control, which has attracted interest from synthetic chemists to this area of solid state chemistry, and has led to a tremendous expansion of the field in recent years.

We can represent a prototypical intercalation of a guest species into a layered inorganic lattice schematically in Figure 2, with the specific example of the reaction of lithium with the layered metal dichalcogenide TiS_2.

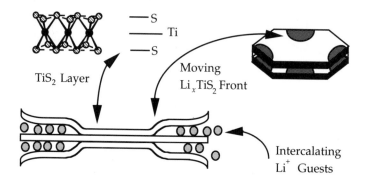

Figure 2 Schematic representation of the intercalation of Li^+ ions into the layered host lattice TiS_2 (adopted from ref [9])

In the reaction shown schematically in Figure 2, the lithium metal reduces the layers of the TiS_2 and the lithium cations are inserted into the vacant interlayer sites to compensate for the negative layer charge caused by the electron transfer. The process can be described by Equation (2). The final product of the reaction, $LiTiS_2$, has a structure which is expanded in the direction perpendicular to the layers.

$$Li \underset{+e^-}{\overset{-e^-}{\rightleftharpoons}} Li^+ + e^- + TiS_2 \rightleftharpoons Li^+[TiS_2]^- \tag{2}$$

The intercalation reactions involving layered host lattices have been much more extensively investigated than those of either framework or linear chain host structures. The structural flexibility of the layer structures, with their ability to adapt to the geometry of the inserted guest species by free adjustment of the interlayer separation, is presumably responsible for the wider occurrence of intercalation compounds for this structure type. It is remarkable that in spite of the differences in composition and detail of the sheet unit structures, the basic chemical reactivity of these phases turns out to be closely related. However, all these layer phases are characterised by strong intralayer covalent bonding and weak interlayer interactions. The layers may be electrically neutral, or possess an overall charge which may be either positive or negative: examples of each category can be found in Table 1. In compounds with neutral layers, the interlayer bonding is often described as *van der Waals'*, and the interlayer space is a connected network of empty lattice sites. In the charged layered systems, the layers are held together by weak electrostatic forces and the interlayer sites are partially or completely filled by ions or by a combination of ions and solvent molecules. Layered host lattices and their intercalates span the entire spectrum of electronic behaviour from insulators (such as clays) through semiconductors and semimetals (graphite and dichalcogenides) to superconducting metals (some dichalcogenides).

The emphasis of this chapter is to describe intercalation into the layered host lattices which reflects the significantly greater body of work which has been published describing intercalation reactions involving this type of host lattice. We do however briefly discuss some of the well known linear chain and framework host lattices which have been observed to undergo intercalation reactions (Sections 4.5 and 4.6 respectively).

Table 1 Examples of layered host structures that exhibit intercalation reactions

Lattice Type	Illustrative Examples	Layer Charge	Ref
Elemental			
Graphite		neutral	[6, 7]
Black-Phosphorus		neutral	[8]
Metal Chalcogenides			
MX_2 (M = Ti, Zr, Hf, V, Nb, Ta, Mo, W; X = S, Se)		neutral	[9–11]
MPX_3 (M= Mg, V, Mn, Fe, Co, Ni, Zn, Cd, In; X = S, Se)		neutral	[12]
AMS_2 (A= Group 1A; M = Ti, V, Cr, Mn, Fe, Co, Ni)		neutral	[11]
Metal Oxides			
M_xO_y (MoO_3, $Mo_{18}O_{52}$, V_2O_5, $LiNbO_2$, $Li_xV_3O_8$)		neutral	[13–15]
$MOXO_4$ (M = Ti, V, Cr, Fe; X = P, As)		neutral	[16]
Metal Oxy-Halides			
MOX (M= Ti, V, Cr, Fe; X = Cl, Br)		neutral	[17]
Titanates $\{K[Ca_2Na_{n-3}Nb_2O_{2n+1}]; 3 \le n < 7\}$		negative	[18]
Niobates ($K_2Ti_4O_9$)		negative	[19]
Metal Halides			
α-$RuCl_3$		neutral	[20]
β-ZrNCl		neutral	[21]
Hydrous Metal Oxides			
Uranium Micas $\{(A^{z+})_{1/z}(H_2O)_y[UO_2XO_4]^{z-}; X=P, As, V\}$		negative	[22]
Tarankite $\{(A^+)_{3+x}(H_2O)_y[Al_5(PO_4)_{2+x}(HPO_4)_{6-x}]^{(3+x)-}\}$		negative	[23]
Double hydroxides $\{X^-(H_2O)_n[Zn_2Cr(OH)_6]^+\}$		negative	[24]
$M(HPO_4)_2$ (M= Ti, Zr, Hf, Ce, Sn)		negative	[25]
Metal Phosphates and Phosphonates			
$Zr(RPO_3)_2$, $Zr(ROPO_3)_2$ (R =Ph, Me, Et, CH_2COOH)		negative	[26, 27]
Smectite Clays and Silicates			
Kaolinites $\{Al_4Si_4O_{10}(OH)_8\}$		negative	[2]
Hectorite $\{Na_{0.6}[Li_{0.6}Mg_{5.4}](Si_{8.0})O_{20}(OH,F)_4\}$		negative	[28]
Montmorillonite $\{Ca_{0.35}[Mg_{0.7}Al_{3.3}](Si_8)O_{20}(OH)_4\}$		negative	[29]
Hydrotalcites			
$LiAl_2(OH)_6OH.2H_2O$, $Zn_2Cr(OH)_6Cl.2H_2O$		positive	[30]
Coordination Compounds			
$Ni(CN)_2$		neutral	[31, 32]
Silicides			
$CaSi_2$		negative	[33]

Table 2 Examples of atomic and molecular ionic guest species that exhibit
intercalation reactions

Guest Type	Illustrative Examples[a]	Host Lattices	Refe
Atomic Ions			
	H^+	MoO_3	[34, 35]
	Li^+, Na^+, K^+, Cs^+	MS_2, FeOCl	[36, 37]
	$Mg^{2+}, Ca^{2+}, Ba^{2+}$	MS_2, $VOPO_4$	[16]
	La^{3+}, Ce^{2+}	TaS_2	[38]
Molecular Ions			
	NH_4^+, NR_4^+	MS_2	[39, 2]
	HSO_4^-	Graphite	[7]
	$[M(Cp)_2]^+$ (M = Co, Cr)	MS_2,MPS_3	[40]
	$[Cr(en)_3]^{2+}$	TaS_2	[41]
	$[Cu(en)_2]^{2+}, [Cu(phen)_2]^{2+}$	$Zr(HPO_4)_2$	[42]
	$[Pt(NH_3)_4]^{2+}$	TaS_2	[43]
	$[M(2,2'-bipyridine)_3]^{2+}$ (M = Fe, Ru)	$MnPS_3$, TaS_2	[41]
	$[Ba(222-crypt)]^{2+}$	MS_2	[43]
	$[Rh(PPh_3)_3]^+$	Hectorite	[44]
	$[Fe(Cp)(diphos)(CO)]^+, [Fe(Cp)(Bz)]^+$	MPS_3	[41]
	$[Co(PMe_3)_3(C_2H_4)]^-$	Graphite	[45]

[a] Cp = η-C_5H_5 , en = ethylenediamine, phen = phenanthroline,
diphos = $Ph_2P(CH_2)_2PPh_2$, Bz = η-C_6H_6,
222-crypt = 4,7,13,16,21,24-hexaoxa-1,10-diazabicyclo[8.8.8]hexacosane

Table 3 Examples of neutral guest species that exhibit intercalation reactions

Guest Type	Illustrative Examples[a]	Host lattices	References
Neutral Molecules			
	NH_3, NR_3, NR_2H, NRH_2, $H_2N(CH_2)_nNH_2$	MPS_3, TaS_2	[46]
	Py, RNC, R_3PO, $RCONH_2$, RCN, RNO_2	MS_2, MoO_3	[2, 47]
	Cl_2, Br_2	Graphite	[7]
	H_2O, RCH_2OH, RCO_2H	$Zr(HPO_4)_2$	[48]
	DMSO, THF, DME, pyrrole	MS_2, FeOCl	[49]
	$[OsF_6]$	Graphite	[50]
	RX (X = Cl, Br)	MS_2	[2]
	Perylene ($C_{20}H_{12}$)	FeOCl	[51]
	$[M(Cp)_2]$ (M = Co, Cr, Fe, Ni), $[Fe(Cp^*)_2]$	FeOCl, MS_2	[52]
	$[Fe(Cp^*)_2]$	FeOCl,	[52]
	$[Mo(Cp)_2H_2]$	ZrS_2, FeOCl	[53, 54]
	$[M(Bz)_2]$ (M = Cr, Mo)	ZrS_2, MPS_3	[55]
	$[Ti(COT)(Cp)]$, $[Cr(CHT)(Cp)]$, $[Cr(Cp)(Bz)]$	ZrS_2, TaS_2	[54]
	TTF, TTN, TTT, TMTTF	FeOCl	[56]
	$[XeOF_4]$, $[XeF_4]$, $[XeF_6]$	Graphite	[7]
	$[Ir(PMe_3)_3H]$	ZrS_2	[53]
	$[Fe_4(\eta\text{-}C_5H_4Me)_4(\mu^3\text{-}S)_4]$	MoO_3	[53]
	$[Fe_6S_8(PEt_3)_3]$	TaS_2	[57]
	$[Fe(Cp(\eta\text{-}C_5H_4CH_2CH_2NH_2))]$	$\alpha\text{-}Zr(HPO_4)_2$	[53]
	$[Fe(Cp(\eta\text{-}C_5H_4CH_2NMe_2))]$	$HUO_2PO_4.4H_2O$	[147]
	$[SnMe_3Cl]$, $[SnCl_4]$	Graphite	[59]
	$[SnMe_3(NMe_3)]$, $[Sn(NMe_2)_4]$	FeOCl	[60]
	K(222-crypt)	Graphite	[61]
	$C_{60}(en)_6$	Fluorohectorite	[131]

[a] Cp = $\eta\text{-}C_5H_5$, Cp* = $\eta\text{-}C_5Me_5$, Bz = $\eta\text{-}C_6H_6$, COT = $\eta\text{-}C_8H_8$, CHT = $\eta\text{-}C_7H_7$,
TTF = tetrathiofulvalene, TTT = tetrathiatetracene,
TMTTF = tetramethyltetraselenafulvalene, TTN = tetrathianaphthalene,
222-crypt = 4,7,13,16,21,24-hexaoxa-1,10-diazabicyclo[8.8.8]hexacosane,
C_{60} = buckminsterfullerene, en = ethylenediamine

4.2. THE KINETICS AND MECHANISM OF INTERCALATION

4.2.1 Introduction

Before we detail the occurrence, characterisation and properties of inorganic intercalation compounds it is worth spending some time discussing our current kinetic and mechanistic understanding of this type of reaction.

4.2.2 Staging

Since the reaction of a potential guest molecule with a prospective inorganic host lattice is a heterogeneous reaction it is important to consider the reaction kinetics, in addition to the thermodynamics of the overall reaction. Intercalation reactions typically involve breaking bonds in the host and the formation of new interactions between the guest and host. The increase in the basal spacing of the unit cell of the layer host lattice upon intercalation imposes one of the most significant energy penalties. Layered host lattices which have stronger interlayer bonding appear to be most difficult to intercalate. This effect can be observed qualitatively in the intercalation reactions of the layered transition metal dichalcogenides (MX_2). The metal disulphides tend to intercalate guest molecules under milder conditions than the equivalent isomorphous transition metal diselenide (MSe_2), and there is no report of intercalation of a layered transition metal ditelluride (MTe_2). This trend is attributed to the decrease in anisotropy down the series S > Se > Te, and the increased covalent bonding between the dichalcogenide layers. This interlayer bond breaking penalty can be minimised by the phenomenon of staging. Staging refers to situations in which certain interlayer regions are totally vacant in the presence of other partially or totally occupied interlayers. The order of the staging is given by the number of layers between successive filled or partially filled layers as shown in Figure 3.

In the reaction between guest and host species two extreme staging behaviours can be observed. The potassium intercalates of graphite can be prepared with specific stoichiometry, each one associated with a specific nth order stage [62]; also for the hydrated phases $\{A_{x/n}^{n+}[MS_2]^{x-}(H_2O)_y\}$ (A^{n+} = alkali, alkaline earth or transition metal ion; M = Ti, Nb and Ta) a sequence of 3rd stage → 2nd stage → 1st stage is observed regularly [43]. The other extreme behaviour is exemplified by the behaviour of Li_xTiS_2 ($0 \leq x \leq 1$), where a single first stage phase is formed over the entire composition range [63]; staging is also not found for VS_2, 1T-TaS_2, or the halide host lattices α-RuCl$_3$ [64], YOCl [65], and FeOCl [66]. However, these two examples are extreme situations and in reality most intercalation reactions lie somewhere between these extremes at restricted composition ranges.

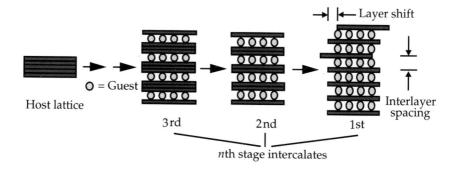

Figure 3 Principal geometrical transitions of layered host lattice matrices upon intercalation of guest species: (i) change in interlayer spacing; (ii) change in stacking mode of the layers; (iii) intermediate phases at low guest concentrations may exhibit staging (from Schöllhorn [67])

Questions have been asked as to whether the origin of this phenomenon is due to structural or electronic reasons. The most simple model explains staging in terms of a smaller loss of lattice energy at low guest concentrations, if the guest species are concentrated in a limited number of van der Waals' gaps rather than being homogeneously distributed over all interlayer sites. A more realistic model proposed by Herold [68] explains staging in terms of a domain scheme with all van der Waals' gaps being involved but with local concentrations of the guest species across the basal plane in certain regions. The size of the guest species may also affect the appearance of higher stage intermediates. Li$^+$ and Cu$^+$ ions with rather low effective ionic radii do not induce stage formation, even at low concentration, due to the small increase in interlayer separation upon intercalation.

The kinetics of incorporation of guest species into host lattices is complex. The reaction seems to be initiated at defects on the host surface, and therefore variations in reactivity often arise from one batch to another of the same host due to microcrystalline or particle size differences of the solid, although relatively few detailed *in situ* studies of intercalation reactions have been carried out. The ^{181}Ta nuclear quadrupole interaction has been monitored during the chemical lithiation of 1T-TaS$_2$ with LiBu^n. Strong evidence was found for a two step reaction leading to a coexistance of three homogeneous phases differing in lithium content. The observed kinetics exhibits a sigmoidal behaviour at the beginning and is dominated by nucleation processes. The data were fitted to a kinetic model involving concentric cylindrical phase boundaries moving towards the crystal centre (Figure 4) [69].

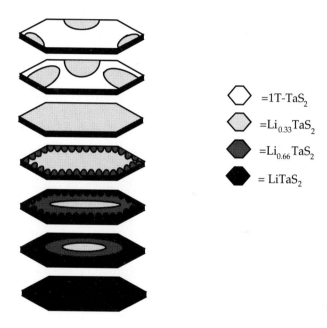

Figure 4 Schematic of the proposed kinetic scheme for the chemical lithiation of 1T-TaS$_2$ (from Ganal *et al.* [69])

Schöllhorn has looked in detail at the mechanism of the electrochemical intercalation of Na$^+$ ions into crystals of 2H-NbS$_2$ by *in situ* X-ray diffraction measurements [67]. He found that during the initial stages of the reaction four different phases can be observed simultaneously: (1) host lattice, (2) hydrated Na$^+$ phase with bilayers of water, (3) hydrated phase with monolayers of water, and (4) 3rd and 2nd stage compound. The X-ray data reveal that (4) is present only in low concentration at the phase boundary to 2H-NbS$_2$ and disappears in the course of the reduction process almost simultaneously with the host lattice. When all the $2H\text{-}NbS_2$ lattice has been reduced only phases (2) and (3) are observed; subsequent diffusion of water into the crystal quantitatively yields phase (2) which represents the final product. This so called 'synchronous multistage mechanism' dramatically illustrates the complexity of intercalation reactions in layered systems. Additional kinetic considerations are the increased activation energy for stage formation with increasing stiffness (elasticity) of the layers or increasing layer thickness.

4.3 SYNTHETIC METHODS

4.3.1 Introduction

There have been numerous advances in the understanding and, particularly, in the synthesis of intercalation systems in recent years [70]. Improved reaction methods have allowed the rapid preparation of many intercalation compounds which had previously remained unexplored owing to their long preparation times using conventional synthetic solid state techniques. These new techniques, in addition to elegant synthetic chemistry, have also led to the discovery of many new host/guest intercalation systems. In the following section I have presented an overview of the wide range of synthetic strategies which have been successfully used to intercalate inorganic lattices. It is hoped this may be useful to synthetic chemists contemplating entering this area of solid state chemistry.

4.3.2 Direct Reaction

The products of intercalation reactions, in common with many other solid state products, are insoluble, therefore they cannot be purified or separated by solution phase techniques, such as recrystallisation, chromatography, or vacuum sublimation. Therefore it is extremely important when designing a synthetic strategy that reactions are clean and the reaction products are free of impurity phases.

The simplest and most widely used method for preparing large amounts of an intercalation compound is the direct reaction of the guest (G) with a potential host lattice (MX_n) as shown in Equation (3).

$$G \ + \ MX_n \quad \longrightarrow \quad G_g MX_n \qquad (3)$$

In the cases where this method is successful it is generally the method of choice. The first alkali metal intercalates of the metal dichalcogenides were prepared in this way by the reaction of the host lattice with the metal vapour at temperatures in the range 600–800 °C. Although direct reaction also appears to be quite a good general method for intercalating the less reactive post transition elements, the more reactive alkali metals have tended to present serious complications such as reactions with the reaction vessel, thermal instability of the product at the reaction temperature, and over reaction to give non-intercalation processes.

Hydrated vanadium oxides containing alkali metal ions can be prepared by hydrothermal treatment of $VOSO_4$ aqueous solutions. These reactions typically involve heating aqueous solutions of $VOSO_4$ with equimolar amounts of the appropriate alkali metal ions in an autoclave to between 100 –280 °C for 10–60 h [71].

The first chemical reagent used extensively for alkali metal intercalation was lithium dissolved in liquid ammonia. While offering many advantages over the direct reaction with alkali metals, it is often accompanied by co-intercalation of ammonia. Heating *in vacuo* to remove the ammonia often resulted in undesired side reactions. The first reagent for lithium intercalation to combine the attributes of simplicity, general applicability, and high purity products was $LiBu^n$ [72, 73]. The reaction proceeds as shown in Equation (4).

$$LiBu^n + MX_n \longrightarrow Li_xMX_n + x/2\ C_8H_{18} \qquad (4)$$

The reaction is heterogeneous, typically involving $LiBu^n$ dissolved in hexane into which is suspended the host lattice. When the reaction is complete, which may be minutes at room temperature or several days or weeks at reflux depending on the host, the reaction product is isolated by filtration, washed with pure solvent to remove any excess $LiBu^n$ and dried *in vacuo*. This widely used reagent does not have an analogy with other cations. However, a class of reagents that appears to be both relatively convenient and general for intercalation of a variety of alkali metal cations is the metal aluminohydrides and borohydrides (M = Li, Na) [74, 75]. The driving force of this reaction is the formation of the nonmetal hydrides and hydrogen. Other chemical reagents, which give lower alkali metal activities are either sodium and potassium naphthalide or benzophenone solutions. Recently two chemical reagents di-nbutylmagnesium and magnesium bis(2,6-di-tert-butylphenoxide) have been used as reagents for the chemical intercalation of Mg^{2+} ions into a range of oxide hosts [76].

A major consideration in the choice of an intercalation reagent is the ease of reduction of the host. Figure 5 shows the relative reduction potentials of some common host lattices and a range of reagents used for lithium intercalation. The importance of using a reagent just on the reducing side of the host is illustrated by a comparison of the reaction products of V_2O_5 with LiI and $LiBu^n$. The reaction of V_2O_5 with LiI [15] is presented in Equation (5). The product of the reaction is $Li_xV_2O_5$ ($0 \le x \le 1$),

$$V_2O_5 \;+\; x LiI \xrightarrow{\hspace{3cm}} Li_xV_2O_5 \qquad\qquad (5)$$

a well characterised intercalation compound, whereas the more reactive $LiBu^n$ reacts with V_2O_5 to give products which are not intercalation compounds.

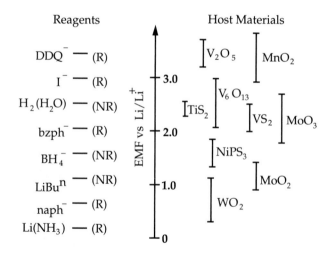

Figure 5 Relative reduction potentials for some common host lattices relative to some common lithium intercalation reagents (from Murphy *et al.* [70])

Direct intercalation of a large range of either organic or organometallic compounds occurs under anhydrous conditions by the direct thermal reaction at temperatures up to 200 °C. Lattice expansions up to *ca* 56 Å have been observed for direct reactions of this type. Potential guests which are either liquids or low melting solids can be used as neat reactants, whereas solid organic and organometallic guests are usually dissolved in polar organic solvents. It is now becoming clear that the choice of solvent can be vitally important for a successful outcome of many of these reactions; unfortunately the choice of the optimum solvent must be determined by experiment. However, as a general guide, solvents such as toluene, acetonitrile, dimethoxyethane, and dimethylformamide have proved successful for a wide range of host/guest systems. Although highly polar solvents often accelerate the intercalation process they can complicate the product distribution by co-intercalating with the guest molecules.

Almost all organometallic compounds known to form intercalation compounds are good reducing agents and form stable cations. The electron transfer mechanism involving oxidation of the guest and electron transfer to the conduction band of the host lattice accounts for the correlation which exists between the reducing power of the organometallic guest (as measured by the first ionisation potential or electrochemical oxidation potential) and its ability to intercalate specific host lattices [77].

Table 4 Examples of intercalation compounds formed by direct reaction of the host lattice with the appropriate guest molecules

Reactants	Products	Reaction Conditions
$0.5\ C_5H_5N\ +\ TaS_2$	$TaS_2(C_5H_5N)_{0.5}$	50 micron, 200 °C, 1 d
$0.25\ [Co(Cp)_2]\ +\ VSe_2$	$VSe_2\{Co(Cp)_2\}_{0.25}$	130 °C, 3 d
$K_{metal}\ +\ C(graphite)$	C_8K	200 °C, two bulb method
$OsF_6\ +\ C(graphite)$	$C_8(OsF_6)$	20 °C
xs Na(naphthalide) + TiS_2	$NaTiS_2$	naphthalene, r.t.
$LiBu^n + V_6O_{13}$	$Li_xV_6O_{13}$	hexane, 20 °C
$HCl/Zn\ +\ WO_3$	H_xWO_3	H_2O, 25 °C

This point is nicely exemplified by the intercalation reactions involving the first row transition metal metallocenes. Cobaltocene [Co(η-$C_5H_5)_2$] with a first ionisation potential of 5.5 eV will intercalate into both TaS_2 and FeOCl, but ferrocene [Fe(η-$C_5H_5)_2$], which is much less reducing (1st I.P. = 6.88 eV) will only intercalate into the more oxidising host lattice FeOCl. Although reducing power appears to be a necessary condition for intercalation, it does not guarantee intercalation, for example, $[Co(\eta\text{-}C_5Me_5)_2]$ (1st IP = 4.7 eV) or $[Cr(\eta^5\text{-}C_9Me_7)_2]$ (1st IP = 4.6 eV) do not intercalate into the metal dichalcogenides (MS_2; M = Nb, Ta, Zr and Sn) [77]. Examples of typical direct intercalation reactions involving organic and organometallic guests are given in Table 4.

4.3.3   Ion Exchange

Once an intercalation compound has been formed the intercalated guest ions can often be replaced by immersing the material in a concentrated solution containing another potential guest ion, for example the ion-exchange reaction of Li^+ ions for Na^+ ions in TiS_2 (Equation (6)).

$$Na_xTiS_2 \; + \; LiPF_6 \longrightarrow Li_xTiS_2 \qquad (6)$$

Some illustrative examples are given in Table 5. These exchange reactions are driven by the great excess of the replacement ion. In some cases pre-intercalating with an alkali metal ion and then ion exchanging the alkali metal cation with the final guest cation provides a useful strategy for intercalating large guest cations that do not intercalate by direct reaction. Clement used ion-exchange techniques extensively to intercalate many potentially reactive guests into $MnPS_3$ [41]. He initially formed the potassium intercalate $Mn_{0.8}PS_3(K)_{0.4}(H_2O)_y$ by treating $MnPS_3$ with an aqueous KCl solution. This phase readily cation-exchanges with added guest ions, for example tris(2,2'-bipyridyl)ruthenium dication can be exchanged for the potassium ions at 50 °C for 3 days.

Table 5 Examples of intercalation compounds prepared by ion-exchange methods

Reactants	Products	Reaction Conditions
$Na_xTiS_2 \; + \; LiPF_6 \longrightarrow$	Li_xTiS_2	dioxolane, 25 °C
$Na_{0.5}(H_2O)_yMoO_3 + 0.25Mg^{2+} \longrightarrow$	$Mg_{0.25}(H_2O)_yMoO_3$	H_2O, 1M in Mg^{2+}
$K[Al_2(OH)_2(Si_3AlO_{10})]^- + Na^+ \longrightarrow$	Na(smec)	H_2O, 1M in Na^+
$Na_{0.33}(H_2O)_{0.66}TaS_2 + [Co(Cp)_2]^+ I^- \longrightarrow$	$TaS_2\{Co(Cp)_2\}_{0.2}$	MeOH, 25 °C
$Na_{0.33}(H_2O)_{0.66}TaS_2 + [Mo(Cp)_2]^+ I^- \longrightarrow$	$TaS_2\{Mo(Cp)_2\}_{0.15}$	MeOH, 25 °C

Intercalation reactions involving the zeolites, pyrochlores, silicates, and clay hosts all generally proceed by ion-exchange processes.

4.3.4 Swelling, Floculation and Refloculation

The ion-exchange reactions discussed previously clearly demonstrate that the high activation energies associated with the deformation of a layered structure to accommodate the incoming guests can be largely overcome if the host lattice can be expanded by pre-intercalation of a small molecule or ion to the point where the interlayer spacing is comparable to the diameter of the reactant molecule. If this reaction is taken to its extreme then the host structure is completely dispersed or exfoliated.

Smectic clays are examples of systems which completely exfoliate in water to form dispersions, and consequently, undergo rapid ion exchange with large cations such as polyoxometallates [28], and metal chelates [78]. In solvents with high dielectric constants the hydrated alkali metal metal dichalcogenides such as $Na_x(H_2O)_yTaS_2$ can also be completely dispersed to give homogeneous dispersions or colloidal solutions of negatively charged MS_2 layers.

These colloidal dispersions can be refloculated by addition of electrolyte solutions of other cations. For example, Jacobson *et al.* has used this procedure to intercalate the large cluster cation $[Fe_6S_8(PEt_3)_6]^{2+}$ in TaS_2 by initially exfoliating $Na_{0.33}TaS_2$ in N-methylformamide/H_2O solution and then restructuring the layers in the presence of $[Fe_6S_8(PEt_3)_6]^{2+}$ [57]. Large aluminium and gallium oxide clusters $([M_{13}O_4(OH)_{24}(H_2O)_{12}]^{7+}; M = Al$ and $Ga)$ can be intercalated into MoO_3 by combining a solution of the guest cations with a dispersion of either $Na_xMoO_3(H_2O)_y$ or $Li_xMoO_3(H_2O)_y$ in water [57].

4.3.5 Electrointercalation Methods

Intercalation reactions can be performed electrochemically. The host lattice serves as the cathode of an electrochemical cell. Some of the advantages of the electrolysis method over conventional techniques are its simplicity, ease of control of stoichiometry, and fast rate of reaction at room temperature. In addition, it provides a convenient method to carry out detailed thermodynamic measurements and an elegant means of studying the staging phenomenon.

The formation of Cu_xTiS_2 ($x_{max} = 0.9$) has been studied with the liquid aprotic electrolyte $CH_3CN/CuCl$ [79]. The current versus cell voltage characteristics (Figure 6) indicate two non-stoichiometric regions $0 \leq x \leq$ 0.5 and $0.7 \leq x \leq 0.9$. The EMF of this cell also provides a direct measure of the free energy of the cell reaction, given in Equation (7), for a given stoichiometry, Cu_xTiS_2.

$$\delta x Cu \ + \ Cu_xTiS_2 \longrightarrow Cu_{(x+\delta x)}TiS_2 \qquad (7)$$

Such measurements have an important practical use, since the measured EMF will correspond to that obtainable from a secondary battery at zero current, the 'open-circuit' voltage. However, these electrolysis methods are not ideal for preparation of bulk samples and they cannot be used to intercalate insulating hosts, or for the intercalation of neutral guest species.

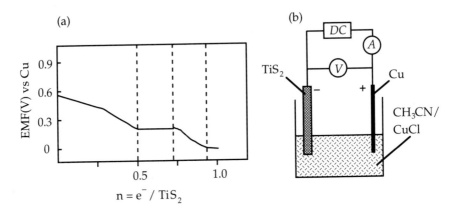

Figure 6   (a) Potential/charge transfer diagram for the cathodic reduction of TiS_2 in $CH_3CN/CuCl$; (b) Schematic of the experimental set-up

4.3.6   Rate Enhancement Methods

One of the major problems synthetic chemists face when carrying out intercalation reactions is that they are often lengthy (weeks or even months) and require elevated temperatures. The development of new synthetic methods which could decrease reaction times and lower reaction temperatures would be useful.

Recently, ultrasonic irradiation of reaction mixtures has been observed to increase the rates of intercalation (by as much as 200 fold) of organic and organometallic compounds into MoO_3, ZrS_2, and TaS_2 [80]. It appears that most of the chemical effects of ultrasound result from acoustic cavitation, which is the creation, expansion, and implosive collapse of bubbles in ultrasonically irradiated liquids. In cavitation near surfaces, the implosive collapse is not spherically symmetric and a localised 'micro-jet' of liquid is driven into the surface at extremely high velocities. The ultrasonic irradiation does not increase intercalation rates through improvement of mass transport, but seems to produce a significant decrease in the particle size and surface damage. Although the rate of the intercalation reaction is, unfortunately, increased by sonication, the crystallinity of the final product is much inferior compared to the material obtained using conventional methods, making structural characterisation using diffraction methods very difficult.

Mingos *et al.* have recently demonstrated that the intercalation of pyridine and substituted pyridines into α-$VO(PO_4)$.$2H_2O$ can be achieved two orders of magnitude more quickly by using microwave dielectric loss heating (compared to conventional thermal techniques) [81]. In contrast to the detrimental effects of ultrasonic irradiation, the intercalates produced so far using microwave heating retain a high degree of crystallinity. The microwave heating causes the reaction described above to achieve temperatures of 200 °C at *ca* 50 atm pressure within a minute, by which time the reaction is complete. It is believed that these short reaction times are responsible for the preservation of the sample crystallinity.

4.4 LAMELLAR HOST LATTICES AND THEIR INTERCALATES

4.4.1 Introduction

In the remainder of this chapter we will review the characteristic features of the most significant host lattice structures and their intercalation reactions and products.

4.4.2 The Metal Dichalcogenides

4.4.2.1 Host Structure

Many of the transition metal dichalcogenides possess layered structures as shown in Figure 7. These lattices consist of two hexagonally close-packed chalcogen layers between which reside the metal ions. The metal ions can be found in sites of either octahedral or trigonal prismatic symmetry

(Figure 7). These six coordinate building blocks are then stacked together to form the overall structure. The bonding within the layers is strong and largely ionic in character, whereas the interaction between the layers is much weaker and is often described as van der Waals' in origin. It is likely that these weak forces are of a different nature, such as p_z overlap between chalcogen atoms which would oppose the electrostatic repulsion between layers. Information about the strength of these intra-and inter-layer forces can be gained from infrared spectroscopy. In MoS_2 the rigid layer shear mode is observed at 32 cm^{-1} as compared to 384 cm^{-1} for the dipolar mode, showing that the interlayer forces are about 100 times weaker than the intralayer forces [82].

The ability of the metal atom to adopt both octahedral and trigonal prismatic coordination and for the X–M–X units to stack in different sequences gives rise to a wide variety of polymorphic and polytypic forms for these materials. The nomenclature devised to describe the relative positions of all the atoms in these layered structures is referred to in the Ramsdell notation [83] and has been used to describe the MS_2 phases in previous reviews [11, 84]. In brief, the different anion sites in the layers are denoted by A, B, C and the metal sites are denoted by a, b, c. Intercalated guest ions are denoted by [a], [b], [c]. Table 6 summarises the commonly occurring stacking arrangements.

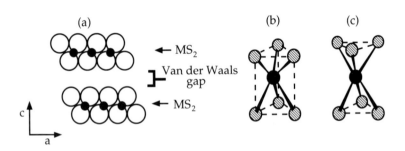

Figure 7 (a) Hexagonally close packed MX_2 layers showing van der Waals' gap. The local metal coordination within the layer can be either (b) Trigonal prismatic or (c) Octahedral

Table 6 Stacking arrangements and crystal structures of layered metal dichalcogenides (from Jacobson [9])

Polytype Designation[a]	Stacking Sequence	Crystal Space Group	Examples	Metal Coordination[a]
1T	AbC	$P\bar{3}m1$	MX_2 (M = Ti, Zr, Hf and V; X = S, Se and Te)	Oct
2Ha	BaB|CaC	$P6_3/mmc$	MX_2 (M = Ta and Nb; X = S and Se)	TP
2Hb	BaB|Cb C	$P\bar{6}m\,2$	$TaSe_2$ and $NbSe_2$	TP
2Hc	BcB|Cb C	$P6_3/mmc$	MX_2 (M = Mo and W; X =S and Se), $MoTe_2$	TP
3R	cB|CaC|AbA	R3m	MX_2 (M = Ta and Mo ; X = S and Se), WS_2	TP
4Hb	aB|CaC|BaCl|BaB|Ca	$P6_3/mmc$	TaS_2 and $TaSe_2$	TP, Oct

[a] T = trigonal, H = hexagonal, R = rhombohedral, Oct = Octahedral, TP = trigonal prismatic

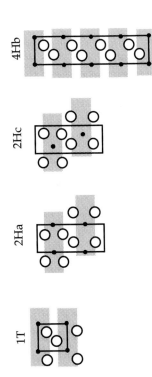

Figure 8 The (110) projections of the common polytypes for the layered transition metal dichalcogenides

4.4.2.2 Metal Ion Guests

The first report of intercalation of a metal ion into a layered dichalcogenide dates from 1959 when Rudorff [85] reported that solutions of alkali metals in liquid ammonia reacted with TiS_2. In many of these reactions the ammonia appears to co-intercalate and can be subsequently removed by heating *in vacuo*. Divalent cations, such as the alkaline earth ions and the lanthanide ions Eu^{2+} and Yb^{2+}, have also been intercalated by this method [86].

The use of $LiBu^n$ has also been extensively used as a mild, efficient means of inserting lithium into the metal dichalcogenides. Hydrated metal intercalate phases with typical composition $A_xMX_2(H_2O)_y$ can be formed by several procedures including electrochemical reduction in aqueous electrolytes, reaction with chemical reducing agents such as borohydrides or dithionite, or simply by exposure of the anhydrous phase to water [9].

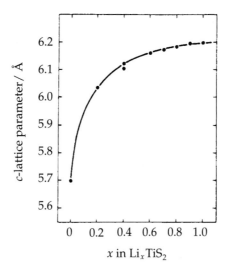

Figure 9 Variation in the *c*-axis lattice parameter for Li_xTiS_2 ($0 \le x \le 1$; from Whittingham [89])

These hydrated compounds exist in many phases depending on the degree of hydration. Typically at high solvation bilayers of water are formed, while at low water partial pressures or for cations with low hydration energies a monolayer of water molecules is observed. In all of these phases both the hydrated metal ions and the water molecules have high mobilities. Consequently these materials have proved useful starting materials for synthesising other new intercalates using ion or solvent exchange routes.

As you might expect, there are significant changes in the lattice parameters of the MX_2 hosts upon alkali metal intercalation. The most extensively studied systems are those involving the lithium intercalated lamellar metal dichalcogenides. In the case of the host TiS_2, a single homogeneous phase has been found for the entire stoichiometry range Li_xTiS_2 ($0 \leq x \leq 1$; Figure 9) [87].

Neutron diffraction studies indicate that lithium occupies the octahedral interlayer sites and the final product, $LiTiS_2$, is isostructural with $LiVS_2$ and $LiCrS_2$ [88]. The radii of these sites are *ca* 0.71 Å, assuming ideal chalcogen octahedra, and this is comparable with the radius of Li^+ so that only a small expansion along the *c*-axis is required to accommodate this cation.

The structures adopted by intercalation complexes formed with other alkali metals are much more varied, as these larger ions can occupy either octahedral or trigonal prismatic interlayer sites. Rouxel *et al.* have extensively investigated the stability limits for the alkali metal intercalation phases of TiS_2 and ZrS_2 [90]. In addition to the two different types of inter-layer sites, staging is observed (Figure 10).

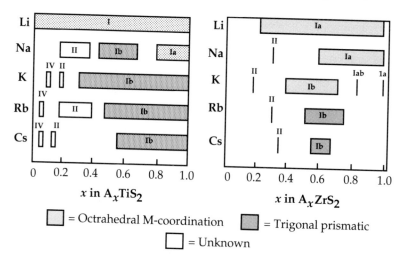

Figure 10 Phase relations for the alkali metal intercalates of TiS_2 and ZrS_2. I, II and IV indicate 1st, 2nd and 4th stage intercalates respectively (from Whittingham [91])

At low alkali metal concentrations (except for lithium) staging results in the formation of compounds with alternating sequences of filled and empty van der Waals' gaps, as shown earlier in Figure 3. The existence of staging indicates that at low concentrations the energy of the system with a homogeneous distribution of cations throughout the interlayer space is higher than that for a phase with the cations concentrated in a fraction of the interlayer sites. For example, for TiS_2, second stage compounds are formed for all the alkali metals except lithium and fourth stage phases occur with K, Rb and Cs.

There is now a wealth of experimental evidence that the alkali metal atoms which enter between the dichalcogenide layers are ionised, with most or all their outer, valence s-electrons going into the conduction band of the host lattice, assuming the rigid band model for the electronic structure of these hosts. An NMR study of Li_xTiS_2 [92] provides one of the clearest pieces of evidence in support of charge transfer from metal ion guest to host lattice. The very small variation of the ^7Li Knight shift (3–12 ppm) for the intercalates, compared with 240 ppm in lithium metal, shows that the lithium is essentially ionised but that the degree of ionisation decreases slightly with increasing lithium content.

These additional filled electronic states in the conduction band produce profound changes in the electronic properties of the host. The electrical and optical properties of the intercalates of the transition metal dichalcogenides have been reviewed extensively by Friend and Yoffe [93] and we are only able to briefly discuss the properties of these materials in Section 4.4.2.5. Table 8 summarises some of the dramatic electronic changes which may occur upon intercalation of alkali metal ions.

4.4.2.3 Organic Guests

Lewis base organic molecules such as amines, hydrazines, acid amides, N-oxides, heterocycles and phosphines can be intercalated into the metal dichalcogenides by direct thermal reaction at temperatures up to *ca* 200 °C (Table 7). The intercalation of ammonia, pyridine and alkylamines have been studied in most detail and we discuss these examples to illustrate the more general phenomenon.

Distillation of anhydrous ammonia onto a layered metal dichalcogenide at – 78 °C followed by warming to room temperature leads to a rapid reaction. The onset of intercalation is marked by swelling of the sample and often a slight colouration of the solution. The reaction proceeds until the limiting first stage composition MX_2NH_3 is achieved. It appears that this phase readily loses NH_3 to go to the second stage material $MX_2(NH_3)_{0.5}$. Both NMR [94] and neutron diffraction [95] experiments have shown that the NH_3 molecules are located in the trigonal prismatic interlayer sites and are orientated with their three-fold axes parallel to the layers. A careful analysis of the reaction stoichiometry for ammonia intercalating into TaS_2 (Equation (8)) has shown that ammonia oxidation is involved in the reaction and that these materials contain ammonium ions solvated by neutral molecules [96].

$$(1 + x/3)\ NH_3\ +\ TaS_2\ \longrightarrow\ (NH_4)_x(NH_3)_{1-x}TaS_2\ +\ x/6\ N_2 \qquad (8)$$

Table 7 Selected examples of intercalation compounds formed by the metal disulphides with different guest ions and molecules

Host	Guest	Guest Occupancy / x Host$\{G\}_x$	Interlayer Spacing, c/Å	Lattice Expansion, Δc/Å
TiS_2		—	5.69	none
	Li	1.0	6.19	0.50
	K	0.8	7.56	1.87
	Cs	0.6	8.36	2.67
TaS_2		—	6.04	none
	Na	1.0	11.87	5.83
	NH_3	1.0	9.10	3.06
	Pyridine	0.5	11.85	5.81
	$[Co(Cp)_2]$	0.25	11.45	5.41
	$[Fe_6S_8(PEt_3)]$	0.05	17.49	11.45
NbS_2		—	6.00	none
	$C_{18}H_{37}NH_2$	0.30	56.50	50.5
ZrS_2		—	5.83	none
	$[Mo(\eta\text{-}C_6H_6)_2]$	0.16	11.64	5.81
	$[Mo(\eta\text{-}C_6H_3Me_3)_2]$	0.08	11.61	5.78
MoS_2		—	6.145	none
	Li	1.0	6.27	0.12
	Rb	0.30	8.60	2.45

Thus the ammonia orientation seems to be determined by the ion-dipole interactions with the NH_4^+ cations (Figure 11). The reactivity of pyridine is closely analogous to that exhibited by ammonia. Direct reaction of $2H\text{-}TaS_2$ with pyridine leads to the formation of the first stage phase with limiting composition $TaS_2(py)_{0.5}$. Neutron diffraction studies on $TaS_2(py\text{-}d^5)_{0.5}$ [97] have determined that the nitrogen-lone pair is directed parallel to the layers (Figure 11). Chemical analysis has also demonstrated that these reactions involve redox chemistry giving a product formulated as $TaS_2(py)_{0.5-2x}(pyH)_x(bipy)_{x/2}$.

The intercalation of n-alkylamines is a very general reaction of the layered metal dichalcogenides, and large systematic changes in interlayer spacing up to *ca* 60 Å have been observed [98]. Reviewing the remaining vast variety of inorganic and organic molecular guests which have been intercalated into the metal dichalcogenides is not possible here. Other examples can be found in Table 7 or by reference to some excellent reviews on the metal dichalcogenide intercalates [91, 99].

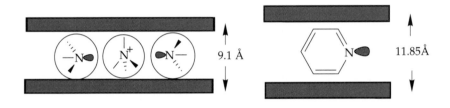

Figure 11 Schematic of the packing and orientation of the guests in
(a) $TaS_2(NH_3)$ (b) $TaS_2(Py)_{0.5}$

4.4.2.4 Organometallic Guests

Redox active organometallic compounds can be intercalated into the metal dichalcogenides by either direct reaction, ion-exchange, or by electrochemical routes. This significant new development was reported in 1975 when Dines described the intercalation of $[Co(Cp)_2]$ and $[Cr(Cp)_2]$ into a range of metal disulphides (MS_2: M = Ti, Zr, Nb, Ta and Sn) [100]. The range of organometallic guests which can be intercalated into these hosts is quite extensive, including metal π-complexes [101], metal clusters [102] and metal phosphine complexes [103].

The electron transfer mechanism from guest to host accounts for the correlation which exists between the reducing power of the organometallic guest and its ability to intercalate into a given host lattice. For example, ferrocene does not appear to intercalate into MS_2 compounds whereas cobaltocene readily intercalates into a wide range of the dichalcogenides. The magnetic susceptibilities of $TaS_2\{Co(Cp)_2\}_{0.25}$

and $TaS_2\{Cr(Cp)_2\}_{0.25}$ are consistent with complete electron transfer from the metallocenes to the hosts. However, X-ray photoelectron studies on $SnS_2\{Co(Cp)_2\}_{0.3}$ indicate that this may not always occur and complex mixed valence behaviour may result [104].

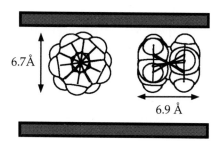

6.7Å

6.9 Å

Figure 12 van der Waals' dimensions of cobaltocene

The orientation of the unsubstituted metallocenes with respect to the host layers has been subject to much debate since Dines' initial report [100]. The complication arises from the fact that the lattice expansion of *ca* 5.31 Å observed for all simple metallocene intercalates does not immediately reveal the orientation of the guest. These molecules have almost a spherical van der Waals' surface as shown in Figure 12. For $SnS_2\{Co(Cp)_2\}_{0.3}$ the debate has been resolved by a combined X-ray and neutron diffraction study, which shows that the cobaltocene molecules are ordered within the van der Waals' gap with their principal axes parallel to the layer planes [105]. In contrast, a variable temperature solid state ^2H NMR study of $2H\text{-}TaS_2\{Co(\eta\text{-}C_5D_5)_2\}_{0.25}$ convincingly suggests that both parallel and perpendicular orientations exist which are in dynamic exchange [106].

4.4.2.5 Electronic Properties

Changes in the electronic properties of the layered metal dichalcogenides after intercalation have been followed by measurements, for example, of the optical absorption spectra, electrical conductivity, Hall constant, magnetic susceptibility and solid state NMR spectra. In general these measurements provide evidence for our premise of charge transfer from the intercalant to the host lattice. For example, the optical absorption spectrum of $2H\text{-}NbSe_2$ shows the Drude reflectivity edge, typical of a metal, at just below 1 eV arising from the half-filled d_{z^2} sub-band. After intercalation with amines this Drude edge moves to lower energies by *ca* 0.2 eV due to partial filling of the $d_z{}^2$ band and the consequent reduction in the number of carriers (holes) and a shrinking of the Fermi surface [93]. There is an excellent comprehensive review of the electronic properties of the intercalation complexes of the transition metal dichalcogenides by Friend and Yoffe [93].

Over the last decade there was great excitement when some of the intercalation compounds of the metal dichalcogenides were found to have significantly higher superconducting critical temperatures (T_c) than the parent host lattices. Today, superconductivity has been widely observed in the intercalation complexes of the metal dichalcogenides. Of the unintercalated group IV, V, and VI transition metal dichalcogenides only the group V dichalcogenides are metallic and only the trigonal prismatic coordination polytypes are superconductors. 2H-NbSe$_2$ has the highest reported superconducting transition temperature (T_c = 7.1 K) for a layered dichalcogenide, Table 8.

Table 8   Examples of the significant electronic changes which may occur to the metal dichalcogenides on potassium intercalation

Host	Host Properties	Intercalate	Intercalate Properties
1T-HfS$_2$	wide band gap semiconductor	K$_x$HfS$_2$	metal
2H-NbSe$_2$	metal superconductor	K$_x$NbSe$_2$	poor metal (x = 1), expect a semiconductor
2H-MoS$_2$	diamagnetic semiconductor	K$_x$MoS$_2$	superconductor

The alkali metal intercalation complexes of the group IV, V, and VI dichalcogenides provide examples of electronic conductivities ranging from semiconductors to superconductors, Table 9. The amine intercalates of 2H-TaS$_2$ have been particularly well studied because these guests substantially enhance the superconducting transition temperature, even when long chain amines are used which produce lattice expansions greater than *ca* 30 Å. In a very few cases layered semiconductors have been transformed to superconductors on intercalation of the appropriate guest. For example, the semiconductors MoS$_2$ and SnSe$_2$ become superconducting on intercalation of alkali metal ions [107] and $[Co(Cp)_2]$ [108] respectively.

A number of fundamental questions have arisen regarding the physical description of the dimensionality of superconductivity inview of the highly anisotropic character of these materials. Could the increased transition temperatures of some of these intercalates suggest a novel mechanism for the superconductivity ?

Table 9 Superconducting metal dichalcogenides hosts and their intercalates.

Host	T_c/K	Intercalate	T_c/K
2H-TaS$_2$	0.6	TaS$_2$(CH$_3$NH$_2$)$_{1.2}$	5.6
2H-TaS$_2$ (high pressure)	4.5	TaS$_2${Co(Cp)$_2$}$_{0.25}$	3.2
		TaS$_2$(py)$_{0.5}$	3.5
		TaS$_2$Gd$_{0.11}$(H$_2$O)$_2$	2.1
2H-NbS$_2$	6.1	TaS$_2$Mn$_{0.16}$(H$_2$O)$_2$	2.6
2H-NbSe$_2$	7.1	NbS$_2$(py)$_{1.2}$	4.1
ZrSe$_2$	none	Li$_{0.5}$ZrSe$_2$	1.8
MoS$_2$	none	Rb$_{0.5}$MoS$_2$	6.9
SnSe$_2$	none	SnSe$_2${Co(Cp)$_2$}$_{0.3}$	8.1

It is now generally agreed, however, that the observed increase in T_c results from suppression of charge density wave formation rather than some exotic quasi-two-dimensional mechanism [109]. However, the effect of the two-dimensional anisotropy of the material is of considerable interest. It has been found that the critical field behaviour is in broad agreement with theoretical predictions based on a model of a layered compound containing two-dimensional superconducting layers weakly coupled via Josephson tunneling [110].

4.4.3 The Metal Oxyhalides

4.4.3.1 Host Structure

Of the small number of metal oxyhalides that form layered structures only the FeOCl structure type (e.g. FeOCl, TiOCl and VOCl) is known to undergo topotactic redox reactions. The crystal structure of FeOCl was first described by Goldsztaub [111]; it belongs to the orthorhombic space group Pmnm, with two formula units per unit cell.

The crystal structure consists of a stack of double sheets of cis-$FeCl_2O_4$ distorted octahedra linked together by shared edges within the crystallographic ac-plane. The neutral layers of FeOCl are orientated perpendicular to the b-direction, with chlorine forming the outermost atoms of each layer (Figure 13). The easy cleavage plane is perpendicular to the b-direction and it is this cell parameter which is greatest affected on intercalation. Other compounds which adopt the FeOCl structure are MOCl (M = Ti, V and Cr) and InOX (X = Cl, Br and I).

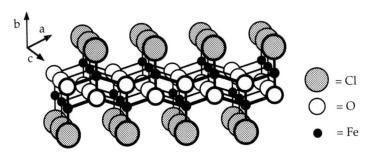

Figure 13 Perspective view of a layer in FeOCl (after Halbert [112])

4.4.3.2 Redox Intercalation

Until recently only electron donors have been intercalated into the metal oxyhalides. The process primarily involves the reduction of the M^{3+} cation. FeOCl represents the most favourable case as shown by the M^{3+}/M^{2+} redox potentials (Table 11).

The first attempts to intercalate FeOCl with lithium were made by treating the host lattice with lithium dissolved in liquid ammonia. The reaction in fact produces FeO(NH)Li probably via the formation of $FeONH_2$ by substitution of Cl^- by NH_2^- ions [113]. The topotactic intercalation of lithium was subsequently achieved by using either $LiBu^n$ or by electrointercalation methods [114]. However, the compounds have low crystallinity, and are not well characterised. The hydrated alkali metal ion intercalates $M_{0.14}FeOCl(H_2O)_y$ (M = Li, Na, K and Cs) can be prepared using $[Fe(CN)_6]^{4-}$ as the reducing agent. They exhibit large interlayer spacings which are dependent on the water content. This behaviour is in contrast to the hydrated alkali metal intercalates of the metal dichalcogenides and indicates much stronger interactions between solvating water molecules with either the chloride layers or with partially hydrolysed layers in which some Cl^- groups have been replaced by OH^-.

Reaction of organic Lewis bases such as ammonia and amines are some of the earliest reported intercalation reactions of compounds with neutral layers [115]. The solid state chemistry is generally similar to that exhibited by the metal dichalcogenides. FeOCl is however much more reactive due to its stronger oxidising power. Consequently, a much wider range of organic and organometallic guest compounds, including pyridines, anilines, organophosphorus and sulphur compounds and metallocenes have been intercalated into these host lattices (Table 10). FeOCl is one of the very few host lattices that will directly intercalate ferrocenes. Even the peralkylated derivatives $[Fe(\eta-C_5Me_5)_2]$ and $[Fe(\eta-C_5Me_4Et)_2]$ have been intercalated by direct reaction with FeOCl [116].

Table 10 Examples of intercalation compounds of MOCl (M = Fe, V and Ti)

Host	Guest	Guest Occupancy/x Host{G}$_x$	Interlayer Spacing, b/Å	Lattice Expansion, Δb/Å
FeOCl		—	7.91	none
	Li^+	0.12	11.71	3.8
	NH_4^+	0.75	11.33	3.42
	NH_3	0.5	10.37	2.46
	Pyridine	0.33	13.45	5.54
	Quinuclidine	0.16	13.36	5.45
	$[SnMe_3(C_5H_4N)]$	0.13	14.11	6.20
	$[Fe_4(\eta-C_5H_4Me)_4(\mu^3-S)_4]$	0.08	17.02	9.11
	Pyrrole	0.34	13.20	5.29
	$[Fe(Cp)_2]$	0.16	13.03	5.11
	$[TTF]^+$	0.12	13.01	5.09
	$[Fe(\eta-C_5Me_4Et)_2]$	0.16	15.46	7.54
	Kryptofix-22 [a]	0.18	12.19	4.28
VOCl		—	7.91	none
	$[Co(Cp)_2]$	0.16	12.77	4.86
	Pyridine	0.33	13.35	5.45
TiOCl		—	8.03	none
	$[Co(Cp)_2]$	0.16	13.16	5.13

[a] Kryptofix-22 = 1,7,10,16,-tetraoxa-4,13-diazacyclooctadecane

Intercalation of organic and organometallic guests is generally carried out by direct reaction in sealed glass ampoules at temperatures up to *circa* 120 °C. Two structural consequences result from intercalation (i) the unit cell expansion along the *b*-axis, depending on the size of the intercalated molecule, (ii) a change in the stacking of successive layer planes. The lattice expansion (Δb) is always smaller than the shortest dimension of the guest molecule which implies there is a significant interpenetration of the guest molecule into the van der Waals' layers defined by the chlorine atoms.

For example, the lattice expansion (Δb = 5.13 Å) observed when ferrocene is intercalated into FeOCl suggests that the metallocenes are accommodated with the η-cyclopentadienyl rings perpendicular to the layers [113]. The stoichiometry $FeOCl\{Fe(Cp)_2\}_{0.16}$ is close to that calculated for close packed metallocene cations in this model. In addition, the powder X-ray diffraction patterns from the intercalates indicate a change in lattice symmetry from primitive to body centred showing that the intercalation produces a relative translation of adjacent layers (Figure 14).

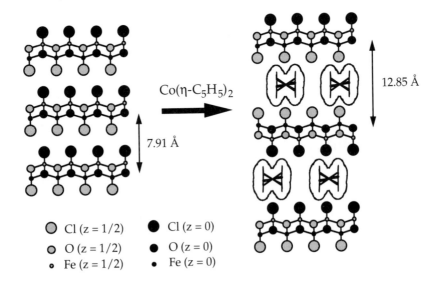

$Co(\eta\text{-}C_5H_5)_2$

12.85 Å

7.91 Å

○ Cl (z = 1/2)   ● Cl (z = 0)
◉ O (z = 1/2)   ● O (z = 0)
∘ Fe (z = 1/2)   • Fe (z = 0)

Figure 14 Schematic representation of cobaltocene in FeOCl (from Halbert [117])

Mössbauer spectroscopy has proved to be a very useful technique for studying electron transfer and mixed valency in the FeOCl intercalates. In the Lewis base intercalation compounds of FeOCl the low temperature ^{57}Fe Mössbauer spectrum shows a resonance assigned to Fe^{2+} which can be used to estimate the degree of charge transfer. For a range of amines and heterocyclic guests the degree of charge transfer is observed to be 0.07–0.14 e^-/FeOCl. The nature of the electron transfer process which occurs on formation of $FeOCl\{Fe(Cp)_2\}_{0.16}$ has been studied in detail by ^{57}Fe Mössbauer spectroscopy. At temperatures in the range 77–100 K the Mössbauer spectrum exhibits resonances characteristic of the ferrocenium cation and two chemically distinct Fe^{3+} sites. The two Fe^{3+} sites are different in the number of nearest neighbour Fe^{2+} ions .

Quantitative measurements confirm that all the ferrocene guest molecules are oxidised and complete electron transfer to the host has occurred. At 300 K only one ^{57}Fe resonance is observed as a consequence of rapid electron hopping between iron ions, both within layers and between layers, presumably via the guest molecules [118].

Pyrrole can be intercalated into FeOCl; the guest polymerises within the van der Waals' gap of the layered lattice giving a novel class of conductive polymer-inorganic hybrid material [119]. Although the electrical conductivity of this intercalate is significantly higher than pristine FeOCl, the conductivity is not higher than doped polypyrrole.

Intercalation compounds are also formed by reaction of FeOCl with organic compounds which are good electron donors. The most extensively studied family of electron donors has been the 'thiofulvalenes'; for example, tetrathiofulvalene (TTF), and tetrathionaphthalene (TTN). These materials are highly crystalline. Powder neutron diffraction data on $FeOCl(TTF)_{0.12}$ shows that the TTF is orientated parallel to the b-axis (interlayer axis) and the central C=C bond is aligned with the c-axis of FeOCl [120]. FeOCl will even intercalate the organic hydrocarbon perylene forming $FeOCl(perylene)_{0.11}$ [121]. The electrical conductivity and infrared data indicate that the intercalant is present as the perylene radical cation.

4.4.3.3 *Electronic and Magnetic Properties*

The layered transition metal oxyhalides are paramagnetic semiconductors. The room temperature conductivity of single crystals of FeOCl is *circa* $10^{-7}\ \Omega^{-1}\ cm^{-1}$. Intercalation of the organic and organometallic guest molecules substantially increases the electrical conductivity of the host lattice; for example, $FeOCl(py)_{0.33}$ has a σ_{rt} *ca* $10^{-1}\ \Omega^{-1}\ cm^{-1}$, and $FeOCl(ET)_{0.25}$ (ET = bis(ethylenedithio)tetrathiafulvalene) has the highest conductivity of any molecular intercalate of

FeOCl (σ_{rt} *ca* 0.26 Ω^{-1} cm^{-1}) [122]. These phases all show a temperature dependence of the conductivity indicative of semiconducting behaviour. In all cases where it has been measured the thermoelectric power is negative, suggesting that the charge carriers are electrons.

Table 11 Summary of some of the electronic and magnetic properties of the lamellar MOCl hosts (M = Fe, V and Cr)

Host	Band Gap/eV	$T_{Néel}$/K	Reduction Potential $(M^{3+} \to M^{2+})$/ V versus SCE
FeOCl	1.9	92	0.77
VOCl	2.1	80	-0.25
CrOCl	3.2	10	-0.41

Pristine FeOCl exhibits a paramagnetic to antiferromagnetic phase transition with a Néel temperature of 91 K (Table 11), which can be conveniently studied by Mössbauer spectroscopy. The major effect on the magnetic properties of FeOCl on intercalation of organic or organometallic guests is the depression of the Néel temperature [123]. The lower temperatures required for the onset of three-dimensional magnetic ordering has been attributed to a decrease in the interlayer magnetic exchange interactions and site disorder effects [118].

4.4.4 The Metal Phosphorus Trisulphides

4.4.4.1 Host Structure

Klingen [124] was the first to describe a series of lamellar semiconductors of the formula MPS$_3$, where M is a wide range of divalent metal ions (M = Mg, V, Mn, Fe, Co, Zn, Ni, Pd and Cd). The selenides (MPSe$_3$) also form similar structures with M = Mg, Mn, Fe, Ni and Cd [125]. These materials can be prepared by standard solid state procedures, i.e. by heating the elements in sealed evacuated quartz ampoules at *ca* 700 °C [126].

The first row transition metal phosphorus trisulphides all form a common monoclinic structure (C2/m) based on the CdCl$_2$ type (Figure 15). This consists of a cubic close-packed anion array with alternate layers of cation sites vacant. Within a layer two-thirds of the cation sites are occupied by M^{2+} cations and one-third by P$_2$ pairs. The structure is related to the CdCl$_2$ lattice, since the M^{2+} ions and the centre of the P–P pairs occupy the Cd positions and the sulphur ions occupy the chloride positions. In this way the transition metal ions and the P–P pairs are approximately octahedrally coordinated in a distorted cubic-close packed lattice.

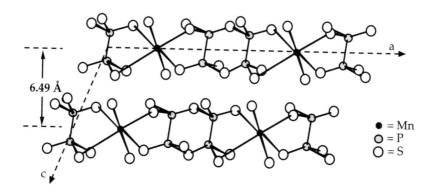

Figure 15 Structure of MPS$_3$ (M = Mg, V, Mn, Fe, Co, Ni, Pd and Cd)

Depending on the particular metal ion, the energy values for the band gaps of these semiconducting MPS$_3$ lattices lie in the range 3.5–1.6 eV. For example, CdPS$_3$ is colourless and transparent (band gap 3.5 eV), MnPS$_3$ appears transparent with bright green colouration (band gap 3.0 eV), whereas FePS$_3$ and NiPS$_3$ are dark with band gaps close to 1.6 eV.

The metal ions are octahedrally coordinated and have high spin electronic configurations where applicable. The MPS$_3$ phases containing paramagnetic M^{2+} ions (Mn^{2+}, S = 5/2; Fe^{2+}, S = 2; Ni^{2+}, S = 1) are antiferromagnets with Néel temperatures of 110, 126 and 253 K respectively. However, the susceptibility maxima are broad indicative of the strong two-dimensionality of these materials.

The metal phosphorus trisulphides exhibit two types of intercalation reaction. Redox intercalation involves electron donating guests using chemical or electrochemical routes and with low band gap MPS$_3$ host lattices. The second reaction type is unique to the MPS$_3$ compounds and involves intercalation of cation species with concomitant loss of the M^{2+} cations from the MPS$_3$ layers into the solution to maintain charge balance. This produces cation vacancies in the MPS$_3$ layers and, as we shall see, is responsible for some of the unique reactivity and potentially important properties of these materials.

4.4.4.2 Redox Intercalation

The intercalation of alkali metal ions into MPS_3 hosts has been much less extensively studied than intercalation of these ions into the metal dichalcogenides. However, the chemical and electrointercalation of lithium into $NiPS_3$ has received particular attention due to its potential application as a battery material since its theoretical energy density output is double that of the Li/TiS_2 battery system [127].

Reaction of $NiPS_3$ with $LiBu^n$ or sodium or potassium naphthalide yields materials with limiting stoichiometry M_xNiPS_3 (M = Li, Na, K; $x \approx 4.4$). These extremely high values of alkali metal incorporation are much greater than the maximum value of $M_{1.5}NiPS_3$ which corresponds to complete occupation of the interlayer sites. X-ray powder diffraction data suggest that degradation of the host lattice occurs and that alkali metal sulphides are formed. This hypothesis has been confirmed by a study of the electrointercalation of Li^+ ions into $NiPS_3$. Electrochemical reduction beyond $x = 1.5$ is irreversible, but in the stoichiometry range $0 \leq x \leq 1.5$ reversible intercalation of lithium occurs giving two structural phases [127].

Organic amines have been intercalated into $NiPS_3$ and $MnPS_3$ by direct reaction [128]. These intercalates contain both protonated and nonprotonated amines between the layers, analogous to the metal dichalcogenides. Pyridine does not intercalate into $NiPS_3$ but will intercalate into $CdPS_3$. A solid state ^2H NMR study of $CdPS_3(py)_{0.5}$ concluded that the pyridine plane is orientated perpendicular and the C_2 symmetry axis parallel to the lattice planes at low temperature (< 285 K); at temperatures above 310 K the guest is dynamically disordered in 3-dimensions [129].

MPS_3 compunds (M = Fe, Ni, Mn and Cd) react directly with cobaltocene in toluene solution to give $MPS_3\{Co(Cp)_2\}_{0.3}$; $CdPS_3\{Co(Cp)_2\}_{0.40}$ has been extensively studied using EPR spectroscopy by Cleary and Francis [40, 130]. The quantitative EPR measurements concluded that complete electron transfer from the $Co(Cp)_2$ to the $CdPS_3$ layers does not occur and both $[Co(Cp)_2]$ molecules and $[Co(Cp)_2]^+$ cations co-exist between the layers.

4.4.4.3 Ion Exchange

Higher band gap materials, for example, $MnPS_3$, $CdPS_3$ and $ZnPS_3$, can be intercalated with cations such as K^+, NH_4^+ and $[Co(Cp)_2]^+$ by treatment with an aqueous solution of the appropriate cation salt. These reactions are believed to occur by an ion-exchange mechanism forming an intercalate phase with general stoichiometry $M_{1-x}PS_3(G)_{2x}(H_2O)_y$ (Table 12). The positive charge of the guest cation (G^+) is counterbalanced by the removal of an equivalent amount of interlayer M^{2+} ions.

Table 12 Examples of intercalates obtained by cation exchange with hydrated potassium or sodium pre-intercalated MPS_3 [132, 133]

Host	Intercalates	Interlayer Spacing, c/Å	Lattice Expansion, Δc/Å
$MnPS_3$		6.50	none
	$Mn_{0.8}PS_3(K)_{0.4}(H_2O)_{0.9}$	9.37	2.87
	$Mn_{0.85}PS_3\{N(Me)_4\}_{0.3}(H_2O)_{0.9}$	11.45	4.95
	$Mn_{0.8}PS_3\{Ru(2,2'\text{-bipy})_3\}_{0.15}(K)_{0.1}(H_2O)_{1.5}$	15.20	8.70
	$Mn_{0.85}PS_3(pyH)_{0.28}(H_2O)_{0.7}$	9.65	3.15
	$Mn_{0.84}PS_3\{Co(Cp)_2\}_{0.32}(H_2O)_{0.3}$	11.82	5.32
	$Mn_{0.83}PS_3(TTF)_{0.42}(K)_{0.1}(H_2O)_{0.4}$	12.15	5.65
$CdPS_3$		6.55	none
	$Cd_{0.75}PS_3(K)_{0.5}(H_2O)_{1.0}$	9.43	2.93
	$Cd_{0.84}PS_3\{Co(Cp)_2\}_{0.32}(H_2O)$	11.87	5.32
$FePS_3$		6.40	none
	$Fe_{0.85}PS_3(NEt_4)(H_2O)_{0.3}$	11.18	4.78
	$Fe_{0.84}PS_3\{Co(Cp)_2\}_{0.32}(EtOH)_{0.3}$	11.72	5.32
	$Fe_{0.84}PS_3(pyH + py)_{0.33}(H_2O)_{0.5}$	9.70	3.30
$NiPS_3$		6.34	none
	$Ni_{0.84}P_{0.9}S_3\{Co(Cp)_2\}_{0.36}Na_{0.03}(H_2O)_{0.5}$	11.90	5.56
	$Ni_{0.84}P_{0.9}S_3Cu_{0.34}Na_{0.06}(H_2O)_{0.6}$	6.05	0.16

This is a unique type of intercalation reaction, and implies that the M^{2+} cations are able to leave their intralamellar site and go into solution. This direct ion-exchange reaction fails if the guest ion is above a critical size, e.g. $[Ru(2,2'\text{-bipy})_3]^{2+}$, however, this cation can be intercalated into $MnPS_3$ by pre-intercalation with K^+ cations giving $M_{1-x}PS_3(K)_{2x}(H_2O)_y$

and then ion exchanging the K^+ cations for the larger $[Ru(2,2'-bipy)_3]^{2+}$ [133]. The planar organic molecule tetrathiofulvalene (TTF) has also been intercalated using this procedure by treatment of $Mn_{0.8}PS_3(K)_{0.4}(H_2O)_y$ with $(TTF)_3(BF_4)_2$. The lattice expansion suggests that the molecular planes of the TTF guests are orientated perpendicular to the $MnPS_3$ layers [134].

Among the MPS_3 series, $MnPS_3$ is by far the most reactive with regard to this ion-exchange reaction and consequently has been the most extensively studied. Other members of the series M = Cd, Zn and Fe undergo this 'ion-transfer' reaction, but the reaction requires assistance usually achieved by complexing the leaving M^{2+} ions (typically 0.1 M EDTA, pH $\approx$ 9–10) [135].

Deintercalation can be easily carried out when the guest species is small. For instance, when $Mn_{0.8}PS_3(K)_{0.4}(H_2O)_y$ is treated with an aqueous solution of $NiCl_2$, the solvated K^+ ions are replaced by large $[Ni(H_2O)_6]^{2+}$ cations. Upon dehydration, the nickel ions lose their aquo ligands and move into the intralayer vacancies forming $Ni_{0.28}Mn_{0.72}PS_3$ which has identical lattice constants and magnetic properties to the phase prepared by direct reaction of the elements at high temperature [136].

4.4.4.4 Electronic and Magnetic Properties

The electronic structure and properties of MPS_3 hosts and their intercalates has been reviewed by Brec [137]. The hosts are all described as broad band semiconductors; a more detailed picture of the electronic structure is given in a recent tight binding band structure calculation [138]. Of the intercalates that can be described by a simple redox process the lithium intercalates of MPS_3 (M = Ni, Fe and Co) have been most extensively studied. The complete ionisation of the lithium upon intercalation was determined by a multinuclear NMR study of Li_xNiPS_3 [139]. The chemical shift of the ^7Li NMR signal indicates that the lithium is intercalated as Li^+ cations throughout the concentration range $0 < x < 1.5$. The Li^+ diffusion coefficients were estimated to be of the order 10^{-14} cm^2 sec^{-1} over the entire composition range. The ^{31}P NMR for the Li_xNiPS_3 series reveals some interesting magnetic behaviour of these phases. It appears that between $x = 0.0$–0.5 a single phase is present with magnetic properties analogous to the host lattice, but between $x = 0.5$–1.5 a mixture of magnetic $Li_{0.5}NiPS_3$ and non-magnetic $Li_{1.5}NiPS_3$ forms. These NMR results imply that the electrons enter a band in the host that does not include either metal 3d or phosphorus orbitals, but consists of either sulphur or metal 4s orbitals [140].

The electronic conductivity of the intercalates formed via the 'ion transfer' route do not appear to be significantly altered relative to the pristine host lattice. For example, the room temperature single crystal electronic conductivity of pristine $MnPS_3$ is 3×10^{-9} Ω^{-1} cm^{-1} compared to 1×10^{-8} Ω^{-1} cm^{-1} for $Mn_{0.8}PS_3(K)_{0.4}(H_2O)_y$. All the MPS_3 hosts and their known intercalates are semiconductors except for $Fe_{0.82}PS_3(TTF)_{0.38}$ has been recently shown to exhibit metallic conductivity [141].

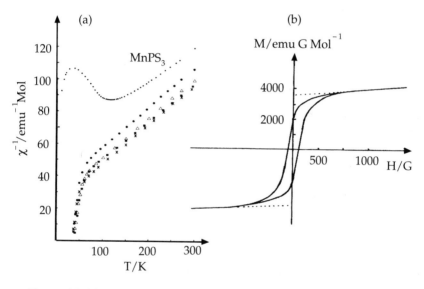

Figure 16 (a) Inverse molar susceptibility versus Temperature for $\triangle = Mn_{0.8}PS_3(K)_{0.4}(H_2O)_{0.9}$, $\times = Mn_{0.84}PS_3$(n-octylamine)$_{0.22}(H_2O)_{0.3}$, $\blacksquare = Mn_{0.84}PS_3[Co(Cp)_2]_{0.32}(H_2O)_{0.3}$, $\bullet = Mn_{0.8}PS_3$(pyH)$_{0.28}(H_2O)_{0.7}$, (b) Molar magnetisation versus applied magnetic field for $Mn_{0.84}PS_3\{Co(Cp)_2\}_{0.32}(H_2O)_{0.3}$

In contrast, the magnetic properties of the strongly coupled two-dimensional Heisenberg antiferromagnet $MnPS_3$ are dramatically changed. Intercalation via the 'ion transfer' route yields intercalation compounds with reduced antiferromagnetic interactions, and the susceptibility of many of these intercalates increases dramatically below 40 K. In fact, the magnetic susceptibility of $Mn_{0.84}PS_3\{Co(Cp)_2\}_{0.32}$-$(H_2O)_{0.3}$ exhibits a spontaneous magnetisation in zero applied magnetic field, which is indicative of weak ferromagnetism (Figure 16) [142]. In addition, the low temperature powder neutron diffraction data show coherent magnetic scattering indicative of long range magnetic order [143]. The origin of the spontaneous magnetisation has not been determined. Two hypotheses have been proposed (Figure 17). The micro-analytical data suggests that there are manganese ion vacancies within the $MnPS_3$ layers, therefore a ferrimagnetic ground state could arise from incomplete compensation of the antiferromagnetically coupled Mn^{2+} spins in the manganese layers if the vacancies are ordered as shown in Figure 17b. Alternatively, if the spins on the Mn^{2+} ions in adjacent layers do not align perfectly antiparallel but are canted they would produce a resultant net magnetisation at 90° to the layers.

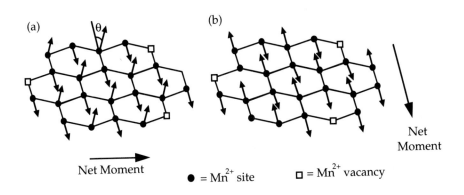

Figure 17 Two proposed spin arrangements for $M_{1-x}PS_3(G)_{2x}(H_2O)_y$ to account for the spontaneous magnetism, (a) canted antiferromagnetism, (b) frustrated antiferromagnetism

4.4.5 The Metal Oxides

4.4.5.1 Host Structure

Simple layered metal oxide structures with true van der Waals' gaps are found only for the oxides of the transition metals in high formal oxidation states. In these metal oxides the layer structure is stabilised by the formation of strong multiple covalent bonds to oxygen, for example MoO_3 and V_2O_5 (Figure 18).

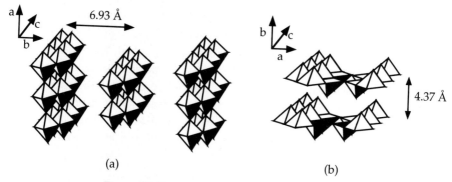

Figure 18 Structures of (a) MoO_3 and (b) V_2O_5

4.4.5.2 *Intercalation into Molybdenum Trioxide*

Molybdenum trioxide possesses a unique layered structure (Figure 18) [144]. It may be thought of as being formed from ReO_3-like chains, with edge and vertex-sharing MoO_3 octahedra which form a puckered layered unit. These layers are separated by a true van der Waals' gap and the oxide undergoes a full range of topotactic redox chemistry similar to that outlined for the metal dichalcogenides.

The intercalation of hydrogen into MoO_3 forming the *oxide bronzes* is the most extensively investigated of its intercalation reactions. Early work by Glemser *et al.* showed that chemical reduction of MoO_3 in aqueous acid results in the formation of a series of compounds which could be formulated as H_xMoO_3 [145]. More recently, a study of the phase diagram of the MoO_3/H system has showen four distinct phases [146]. The lattice parameters of these intercalates are very similar to those of the host MoO_3 lattice. Detailed structural information regarding the location of the hydrogen in the lattice has been obtained via neutron diffraction experiments on $D_{0.36}MoO_3$ and $D_{1.68}MoO_3$ [35]. At low hydrogen concentrations the hydrogen is attached to bridging oxygen atoms as –OD groups in interlayer sites and is not involved in hydrogen bonding between layers. The neutron diffraction data on the more highly reduced phase suggests that the deuterium atoms are attached to the terminal oxygens to form –OD_2 groups projecting into the interlayer space.

Wide line ^1H NMR measurements on $H_{0.36}MoO_3$ and $H_{1.7}MoO_3$ indicate rapid hydrogen mobility with room temperature diffusion rates

$D(300) = 10^{-8} \, cm^2 \, sec^{-1}$ and $10^{-11} \, cm^2 \, sec^{-1}$ respectively [148]. The difference in the hydrogen mobility between $H_{0.36}MoO_3$ and $H_{1.7}MoO_3$ has been attributed to a change in structure.

Li_xMoO_3 can be prepared by reaction of MoO_3 with either $LiBu^n$, $LiBH_4$ or electrochemically. The maximum value of x obtained by chemical lithiation is 1.55. An extensive investigation by Schöllhorn *et al.* [149] has shown that a wide range of guest species can be intercalated into MoO_3 by chemical or electrochemical reduction. Suspensions of MoO_3 in neutral aqueous electrolyte solutions in the presence of strong reducing agents yield the hydrated molybdenum oxide bronzes $A^{x+}(H_2O)_y[MoO_3]^{x-}$ (A = Na, K, Cs, NH_4 and Mg). The alkali cations are highly mobile within the interlayer solvate and undergo exchange rapidly with other organic and inorganic cations.

The oxide bronze $H_{0.5}MoO_3$ reacts with Lewis bases such as pyridine to give $H_{0.5}MoO_3(py)_{0.3}$ with an interlayer separation increased by 5.84 Å [151]. Only some of the interlayer pyridine molecules are protonated and these pyridinium cations are stabilisied by hydrogen bonding to neutral pyridine molecules in an arrangement analogous to that described previously for the metal dichalcogenides.

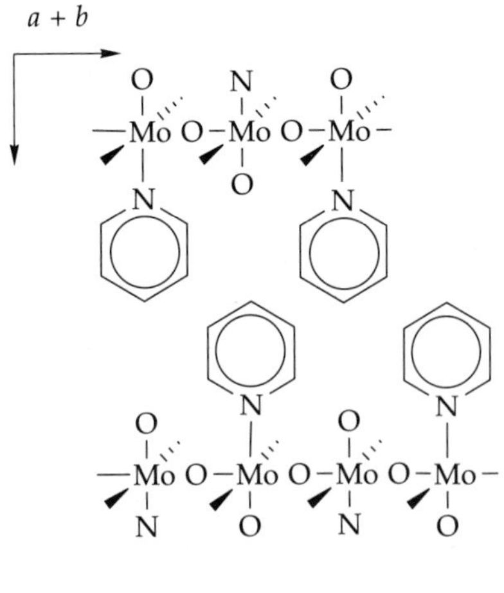

Figure 19 Schematic structure of MoO_3(pyridine) viewed parallel to the layers (from Johnson *et al.* [150])

The intercalation of pyridine under strict anhydrous conditions gives a different phase in which some of the oxide ligands have been replaced by pyridine ligands (Figure 19).

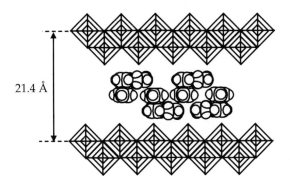

Figure 20 Schematic diagram of the proposed bilayer arrangement of $[Fe(Cp)(\eta-C_5H_4CH_2CH_2NH_2)]$ intercalated in MoO_3

MoO_3 will also accommodate quite large organometallic guests. For example, the cubane cluster $[Fe_4(\eta-C_5H_4Pr^i)_4(\mu^3-Se)_4]$ has been intercalated into MoO_3, producing an observed lattice expansion $\Delta b = 15.2$ Å [103]. Although ferrocene does not directly intercalate into MoO_3 the ring functionalised amino-ferrocene derivative $[Fe(Cp)(\eta-C_5H_4CH_2CH_2NH_2)]$ readily intercalates into MoO_3 forming $MoO_3\{Fe(Cp)(\eta-C_5H_4CH_2CH_2NH_2)\}_{0.5}$. The success of this reaction appears to result from the affinity of MoO_3 to intercalate amine guests. The amino-ferrocene is proposed to adopt a bilayer structure as shown in Figure 20 in common with many other amine intercalates of MoO_3 [103].

4.4.5.3 Intercalation in Vanadium Pentoxide

The V_2O_5 structure can be considered as being formed of V_2O_5 ribbons formed by shearing two ReO_3-like chains (Figure 18) [152]. The structure is layered, however the interlayer interactions are sufficiently strong that it only intercalates the smallest cations, H^+ and Li^+.

The hydrogen intercalation compounds of V_2O_5 can be prepared by the same techniques described for MoO_3. Electrochemical synthesis using a solid proton conducting electrolyte indicates that three distinct phases in the $H_xV_2O_5$ system exist, with compositions $x \leq 0.5$, $1.3 < x < 2.3$, and $3.0 < x < 3.8$. They appear to be structurally similar to the molybdenum oxide bronzes. Lithium may be intercalated chemically or electro-chemically into the V_2O_5 lattice to give $Li_xV_2O_5$ ($0 < x < 2.0$) and V_2O_5 has been used reversibly in high power lithium cells. When $x < 1.0$, the compounds are crystalline and electrochemical measurements show three single phase regions with $x < 0.2$, $0.35 \leq x \leq 0.5$, and $0.9 \leq x \leq 1.0$. At low lithium concentrations, $x < 1.0$, the lattice constants are close to the parent oxide host, indicating topochemical intercalation of lithium. The latter two phases have structures which differ considerably from the host structure. V_2O_5 also reacts with either lithium iodide in acetonitrile solution or $LiBu^n$ in hexane to give $Li_xV_2O_5$. The more strongly reducing $LiBu^n$ gives reduction beyond $x = 1.0$, but the products are poorly crystalline and not well characterised.

Table 13 Examples of intercalation compounds formed by the layered metal oxides

Host	Intercalates[a]	Interlayer Spacing, b/Å	Lattice Expansion, Δb/Å
MoO_3		6.93	none
	$MoO_3H_{0.31}$	7.05	0.12
	$MoO_3(NH_3)H_{0.3}$	9.23	2.3
	$MoO_3Li(H_2O)$	11.48	4.55
	$MoO_3(NH_4)(H_2O)$	12.18	5.25
	$MoO_3(Py)$	11.48	4.55
	$MoO_3\{Co(Cp)_2\}_{0.66}$	12.73	5.80
	$MoO_3\{Fe(Cp)(Cp'')\}_{0.6}$	21.43	14.5
	$MoO_3\{Fe_4(Cp')_4(\mu^3\text{-}S)_4\}_{0.15}$	17.23	10.3
	$MoO_3\{(Al_{13})_{0.047}Na_{0.003}(H_2O)_{1.0}\}$	17.35	10.42
V_2O_5		4.37	none
	V_2O_5Li	9.90	5.53
	$V_2O_5.1.5H_2O$ (Xerogel)	11.55	—
	$V_2O_5(py)0.5.H_2O$ (Xerogel)	13.0	1.45
	$V_2O_5(C_6H_4NH)_{0.44}.0.5H_2O$	13.94	2.39

[a] $Cp' = \eta\text{-}C_5H_4CH_3$; $Cp'' = \eta\text{-}C_5H_4CH_2CH_2NH_2$

V_2O_5 gels prepared by polycondensation of decavanadic acid have an average molecular weight of *ca* 10^6 and approximate composition $V_2O_5.1.6H_2O$. X-ray diffraction studies of orientated films indicate that the structure is layered with an interlayer spacing of 11.55 Å. Intercalation compounds of these gels with ammonium, cobaltocenium and ferrocenium cations have been reported [153]. Complete removal of water at *ca* 300 °C from vanadium pentoxide gels leads to the crystallisation of orthorhombic V_2O_5.

4.4.5.3 *The Layered* A_xMO_2 *Oxides*

The A_xMO_2 consist of a sheet of guest alkali metal ions (A) sandwiched between MO_2 layers. All the structures can be described by layers of edge-sharing (MO_6) octahedra with alkali ions inserted between these layers. The minimum value of x required to form a stable phase is 0.5. Therefore the pure MO_2 phases cannot be prepared because some alkali metal content is necessary to stabilise the layer structure; staging is not found because the interlayers cannot be completely emptied of all cations.

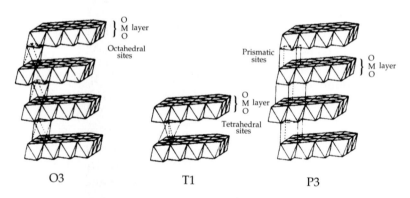

Figure 21 Various oxygen packing types in A_xMO_2 layer oxides (from Delmas *et al.* [154])

The main structure types are denoted by O3, T1 and P3 and result from differing oxygen packing. The letters O, T and P refer to the alkali ion environment of either octrahedral, tetrahedral or prismatic coordination respectively, and the number indicates the number of $(MO_2)_n$ sheets within the unit cell. Between the sheets the bonds are relatively weak and allow ionic transport and facile layer gliding. Consequently these materials undergo a range of classical ion exchange, deintercalation, and re-intercalation reactions as shown schematically in Figure 22.

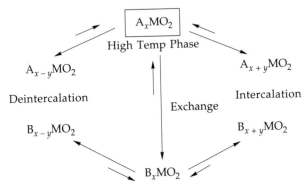

Figure 22 Low temperature reactions involving A_xMO_2 sheet oxides

4.4.5.4 Physical Properties

Molybdenum trioxide is insulating and white when fully oxidised. Reductive intercalation produces a large increase in the electronic conductivity and a dramatic change in colour to metallic blue-black even on small amounts of reduction. All the molybdenum trioxide hydrogen bronzes show metallic conductivity and weak temperature dependent paramagnetism.

V_2O_5 xerogels can be deposited as thin films that exhibit quite high electrical conductivity. For example, the polyaniline intercalation compound $(V_2O_5(C_6H_5NH)_{0.44}\cdot0.5H_2O)$ has a conductivity as a free standing film of *ca* 0.5 Ω^{-1} cm^{-1} [155]. The temperature dependence of the conductivity is characteristic of thermally activated semiconducting behaviour, similar to that observed for protonated polyaniline polymer.

4.4.6   The Metal Phosphates, Hydrogen Phosphates and Phosphonates

4.4.6.1 The Host Structures

The group 4 elements (except carbon) and group 14 elements form a series of isostructural hydrogen phosphates $\{M^{IV}(HPO_4)_2\}$ [156]. We discuss the structure and intercalation chemistry of zirconium hydrogen phosphate $\{\alpha\text{-}Zr(HPO_4)_2\cdot H_2O\}$ as an illustrative example of this class of lamellar host lattice. The structure of $\alpha\text{-}Zr(HPO_4)_2\cdot H_2O$ has been elucidated by Clearfield *et al.* [157]. It crystallises in the monoclinic system with each layer consisting of zirconium atoms lying in a plane and linked by bridging phosphate groups. Three oxygens of each tetrahedral phosphate are linked to three zirconium atoms so that each zirconium is octahedrally coordinated with six oxygens from six different phosphate groups as shown in Figure 23.

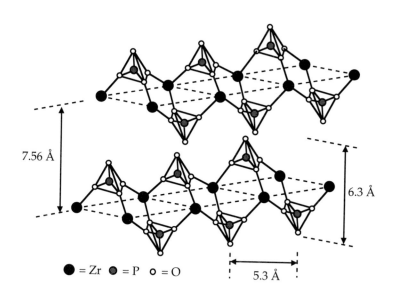

7.56 Å

6.3 Å

● = Zr ◉ = P ○ = O

5.3 Å

Figure 23 Structure of α-Zr(HPO$_4$)$_2$.H$_2$O (from Alberti and Costantino [159])

Another series of layered complex metal oxides have general formula MOXO$_4$ (M = V, Nb, Ta and Mo; X = P, As and S). The α form of VOPO$_4$ is a member of this series; it has the richest intercalation chemistry of any member of this series, and so has been chosen as an illustrative example. The tetragonal layer structure of these compounds is made up of distorted VO$_6$ octahedra and PO$_4$ tetrahedra linked by corner-sharing oxygen atoms. The oxygen atoms in each VO$_6$ unit form a near regular octahedron, with the V atoms displaced along the c-axis [158]. The vanadium coordination is similar to that observed for V$_2$O$_5$. However, the interlayer interactions in α-VOPO$_4$ are much weaker due to the fact that the number of V=O···V interactions is reduced by the presence of PO$_4$ groups.

The related uranium mica compounds with general formula M(UO$_2$PO$_4$)$_2$.nH$_2$O, where M is a monovalent or divalent cation, possess

212   *Dermot O'Hare*

layer structures. They contain negatively charged layers of $(UO_2PO_4)_n^{n-}$ separated by staggered layers of water molecules and compensating cations [160]. The proton exchanged form $HUO_2PO_4.4H_2O$ crystallises in the tetragonal system with an interlaminar spacing of 8.8 Å and has proved to be the most common starting material for preparing intercalation complexes of this lattice type.

4.4.6.2 Intercalates of Zirconium Hydrogen Phosphate

The majority of investigations on the intercalation chemistry of layered acid salts have been performed with α-zirconium hydrogen phosphate. This phase exhibits a rich and varied intercalation chemistry [161]. Its reactivity arises from the high acidity in the protonic form and the weak forces between the layers.

N-alkylamines are readily intercalated into α-$Zr(HPO_4)_2.H_2O$ from aqueous solution, the gas phase or from organic solvents. A maximum of 2 moles of amine per formula weight is intercalated forming a bilayer of guest molecules in the van der Waals' gap [162]. The experimental increase of 2.21 Å in interlamellar separation for each additional carbon atom added to the alkyl chain of the n-alkylmonoamine suggests that the molecules are not perpendicular to the layers but are inclined to the layers by ca 56°, as illustrated schematically in Figure 24. Since alkylamines are relatively strong bases, it is probable that these molecules are present as alkylammonium ions.

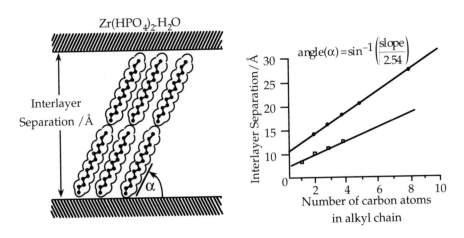

Figure 24   Schematic diagram of the structure of linear chain amines and alcohols in $Zr(HPO_4)_2.H_2O$ (after Costantino [163])

α-Zr(HPO$_4$)$_2$.H$_2$O does not directly intercalate alcohols and the crystals do not swell, either in water or alcohol. However, when the half-sodium exchanged phase Zr(NaPO$_4$)(HPO$_4$).5H$_2$O, which has an interlayer spacing of 11.8 Å, is treated with methanol in acid solution the methanol is intercalated and the Na$^+$ cations are exchanged with protons giving Zr(HPO$_4$)$_2$.MeOH. The methanol in this phase is easily displaced by other alcohols, polar solvents and other potential guests and is a very useful starting material for exploratory synthesis.

Other organic guests which have been intercalated by both direct and indirect methods are urea, amides, biopharmaceuticals such as aminoacids, and heterocyclic bases such as imidazole [163, 164]. Organometallic guests such as cobaltocene [165], and [FcCH$_2$CH$_2$NH$_2$] {Fc = Fe(Cp)(η-C$_5$H$_4$)} have also been intercalated [166].

Since α-Zr(HPO$_4$)$_2$.H$_2$O is optically transparent into the ultraviolet, several studies have been performed on the photochemical reactivity of inorganic complexes intercalated into this host lattice [167]. For example, photolysis of the dimeric manganese tricarbonyl complex [Mn(CO)$_3$(η-C$_5$H$_3$(CH$_3$)CH$_2$–]$_2$NH intercalated in α-Zr(HPO$_4$)$_2$.H$_2$O leads to a steady decrease of the intensity of ν$_{CO}$ bands of the guest and the growth of absorptions assigned to a phosphinol-bound manganese dicarbonyl complex [168].

4.4.6.3 Intercalates of Vanadyl Phosphate

Vanadium phosphate shows a wide range of intercalation reactions which parallel, in many respects, the chemistry of the covalent oxide systems discussed in Section 4.4.5. VOPO$_4$ readily forms reversible hydrated phases; the dihydrate is an intercalate where one water molecule is coordinated to the vanadium and the second located in the interlayer space. Similar reactions have been described for larger donor ligands such as pyridine and bipyridines [169].

Reaction of either VOPO$_4$ or VOPO$_4$.2H$_2$O with pyridine leads to the formation of VOPO$_4$.py in which all the vanadium sites in the metal oxide layer are coordinated by pyridine. Unlike other layer host systems, no reduction of the metal to V^{4+} or protonation of the guest to pyH$^+$ is detected. When 4,4'-bipyridine is used a cross-linked or 'pillared' structure is obtained [169]. These coordination intercalation compounds are generally similar to the corresponding pyridine and bipyridine derivatives of MoO$_3$ [150].

VOPO$_4$.2H$_2$O readily undergoes redox intercalation reactions with group 1A and 2A metal cations in the presence of a reducing agent. The oxidising power of V^{5+} is such that only mild reducing agents such as the metal iodides are required [170]. The iodide is oxidised to iodine, V(V) is reduced to V(IV) and the metal cations are inserted between the layers. An important difference between $VOPO_4.2H_2O$ and other oxide lattices is that redox intercalation of the type discussed above gives materials with low electrical conductivity.

Table 14 Examples of organic and organometallic intercalation compounds formed by the layered metal hydrogen phosphates and phosphates

Host	Guest	Stoichiometry Host$\{G\}_x$	Interlayer Spacing, c/Å	Lattice Expansion, Δc/Å
α-Zr(HPO$_4$)$_2$.H$_2$O			7.54	none
	MeOH	1.0	9.3	1.76
	Urea	—	9.4	1.86
	DMF	—	11.2	3.66
	Decylamine	—	32.0	24.46
	Octan-1-ol	—	26.7	19.16
α-Sn(HPO$_4$)$_2$.H$_2$O			7.8	none
	Pyridine	0.41	11.48	3.68
	Butylamine	2.0	19.5	11.7
	Heptylamine	2.0	26.6	18.8
	Aniline	2.0	19.1	11.3
	DMDMFc	0.81	20.7	12.9
α-VOPO$_4$			4.11	none
	H$_2$O	2.0	7.41	3.30
	Ethanol	2.0	13.17	9.06
	Pyridine	1.0	9.59	5.48
	DMDMFc	0.11	8.5	4.39
NbOPO$_4$.3H$_2$O			8.04	none
	Butylamine	1.5	17.1	9.06
	Aniline	—	26.0	17.96
HUO$_2$PO$_4$.4H$_2$O			8.69	none
	Pyridine	1.0	9.3	0.61
	Piperidine	0.62	12.56	3.87
	DMDMFc	0.8	18.8	10.11

[a]DMDMFc = Dimethylaminomethylferrocene

4.4.6.4 Intercalates of Hydrogen Uranyl Phosphate

Hydrogen uranyl phosphate ($HUO_2PO_4.4H_2O$) readily undergoes intercalative ion-exchange reactions with alkali metal (M^+) and alkaline earth (M^{2+}) cations. Typically, the protons residing between the $(UO_2PO_4)_n{}^{n-}$ sheets can be exchanged by addition of the host lattice to an aqueous solution of the guest cation [22]. Divalent transition metal cations can also be intercalated by the pre-expanded lattice $HUO_2PO_4.4H_2O$(butylamine) [171]. The concept of using the butylamine intercalate precursor to facilitate exchange in layer hydrogen phosphates was first demonstrated by Clearfield and Tindwa [162], although Weiss *et al.* were the first to prepare *n*-alkylammonium derivatives of hydrogen uranyl phosphate [172]. Many other neutral guest molecules such as amides, pyridines, and (dimethylaminomethyl)ferrocene have been intercalated into $HUO_2PO_4.4H_2O$ by direct reaction in methanol or aqueous solutions [147].

4.4.6 Layered Metal Phosphonates and their Intercalates

An important advance in the chemistry of metal phosphates was made when Alberti *et al.* [173] showed that layered hosts could be prepared using phosphonic acids (as well as organic phosphates). Initially the phenyl derivative $Zr(O_3PPh)_2$ was prepared as shown in Figure 25, but subsequently a wide range of derivatives have been prepared of general class $Zr(O_3PR)_2$ [161].

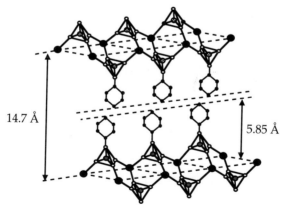

Figure 25 Schematic representation of $Zr(O_3PPh)_2$ (from Alberti *et al.* [173])

These materials have rapidly been exploited as sorbents, catalysts and ion exchangers. Their properties are related to the easy access of molecules to the large internal surface area of the layer structure. These materials contain voids and coordination sites which are well-defined both chemically and spacially. These properties have been dramatically exploited to give shape-selective host lattices. The vanadyl phosphonates $\{VO(C_nH_{2n+1}PO_3).H_2O\}$ can distinguish between groups of alcohol molecules. This selectivity can be controlled by the steric constraint around the absorption site, which is in turn determined by the nature of the organic group bound to phosphorus. For example, the hexylphosphonate hydrate, $VO(C_6H_{13}PO_3).H_2O$ intercalates the primary alcohols n-butanol and isobutyl alcohol but does not react with sec-butanol or tert-butyl alcohol [165].

4.4.7 Graphite

4.4.7.1 Host Structure

Graphite is the thermodynamically most stable form of carbon at room temperature and has one of the simplest of layered structures as shown in Figure 26. The carbon-carbon distance in the planar hexagonal sheets is 1.42 Å, compared with the d(C–C) = 1.35 Å found in benzene. Each layer is separated by 3.35 Å, which is approximately double the van der Waals' radius of carbon. Successive layers are displaced so that the centroids of the hexagons of one layer lie above a carbon in the next layer.

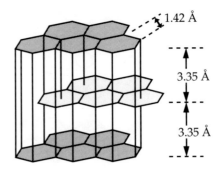

1.42 Å

3.35 Å

3.35 Å

Figure 26 Layer structure of graphite

The overlap of the valence and conduction energy levels results in graphite being described as a semimetal rather than a semiconductor

[174]. It also explains one of the unique features of graphite chemistry in that it can intercalate both electron acceptors and electron donors.

4.4.7.2 Graphite Intercalation Compounds

Graphite sulphate is believed to have been the first intercalation compound reported in the scientific literature. It was made by Schafhäutl in 1840 during his attempts to find a solvent for graphite [3]. He reported that on attempting to dissolve graphite in boiling concentrated sulphuric acid the graphite flakes appeared to swell up. Fredenhagen and Cadenbach in 1926 [175] were the first to report the intercalation of alkali metals in graphite by exposure to the metal vapour at 400 °C. Since then graphite intercalation chemistry has been extensively investigated and numerous review articles have appeared through the years [176–179].

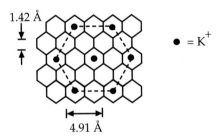

Figure 27 Schematic diagram of the packing of the alkali ions in the basal planes of C_8K (from Krebs [180])

The alkali metal intercalates show classical staging phenomena. The final first stage phase for potassium intercalated into graphite has an elemental composition C_8K; the packing of the atoms in the graphite layers is illustrated in Figure 27. The heavier alkali metals form similar compounds, but little is known about the sodium complexes. The lithium intercalates have been found to have composition $C_{6n}Li$ ($n = 1, 2$ and 3). The alkali metal ions are highly mobile within the layers at ambient temperatures with only short range order so that the intercalate can be thought of as a 2-dimensional liquid within a solid matrix [181]. On lowering the temperature ordering of the ions occurs at -175, -114 and -110 °C for K, Rb and Cs respectively .

Just as with other alkali metal intercalates, those of graphite also readily take up polar molecules such as ammonia, tetrahydrofuran, dimethoxyethane, and heterocyclic molecules such as hexamethyl-phosphoramide. In fact the potassium ion in $C_{36}K$ can react with the 3-dimensional 222-crypt (222-crypt = 4,7,13,16,21,24-hexaoxa-1,10-diazabicyclo[8.8.8]hexacosane) to yield a first stage intercalation compound with an interlayer spacing of 15.5 Å in which the K^+ ions are probably within the macroheterobicyclic cage [182]. Multicharged ions are generally only intercalated as a solvated complex, which presumably reduces the electrostatic interactions between guest and host.

Graphite also forms compounds with electron acceptors such as Br_2, AsF_5, HNO_3, and XeF_4. The host loses its electrons from the valence band and becomes a p-type electronic conductor. With bromine the limiting composition C_8Br corresponds to a second stage compound with bromine atoms arranged in chains [183]. The mineral acids react with graphite giving compounds such as $C_{24}{}^+X^-.2HX$ (X = $ClO_4{}^-$, $HSO_4{}^-$) and $C_{24}{}^+NO_3{}^-.3HNO_3$.

4.4.7.3 Electronic Properties of Graphite Intercalates

The increase in the electrical conductivity by an order of magnitude in the *ab*-plane and approximately twenty orders of magnitude along the *c*-axis for 1st stage intercalation compounds containing the alkali metals is consistent with electron donation from the alkali metal to the graphite lattice. Hall measurements on $C_{36}K$ showed that R_H is negative in sign, indicating that electrons are the carriers. Dramatic colour changes are observed, from brass-like for the first stage through copper red, steel blue, blue-green, bluish-black to black, which may be associated with the change in electron concentration in the conduction band of the host.

The sizeable diamagnetic susceptibility of the pristine graphite is lost on reductive intercalation and a small Pauli paramagnetism results [184]. Raman studies of the carbon-layer vibrations and ^{133}Cs NMR Knight shifts suggest there is *ca* 55% ionisation of the alkali metal in C_8Cs and 100% in $C_{24}Cs$. In the partially intercalated phases the electrons transferred to the host are largely confined to the graphite layers adjacent to the metal atom layers.

4.4.8   Layered Clay Minerals

4.4.8.1 Host Structures

The use of the geological term *clay* refers to any naturally occurring material with a particle size less than 2 microns. Clay minerals are typically layered oxides or silicates and are the products of the weathering of primary oxides [185]. Many members of this group are able to undergo intercalation reactions and they can be divided into two subgroups, the kandite minerals and the smectites and vermicalites.

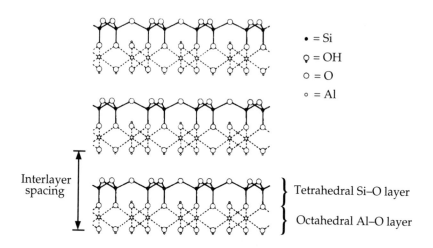

• = Si

◌ = OH

o = O

◦ = Al

Interlayer
spacing

Tetrahedral Si–O layer

Octahedral Al–O layer

Figure 28 Schematic of the structure of kaolinite

The kandite minerals, e.g. kaolinite, nacrite, dickite, and halloysite, have a layered structure consisting of single tetrahedral Si/O layers which are connected to an edge-shared octahedral M/O, OH (M = Al, Mg) sheet (Figure 28). The sheets in kanadite minerals are electrically neutral and are linked head to tail via H-bonding .

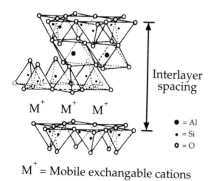

Interlayer
spacing

M^+ M^+ M^+

• = Al

• = Si

o = O

M^+ = Mobile exchangable cations

Figure 29 Schematic structure of the smectic aluminosilicate mineral, Argile

The smectite clays e.g. montmorillonite, hectorite, beidellite and the mica minerals e.g. talc and pyrophillite possess a structure which is exemplified in Figure 29. The host lattice is formally composed of three sublayers; two tetrahedral Si/O and one octahedral M/O, OH (M = Al, Mg) central layer. These layers bear an excess negative charge and compensation of these charges is achieved by the presence of interlayer cations.

4.4.8.2 Intercalation Compounds of Layered Clay Minerals

Kaolinite is one of the few clay minerals which is able to intercalate neutral guest species. However, the guest molecules must be able to form strong hydrogen bonds with the host e.g urea, formamide, acetamide, and hydrazine. Diffraction studies have revealed significant guest/host as well as guest/guest interactions. Alkali metal or ammonium salts of the short chain fatty acids can also be intercalated. The large monovalent cations are favoured because of their polarisability and low solvation energy. If a potential guest species does not intercalate directly then ion-exchange routes can be employed. Most guests can be easily removed from kaolinite by washing with water since this leads to formation of an unstable kaolinite hydrate.

Table 15 Examples of cation-exchanged clay minerals

Host	Intercalate	Approx. Formula	Interlayer Spacing/Å
Montmorillonite[a]			
	Na^+-Mont hydride	$Na_{0.5}M(H_2O)_{3.5}$	12.40
	Na^+-Mont-THF	$Na_{0.5}M(C_4H_8O)_{1.3}$	14.99
	Cu^{2+}-Mont-azobenzene	$Cu_{0.25}M(C_{12}H_{10}N_2)_{0.8}$	20.50
	Sr^{2+}-Mont-γ-butyrolactone	$Sr_{0.25}M(C_4H_6O_2)_{4.8}$	23.10

[a] Montmorillonite = $Ca_{0.35}[Mg_{0.70}Al_{3.3}](Si_{8.0})O_{20}(OH)_4$

Mica type silicates may reversibly intercalate a large variety of polar solvents which seem to interact with the cations in the van der Waals' layer. The observed interlayer separation between tetrahedral sheets is dependent on the size and charge of the interstitial cation, the negative charge density of the layers and the distribution of the charge in the layers. A large range of intercalates can be obtained by ion-exchange with main group cations, transition metal cations and organic cations especially the tetraalkylammonium ions. The cationic hydrogenation catalyst precursor $[Rh(PPh_3)_3]^+$ can be intercalated in the smectite clays montmorillonite and hectorite [186]. This material is a hydrogenation catalyst which reduces terminal olefins without isomerisation whereas with the analogous homogeneous system isomerisation to an internal olefin occurs.

4.4.9 Other Layered Host Lattices

4.4.9.1 *α-RuCl₃*

Among the layered metal halide lattices only α-$RuCl_3$ has been shown to undergo a large variety of intercalation reactions with alkali, alkaline earth and transition metal cations (Co^{2+}, Ni^{2+}) [187]. For example, the layered hexagonal semiconductor α-$RuCl_3$ can be easily converted by cathodic reduction in aqueous alkali metal electrolyte solution to a hydrated black hexagonal phase e.g. $Na_{0.5}(H_2O)_y RuCl_3$. Further cathodic reduction gives the yellow $Na(H_2O)RuCl_3$. The cation intercalates $Cu_x RuCl_3$ and $Ag_x RuCl_3$ have also been prepared electrochemically. The maximum stoichiometry is $x = 1$ and 0.4 for Cu and Ag respectively. The phase $CuRuCl_3$ corresponds to reduction of all the Ru^{3+} ions to Ru^{2+} [188].

4.4.9.2 *Ni(CN)₂*

The layered transition metal cyanide, $Ni(CN)_2$ has been reported to intercalate H_2O and other polar Lewis bases [189]. These intercalates have a general stoichiometry $(G)_{1-y}[Ni(CN)_2]$ ($0 < y < 0.2$) where y represents a number of intercalated guest molecules which are not coordinated to the nickel ions in the host layers [190].

4.4.9.3 *β-ZrNCl*

β-ZrNCl has a layer structure consisting of zirconium nitride double layers sandwiched between two close-packed chloride ion layers. These ZrNCl (Cl–Zr–N–N–Zr–Cl) layers are then stacked on each other giving the overall structure [191]. Lithium can be intercalated in β-ZrNCl by treatment with LiBu^n in donor solvents such as THF, propylene carbonate or DMF [192]. Reaction of β-ZrNCl with Li naphthalide in THF leads to co-intercalation of THF and Li^+ giving two phases which have differing

THF arrangements within the layer structure [193]. The β-ZrNCl structure contains two types of Van der Waals gaps, between Cl···Cl layers and in between N···N layers. The one-dimensional electron density synthesis of $Li_{0.18}(THF)_{0.28}ZrNCl$ clearly shows that the THF molecules are intercalated between the Cl···Cl layers (Figure 30).

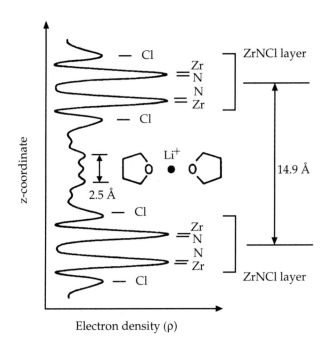

Figure 30 One-dimensional electron density map along the direction normal to the ZrNCl layers for $Li_{0.18}(THF)_{0.28}ZrNCl$ (from Ohashi *et al*. 193])

4.5 CHAIN STRUCTURE HOST LATTICES AND THEIR INTERCALATES

4.5.1 Introduction

Many inorganic materials are known to form chain structures which consist of one-dimensional stacks separated by van der Waals' channels. In common with layer structures they have the potential for a high degree of structural flexibility and could accommodate guest species with different size and geometry. In practice, these materials have generally low structural stability and are much more prone to lattice defects and disorder than the layer lattices. Thus relatively few studies have been carried out on the intercalation chemistry of one-dimensional host lattices.

4.5.2 Niobium Triselenide

The structure of $NbSe_3$ is built up of trigonal prismatic $NbSe_6$ units stacking along the crystallographic b-axis sharing Se_3-triangular bases, as shown in Figure 31.

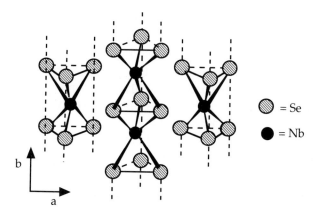

$\bigcirc$ = Se

$\bullet$ = Nb

Figure 31 Structure of $NbSe_3$ (from Schöllhorn [194])

Up to three lithium ions per $NbSe_3$ can be intercalated into this structure. The structure is preserved for lithium concentrations ≤ 1 [195]. At higher concentrations of lithium the structure seems to change drastically with many X-ray lines disappearing and the remaining lines broadening. However, the powder X-ray powder pattern of final phase Li_3NbSe_3 has been indexed on a *ca* 30% larger unit cell than the host, and a structure has been proposed for localisation of lithium ions around the $NbSe_3$ chains [196].

4.5.3 Other One-Dimensional Systems

The ternary molybdenum chalcogenides AMo_3X_3 (A = alkali metal, In, Tl; X = S, Se) are highly anisotropic and can be considered as resulting from the condensation and stacking of the Chevrel clusters Mo_6X_8 along the c-axis. This forms infinite $(Mo_3X_3)_\infty$ chains separated by parallel stacks of A atoms in a hexagonal array [197]. The intercalated A atoms help to stabilise this structure by giving their s-electrons to the conduction band formed by Mo d-orbital overlap. The poor interchain coupling leads to physical properties with a pronounced quasi-one-dimensional character. The alkali metal derivatives are all semiconducting and show classical Peierls instability. Ion exchange of the electropositive alkali metals for In and Tl gives greater interchain coupling, suppresses Peierls type instabilities, and yields superconducting phases $InMo_3Se_3$ (T_c = 3 K) and $TlMo_3Se_3$ (T_c = 6.5 K) [198].

The one-dimensional inorganic polymer polythiazyl $(SN)_x$ is a metallic conductor with a superconducting transition temperature T_c *ca* 0.26 K. Polythiazyl intercalates bromine to give $[(SNBr)_y]Br_x$ ($0.4 \leq x \leq 1.5$) which have room temperature electrical conductivities *ca.* tenfold larger along the chain direction but show almost no increase in the superconducting temperature. The $(SN)_x$ chains remain intact on intercalation and the increased conductivity of the intercalate is believed to arise from electron transfer from the chain matrix and the formation of polyhalide anions $[(SN)_x]^{n+}Br_z^{n-}$ [199].

Ruthenium tribromide has an orthorhombic structure of the TiI_3 type, consisting of $(RuBr_3)_\infty$ chains of face-sharing $RuBr_6$ octahedra. It is believed to be the only known one-dimensional metal halide to undergo reversible topochemical intercalation reactions [187]. For example, electrochemical or chemical reduction in electrolytes containing Ag(I) or Cu(II) cations leads to an increase of the interchain separation with a limiting stoichiometry $M_{0.5}RuBr_3$ (M = Ag, Cu).

4.6   FRAMEWORK HOSTS AND THEIR INTERCALATES

4.6.1   Introduction

Framework host structures that have been reported to undergo intercalation reactions contain a network of interconnecting vacant channels. These lattice channels put severe restrictions on the size of any potential guest in contrast to the intercalation reactions of the layered hosts.

4.6.2   Zeolites

4.6.2.1   Host Structures

A zeolite is typically a crystalline aluminosilicate with the formal composition $M_{2/n}O.Al_2O_3.xSiO_2.xH_2O$ *(n* = valency of the mobile cation,

M^{n+}; $x \geq 2$). There are 34 naturally occurring zeolites and nearly 100 synthetic types have been prepared under hydrothermal conditions from aluminosilicate gels. They are structurally characterised by a rigid open network matrix of SiO_4 and AlO_4 tetrahedra as shown in Figure 32. In the sodalite, faujasite and A, X and Y zeolites the silica and alumina tetrahedra are linked together to form cubeoctahedra building blocks, which are linked together in a variety of ways to give the individual zeolites. The ZSM-5 and mordenite structures contain different building blocks which are linked together to give a unique pore structure that consists of intersecting channel system.

(a) (b) (c)

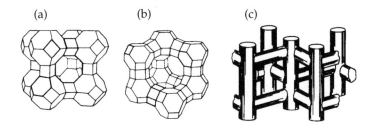

Figure 32 Schematic structures of (a) zeolite A, (b) faujasite and (c) channel network of ZSM-5

4.6.2.2 Intercalation Chemistry

The water present in all genuine zeolites can be reversibly removed with only small changes in the lattice constants. Other molecular inorganic and organic guests can be either directly intercalated into these voids or by solvent exchange reactions. These intercalation reactions are selective and depend upon the critical cross section of the guest ion in the channels. This critical cross section is variable between *ca* 3 Å and *ca* 10 Å which is the origin of the selectivity or 'molecular sieve' nature of these materials.

Intercalation of guest ions can be achieved by ion-exchange methods. Apart from metal ions, molecular cations such as NH_4^+ can be intercalated provided they are below a critical maximum cross section for the particular zeolite host.

Many zeolites are now of major industrial importance especially the tubular-pore zeolite ZSM-5 and it homologues. In 1960 Weisz and Frilette reported the first shape selective catalytic reaction involving zeolites [200]. Since then many application have been found in the petroleum and chemical industry including catalytic cracking, olefin and paraffin isomerisation and aromatic alkylation [201].

4.6.3   Sodium β-Alumina

4.6.3.1 Host Structure

Sodium β-alumina is a nonstoichiometric phase belonging to the Na_2O/Al_2O_3 phase diagram. It can be described by the nominal formula $Na^+[Al_{11}O_{17}]^-$. It has a layer structure comprising close packed blocks of oxygen and aluminium, four layers thick with the γ-Al_2O_3 spinel structure, held together by Al–O–Al bridging links. The sodium ions are located in these bridging layers with one Na^+ cation to every O^{2-} ion per layer in the unit cell as shown in Figure 33.

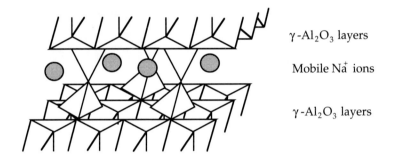

γ-Al_2O_3 layers

Mobile Na^+ ions

γ-Al_2O_3 layers

Figure 33 Schematic structure of β-alumina ($NaAl_{11}O_{17}$) (from Schroder [202])

The packing in this layer is very loose, and there is sufficient room to accommodate up to three more ions in the unit cell. These additional unoccupied lattice sites in the Na^+ plane between the Al/O columns give the alkali ions a high mobility. This and the high thermal stability are the reasons for the technological importance of this material as a solid electrolyte.

4.6.3.2 Intercalation Chemistry

Unlike most other layered structures, the lattice cannot readily change dimensions to accommodate different size cations due to the rigid Al–O–Al cross links. Therefore ion-exchange reactions do not normally permit a change in the total mobile ion composition, in contrast to the situation with the electronically conducting intercalation compounds.

The normally resident sodium ions can be exchanged for almost any other singly charged cation, including H^+, Ga^+ and NO^+ but the upper critical radius for ion-exchange appears to be *ca* 2.8Å.

4.6.4 Tungsten Trioxide

Hexagonal WO_3 is another example of a host lattice with sufficiently large isolated channels to undergo chemical intercalation reactions, as shown in Figure 34. It can be obtained either by dehydration of $WO_3(H_2O)_{0.33}$ or by chemical oxidation of $(NH_4)_xWO_3$.

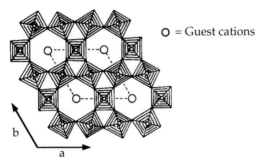

O = Guest cations

Figure 34 Structure of hexagonal WO_3 viewed perpendicular to the channel axes

The 'tungsten bronzes' H_xWO_3 can be prepared by chemical reduction of WO_3 with Zn/HCl or by electrochemical reduction in 0.1 M H_2SO_4. The term 'bronze' arose from their intense colour and metallic lustre, high electronic conductivity and variable hydrogen content. The composition range found for H_xWO_3 corresponds to $0 \leq x \leq 0.6$. The proton mobility of the hydrogen-bronze is high, and has been shown recently to be strongly one-dimensional. Chemical intercalation of the alkali metal cations can be acheived using either $LiBu^n$, NaNp or KNp ($Np^- =$ naphthalide) giving $Li_{0.33}WO_3$, $Na_{0.23}WO_3$ and $K_{0.3}WO_3$ respectively.

4.7 CONCLUSIONS

In a single chapter it has proved impossible to comprehensively review the intercalation chemistry of all the inorganic host lattices and guests that exhibit this reaction. I have tried to highlight the historical perspective of the development of the subject and to illustrate how the subject has evolved with modern developments in solid state chemistry and spectroscopy. I have selected systems for inclusion in the text which I believe have been well studied and illustrate the general concepts underlying our current understanding of this remarkable and wide ranging reaction.

Intercalation chemistry is now well established as an important discipline interconnecting with solid state chemistry, material science, analytical, organic, inorganic and physical chemistry. The future for basic exploratory synthesis in this area is still as unexpected and challenging as it was when Schafhäutl attempted to dissolve graphite in concentrated sulphuric acid in 1840. Today the mechanistic details of the chemistry and the characterisation of intercalation compounds remains one of the challenges for our battery of modern spectroscopic techniques. Consequently, a complete understanding of all aspects of the molecular and electronic structure exists for relatively few intercalation systems.

4.8   REFERENCES

1.   B Laufer, *Beginnings of Porcelain in China*, ARC Press (1917).
2.   A Weiss and A Weiss, *Angew Chem*, **72**, 413 (1960).
3.   C Schafhäutl, *J Prakt Chem*, **3** 129 (1840).
4.   K Fredenhagen and G Cadenbach, *Z Anorg Allg Chem*, **158**, 249 (1926).
5.   R Schöllhorn, *Angew Chem Int Ed Engl*, **19**, 983 (1980).
6.   R Csuk, B I Glanzer and A Furstner, *Adv Organomet Chem*, **28**, 85 (1990).
7.   H Selig and L B Ebert, *Adv Inorg and Radiochem*, **23**, 281 (1980).
8.   T Nishii, I Shirotani, T Inabe and Y Maruyama, *Synth Met*, **18**, 559 (1986).
9.   A J Jacobson, in: *Intercalation Reactions of Layered Compounds*, In *Solid State Chemistry: Compounds*, A K Cheetham and P Day, Eds, OUP, Oxford (1992).
10.   C Riekel, H G Reznik and R Schöllhorn, *J Solid State Chem*, **34**, 253 (1980); A Lerf and R Schöllhorn, *Inorg Chem*, **16**, 2950 (1977).
11.   M S Whittingham, *Prog Solid State Chem*, **12**, 41 (1978).
12.   R Brec, *Solid State Ionics*, **22**, 3 (1986).
13.   R Schöllhorn, R Kuhlmann and J C Besenhard, *Mat Res Bull*, **11**, 83 (1976).
14.   P Dickens, S J French, A T Hight and M F Pye, *Mat Res Bull*, **14**, 1259 (1979).
15.   D W Murphy, P A Christian, F J Disalvo and J V Waszczak, *Inorg Chem*, **18**, 2800 (1979).
16.   A J Jacobson, J W Johnson, J F Brody, J C Scanlon and J T Lewandowski, *Inorg Chem*, **24**, 1782 (1985).

17. J Rouxel and P Palvadeau, *Rev Chimie Miner*, **19**, 317 (1982).
18. A J Jacobson, J W Johnson and J T Lewandowski, *Inorg Chem*, **24**, 3727 (1985).
19. M Tournoux, R Marchand and L Brohan, *Prog Solid State Chem*, **17**, 33 (1986).
20. R Schöllhorn and H D Zagefka, *Angew Chem Int Ed Engl*, **16**, 199 (1977).
21. M Ohashi, K Uyeoka, S Yamanoka and M Hattori, *Chem Lett*, 93 (1990).
22. G L Rosenthal and A B Ellis, *J Am Chem Soc*, **87**, 3157 (1987).
23. A Weiss, K Hartl and A Weiss, *Angew Chem*, **73**, 707 (1961).
24. M Lal and A T Lowe, *J Chem Soc, Chem Commun*, 737 (1980).
25. A Clearfield, *Inorganic Ion Exchange Materials*, CRC Press (1982).
26. A Clearfield, *Comments Inorg Chem*, **10**, 89 (1990).
27. M B Dines, P M Digiacomo, K P Callahan, P C Griffith, R H Lane and R E Cooksey, *ACS Symp Series*, **192**, 223 (1982).
28. T J Pinnavaia, *NATO ASI SER, Ser B*, **172**, 233 (1987).
29. J M Adams, *J Chem Soc, Dalton Trans*, 2286 (1974).
30. M A Drezdzon, *Inorg Chem*, **27**, 4628 (1988).
31. G F Walker and D G Hawthorne, *Polym Lett*, **6**, 593 (1968).
32. A Weiss, *Chem Ber*, **91**, 487 (1958).
33. F Liebau, *Structural Chemistry of Silicates*, Springer Verlag, Berlin (1985).
34. B Schlsche and R Schöllhorn, *Rev Chim Miner*, **19**, 534 (1982).
35. P G Dickens, S Crouch-Baker and M T Weller, *Solid State Ionics*, **18**, 89 (1986).
36. J A Maguire and J J Banewicz, *Mater Res Bull*, **19**, 1573 (1984).
37. R Schöllhorn and H Meyer, *Mater Res Bull*, **9**, 1237 (1974).
38. K Suzuki, N Kojima, T Ban and I Tsujikawa, *Synth Met*, **19**, 893 (1987).
39. G Lagaly, *Solid State Ionics*, **22**, 43 (1986).
40. R Clement, *J Chem Soc, Chem Commun*, 647 (1980).
41. R Clement, *J Am Chem Soc*, **103**, 6998 (1981).
42. C Ferragina, M A Massucci, P Patrono, A L Ginestra and A A G Tomlinson, *J Chem Soc, Dalton Trans*, 265 (1986).
43. C Riekel, H G Reznik and R Schöllhorn, *J Solid State Chem*, **84**, 253 (1980).
44. T J Pinnavaia, R Raythatha, J G S Lee, L J Halloran and J F Hoffman, *J Am Chem Soc*, **101**, 6891 (1979).
45. J O Besenhard, I Kain, H F Klein and H Witty, *Mater Res Soc Symp Proc*, **20**, 221 (1983).
46. P J S Foot and N G Shaker, *Mat Res Bull*, **18**, 173 (1983).
47. J W Johnson, A J Jacobson, S M Rich and J F Brody, *J Am Chem Soc*, **103**, 5246 (1981).

48. U Costantino, *J Chem Soc, Dalton Trans*, 402 (1979).
49. M G Kanatzidis, L M Tonge, T J Marks, H O Marcy and C R Kannewurf, *J Am Chem Soc*, **109**, 3797 (1978).
50. N Bartlett, R N Biagioni, B W McQuillan, A S Robertson and A C Thompson, *J Chem Soc, Chem Commun*, 200 (1978).
51. J F Bringley and B A Averill, *J Chem Soc, Chem Commun*, 399 (1987).
52. T R Halbert, D C Johnston, L E McCandlish, A H Thompson, J C Scanlon and J A Dumesic, *Physica B*, **99**, 128 (1980).
53. K Chatakondu, M L H Green, J Qin, M E Thompson and P J Wiseman, *J Chem Soc, Chem Commun*, 223 (1988).
54. R P Clement, W B Davies, K A Ford, M L H Green and A J Jacobson, *Inorg Chem*, **17**, 2754 (1978).
55. Y Mathey, R Clement, C Sourisseau and G Lucazeau, *Inorg Chem*, **19**, 2773 (1980).
56. S M Kauzlarich, J F Ellena, P D Stupik, W M Reiff and B A Averill, *J Am Chem Soc*, **109**, 4561 (1987).
57. L F Nazar and A J Jacobson, *J Chem Soc, Chem Commun*, 570 (1986).
58. R C Morris and R V Coleman, *Phys Rev B*, **7**, 991 (1973).
59. P Bowen, W Jones, J M Thomsa, and R Schlogl, *J Chem Soc, Chem Commun*, 677 (1981).
60. J E Phillips and R H Herber, *Inorg Chem*, **25**, 3081 (1986).
61. R Setton, F Beguin, L Facchini, M F Quinton, A P Legrand, B Ruisinger and H P Boehm, *J Chem Soc, Chem Commun*, 36 (1983).
62. S Aronson, F J Salzano and D Bellafiore, *J Chem Phys*, **49**, 434 (1968).
63. M S Whittingham, *J Electrochem Soc*, **123**, 315 (1976).
64. R Steffen and R Schöllhorn, *Solid State Ionics*, **22**, 31 (1986).
65. J E Ford and J D Corbett, *Inorg, Chem*, **24**, 4120 (1985).
66. H Meyer and A Weiss, *Mater Res Bull*, **13**, 913 (1978).
67. R Schöllhorn, *NATO ASI SER, Ser B*, **172**, 149 (1987).
68. A Herold, *Intercalated Layered Materials*, Reidel, Dordrecht, p 321 (1979).
69. P Ganal, T Butz and A Lerf, *Synth Met*, **34**, 641 (1989).
70. D W Murphy, S A Sunshine and S M Zahurak, *NATO ASI Ser, Ser B*, **172**, 173 (1987).
71. Y Oka, T Yao and N Yamamoto, *Nippon Seramikkusu Kyokai Gakujutsu Ronbunshi*, **98**, 1366 (1990).
72. M B Dines, *Mater Res Bull*, **10**, 287 (1975).
73. D W Murphy, P A Christian, F J Disalvo and J V Waszczak, *Inorg Chem*, **15**, 17 (1976).
74. S J Hibble, P G Dickens and J C Evison, *J Chem Soc, Chem Commun*, 1809 (1985).
75. M G Kanatzidis and T J Marks, *J Am Chem Soc*, **26**, 783 (1987).

76. P G Bruce, F Krok, J Nowinski, V C Gibson and K Tavalloli, *J Mater Chem*, **1**, 705 (1991).

77. W B Davies, M L H Green and A J Jacobson, *J Chem Soc, Chem Commun*, 781 (1976).

78. T J Pinnavaia, M S Tzou and S D Landau, *Inorg Chem*, **17**, 1090 (1978).

79. R Schöllhorn, *Physics of Intercalation Compounds*, Springer Verlag, Berlin, p 33 (1981).

80. K Chatakondu, M L H Green, M E Thompson and K S Suslick, *J Chem Soc, Chem Commun*, 900 (1987).

81. K Chatakondu, M L H Green, D M P Mingos and S M Reynolds, *J Chem Soc, Chem Commun*, 1515 (1989).

82. B L Evans, *Physics and Chemistry of Compounds with Layered Structures*, Reidel, Dordrecht, Vol 4, p 1 (1976).

83. L S Ramsdell, *Am Mineralogist*, **32**, 64 (1947).

84. F Hulliger, *Structural Chemistry of Layer-Type Phases*, in: *Physics and Chemistry of Materials with Layer Structures*, F Levy, Reidel, Ed, Dordrecht, 1976

85. W Rudorff, *Angew Chem*, **71**, 487 (1959).

86. W Rudorff, *Chimia*, **19**, 489 (1965).

87. J Bichon, M Danot and J Riuxel, *Compt Rend*, **276C**, 1283 (1973).

88. B van Laar and D J W Ijdo, *J Solid State Chem*, **3**, 590 (1971).

89. M S Whittingham, *J Electrochem Soc*, **123**, 315 (1976).

90. J Rouxel, M Danot and J Bichon, *Bull Soc Chim Fr*, 3930 (1971).

91. M S Whittingham, *Prog Solid State Chem*, **12**, 41 (1978).

92. B G Silbernagel and M S Whittingham, *J Chem Phys*, **64**, 3670 (1976).

93. R H Friend and A D Yoffe, *Adv Phys*, **36**, 1 (1987).

94. B G Silbernagel, M B Dines, F R Gamble, L A Gebhard and M S Whittingham, *J Chem Phys*, **65**, 1906 (1976).

95. C Riekel and R Schöllhorn, *Mater Res Bull*, **11**, 369 (1976).

96. R Schöllhorn and H D Zagefka, *Angew Chem Int Ed Engl*, **16**, 199 (1977).

97. C Riekel, D Hohlwein and R Schöllhorn, *J Chem Soc, Chem Commun*, 863 (1976).

98. A Weiss and A Weiss, *Angew Chem*, **72**, 413 (1960).

99. M S Whittingham and M B Dines, *Surv Prog Chem*, **9**, 55 (1980).

100. M B Dines, *Science*, **188**, 1210 (1975).

101. R P Clement, W B Davies, K A Ford, M L H Green and A J Jacobson, *Inorg Chem*, **17**, 2754 (1978).

102. L F Nazar and A J Jacobson, *J Chem Soc, Chem Commun*, 570 (1986).

103. K Chatakondu, M L H Green, J Qin, M E Thompson and P J Wiseman, *J Chem Soc, Chem Commun*, 223 (1988).

232   *Dermot O'Hare*

104. D O'Hare, W Jaegermann, D L Williamson, F S Ohuchi and B A Parkinson, *Inorg Chem,* **27,** 1537 (1988).
105. D O'Hare, J S O Evans, C K Prout and P J Wiseman, *Angew Chem Ind Ed Engl,* **30,** 1156 (1991).
106. S J Heyes, N J Clayden, C M Dobson, M L H Green and P J Wiseman, *J Chem Soc, Chem Commun,* 1560 (1987).
107. J A Woolam and R B Somoano, *Mat Sci Eng,* **31,** 289 (1977).
108. D O'Hare, C Formstone, J Hodby, M Kermoo, E FitzGerald and P A Cox, *J Chem Soc, Chem Commun,* **11** (1990).
109. R C Morris and R V Coleman, *Phys Rev B,* **7,** 991 (1973).
110. D E Prober, R E Schwall and M R Beasly, *Phys Rev B,* **21,** 2717 (1980).
111. S Goldsztaub, *Bull Soc Fr Mineral Crystallogr,* **58,** 6 (1935).
112. T R Halbert, D C Johnston, L E McCandlish, A H Thompson, J C Scanlon and J A Dumesic, *Physica B,* **99,** 128 (1980).
113. J Rouxel and P Palvadeau, *Rev Chimie Miner,* **19,** 317 (1982).
114. P Palvadeau, L Coic, J Rouxel and J Portier, *Mater Res Bull,* **13,** 221 (1978).
115. P Hagenmuller, J Portier, B Barbe and P Bouclier, *Z Anorg Allg Chem,* **355,** 209 (1967).
116. H Stahl, *Inorg Nucl Chem Lett,* **16,** 271 (1980).
117. T R Halbert, *Intercalation Chemistry,* A J Jacobson and M S Whittingham, Eds, Academic Press, New York, p 375, Chapter 12 (1982).
118. R H Herber, *Z Phys Chim Neu Fol B,* **151,** 69 (1987).
119. M G Kanatzidis, L M Tonge, T J Marks, H O Marcy and C R Kannewurf, *J Am Chem Soc,* **109,** 3797 (1978).
120. S M Kauzlarich, B K Teo and B A Averill, *Inorg Chem,* **25,** 1209 (1986).
121. J F Bringley and B A Averill, *J Chem Soc, Chem Commun,* 399 (1987).
122. J F Bringley, J M Fabre and B A Averill, *J Am Chem Soc,* **112,** 4577 (1990).
123. R H Herber and Y Maeda, *Inorg Chem,* **20,** 1409 (1981).
124. W Klingen, R Ott and H Hahn, *Z Anorg Allg Chem,* **396,** 271 (1973).
125. G Ouvrard, R Brec and J Rouxel, *Mater Res Bull,* **20,** 1181 (1985).
126. B E Taylor, J Steger and A Wold, *J Solid State Chem,* **7,** 461 (1973).
127. A H Thompson and M S Whittingham, *Mater Res Bull,* **12,** 741 (1977).
128. P J S Foot and N G Shaker, *Mat Res Bull,* **18,** 173 (1983).
129. P L McDaniel, G Liu and J Jonas, *J Phys Chem,* **92,** 5055 (1988).
130. D A Cleary and A H Francis, *J Phys Chem,* **89,** 97 (1985).
131. V Mehrotra, E P Giannelis, R F Ziolo and P Rogalskyj, *Chem Mater,* **4,** 20, 1992.

132. R Clement, *J Chem Soc, Chem Commun,* 647 (1980).
133. R Clement, *J Am Chem Soc,* **103,** 6998 (1981).
134. P Lacroix, J P Audière and R Clement, *J Chem Soc, Chem Commun,* 536 (1989).
135. R Clement, M Doeuff and C Gledel, *J Chimie Phy,* **85,** 1053 (1988).
136. R Clement and A Michalowicz, *Rev Chim Miner,* **21,** 426 (1984).
137. R Brec, *Solid State Ionics,* **22,** 3 (1986).
138. H Mercier, Y Mathey and E Canadell, *Inorg Chem,* **26,** 963 (1987).
139. A LeMehaute, G Ourvard, R Brec and J Rouxel, *Mater Res Bull,* **12,** 1191 (1977).
140. C Berthier, Y Chabre and M Minier, *Solid State Commun,* **28,** 327 (1978).
141. L Lomas, P Lacroix, J P Audière and R Clement, *J Mater Chem,* **1,** 475 (1991).
142. R Clement, J J Girerd and I Morgenstern-Badarau, *Inorg Chem,* **19,** 2852 (1980).
143. D O'Hare, J S O Evans and R Clement, unpublished results.
144. L Kihlborg, *Ark Kemi,* **21,** 357 (1963).
145. O Glemser, G Lutz and G Meyer, *Z Anorg Allg Chem,* **285,** 173 (1956).
146. J J Birtill and P J Dickens, *Mater Res Bull,* **13,** 311 (1978).
147. R Pozas-Tormo, L Moreno-Real, M Martinez-Lara and S Bruque-Gamez, *Can J Chem,* **64,** 30 (1986); L Moreno Real, R Pozas Tormo, M Martinez Lara and S Bruque, *Mat Res Bull,* **22,** 19 (1987).
148. R C T Slade, T K Halstead and P G Dickens, *J Solid State Chem,* **34,** 183 (1980).
149. R Schöllhorn, R Kuhlmann and J C Besenhard, *Mater Res Bull,* **11,** 83 (1976).
150. J W Johnson, A J Jacobson, S M Rich and J F Brody, *J Am Chem Soc,* **103,** 5246 (1981).
151. R Schöllhorn, T Schulte-Nolle and G Steinhoff, *J Less Common Metals,* **71,** 71 (1980).
152. H G Bachmann, F R Ahmed and W H Barnes, *Z Kristallogr,* **115,** 110 (1961).
153. P Aldebert, N Baffier, J J Legendre and J Livage, *Rev Chim Miner,* **19,** 485 (1982).
154. C Delmas, J J Braconnier, A Maazaz and P Hagenmuller, *Rev Chim Miner,* **19,** 343 (1982).
155. M G Kanatzidis, C G Wu, H O Marcy and C R Kannewurf, *J Am Chem Soc,* **111,** 4139 (1989).
156. A Clearfield, *Inorganic Ion Exchange Materials,* CRC Press, Boca Raton, Fl, U S A (1982).
157. A Clearfield and G D Smith, *Inorg Chem,* **8,** 431 (1969).
158. B Jordan and C Calvo, *Can J Chem,* **51,** 2621 (1973).

159. G Alberti and U Costantino, *Intercalation Chemistry*, A J Jacobson and M S Whittingham, Eds, Academic Press, p 147, Chapter 5 (1982).
160. B Morosin, *Acta Crystallogr, Sect B*, **34**, 3733 (1978).
161. A Clearfield, *Comments Inorg Chem*, **10**, 89 (1990).
162. A Clearfield and R M Tindwa, *J Inorg Nucl Chem*, **41**, 871 (1979).
163. U Costantino, *J Chem Soc, Dalton Trans*, 402 (1979).
164. D Behrendt, K Beneke and G Lagaly, *Angew Chem Int Ed Engl*, **15**, 544 (1976).
165. J W Johnson, *J Chem Soc, Chem Commun*, 263 (1980).
166. D O' Hare, M Kermoo, C Formstone, E FitzGerald and P A Cox, *J Mater Chem*, **1**, 51 (1991).
167. G L Rosenthal and J Caruso, *J Solid State Chem*, **93**, 128 (1991).
168. C F Lee and M E Thompson, *Inorg Chem*, **30**, 4 (1991).
169. J W Johnson, A J Jacobson, J F Brody and S M Rich, *Inorg Chem*, **21**, 3820 (1982).
170. A J Jacobson, J W Johnson, J F Brody, J C Scanlon and J T Lewandowski, *Inorg Chem*, **24**, 1782 (1985).
171. R Pozas-Tormo, L Moreno-Real, M Martinez-Lara and S Bruque-Gamez, *Can J Chem*, **64**, 30 (1986).
172. V A Weiss, K Hartl and U Hofmann, *Z Naturforsch*, **B12**, 351 (1957).
173. G Alberti, U Costantino, S Alluli and N Tomassini, *J Inorg Nucl Chem*, **40**, 113 (1978).
174. D E Soule, J W McClure and L B Smith, *Phys Rev*, **134A**, 453 (1964).
175. K Fredenhagen and G Cadenbach, *Z Anorg Allg Chem*, **158**, 249 (1926).
176. R Csuk, B I Glanzer and A Furstner, *Adv Organomet Chem*, **28**, 85 (1990).
177. H Selig and L B Ebert, *Adv Inorg and Radiochem*, **23**, 281 (1980).
178. F Hulliger, *Phys Chem Mater Layered Struct*, **5**, 52 (1976).
179. G R Hennig, *Prog Inorg, Chem*, **1**, 125 (1959).
180. H Krebs *Fundamentals of Inorganic Crystal Chemistry*, McGraw-Hill, New York (1968).
181. D E Nixon and G S Parry, *J Phys D*, **1**, 291 (1968).
182. R Setton, F Beguin, L Facchini, M F Quinton, A P Legrand, B Ruisinger and H P Boehm, *J Chem Soc, Chem Commun*, 36 (1983).
183. W T Eeles and J A Turnbull, *Proc R Soc Lond, Ser A*, **283**, 179 (1965).
184. P Delhaes, J C Rouillon, J P Mareche, D Guerard and A Herold, *J Phys Lett*, **37**, L127 (1976).
185. T J Pinnavaia, *NATO ASI SER, Ser B*, **172**, 233 (1987).
186. T J Pinnavaia, R Raythatha, J G S Lee, L J Halloran and J F Hoffman, *J Am Chem Soc*, **101**, 6891 (1979).

187. R Schöllhorn, R Steffen and K Wagner, *Angew Chem Int Ed Engl*, **22**, 555 (1983).

188. R Schöllhorn, *NATO ASI SER, Ser B*, **172**, 149 (1987).

189. Y Mathey and C Mazieres, *Can J Chem*, **52**, 3637 (1974).

190. Y Mathey, R Setton and C Mazieres, *Can J Chem*, **55**, 17 (1977).

191. R Juza and H Friedrichsen, *Z Anorg Allg Chem*, **332**, 173 (1964).

192. M Ohashi, S Yamanaka, M Sumihara and M Hattori, *J Inclusion Phenomena*, **2**, 289 (1984).

193. M Ohashi, K Uyeoka, S Yamanoka and M Hattori, *Chem Lett*, 93 (1990).

194. R Schöllhorn, *Inclusion Compounds*, Academic Press, London, Vol 1, p 249, Chapter 7 (1984).

195. D W Murphy and F A Trumbore, *J Crystal Growth*, **39**, 135 (1977).

196. R R Chianelli and M B Dines, *Inorg Chem*, **14**, 2417 (1975).

197. M Potel, R Chevrel and M Sergent, *Acta Cryst B*, **36**, 1545 (1980).

198. R Brusetti, P Monceau, M Potel, P Gougeon and M Sergent, *Solid State Commun*, **66**, 181 (1988).

199. G B Street, R H Geiss, W D Gill, J Kuyper and R D Smith, *Synthesis and Properties of Low-Dimensional Materials*, The New York Acad Sciences, New York (1978).

200. P B Weisz and V J Frilette, *J Phys Chem*, **64**, 382 (1960).

201. S M Csicsery *Zeolite Chemistry and Catalysis*, ACS, Washington, D C, p 680 (1976).

202. F A Schroder and J Scherle, *Acta Cryst*, **B31**, 531 (1975).

5 Biogenic Inorganic Materials

Stephen Mann

Inorganic Materials. Edited by Duncan W Bruce and Dermot O'Hare
© 1992 John Wiley & Sons Ltd

5.1   INTRODUCTION

The natural world is full of specialised inorganic materials. Oyster shells, corals, ivory, sea urchin spines, cuttlefish bone, limpet teeth, and magnetic crystals in bacteria are just a few of the vast variety of biological minerals engineered by living creatures. Many of these biological building materials consist of inorganic minerals intricately combined with organic polymers. Together, these materials are fashioned into a fascinating variety of shapes and forms which serve many different functions. Not only are biominerals highly refined but the extent of their production is enormous. For example, geological deposits such as the white cliffs of Dover contain billions of exquisite $CaCO_3$ scales formed within the cells of marine algae. Similarly, large scale geographic features such as coral reefs are the result of immense biological activity.

Chemists are interested in how biominerals are made because the ability to produce well-defined inorganic materials is of great value in fields such as catalysis, electronics, magnetism and ceramics. For materials scientists, biomineralisation provides a unique opportunity to study solutions to key problems in mechanical design. Materials such as bones, teeth and shells are synthesised as complex composites and the organisation and interfacial chemistry of the components are optimised for functional use. Mimicking such structures would be a significant step towards so-called 'smart' materials [1, 2]. For instance, the fabrication of high strength, 'macro-defect-free' cements [3] was much inspired by the high strength and toughness of the nacre (mother-of-pearl) layer of sea shells, and a similar perspective could be applied to the recent production of cheap, tough ceramics [4]. The crucial factor in these developments was the realisation that brittle materials can become strong if cracks generated under tension are unable to propagate through the material because of an absence of pores or defects in the structure. Furthermore, weak interfaces between high compression strength components act to dissipate the energy of crack propagation. In nacre, this is accomplished by a laminar structure of aragonitic crystal blocks sandwiched between thin layers of proteinaceous material (Figure 1). For high strength in cements this interface is mimicked by the adsorption of an organic polymer between calcium silicate grains, and in ceramics by coating thin sheets of silicon carbide with graphite interlayers.

Figure 1 Cross-section of a mollusc shell showing the brick wall arrangement of oriented tabular aragonite (CaCO₃) crystals. Thin sheets of protein are sandwiched between adjacent layers of the crystals. Scale bar = 100 μm (courtesy of Prof. S. Weiner, Weizmann Institute of Science, Israel)

Clearly, there is much to learn from the mechanical properties of biogenic inorganic materials. In addition, the micro-architecture of these materials depends ultimately on the biological regulation of molecular processes involving the nucleation, crystal growth and organisation of mineral structures and it is at this level that chemists and biochemists have made a major contribution in recent years. Indeed, for many purposes the term 'biomineralisation' is synonymous with 'solid state bioinorganic chemistry' and it is towards this perspective that this chapter is focused.

We start with a discussion of the types and associated functions of biogenic minerals and attempt to highlight the profound differences between these materials and their inorganic or geological counterparts. We then review the control mechanisms involved in biomineralisation through a series of selected examples including iron storage proteins, magnetic bacteria, coccolithophorids, invertebrate shells, bones and teeth. The importance of model systems in elucidating molecular mechanisms of biomineralisation is also addressed.

5.2   BIOMINERALS; TYPES AND FUNCTIONS

Table 1 presents a list of the major bioinorganic solids formed in living organisms [5, 6]. Insoluble calcium salts such as carbonates and phosphates are widespread throughout the biological world. Many of these deposits are utilised as structural support systems or as specialised hard-parts (Figure 2).

Figure 2 Scanning electron micrograph of a fractured section of a cuttlefish bone. The cuttlebone is a modified shell that is used as a buoyancy device. It is constructed of calcium carbonate (aragonite) and the polysaccharide, β-chitin. The biomineral is deposited in two distinct but interdependant structures; viz. thin sheets and S-shaped pillars. The mineral lamellae is propped up by the crystalline pillars such that a network of chambers are formed throughout the cuttlebone. The number of chambers filled with liquid or gas dictates the density of the material and enables the cuttlefish to maintain a constant depth without the expenditure of metabolic energy. Scale bar is 100 μm (courtesy of Mr J. B. A. Walker, University of Bath, UK)

Hydroxyapatite $[Ca_{10}(PO_4)_6(OH)_2]$ is the major inorganic component of bone and teeth whilst the shells (exoskeletons) of invertebrates (e.g. molluscs) consist of calcite and aragonite. Why the garden snail uses calcium carbonate to support soft tissue but the skeletons of vertebrates consist of calcium phosphate is unclear. Presumably there are specific chemical, biological and mechanical properties that favour one mineral over another depending on the evolutionary status of the organism.

5.2.1 Calcium Carbonate

There are a range of functions, besides structural support, that are exhibited by the carbonate biominerals. For example, inside the inner ear there are hundreds of small calcite crystals which act as an inertial mass for the detection of changes in linear acceleration [7] (Figure 3).

Figure 3 Inner ear calcite crystals (otoconia). Note the characteristic rounded bodies and crystallographic end faces. Scale bar = 8 μm

These gravity devices function in a similar way as the fluid in the semi-circular canals (which detect changes in angular momentum). The crystals are sited on a membrane under which are located sensory cells. During a change in linear acceleration, the movement of the crystal mass relative to the delicate hair-like extensions of the cells results in the electrical signalling of the applied force to the brain.

Table 1 The types and functions of the main inorganic solids found in biological systems

Mineral	Formula	Organism/function
Calcium carbonate:		
Calcite	$CaCO_3^*$	Algae/exoskeletons Trilobites/eye lens
Aragonite	$CaCO_3$	Fish/gravity device Molluscs/exoskeleton
Vaterite	$CaCO_3$	Ascidians/spicules
Amorphous	$CaCO_3.nH_2O$	Plants/Ca store
Calcium phosphate:		
Hydroxyapatite	$Ca_{10}(PO_4)_6(OH)_2$	Vertebrates/endoskeletons teeth, Ca store
Octa-calcium phosphate	$Ca_8H_2(PO_4)_6$	Vertebrates/precursor phase in bone
Amorphous	?	Mussels/Ca store Vertebrates/precursor phases in bone
Calcium oxalate:		
Whewellite	$CaC_2O_4.H_2O$	Plants/Ca store
Weddellite	$CaC_2O_4.2H_2O$	Plants/Ca store
Group IIA metal sulphates:		
Gypsum	$CaSO_4$	Jellyfish larve/gravity device
Barite	$BaSO_4$	Algae/gravity device
Celestite	$SrSO_4$	Acantharia/cellular support
Silicon dioxide:		
Silica	$SiO_2.nH_2O$	Algae/exoskeletons
Iron oxides:		
Magnetite	Fe_3O_4	Bacteria/magnetotaxis Chitons/teeth
Geothite	α-FeOOH	Limpets/teeth
Lepidocrocite	γ-FeOOH	Chitons (Mollusca) teeth
Ferrihydrate	$5Fe_2O_3.9H_2O$	Animals and plants Fe storage proteins

*A range of magnesium-substituted calcites are also formed

A further spectacular role is the use of biogenic calcite in the compound eyes of the extinct trilobites. These creatures are well-preserved as fossils such that the structure and organisation of the corneal lenses have been determined. The eyes consist of hexagonally-packed calcite single crystals. Interestingly, single crystals of calcite are renown for their ability to doubly refract white light suggesting that the trilobites suffered a life of continual double vision! However, studies of well-preserved fossilised material show that each crystal is aligned in the eye such that the unique (non-refracting) *c*-axis is perpendicular to the surface of each lens [8]. In this orientation, the calcite lens behaves isotropically like glass and a single well-defined image is formed. The sophisticated design of bioinorganic solid state assemblies in this manner exemplifies the fundamental interdependence of structure, organisation and function in many biomineralisation processes.

Although most of the calcium carbonates formed in biological systems have structures of calcite or aragonite, some organisms deposit vaterite. Vaterite is the least thermodynamically stable of the three non-hydrated crystalline polymorphs and rapidly transforms to calcite or aragonite in aqueous solution. It occurs as spicules in a few marine sponges [9] (the majority of calcareous sponges have magnesium-rich calcite spines), where it possibly acts as a structural support or as a deterrent against predators. The mineral has also been observed in the inner ears of two fish species [7].

Finally, we note that amorphous calcium carbonate is deposited in the leaves of many plants where it acts as a calcium store [10]. Although this material is exceedingly unstable in inorganic systems due to rapid phase transformation in aqueous solution, the biogenic mineral appears to be stabilised through the adsorption of biological macromolecules such as polysaccharides at the solid surface.

5.2.2 Calcium Phosphate

Perhaps bone, more than any other biomineral, reflects the greatest distinction between an inorganic and bioinorganic solid. For many purposes the calcium phosphate of bone is best thought of as a 'living mineral' since it undergoes continual growth, dissolution and remodelling. The dynamic property of bone indicates that the mineral is more than just a structural support. The mineral acts as an important calcium store in homeostasis as well as a supply of calcium in times of high demand.

The structure and mechanical properties of bone are derived from the organised mineralisation of hydroxyapatite (often as carbonated apatite) within a matrix of collagen fibrils, proteoglycans and many other proteins. The initial sites of mineralisation are located within gap regions between collagen molecules [11], and the plate-like crystals are crystallographically orientated with respect to the organic matrix [12] (Figure 4). Recent observations have shown that the crystals are aligned in parallel arrays across the fibrils and this level of organisation is often coherent across adjacent fibrils suggesting that this long range order could be responsible for much of the unusual fracture properties of bone [13].

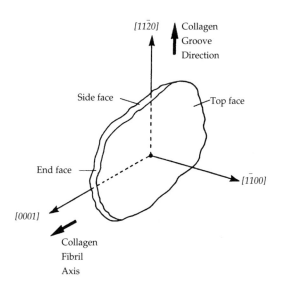

Figure 4 Schematic showing the morphology and crystallographic orientation of hydroxyapatite crystals in the collagen matrix of turkey tendon

The non-stoichiometric nature of bone mineral may be responsible for the piezo-electric response observed in this calcified tissue. Although the precise mechanisms are not yet determined, it has been known for a long time that the application of pressure stimulates the growth of bone mineral, (to put this principle into a vivid historical context soldiers of the First World War had their broken legs stimulated by a helpful matron wielding a wooden mallet!). There may be, therefore, processes inherent in biological mineralisation which are not immediately apparent from simulated studies done in the test tube. This note of caution implies that the growth of bioinorganic solids is not only intriguing but that modelling such processes in the laboratory is notoriously difficult.

The structure and organisation of tooth enamel, like bone, derives from a highly complex system designed to withstand specific types of mechanical stress. Enamel is unique in that it consists of long ribbon-like hydroxyapatite crystals that make up a very high weight percentage of the mature tissue (95 wt%; cf. bone, 65 wt% (average)). This is achieved by depletion of the organic components during tooth maturation and crystal growth.

5.2.3   Other Calcium Biominerals

Calcium oxalates and sulphates are also formed, to a limited extent, in biological systems. The monohydrate and dihydrate oxalates are formed in many plants and act as deterrents against insects, as structural supports and as important calcium stores in plants. Calcium sulphate deposits are effective stores of calcium and sulphur for metabolic use in plants. They are also present as gravity devices (statoliths) in jellyfish larvae [14].

5.2.4   Iron Oxides

Bioinorganic iron oxides are widespread and serve several different functions (Table 1). These solids have important inorganic counterparts that are extensively used in catalytic and magnetic devices. The mixed valence compound, magnetite (Fe_3O_4), is of particular biological relevance. It is synthesised in the form of discrete crystallographically oriented inclusions in a wide range of magnetotactic bacteria [15] (Figure 5). These organisms are aligned in the earth's magnetic field such that in the Northern Hemisphere they swim downwards (north-seeking) towards the oxygen-depleted zone at the sediment-water interface of freshwater and marine environments. The same species in the Southern Hemisphere have a similar arrangement of magnetite crystals but of the opposite polarity (south-seeking) such that they also swim exclusively downwards. Details of structural studies on these biogenic materials are described in Section 5.4.3.1.

Other iron oxides such as goethite (α-FeOOH) and lepidocrocite (β-FeOOH) are deposited in the teeth of certain molluscs. For example, the common limpet is armed with sabre-like rust-coloured goethite teeth [16]. During feeding, these hardened structures are rasped across rocks encrusted with algae. In certain species, the chitons, the teeth, comprise both lepidocrocite and magnetite [17] and are therefore magnetic! An important and widespread iron oxide is the hydrated mineral called ferrihydrite. This is the brown gelatinous precipitate which is readily formed in a test tube by the addition of sodium hydroxide to an Fe(III) solution, the so-called amorphous 'ferric hydroxide'. The iron storage protein, ferritin, (see Section 5.4.2.1 for details) contains a 5 nm central core of this mineral wrapped up in a protein coat. Encapsulation of the mineral in this way protects the organism from the all-familiar problem of rusting as well as providing a means of cellular protection from the deleterious effects of labile iron. Ferritin iron may also be utilised in biochemical processes such as haemoglobin synthesis. Note that, unlike the above iron oxides, ferrihydrite is a disordered material with relatively high solubility and this is commensurate with the dynamic behaviour of this iron store. More thermodynamically stable oxides such as goethite or magnetite would be chemically unsuited for such a function.

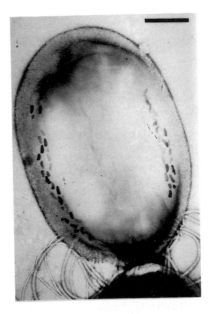

Figure 5 Transmission electron micrograph of a magnetotactic bacterial cell containing chains of bullet-shaped magnetite (Fe_3O_4) crystals. Scale bar = 0.5 μm (courtesy of Dr R. P. Blakemore, University of New Hampshire, USA)

5.2.5 Iron Sulphides

Until very recently it was not considered that organised iron sulphide biominerals were generated by living systems. Many iron sulphide minerals are known to be associated with sulphate-reducing bacteria [18] but they show none of the characteristics of genetically-controlled biomineralisation. Most of these products are adventitious and arise from the reaction of metabolic products such as H_2S with Fe(III) species in the surrounding environment. However, recent studies have shown that certain types of magnetotactic bacteria present in sulphide-rich environments synthesise and organise crystals of the ferrimagnetic mineral, greigite (Fe_3S_4) [19, 20]. The inclusions are discrete single crystals of narrow size distribution. They have species-specific morphologies and appear to be crystallographically aligned within chains (Figure 6). In one species, pyrite (FeS_2) crystals are associated with the greigite particles [19]. Clearly, these processes are under strict biological control. Moreover, as sulphide rather than oxide chemistry was likely to be dominant in the early stages of the earth's history, the formation of intracellular bacterial iron sulphides could be representative of an ancient process in which inorganic materials were adapted to specific biological functions.

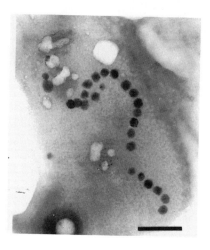

Figure 6 Transmission electron micrograph of a magnetotactic bacterial cell containing a single chain of cubo-octahedral greigite (Fe_3S_4) crystals. Scale bar = 0.4 μm (courtesy of Dr B. R. Heywood, University of Bath, UK)

5.2.6 Silica

In contrast to the other bioinorganic solids described in Table 1, biogenic silica is always deposited in the amorphous state. A wide range of unicellular organisms such as diatoms synthesise elaborate exoskeletons from this biomineral [21] (Figure 7). In plants, the presence of a large number of silica spines and nodules presents an unpalatable meal to a discerning predator. In some plants such as the horsetail the level of silica is extremely high being of the order of 20–25% of the dry weight. The dried plant is essentially akin to sandpaper and was used (so the story goes) by the early American pioneers as an effective means of cleaning teeth.

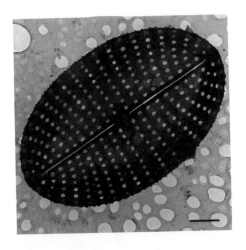

Figure 7 Transmission electron micrograph showing the silica shell of a unicellular diatom. Note the exquisite patterning of the perforated shell. Scale bar = 2 μm

The biological deposition of exclusively non-crystalline silica raises an important question; what are the functional advantages of using an amorphous material rather than a crystalline mineral such as calcium carbonate? One possibility is that because the fracture and cleavage planes inherent to crystalline structure are missing, the amorphous biomineral can be subsequently moulded, without loss of strength, into a wide variety of shapes. Biogenic silica will not be discussed in detail in this article. See references [22, 23] for reviews of chemical and biochemical studies of silicification.

5.3   CRYSTAL ENGINEERING

It is clear from the above discussion that the distinction between inorganic and bioinorganic minerals lies with the extent of crystal chemical specificity exhibited by these materials. Although biominerals have relatively simple inorganic structures compared with those frequently obtained in synthetic solid state chemistry, the level of engineering of the solid phase is often extremely high. For example, Figure 8 shows a single crystal of inorganic calcite ($CaCO_3$) formed in the laboratory. The crystal habit reflects the rhombohedral symmetry of the unit cell and the relatively low surface energy of the exposed {104} faces.

Figure 8 Single crystal of calcite grown under room temperature laboratory conditions. Scale bar = 10 μm (courtesy of Prof. S. Weiner, Weizmann Institute of Science, Israel)

In contrast, Figure 9 shows a single crystal of biogenic calcite that constitutes the individual spines of an adult sea urchin. Clearly, the spine shows none of the underlying crystallographic symmetry although the *c*-axis is aligned parallel to the morphological long axis. Furthermore, the spine is fabricated with an open porous structure with curved edges and contains specialised features such as the spherical end that serves as a rotational joint within the body of the central shell of the organism.

Recent chemical analyses of sea urchin spines and tests indicate that the crystals contain 0.02 wt% protein [24], i.e. of the order of one protein molecule per 10^5 unit cells. This is sufficient to explain the conchoidal fracture of the biomineral and associated mechanical properties. How these intracrystalline proteins are incorporated into the crystal structure whilst maintaining the high degree of long range order as shown by synchrotron X-ray diffraction studies [25] is a crucial question in understanding the biological engineering of these materials.

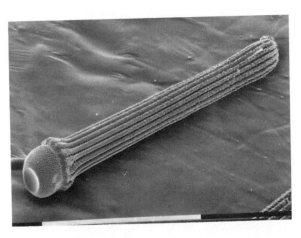

Figure 9 Single crystal of biogenic calcite. The crystal constitutes a single sea urchin spine. Scale bar = 1mm (courtesy of Prof. S. Weiner, Weizmann Institute of Science, Israel)

The interplay between the precision of crystallochemical properties such as size, structure, composition, morphology and orientation, and the evolutionary requirements of organisms as expressed through functional properties is relevant within a technological context. Since many of our new technologies rely on the fabrication of well-characterised materials of reproducible, high performance capability, understanding the underlying principles of synthesis, control and organisation of biogenic inorganic materials will aid the elucidation of principles which can be utilised in the generation of synthetic materials. The possibilities of crystal engineering are far-ranging; catalyst particles of controlled morphology and size, magnetic materials of anisotropic shape, high porous materials, shape-specific amorphous solids, complex composites and organised crystalline assemblies are all features of biomineralisation that might be replicated in the laboratory. The success of this approach depends on a detailed knowledge of the underlying mechanisms of crystal synthesis and the control of these processes at the molecular level, and it is these aspects of biomineralisation that we will now explore.

5.4 CONTROL MECHANISMS

The formation of a complex material such as a sea urchin spine involves many features of control at the site of mineralisation. The site must be delineated from the surrounding biological environments, activated at specific times in the life of the organism, constrained in size and shape, and highly regulated with respect to the chemistry of the mineralisation process. The principal processes to be controlled are nucleation and crystal growth both of which are critically dependent on the ionic composition (supersaturation) of the medium and the nature of interfaces (mineral-matrix and mineral-environment) present in the system (Figure 10).

Details of these processes are given elsewhere [26, 27]. In general, the chemical processes are regulated by transport and/or reaction-mediated mechanisms. The delineation of the mineralisation site by a surrounding organic membrane enables the supersaturation levels to be precisely regulated through facilitated ion-flux, complexation/decomplexation switches, local redox and pH modifications and changes in local ion activities via ionic strength and vectorial water fluxes. Reaction mechanisms influence the kinetics of surface-mediated processes such as cluster formation in nucleation, expression of crystal habit and phase transformations and aggregation in crystal growth.

The presence of macromolecular substrates is a key feature in many biomineralisation processes and many systems are characterised as being 'organic-matrix-mediated' [28]. The underlying concept lies with the putative rôle of an organic surface to induce nucleation through the lowering of the activation energy of cluster formation. In many cases, the matrix is assumed to have an orientational rôle such that specific molecular interactions at the nuclei-matrix interface result in crystallographic alignment of the forming crystals. This is a profound idea, supported by much circumstantial evidence, and if substantiated could give rise to important technological developments. Model systems based on the hypothesis of molecular recognition at inorganic-organic interfaces are discussed in Section 5.4.6.2.

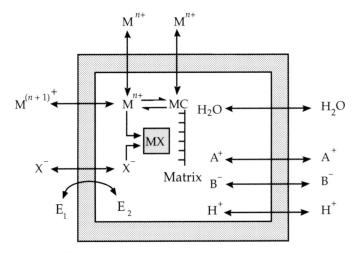

M^{n+} = metal cations; X^- = anions; MC = cation complex

$E_1 E_2$ = enzymes; MX = biomineral; A^+, B^- = extraneous ions

Figure 10 Generalised strategies for controlling supersaturation in biomineralisation. Direct mechanisms include; (i) cation-gated systems via membrane-bound pumps and molecular channels, e.g. Ca^{2+}. Facilitated transport may involve redox changes, e.g. Fe(II)/Fe(III); (ii) controlled destabilisation of ion-complexes, e.g. citrate, pyrophosphate; (iii) enzymatic regulation, e.g. removal of H_2CO_3 in calcification by carbonic anhydrase. Indirect mechanisms; (i) vectorial water movement inducing condensation reactions, e.g. silica deposition; (ii) pH changes induced by cellular metabolism; (iii) changes in ionic strength and activities through differential pumping of extraneous ions.

5.4.1 Mineralisation Sites

The isolation of the mineralisation zone from the cellular environment is a prerequisite for the controlled deposition of biogenic inorganic materials. This space can be generated in several ways [29] (Figure 11); (a) epicellularly, e.g. on or in the outer membrane wall of bacterial cells, (b) extracellularly, e.g. in the collagen matrix of bone and dentine, and non-collagenous tubules of enamel, (c) intracellularly, e.g. in membrane-bound vesicles of unicellular organisms. The vesicles and their mineral products may remain in the cell or undergo translocation to the external surface. In some systems, e.g. the tests and spines of adult sea urchins, the vesicles are produced by the fusing of individual cells to form an extensive collaborative framework (syncytium); and (d) intercellularly, e.g. in spaces generated by algal cells in coral formation.

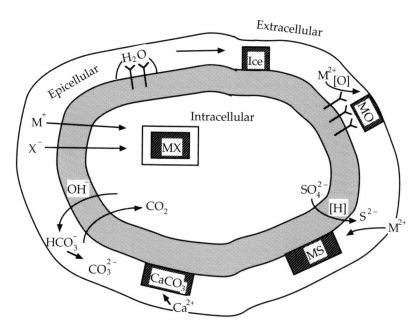

Figure 11 Possible mineralisation sites. In unicellular organisms the sites are confined to inside and on the cell wall. The extracellular space can be utilised in multicellular organisms

Although the mineralisation zone may be microscopic, and in some cases macroscopic, the size of the initial nucleation space may be exceedingly small, often on the nanometre scale. For example, calcium phosphate deposition in the collagen extracellular matrix occurs initially within the 40 x 3 nm gap regions between individual fibrils [11]. Similarly, iron oxide mineralisation in the storage protein, ferritin, is confined to a cavity 8 nm in diameter [30]. The ability to deposit biominerals in such confined spaces may have relevance to the production of inorganic clusters and nanoparticles in synthetic systems. We illustrate this potential through a discussion of the mineralisation processes involved in the formation of iron oxide cores in ferritin.

5.4.2 Solid State Chemistry in a Nanospace

5.4.2.1 *Ferritin*

The storage and mobilisation of surplus iron in eukaryotes and some prokaryotes is regulated by the iron storage protein, ferritin. Ferritin isolated from horse spleen consists of a hollow spherical shell of 24 symmetrically related protein subunits (*ca* 18–20 kDa per subunit) surrounding a core of inorganic hydrated Fe(III) oxide [30] (Figure 12).

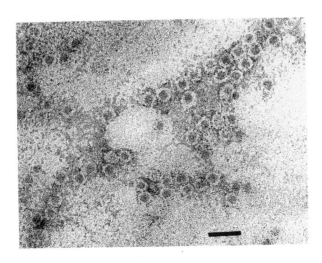

Figure 12 Transmission electron micrograph showing stained molecules of ferritin. The iron oxide core is imaged as a dark centre and the protein shell as a surrounding light halo. Scale bar = 25 nm (courtesy of Ms F. C. Meldrum, University of Bath, UK)

Most ferritins are heteropolymers of two types of subunits, designated H and L. Phosphate may be associated with the surface of the iron oxide core in horse ferritin but does not appear to be a critical factor for core formation in reconstituted ferritins [31]. The diameter of the cavity set by the protein shell is of the order of 70–80 Å resulting in an upper limit of 4500 iron atoms (*ca* 30% wt/wt Fe) which can be stored within the molecule.

The uniqueness of the ferritin structure arises from the metabolic requirement to organise and utilise dissolved iron at concentrations and pH levels which induce precipitation of potentially toxic solid phases. Not only is iron solubilised by micelle encapsulation, but homeostatic control is also maintained. The study of ferritin therefore provides an important example of the biological control of solid state reactions involved in the formation and organisation of nanometre size inorganic materials.

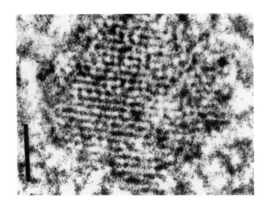

Figure 13 Lattice image of an individual iron oxide core of human spleen ferritin. The fringes correspond to the {110} lattice planes of ferrihydrite. Scale bar = 2 nm.

The structure of ferritin cores has been studied by X-ray diffraction [32], electron diffraction [33] and high resolution electron microscopy (HRTEM) [34, 35]. The results indicate that the iron-containing cores of mammalian ferritins are crystalline with a unit cell based on a four layer repeat of hexagonally close-packed oxygen atoms with variable octahedral occupancy of Fe(III) ions. The analogous inorganic mineral is ferrihydrite $(5Fe_2O_3.9H_2O)$. Recently, investigations using HRTEM have been undertaken to determine the structural similarities and differences of individual ferritin cores isolated from vertebrate (human), invertebrate (limpet, *Patella vulgata*) and bacterial (*Pseudomonas aeruginosa*) sources [35]. Whereas a predominance of single crystalline cores was observed for human ferritin cores (Figure 13), limpet and bacterial ferritins gave very few lattice images. Limpet cores imaged with resolvable lattice fringes contained crystalline domains with dimensions in the range of 30–50 Å.

Bacterial ferritin cores, on the other hand, showed only incoherent fringe patterns suggesting lamella-like structures in which the ordering was extremely short-range (often 10–20 Å). These differences were also reflected in the different ^{57}Fe Mössbauer spectra obtained from these samples (Figure 14) [36]. Human ferritin cores were superparamagnetic (i.e. antiferromagnetically coupled particles of small dimension) with a blocking temperature of *ca* 40 K. Limpet ferritin had a lower blocking temperature (*ca* 20 K) and there was also evidence for magnetic disorder. Bacterial ferritins, in contrast, were magnetically disordered down to a temperature of 4 K.

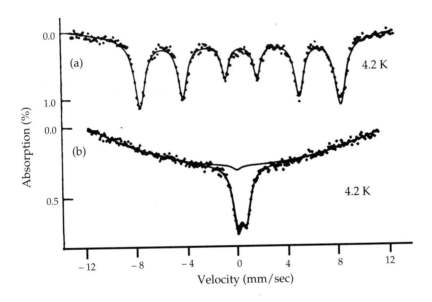

Figure 14 ^{57}Fe Mössbauer spectra at 4.2 K of (a) human spleen ferritin and (b) bacterial ferritin. The human ferritin exhibits a magnetic hyperfine split sextet whilst the bacterial protein shows a quadrupole doublet corresponding to magnetically disordered high spin Fe(III)

These results show unequivocally the fundamental difference in the crystallographic nature of ferritin mineral cores isolated from human, limpet and bacterial sources. Factors governing this change in crystal chemistry include the rate of oxidation of Fe(II) on entry into the protein cavity and the subsequent mechanisms of growth of the Fe(III) solid phase. These factors will be dependent on local redox and pH conditions, ionic concentrations (influx and efflux rates) and binding sites at the protein interface. For example, rapid oxidation of Fe(II) will kinetically favour amorphous and poorly-ordered ferrihydrites. Another factor which may be important is the much higher inorganic phosphate level in the bacterial iron-containing cores. This protein may contain hydrated Fe(III) oxy-phosphate cores rather than the Fe(III) oxide phases of invertebrate and vertebrate proteins.

The growth of single crystal cores within the protein cavity of mammalian ferritin molecules indicates that nucleation proceeds at one location at the protein surface. Spectroscopic evidence and chemical modification studies suggest that the important sites are carboxylic residues lying close to the subunit dimer interface (see [30] for details). Recent work using recombinant human mutant ferritins has shown that the H-chain but not the L-chain has ferroxidase activity [37, 38], and crystallographic studies have identified the ferroxidase centre ligands as two neighbouring glutamates and a histidine close to the inner surface [39] which are not present in the L-chain. Although oxidation can proceed at these sites, it is unlikely that they act as unique nucleation centres since a critical cluster involving many Fe(III) ions must be stabilised. However, the clustering of glutamate residues at the subunit dimer interface [30] could provide the necessary charge density to act as the nucleation zone [35] (Figure 15).

In order to elucidate the factors which may be important in determining the structural differences of ferritin iron cores isolated from different biological sources, experiments involving the structural and magnetic characterisation of reconstituted horse spleen and bacterioferritins have been investigated [40]. Reconstitution to a Fe loading of *ca* 3000 atoms was undertaken in the absence of phosphate at pH 7.0 and the cores were compared with those of native proteins. The initial rate of core reconstitution was significantly faster for bacterial (*A. vinelandii*) compared with horse ferritin.

The reconstituted bacterial ferritin cores were pseudo-crystalline ferrihydrite compared with the amorphous cores of the native protein. A similar increase in crystallinity was also observed in the reconstitution of *Ps. aeruginosa* bacterioferritin. These structural modifications imposed by *in vitro* reconstitution suggest that the bacterial proteins do not function primarily as crystallochemically-specific interfaces for core mineralisation *in vivo*. The absence of phosphate during reconstitution suggests that a major influence determining core structure may lie in differences in cellular physiology that lead to co-availability of Fe and inorganic phosphate in the bacteria and compartmentalisation of these components in higher organisms.

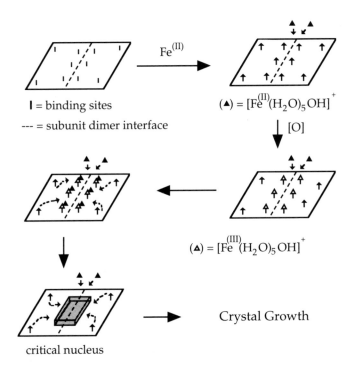

Figure 15 Possible mechanism of mineralisation in ferritin. The subunit dimer interface contains a cluster of possible binding sites for Fe(II). Oxidation of Fe(II) at the ferroxidase site creates a localised concentration of Fe(III). This region is now activated such that it becomes the focus of additional Fe(II) diffusion and oxidation. The activation energy of nucleation is overcome when the number of localised Fe(III) species attains a critical size. Crystal growth continues by further oxidation of Fe(II) within the nucleation zone

The above studies indicate that although the specificity of ferritin in iron uptake and deposition is in part determined by the functionality of the inner surface of the cavity, a major feature lies with the ability of the quaternary structure to encapsulate a nanovolume of aqueous reaction space. For example, several recombinant human H-chain mutants comprising site-directed modifications in the channel or various inner surface amino acids have been successfully reconstituted *in vitro* [41]. However, an intact mutant protein modified in both the ferroxidase ligands and cavity surface glutamates was inactive with regard to core formation suggesting a cooperative effect of these sites on iron oxidation and nucleation [41].

The study of solid state reactions in such small volumes is relatively unexplored although it could be a viable route to monodisperse nanometre-size particles for use in catalysis or semiconductor devices. In the next section we describe an approach to modelling the types of reactions taking place in ferritin through the study of precipitation of inorganic materials in synthetic unilamellar phospholipid vesicles.

5.4.2.2 *Phospholipid Vesicles*

Unilamellar phosphatidylcholine vesicles, *ca* 30 nm in diameter, can be readily prepared by sonicating dispersions of the lipid in aqueous solution at a temperature above the gel-liquid transition point. When formed in the presence of metal ions, the internal space contains encapsulated species which can subsequently undergo crystallisation reactions with membrane permeable species such as OH^- and H_2S (Figure 16).

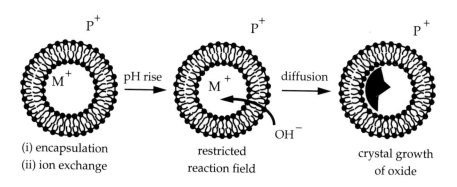

(i) encapsulation   restricted   crystal growth
(ii) ion exchange   reaction field   of oxide

Figure 16 Reaction scheme illustrating the use of phospholipid vesicles in the membrane-mediated precipitation of metal oxides. Cations (M^+) are encapsulated by sonication and replaced in the external phase by inert cations (P^+) by ion exchange chromatography. Increases in the extravesicular pH result in the slow OH^- influx and subsequent nucleation of the oxide on the inner membrane surface

Alternatively, co-reactants can be transported into the interior of the vesicles via ionophores sited in the lipid bilayer. The following materials have been investigated; Ag_2O [42, 43], $CoSiO_3$ [44], Ca phosphates [45], $Fe_2O_3.nH_2O$ and Fe_3O_4 [46, 47], Al_2O_3 [48], and $BaHPO_4$ and SiO_2 [49]. The aim of much of this work was to investigate the types of precipitate which could be formed in confined vesicular environments and determine the level of control exerted by these chemically well-defined supramolecular assemblies. For example, the chemistry of formation of intravesicular materials may be very different from the corresponding bulk solution reactions normally encountered in inorganic chemistry. In particular, the chemical and electrical fields generated within the microvolume may be responsible for changes in redox, kinetic and structural behaviour.

Before reaction, encapsulated cations such as Fe(II), Fe(III) and Co(II) bind strongly to the headgroup phosphates of the phospholipid bilayer so that nucleation is localised at the organic surface. Reaction with OH^- results in finely divided intravesicular particles. Electron diffraction patterns arising from the particles obtained from Fe(III) solutions had *d*-spacings corresponding to poorly ordered goethite (α-FeOOH) whilst similar reactions with entrapped $Fe(II)_{(aq)}$ and Fe(II)/Fe(III) solutions gave intravesicular particles of spherulitic magnetite (Fe_3O_4) and ferrihydrite ($Fe_2O_3.nH_2O$), respectively. These products were different from those formed under identical starting conditions in the absence of vesicles. For example, precipitation of Fe(III) solutions resulted in extended aggregates of ferrihydrite, Fe(II) solutions gave acicular needles of lepidocrocite (γ-FeOOH) and goethite, and Fe(II)/Fe(III) solutions gave irregular-shaped 10–50 nm magnetite particles. Thus there are distinct modifications in structure, morphology, and particle size for precipitation reactions undertaken within unilamellar vesicles. These differences can be attributed primarily to the kinetic control exerted by the vesicle membrane on the rate of OH^- diffusion into the intravesicular space although the charged organic surface may be important in stabilizing the accumulation of ionic charge and subsequent formation of the embryonic crystallites.

Similar studies involving the intravesicular precipitation of Ag_2O from encapsulated Ag(I) solutions reacted with hydroxide ions gave discrete crystals of cubic Ag_2O (Figure 17). The formation of intravesicular single crystallites indicates that nucleation proceeds at a single site on the vesicle membrane. The kinetics of these reactions have been studied by light scattering [43]. There was a linear relationship between the initial rate of precipitation and trapped Ag(I) concentration at constant hydroxide-ion concentration.

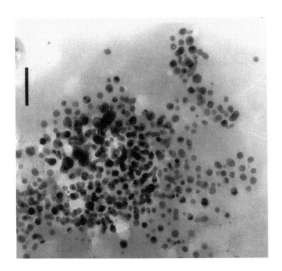

Figure 17 Transmission electron micrograph showing discrete intravesicular crystals of Ag_2O. Scale bar = 80 nm

Thus over the concentration range investigated the initial kinetics were first order with respect to intravesicular Ag(I) concentration ($[Ag]_{in}$). The dependency on extravesicular pH ($[OH^-]_{out}$) was more complex; at pH values < 10 no changes in turbidity were observed, at pH's 11–12 the reaction was strongly dependent on $[OH^-]_{out}$ and above pH 12, the initial rate was essentially independent of $[OH^-]_{out}$. These data suggest a two step reaction mechanism for intravesicular Ag_2O precipitation.

$$[OH^-(aq)]_{out} \xrightarrow{K_{diff}} [OH^-(aq)]_{in} \tag{1}$$

$$2[OH^-(aq)]_{in} + 2[Ag^+(aq)]_{in} \xrightarrow{K} [Ag_2O(aq)]_{in} + H_2O \tag{2}$$

The first step (Equation (1)) is diffusion controlled and depends on the rate of passage of OH^- ions through the lipid membrane. At pH_{out} values below 11.0 the rate of OH^- influx is very small and supersaturation is never attained within the vesicles. At pH_{out} values above 12.0 the rate of crystal growth becomes less dependent on initial pH gradients across the membrane because the limiting rate of diffusion through the membrane is attained.

The relationship between the intra- and extravesicular pH in the above experiments has been studied by ^{31}P NMR spectroscopy [50]. Vesicles containing a mixture of NaH_2PO_4 and $NaNO_3$ were studied to determine whether the presence of the diffusable NO_3^- ion would permit the influx of hydroxide as observed for intravesicular Ag_2O formation. The intravesicular phosphate resonance, initially at -14.98 ppm, did not shift significantly until above pH_{out} 11.0, after which it shifted steadily downfield due to OH^- influx. Similar experiments undertaken with phosphate in the absence of nitrate showed that a pH gradient of *ca* 6 units could be maintained across the bilayer membrane at an external pH of 12.5.

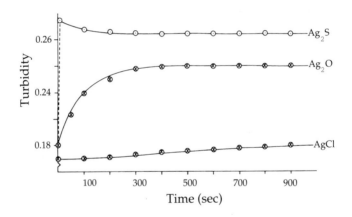

Figure 18 Plot of turbidity measured at 600 nm against time for the formation of various intravesicular deposits. (a) AgCl, (b) Ag_2O, and (c) Ag_2S. The Ag^+ concentration was the same in each experiment

The reduction of the pH gradient across the vesicle membrane in these experiments requires the net transport of OH^- ions into the vesicles. In order to preserve electroneutrality this can only occur if there is an equivalent net migration of anions out of the vesicles. In the experiments with vesicles containing NO_3^- the internal pH responded to a change in pH_{out} above 11.0, whereas the vesicles containing only the more highly charged $[HPO_4]^{2-}$ and $[PO_4]^{3-}$ ions showed no changes in internal pH. Thus the presence of encapsulated diffusible anions is important in determining the rôle played by the membrane in controlling intravesicular pH, and hence the rate of inorganic oxide precipitation within the internal cavity of the vesicles.

The dependency of the rate at which supersaturation is maintained within the vesicles on anion diffusion rates is clearly revealed when anions of different permeabilities are substituted for OH^- and the corresponding intravesicular precipitation is followed by turbidity measurements [43] (Figure 18). In the case of intravesicular AgCl formation the increase in turbidity with time due to the addition of Cl^- was much slower than for Ag_2O formation indicating that diffusion of Cl^- across the lipid membrane was reduced compared with OH^-. The formation of intravesicular Ag_2S, on the other hand, occurred instantaneously after the addition of $(NH_4)_2S$ due to diffusion of free molecular H_2S across the vesicle membrane.

5.4.3 Shaping the Mineralisation Space

We have described how the confinement of inorganic precipitation within reaction volumes of nanometre dimension can result in the controlled biomineralisation of ferritin. These processes can be successfully modelled using supramolecular assemblies of phospholipid molecules. However, both ferritin and synthetic vesicles have no influence on the crystallographic orientation or habit of the encapsulated inorganic materials. This contrasts with many biomineralisation processes that result in highly reproducible crystal morphologies often unknown in inorganic systems. For example, whereas inorganic magnetite (Fe_3O_4) adopts crystal habits (cube, octahedral etc.) based on the cubic space symmetry of the unit cell, bacterial magnetites are usually elongated along preferred crystallographic axes. The main structural and organisational factors responsible for the formation of these anisotropic ultrafine magnetic biogenic particles are reviewed in the following section.

5.4.3.1 Magnetotactic Bacteria

Life on earth has evolved under the influence of the earth's magnetic field and, perhaps not surprisingly, some organisms are thought to have adapted such that they can detect and exploit this phenomenon. In this regard, the discovery of a very simple magnetotactic response in certain species of bacteria was of profound importance [51]. Electron microscopy and electron diffraction studies [52–55] and [57]Fe Mössbauer spectroscopy

[56, 57] have shown that these bacteria contain intracellular single crystals of magnetite, Fe_3O_4 (Figure 19).

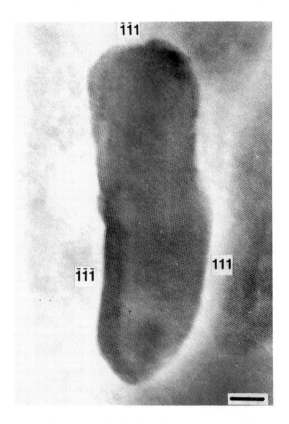

Figure 19 Lattice image of a bullet-shaped magnetite crystal from wild type bacterial cells. The inclusion is a well-ordered single crystal of unusual morphology. The crystal is imaged along the *[110]* axis. The top and side edges correspond to faces of *{111}* form. Lattice fringes correspond to the *{111}* and *{200}* interplanar spacings. Scale bar = 10 nm

Each individual crystal is enveloped by an organic membrane and localised in chains in close proximity to the inner surface of the cytoplasmic membrane [28, 29]. The particles are in the single magnetic domain size range such that a chain of the crystals imparts a sufficiently strong permanent magnetic moment to orient the bacteria in the geomagnetic field [58].

Low-magnification electron micrographs of bacterial magnetite indicate a variety of morphological forms which are species specific [59].

Viewed in projection, cubic, rectangular, hexagonal and bullet-shaped particles have been observed. The idealised three-dimensional morphologies have been established through the detailed analysis of lattice imaging data. The simplest morphological form, cubo-octahedral, is exhibited by crystals synthesised in *A. magnetotacticum* [54] (Figure 20a).

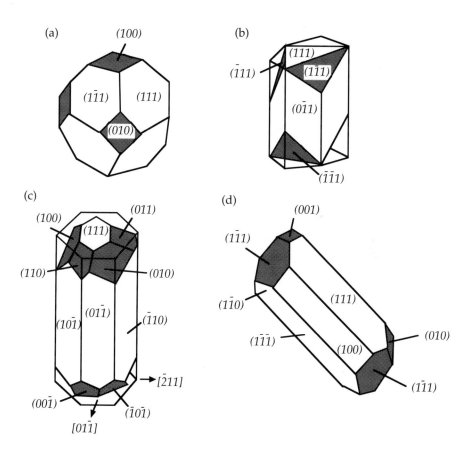

Figure 20 Idealised crystal morphologies of bacterial magnetite; (a) cubo-octahedral, (b) and (c) hexagonal prisms and (d) elongated cubo-octahedral

This crystal habit is the equilibrium form common in inorganic magnetite and reflects the stability of the close-packed octahedral {111} faces and the strongly bonded cubic {100} faces. A similar analysis of magnetite crystals formed in coccoid cells showed that the rectangular morphology of these crystals, when viewed side-on at low magnification, is the projection of a truncated hexagonal prism. In one coccus type [52], the hexagonal prism is capped by only one of the four symmetry-related {111} sets and the other {111} faces are not expressed in the crystal morphology (Figure 20c). Furthermore, the crystal is preferentially elongated along one of the four [111] axes. A similar discrimination is made with regard to faces of index {110}. Six of the 12 symmetry-related {110} faces are extensively formed as the elongated sides of the hexagonal prism whereas the remaining six are expressed as small truncated faces at the ends. A related hexagonal habit has been determined for magnetite crystals synthesised in the cultured marine vibroid MV-1 [53]. The lattice images clearly indicated that both the top and truncated side faces are of {111} form and the crystals had a hexagonal cross-section comprising six {110} faces (Figure 20b). Matsuda *et al.* [60] have obtained similar results in an unspecified coccus cell type.

The bullet-shaped morphology of certain wild type magnetotactic bacteria is based on an elongated cubo-octahedral form comprising a hexagonal prism of {111} and {100} faces, capped by {111} faces and with associated {111} and {100} truncations (Figure 20d) [61, 62]. In many crystals the axis of elongation lies parallel to the [112] crystallographic axis and, again, there is differentiation of symmetry-equivalent faces.

The cubo-octahedral and bullet-shaped crystals are closely related even though they appear to be very different in the final product. Development of the elongated crystals takes place in two distinct stages [62]. The first stage involves the development of isotropic magnetite crystals, of cubo-octahedral morphology, which grow to a size of 20 nm. There is no spatial constraint imposed on crystal growth at this stage since the crystal grows out from the membrane wall equally in all directions. The second stage involves anisotropic growth along the [112] direction, resulting in three of the {111} and {100} planes becoming elongated. One can envisage that the crystals grow either within preformed elongated vesicular sacs or in vesicles that are continuously being extended along a preferential direction (perhaps parallel to the cell membrane) during crystal growth. Thus the cytoskeletal organisation of the vesicular system is of primary importance in morphological specificity in these biogenic magnetic particles. However, the fact that the crystallographic orientation of the particles is established at the nucleation stage through specific formation of the (111) face on the immobile wall of the expanding vesicle indicates that the molecular interactions involved at this stage must be precisely coupled to the axis of unidirectional growth and vesicle elongation.

A different mechanism of anisotropic growth appears to be present in the elongated coccoid and MV-1 crystals. The crystal habit of these particles (Figures 20b and c) appears to be established not via the spatial restriction of a growing isotropic crystal, but as an intrinsic property of the initial crystals nucleated within the magnetosome vesicles. Although specific molecules present within this environment could give rise to novel crystal habits, one would predict that these would remain isotropic since the activity of growth mediators is equivalent on symmetry-related surfaces. One possibility is that the spatial organisation of ion-transport centres on the magnetosome membrane generates different growth rates of symmetry-equivalent directions because the flux of ions to the crystal surfaces is highly directional. Thus the hexagonal prism morphology could be related to a three-fold symmetry of transport centres on the membrane and elongation could arise by a greater flux rate in the axial direction. Moreover, vectorial crystal growth would be enhanced if the membrane and crystal surfaces were in direct contact throughout crystal growth such that the lateral diffusion of ions in solution was minimised.

Investigations of the processes of crystal growth of bacterial magnetite have focused on structural studies of immature crystals since the crystallochemical properties of these crystals reflect the intrinsic mechanisms of crystal synthesis. Crystals at early stages of growth have been studied *in situ* by HRTEM [52, 54, 62]. Lattice imaging of the immature irregular particles showed the presence of contiguous crystalline and non-crystalline regions within the magnetosomes. The crystalline zone was always observed to be single domain with well-ordered lattice planes of magnetite. This data is in agreement with ^{57}Fe Mössbauer spectroscopy studies which showed the presence of a high spin Fe(III) component (ferrihydrite) in addition to magnetite in spectra recorded at room temperature from frozen whole cells [56]. Crystals at intermediate stages of growth in coccoid and wild type cells showed characteristic rounded edges and irregular, structurally disordered surfaces. These observations suggest that growth of the crystals occurs through surface-mediated reactions involving the phase transformation of ferrihydrite to magnetite.

On the basis of HRTEM, Mössbauer spectroscopy and biochemical results, a sequence of events leading to bacterial magnetite can be proposed. These involve: (i) uptake of Fe(III) from the environment via a reductive step in membrane transport [63]; (ii) transport of Fe(II) (or

Fe(III) as ferritin?) to and across the magnetosome membrane; (iii) precipitation of hydrated ferric oxide within the magnetosome vesicles; and (iv) phase transformation of the amorphous Fe(III) phase to magnetite both at the nucleation stage and during surface-controlled growth.

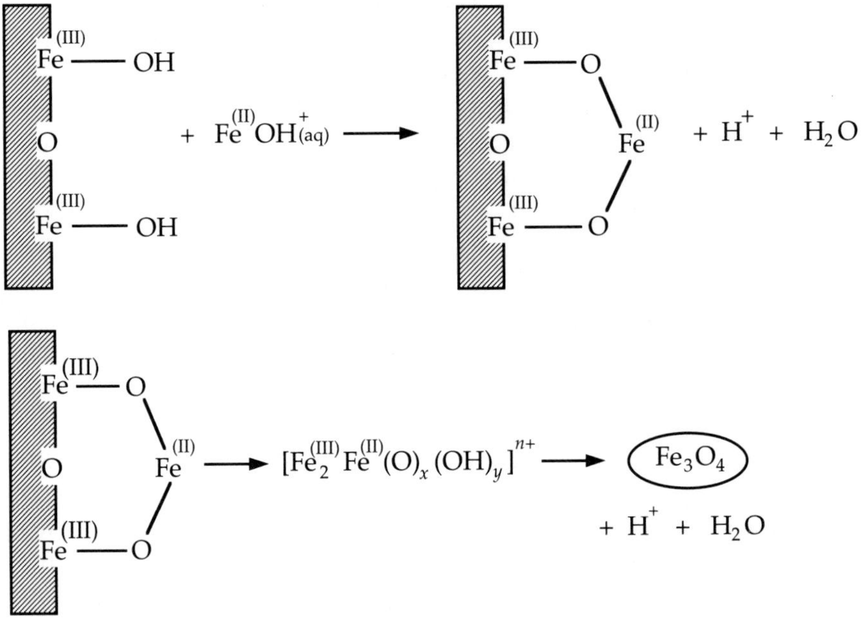

Figure 21 Scheme for the formation of magnetite from reaction of amorphous Fe(III) oxide with Fe(II) under anaerobic conditions

The phase transformation of amorphous hydrated ferric oxide to magnetite can occur at neutral pH provided the redox potential of the reaction environment is of the order of -100 mV [64]. The redox potential will be extremely sensitive to pH such that small changes in $[H^+]$ could have a marked influence on the phase transformation processes. Investigations of the transformation of ferric oxides to magnetite under aqueous conditions in inorganic systems have shown that the critical step is the involvement of aqueous Fe(II) at the ferric oxide surface [65]. A two-step process is postulated (Figure 21).

The rate of magnetite formation appears to be essentially first order with respect to the concentration of the surface intermediate formed in step 1. Although the composition of the intermediate released into solution is unknown, the formation of Fe_3O_4 in the second step involves the release of one further proton. The resultant lowering in the reaction pH and subsequent increase in redox potential implies that for the phase transformation to magnetite to proceed to any significant extent within the magnetosome vesicle, there must be precise regulation of the pH and hence redox potential within the localised mineralisation zone.

The organisation and ultrastructure of the surrounding magnetosome membrane is fundamental to the control of magnetite biomineralisation. This membrane, which has an overall composition similar to other cell membranes, contains two proteins [66] that may be specific to the nucleation and growth of magnetite. The formation of the enclosed vesicle appears to occur prior to mineralisation and provides both a spatial constraint for growth and a defined chemical reaction volume via selective ion transport. Furthermore, the presence of a charged organic surface may be an important factor in determining the kinetics and structural characteristics of the nucleation event.

5.4.4 Organised Assemblies

One of the key functional features of the bacterial synthesis of magnetic crystals is the organisation of crystals of unusual morphology into chains aligned within the bacterial cell (Figures 5 and 6). Although such an assembly may be far removed from the complex composites of bone and teeth, it does indicate that the ability to organise biominerals into suprastructures is established within the relatively undifferentiated cells of prokaryotes (bacteria). The level of organisation of the cellular components is markedly increased in single-celled eukaryotes such as algae and protozoa; consequently the degree of organisation of biomineralised structures is vastly increased. The complexity of the sculptured products far exceeds that currently possible in man-made inorganic materials. Furthermore, the assembly of shaped particles into organised arrays has relevance in a range of fields concerned with multi-interactive particle systems (magnetism, optical effects, catalysis).

Here we describe two systems; (i) silica formation in unicellular organisms termed choanoflagellates, and (ii) calcite formation in marine algae (coccolithophorids).

5.4.4.1 Silica Assemblies in Protozoa

Amorphous materials are isotropic in structure and hence limited in the geometry of their external form. The absence of anisotropy in their molecular structure results in spherical particles that are not influenced by conventional growth additives in solution. Thus the formation of shaped amorphous materials in biological systems is of immediate interest to colloidal and materials scientists.

Stephanoeca diplocostata Ellis is a unicellular loricate choano-flagellate commonly found in coastal waters around Europe and the Mediterranean [67]. The cells comprise a colourless protoplast with a single anteriorly directed flagellum surrounded by a ring of tentacles (the collar). The protoplast is lodged in an open-ended basket-like casing (lorica) constructed of 150–180 silica strips (costal rods) (Figure 22).

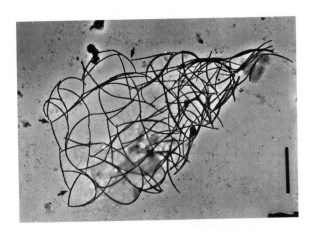

Figure 22 Transmission electron micrograph showing an intact basket of assembled curved silica rods in the unicellular protozoan, *S. diplocostata*. Arrows highlight junctions between individual rods. Scale bar = 8 μm

New costal rods are produced in advance of mitosis within long thin vesicles in the peripheral cytoplasm and then released sideways through the plasmalemma so that on cell division the juvenile, taking the supernumerary strips with it as it leaves the parent lorica, is able to assemble its own basket within 2–3 minutes [68].

HRTEM images of individual costal rods showed irregular incoherent fringes and no evidence of periodicity [69]. No short-range order could be determined extending above 1 nm (approximately three Si–O–Si units)

indicating that the silica comprised a continuous disordered gel-like structure. Moreover, the images indicated a structure based on a random network of (SiO_4) units connected through Si-O-Si bonds of variable bond angle rather than a microcrystalline/cluster structure composed of a random array of microcrystalline polyhedra.

A final process in the assembly of the basket is the joining together of costal rods in a manner which allows the intact lorica to be resistant to forces arising in the marine environment. Figure 23 shows a junction between two costal rods and clearly indicates that the rods are glued together during the construction process. The connective material is generally less electron dense than the adjacent siliceous material of the costal strips. High resolution electron micrographs have indicated that silica and probably organic material constitute the join [69]. This suggests that the inorganic polymer has some residual flow properties which enable silica to move into the junction prior to hardening of the join.

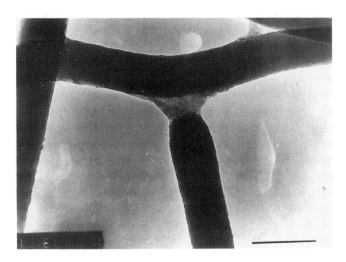

Figure 23 Junction between two silica rods in the intact basket of *S. diplocostata*. The rods are joined by a glue of silica and organic material. Scale bar = 100 nm

Interestingly, the silica-containing costal rods are metastable and slowly redissolve in the aqueous medium. The process of demineralisation depends on the conditions of physical growth; at 20 °C in an agitated solution, demineralisation is complete within 10 days [70]. Studies of silica demineralisation have been important in revealing differences in the local chemical environment of silica within the costal rods. The initial stage of demineralisation takes place at localised centres along the central axis of the strip suggesting that these regions comprise silica which is more hydrated (soluble). Alternatively, the surface silica may be preferentially protected by an organic membrane. The initial centres become increasingly demineralised with time until they extend and join together along the central axis, forming tubular, brittle rods. As demineralisation increases the hollowing becomes more extensive and only at the later stages do the outer edges of the costal rods show signs of demineralisation, becoming rough and pitted. An interesting implication of this preferential dissolution is that the rods become hollow without significant reduction in their mechanical strength (compare the use of tubular steel rods in building scaffolding). Thus the silica basket remains functional, i.e. intact, even though the costal strips have undergone extensive demineralisation. Only when the tubular walls become very thin does fracture and buckling occur. The mechanical design of biogenic silicas is therefore an important consideration in relating structural properties to biological function.

One of the major unresolved questions is how this molecularly isotropic material is fabricated into elongated curved rods. During mineralisation, the rods remain surrounded by a vesicle membrane that lies in association with two microtubular filaments [70] suggesting that the shaping of the vesicle by cellular stresses may be responsible for the elaborate morphology of the mature mineral particles. Deposition could take place in several ways. Silicification could be the result of a specific binding mode for silicic acid within the vesicles, followed by localised polymerisation. Binding of silicic acid to organic molecules can occur through hydrogen bonding, ionic interactions or condensation of OH groups [71]. Alternatively, an energised dissolved silica concentration gradient within the vesicles could initiate silicification through changes in osmotic pressure, pH or concentration. Concurrent with silica deposition must be biochemical processes involving regulation of silicic acid metabolism and control over the organic components involved in vesicle development.

5.4.4.2 *Organised Calcite Crystals in Algae*

In the previous section we described how amorphous silica can be shaped and assembled by biological processes. In principle, the engineering of crystalline materials could be potentially more difficult since any deviation from the crystallographic symmetry requires significant energy. This can be readily overcome in biological systems through structural elements such as microtubules which establish and maintain stress fields within the cells. In many cases, the final form of the biomineral is a compromise between the crystallographic forces and 'organic' symmetry requirements of individual species. As the latter can vary widely, a large range of sculptured structures can be expressed using the same biomineral. Here we describe the formation of complex calcite plates in the unicellular marine alga, *Emiliania huxleyi*.

Cells of *E. huxleyi* synthesise an elaborate calcitic extracellular shield comprising an intricate network of delicately sculptured oval-shaped calcite plates (coccoliths). The morphology of these plates varies in different species. In *E. huxleyi* the coccoliths are composed on a radially arranged array of crystalline segments each consisting of a flattened lower element, a hammer-shaped upper element and a vertical central element (Figure 24). The vertical element connects the other two regions and forms part of the wall of a central cylinder in each coccolith.

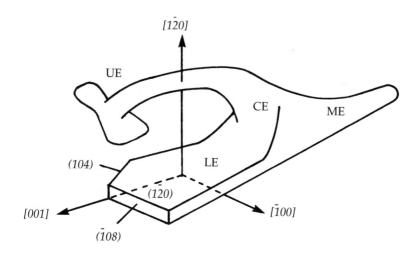

Figure 24 Drawing showing the morphology and crystallographic faces and directions of an individual calcite single crystal radial segment in the coccoliths of *E. huxleyi*. Approximately thirty of these structures are assembled into one coccolith plate. UE = upper element, CE = central element, ME = medial element, LE = lower element

Electron diffraction and lattice imaging studies [72, 73] have shown that each coccolith element is a highly ordered single crystal. Single-crystal electron diffraction patterns and lattice images were obtained from individual lower elements and from local regions within these segments, and the crystallographic directions in these patterns were correlated with their associated TEM images. Viewed from above the coccolith plate, the diffraction patterns for the lower elements corresponded to the *[1$\bar{2}$0]* crystallographic direction (*a*-axis) of calcite (Figure 24) indicating that the top face of the lower element corresponds to the *(1$\bar{2}$0)* face of calcite and that the bottom face corresponds to the *($\bar{1}$20)* crystal face. Furthermore, the patterns showed that the crystallographic *[001]* direction (*c*-axis) was orientated parallel to the direction of elongation of the lower element and that the faceted outer edges of each lower element corresponded to the *($\bar{1}$08)* and *(104)* faces of calcite viewed end-on. Interestingly, the *($\bar{1}$08)* face was always more extensively developed and positioned to the left when viewed from above the base plate (Figure 24).

Electron diffraction patterns recorded from isolated upper elements showed that these complex units were well defined single crystals. The *c*-axis was orientated parallel to the long axis of the upper element and the hammer-head extension at 90° to the stem corresponded to the *[100]* direction. The alignments of the crystallographic axes were coincident with those determined on the base plate indicating that the whole segment was a continuous single crystal.

The first stage in coccolith biosynthesis in *E. huxleyi* involves the formation of a polysaccharide-rich organic base plate apposed to the nuclear envelope [75, 76]. The nucleation sites for calcification are at the rim of this plate. Crystallites are initially formed which subsequently grow in lateral, medial and distal directions to form the complete segments of each coccolith. The precise orientation of crystal nuclei on the surface of the organic base plate indicates that the molecular nature of the underlying organic surface must be of paramount importance. One possibility is that there is a close stereochemical and geometrical correspondence between Ca-binding sites on the organic substrate and lattice sites in the *($\bar{1}$20)* crystal face of the calcite nuclei. It is important to note that it is the *($\bar{1}$20)* face and not the symmetry related *(1$\bar{2}$0)* face that is in direct contact with the organic surface. The nature of the molecular recognition processes active during the nucleation stage is therefore extremely precise.

The Ca atoms coplanar within the *(1$\bar{2}$0)* face lie in rows parallel to the *(104)* edge and are separated by distances of 4.03 Å in each row and 6.4 Å between adjacent rows. These spacings must correspond to commensurate configurations of binding sites on the surface of the polysaccharide base plate. The composition of the polysaccharide isolated from the coccoliths of *E. huxleyi* (strain F61) is known and includes at least 13 different monosaccharides among which uronic acid and methylated and dimethylated sugars are present [76]. Borman *et al.* [77] have shown that the uronic acid moieties interfere with the *in vitro* crystallisation of calcium carbonate. It is feasible that Ca binding to the carboxylate groups, as well as the sulphate residues, could occur in an organised fashion if the polysaccharide molecules of the base plate were themselves ordered. Three major constraints relating to the mode of interaction of Ca atoms and carboxylate groups on the rigid polysaccharide are required for nucleation to occur on the $(\bar{1}20)$ face. Firstly, the O–C–O binding unit of the ligand must lie approximately perpendicular to the surface of the organic base plate if it is to mimic the stereochemical requirements of carbonate groups in this face; furthermore, the coincidence of the *[001]* axis and the direction of elongation of the lower elements implies that the O–C–O binding unit must be also orientated perpendicular to this direction. Secondly, the arrangement of and distance between Ca binding sites at the interface should match those in the plane of the $(\bar{1}20)$ face, i.e. linear arrays of sites separated by approximately 4 Å within and 6.4 Å between each row. Thirdly, the arrangement of the sites across the interface must be such that the symmetry elements describing this organisation are compatible with the deposition of the $(\bar{1}20)$ face and not the $(1\bar{2}0)$ face; this selectivity occurs because these faces are chiral such that the polysaccharide binding sites can be assembled in a mirror-image configuration specific to the Ca atoms in the $(\bar{1}20)$ face. These criteria could be satisfied, for example, if linear chains of polysaccharide molecules separated by a distance of approximately 6.4 Å were preferentially aligned at the nucleation site.

These investigations highlight the precise interplay that exists in organisms between the processes of biological organisation and the inanimate crystallochemical forces present in inorganic minerals. The matching of specific crystallographic axes with the morphological requirements of biological structures and the tailoring of crystallography in biomineralisation must ultimately reflect the evolution of molecular specificity between organic macromolecules and inorganic crystal surfaces. The nature of these interactions has been elucidated through the use of model systems as described in Sections 5.5 and 5.6.

5.4.5   Morphology and Crystal Surfaces

The above examples indicate that a primary strategy for shaping and assembling biogenic materials is to utilise stressed supramolecular assemblies. An analogous process in synthetic systems has yet to be realised. There are, however, other possibilities.

One particularly promising approach is to use tailor-made additives that modify crystal habits through specific interactions at the crystal faces. Many of the soluble macromolecules associated with biominerals are considered to act in this manner. For example, acidic glycoproteins extracted from adult sea urchin tests have been shown to bind specifically to the $\{1\bar{1}0\}$ faces of synthetic calcite crystals [24]. In the past, many additives have been used on an empirical basis and it is only very recently that stereochemical arguments have been developed to explain their effects on inorganic crystallisation [78, 79]. As many biological molecules contain oxyanion functional groups (carboxylate, phosphate and sulphate esters), we have recently been involved in *in vitro* crystallisation experiments using low molecular weight analogues of these macromolecules. Here we describe some of our results on modified calcite crystallisation.

Figure 25 Scanning electron micrograph of a calcite crystal modified in the presence of the additive, γ-carboxyglutamic acid [(HOOC)$_2$CHCH$_2$CH(NH$_2$)COOH]. The molecule is highly charged at pH 6 and interacts specifically with the $\{1\bar{1}0\}$ prismatic faces of calcite. The smooth $\{104\}$ surfaces in the photograph are the only faces present in unmodified crystals, whereas the textured faces correspond to those stabilised by the adsorbed additive molecules. Scale bar is 10 μm

Calcite crystals grown in aqueous solutions containing variable levels of functionalised and non-functionalised α,ω-dicarboxylates have been studied [79]. Crystals formed in the presence of malonate were spindle-shaped at a mole ratio of Ca/malonate = 3. The crystals were elongated along the *c*-axis and exhibited curved {1$\bar{1}$0} prismatic faces. Significantly, these crystals were very similar to the calcite crystals deposited in the inner ear gravity devices of mammals and reptiles described in Section 5.2.1, Figure 3. The biological crystals are formed in the presence of acidic organic molecules including proteins containing the malonate residue of the specialised amino acid, γ-carboxyglutamic acid [80], suggesting that these functional groups are responsible for the characteristic shape of the biogenic deposits.

The formation of spindle-shaped crystals was reduced with increasing chain length between the α,ω-carboxylates. The unsaturated derivative, maleate, was also less potent compared with the saturated malonate compound whereas the *trans*-isomer, fumarate, had negligible morphological effect. Functionalisation of the lower chain acids had a marked influence on crystal morphology. Crystals grown in the presence of aspartate (α-aminosuccinate) exhibited well-defined {1$\bar{1}$0} prismatic faces at Ca/additive = 17, whilst γ-carboxyglutamate had a pronounced effect at ratios as high as 85 (Figure 25).

The affinity of dicarboxylate molecules for the prismatic faces can be attributed to electrostatic, geometric and stereochemical recognition at the crystal/additive interface. Incorporation of carbonate into these faces is via bidentate binding to Ca atoms and this can be closely simulated by dicarboxylate interactions provided that the spacing between carboxylate groups is close to 4.0 Å. Both malonate and maleate fit this criterion but the increased rigidity of the latter reduces the binding affinity. The marked effect observed on substitution of a α-hydrogen of succinate for $[NH_3]^+$ of aspartate suggests that the amine moiety may be able to substitute for a Ca site cooperatively with carboxylate binding via an extended molecular conformation across several layers of the {1$\bar{1}$0} crystal surface. A similar effect, but involving an additional carboxylate, is possible for γ-carboxyglutamate binding (Figure 26).

The importance of stereochemical factors in additive-crystal interactions was also observed for calcite grown in the presence of phosphorus-containing oxyanions. Inorganic phosphate, as $[H_2PO_4]^-$, at Ca/additive ratios as high as 10,000:1, induced the formation of elongated calcite crystals with stabilised {1$\bar{1}$0} prismatic faces [81], analogous to the experiments involving carboxylate binding. Infrared spectroscopy indicated that the phosphate was bound to the surface in a bidentate manner. Thus the general rule appears to be that the prismatic faces will be selectively stabilised by anionic molecules provided that they can compete for surface sites involving bidentate coordination.

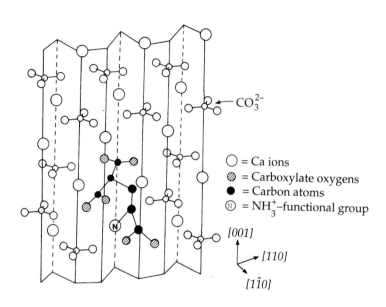

O = Ca ions
⊘ = Carboxylate oxygens
● = Carbon atoms
Ⓝ = NH_3^+–functional group

[001]

[110]

[1$\bar{1}$0]

Figure 26 Perspective drawing of the *(1$\bar{1}$0)* face of calcite showing possible binding mode for γ-carboxyglutamic acid

The importance of charge density and the hydrophobic nature of the substituent attached to the active functional group was also investigated through the use of phosphate esters. In general, a marked reduction in the morphological stabilisation of the prismatic faces was observed as the hydrophobic group increased in size (H > butyl > phenyl > naphthyl). Thus, an increased level of hydrophobicity reduces the strength of the additive/crystal interaction presumably due to repulsive forces at the crystal/water interface. Interestingly, the phenyl phosphoester, $[C_6H_5OPO_3H]^-$, was slightly more potent than its phosphonate counterpart $[C_6H_5PO_3H]^-$ but the latter produced a very different morphology. Whereas phenyl phosphate gave spindle-shaped calcite, crystals grown in the presence of phenyl phosphonate exhibited rhombohedral {018} faces. In the phosphonate, the bulky phenyl group attached directly to the P atom is at a maximum distance from the crystal surface when the molecule is bound in a tridentate manner. The {018} faces expressed by phenyl phosphonate have carbonates arranged at an angle of 26.3° to the plane of the face such that all three oxygens are partially exposed. Thus, insertion of phosphate-containing anions into the surface can proceed essentially in a tridentate manner. By comparison, rotation about the C–O–P linkage in phenyl phosphate enables the hydrophobic group to be at a maximum distance from the surface for a bidentate mode of binding on the prismatic {1$\bar{1}$0} faces.

In this section we have described how crystals can be shaped through the specific molecular interaction of surface bound anions. These interactions are governed by charge, stereochemical and structural matching of anion binding with packing motifs in the crystal lattice. In this regard, the synthesis of specific types of acidic macromolecules associated with many biomineralisation processes may have an important rôle in determining the architecture of these biogenic materials. An alternative function for these macromolecules is in the design of molecular templates for orientated nucleation. It is this aspect of biomineralisation that we consider in the following section.

5.4.6 Matrix-mediated Nucleation

The nucleation of an inorganic mineral at the surface of an organic matrix can be considered as a phase transformation reaction involving surface and bulk processes. Here we will not consider the influence of solution thermodynamics and kinetics (see [26, 82] for details) but will limit our discussion to the interfacial processes that can modify the reaction profile. In many ways we can liken the rôle of the matrix to an enzyme with the incipient nuclei acting as the substrate. A difficulty arises in the choice of description with regard to the forming clusters — in no cases do we have knowledge of their structure (periodic, amorphous, polyhedral), size or composition and we do not know whether the initial interactions involve ion-binding or larger scale polynuclear events. Whatever the details, the role of the matrix is primarily to lower the activation energy of nucleation by decreasing the time between ionic collisions, thereby stabilizing and perhaps catalysing the transition state of the phase transformation reaction.

We know several general features about the matrix [83];

(i) *Primary Structures.* The matrix is matched to the coordination chemistry of the ionic species comprising the mineral through the choice of ligands exposed at the interface. For example, calcified invertebrate tissues contain macromolecules rich in carboxylate residues [84] such that Ca-binding mimics Ca-[CO_3] interactions. Similarly, Ca-[PO_4] motifs in bone, dentine and bacterial deposits can be established through Ca-[$ROPO_3$] binding (phosphoproteins in bone [11] and dentine [85], proteolipids in calcifying bacteria [86]) and ice nucleation is activated [87] and suppressed [88] by hydroxy-rich macromolecules.

(ii) *Secondary and Tertiary Structures.* There are only two options; nucleation at a planar organic surface or at one which is curved. The former can be generated by antiparallel β-pleated sheets and there is X-ray diffraction evidence [89] to support the role of such structures in biomineralisation. Other planar surfaces could be derived by elongation of phospholipid membranes or by crystallisation such as the proteinaceous S-layer of bacterial membranes. Curved surfaces are more common; localised protein pockets and grooves, membrane-bound vesicles, α-helical (antifreeze proteins) and triple-helical (collagen) conformations are all possible.

(iii) *Quaternary Structures.* Quaternary structures may play a fundamental role in biomineralisation. In evolutionary terms, ferritin (as bacterioferritin) is an ancient matrix and functions primarily through its ability to self-assembly into a 24-mer capable of sequestering and nucleating Fe(III) oxide (see Section 5.4.2.1.). Similarly, it is known that the membrane aggregation of proteins is crucial to the efficacy of ice nucleating bacteria [90]. It seems feasible that the 'two-component' model of shell mineralisation may also rely on the ordered assembly of the nucleator macromolecules on the relatively inert framework proteins. For example, the EDTA soluble macromolecules are considered to adopt the β-sheet conformation of the underlying structural proteins [91].

In summary, there are two key structural factors in the use of organic matrices in controlled nucleation. Firstly, the matrix is preorganised with respect to nucleation through processes such as self-assembly, aggregation, membrane vesiculation and controlled polymerisation (cross-linking) which impart spatial regulation of functional groups. Secondly, nucleation at the matrix surface is regiospecific with a limited number of sites being confined to discrete loci. These two factors may be temporally linked since ion-binding could result in specific conformational changes in the preorganised matrix such that nucleation is activated within localised domains.

5.4.6.1 Molecular Complementarity

Some of the possible modes of molecular complementarity existing at the mineral-matrix interface are shown in Figure 27 [27, 92].

It is likely that several of these factors act cooperatively in real systems. The most fundamental aspect of recognition involves the matching of charge and polarity distributions and it seems feasible that the earliest biological approach to regulate nucleation was based on the clustering of charged centres, particularly if these sites were also redox active. The localisation of Mn-oxidizing proteins in the cell walls of bacteria such as *Bacillus* [93] and ferroxidase centres in ferritin [39] are typical examples. The primary role of charge matching is to favour electrostatic accumulation and hence increase the time of encounter of ions in embryonic clusters. In this regard, the topography of the matrix surface may play a significant role (Figure 28).

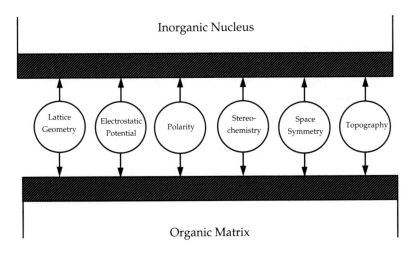

Figure 27 Possible modes of molecular complementarity at inorganic-organic interfaces

For example, it is clear that localised pockets or grooves can give rise to high spatial charge densities over dimensions commensurate with stable nuclei (1–5 nm). Note that high affinity binding may be disadvantageous since it restricts the structural rearrangement of the dispersed hydrated ionic clusters to the stable nuclei. Thus ion-pair accumulation may be important.

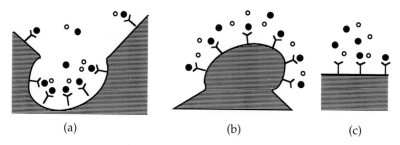

Figure 28 Spatial charge distribution of functional groups (Y) on organic matrices of different topography; (a) concave, (b) convex and (c) planar. Only concave localities give rise to a significant local three-dimensional clustering of ionic charge (● , o)

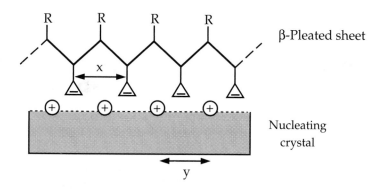

Figure 29 The concept of epitaxy as applied to biomineralisation. Geometric matching exists at the interface between a structured organic surface and nuclei of the inorganic crystal. Cation-cation distances in one specific crystal face (y) are commensurate with the spacing of negatively charged binding sites on, for example, a β-pleated sheet protein (x). Matching can be in one or two dimensions and results in orientated nucleation.

The curvature of molecular cavities provides three-dimensional control over nucleation and a limit on the size of the nucleation site. Planar surfaces, in contrast, provide a nucleation site only in two dimensions and, like convex surfaces, in the absence of other factors, would be less active since the binding sites are not constrained in close proximity to each other as for concave localities. The advantage of planar surfaces, however, is that long range structural matches can be readily established in principle. This leads to the concept of epitaxy which has been a central hypothesis of controlled biomineralisation (Figures 29 and 30) [94, 95].

It is important to make a distinction between inorganic epitaxy and 'biological epitaxy'. The former is well documented and, in many cases, geometric matching at the unit cell level is apparent. However, large degrees of mismatch can occur and ordered aggregation of non-orientated nuclei to give orientated overgrowths has been observed at the post-nucleation stage. Biological substrates differ from analogous inorganic templates in that they do not show the molecular smoothness or rigidity implicit in epitaxial mechanisms. On the other hand, they exhibit surface stereochemistry due to exposed functional groups and this may be advantageous since it provides for variability in surface conformation depending on the binding/ionisation state of the residues. Thus, there may be several potential geometric matches for a given crystal face.

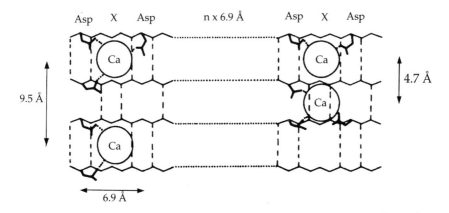

Figure 30 Proposed molecular correspondence at the inorganic-organic interface in the nacreous shell layer of *Nautilus repertus* [91]. There is a close match between the Ca-Ca distance along the aragonite *a*-axis (4.96Å) and periodic spacings in the β-pleated sheet protein (4.7 Å). Sequences of Asp-X-Asp or polyAsp residues along the sheet provide the appropriate binding sites for orientated nucleation

An additional possibility is that geometric epitaxy is superceded by stereochemical correspondence as determined in model systems [78, 96, 97]. This is a crucial concept since it relegates the need for matrix periodicity and is applicable to both planar and non-planar surfaces. A corollary of a stereochemically-driven mechanism is that it may select general rather than specific features of a crystal face. This is important because the putative crystal faces of nuclei are likely to be extensively reconstructed, non-stoichiometric and high in defect sites and general features such as the orientation of carbonate groups may be the only discernable property. In crystallographic terms, nucleation on organic surfaces may be selective for specific zone axes rather than particular crystal faces.

In summary, electrostatic, geometric and stereochemical recognition processes can occur at mineral-matrix interfaces. Understanding these interactions depends very much on determining the nature of assembled organic surfaces and in elucidating the surface structure of inorganic clusters. Furthermore, application of these concepts in synthetic systems will lead to new approaches to the fabrication of organised, orientated assemblies of inorganic materials. One promising approach involves the use of compressed monomolecular surfactant films and recent developments in this area are described in the following section.

5.4.6.2 Controlled Crystallisation under Langmuir Monolayers

Organised organic surfaces can be readily prepared by the spreading and compression of surfactant monolayers at air/water interfaces in a Langmuir trough. The influence of these charged or polar molecules on the orientated nucleation of amino acid and NaCl [98, 99], and $CaCO_3$ [100–102] crystals forming from supersaturated subphases has been reported. Recent work on the crystallisation of $SrSO_4$ on Langmuir-Blodgett multilayer assemblies has been undertaken [103]. Here we summarise our recent results on orientated $CaCO_3$ crystallisation.

Crystals of $CaCO_3$ were grown from supersaturated calcium bicarbonate solutions (pH = 6.0) freshly prepared by purging suspensions of calcite with CO_2 gas for 1 hour. The control experiments gave randomly dispersed aggregates of non-orientated rhombohedral calcite crystals at the air/water interface. In contrast, when compressed monomolecular films of stearic acid [$CH_3(CH_2)_{16}COOH$] or octadecylamine [$CH_3(CH_2)_{17}NH_2$] were spread at the air/water interface, the crystals were discrete and crystallographically orientated. For stearic acid at total [Ca] = 10 mM, rhombohedral calcite plates aligned with the $[1\overline{1}0]$ direction perpendicular to the organic surface were deposited (Figure 31).

At lower concentration ([Ca] = 5 mM), a marked structural modification was observed with the metastable polymorph, vaterite, being exclusively nucleated. These crystals were orientated with the c-axis perpendicular to the monolayer. In comparison, orientated crystals grown under compressed octadecylamine films were exclusively vaterite, independent of Ca concentration.

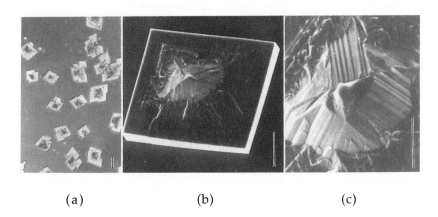

(a)   (b)   (c)

Figure 31 Orientated calcite crystals nucleated under stearic acid films. (a) Optical micrograph, viewed from above the monolayer showing plate-like crystals with central elevations, bar = 10 μm. (b) Scanning electron micrograph showing *[1Ī0]* orientated crystal with roughened upper surface and pyramidal central elevation, bar = 10 μm. (c) High magnification image of central elevation and upper surface of basal plate. Rhombohedral steps are observed on the three inclined faces, bar = 5 μm

Moreover, two distinct orientations of these crystals were observed (Figure 32). One type were morphologically and crystallographically equivalent to the *vaterite* crystals formed under stearic acid films whereas the other type were unique to the octadecylamine monolayers and orientated with the *a*-axis perpendicular to the monolayer surface.

These results indicate that both negatively and positively charged compressed monolayers can induce the orientated nucleation of $CaCO_3$ crystals. Significantly, $CaCO_3$ crystallisation under neutral octadecanol $[CH_3(CH_2)_{17}OH]$ monolayers was severely inhibited, and under compressed cholesterol $[C_{27}H_{45}OH]$ films was no different from the control experiments. This suggests that ion-charge accumulation at the ionised organic surfaces is a critical factor in regulating crystallisation at the organic surface.

(a)   (b)

Figure 32 Scanning electron micrographs of mature vaterite crystals grown under compressed octadecylamine monolayers. Views from above the monolayer surface. (a) Type II crystal orientated along *(110)* (*a*-axis), bar = 10 μm; (b) type I crystal orientated along *(001)* (*c*-axis), bar = 10 μm

The presence of both orientated vaterite and calcite crystals on charged monolayers indicates that the electrostatic interactions between nuclei and the organic surface are influenced by structural relationships at the interface. The nucleation of vaterite on films of positive and negative charge indicates that Ca binding is not a prerequisite for stabilisation of this metastable phase. Oriented calcite, on the other hand, requires Ca binding at the carboxylate headgroups.

Nucleation of calcite on the *(1\overline{1}0)* face under stearate monolayers can be rationalised in terms of charge, stereochemical and geometric complementarity at the inorganic/organic interface. Monolayers of simple fatty acids form hexagonal or pseudo-hexagonal nets of interheadgroup spacing *ca* 5 Å when compressed on aqueous subphases. Thus the first layer of the crystal will be determined primarily by the two-dimensional spacings of the carboxylate binding sites. A comparison of these distances with those between coplanar Ca atoms in the *(1\overline{1}0)* face of calcite (Figure

33) reveals a very close epitaxial match in two dimensions but there is no such match for the *(001)* face of vaterite.

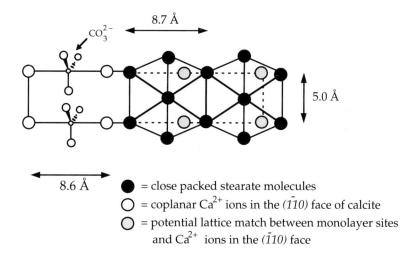

Figure 33 Diagram showing the superimposition of the coplanar Ca atoms of the *(1̄10)* face of calcite on a hexagonal net of stearate molecules with interheadgroup spacing = *ca* 5 Å

Geometric correspondence cannot, however, be solely responsible for [1̄10] orientated calcite nucleation. For example, the *(001)* face of calcite comprises a hexagonal lattice of coplanar Ca atoms of 4.96 Å periodicity and such an arrangement matches the monolayer binding sites almost exactly. A significant difference between the *(1̄10)* and *(001)* faces is the orientation of the carbonate anions; they lie perpendicular to the *(1̄10)* surface but parallel to *[001]*. Thus the stereochemistry of the carboxylate headgroups mimics that of the anions in the *(1̄10)* crystal face but not *(001)* (Figure 34).

The selectivity of the *(001)* vaterite face on stearic acid monolayers and both *(001)* and *(110)* on octadecylamine films suggests that stereochemical factors at the interface are important as Ca binding is not a prerequisite for orientated nucleation. Both these faces contain anions which lie perpendicular to the crystal surfaces and this motif can be mimicked by carboxylate alignment on stearic acid monolayers or bidentate binding of $[HCO_3]^-$ to octadecylamine headgroups (Figure 34).

All carbonate anions in the *(001)* face are perpendicular to the surface whilst in the *(110)* face only a subset of anions have this orientation. Thus, the *(001)* face will be favoured under carboxylate films because of the stereochemical equivalence of all the headgroups. For octadecylamine monolayers, there is no direct stereochemical correspondence, and orientated nucleation must be governed by the structure of the underlying ion-containing boundary layer. In this respect, bidentate binding of $[HCO_3]^-$ at the amine headgroups will be essentially orthogonal to the monolayer surface and at least two nucleation orientations are possible.

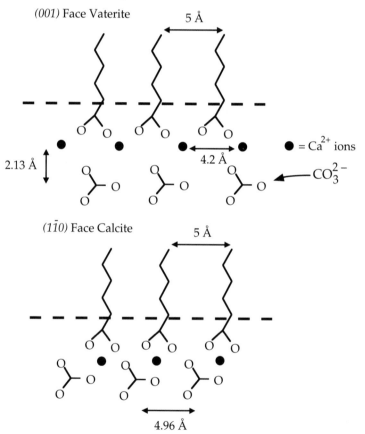

Figure 34a Electrostatic, stereochemical and geometric relationships at monolayer-crystal interfaces. Ca binding (filled circles) is dominant on stearate films and carboxylate stereochemistry is critical

The change from calcite to vaterite nucleation on stearate films at low [Ca] suggests that the extent of Ca binding is important for polymorph selection. The nucleation of calcite is favoured by the formation of a well defined Ca-carboxylate layer which mimics the first layer of the $(1\bar{1}0)$ face of the unit cell. By contrast, the structural requirements for vaterite formation are less precise. This is consistent with vaterite being the dominant phase on amine monolayers where no Ca binding is present and suggests that kinetic factors of charge accumulation and cluster stabilisation overide any structural factors in these systems.

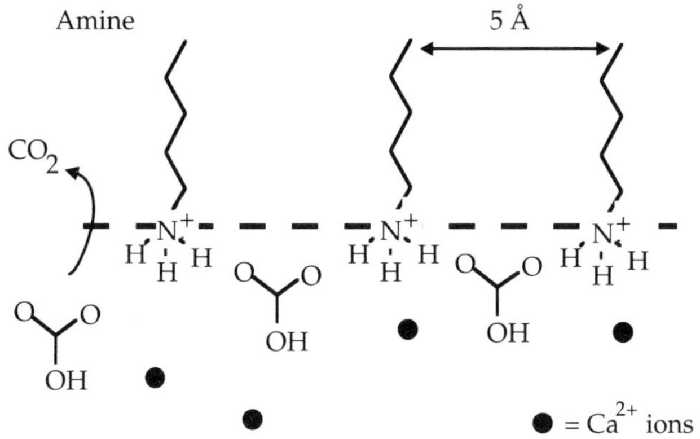

Figure 34b For amine monolayers, orthogonal bicarbonate binding favours the nucleation of specific vaterite faces

5.5   CONCLUSIONS

The development of advanced inorganic materials is now receiving serious attention amongst chemists, physicists and materials scientists. There are two major approaches. The first involves the development of synthetic pathways to new materials with novel structures and properties. An alternative strategy is to design new approaches to the controlled fabrication of known materials in order to enhance their potential in chemical, magnetic and electronic devices. In this respect, the study of biomineralisation is of key importance. We have described above how organisms can synthesise relatively simple inorganic materials in precisely defined forms. Crystal size, shape, structure, orientation and assembly can all be controlled and replicated. Moreover, these processes are characterised by specific molecular interactions that involve recognition between inorganic and organic surfaces.

The modelling of biomineralisation is a promising approach to controlled materials synthesis. In particular, the use of enclosed supramolecular assemblies appears to be a viable route to nanophase inorganic materials. A complementary approach, involving planar assemblies of organic macromolecules, has potential in the orientated nucleation of organised inorganic materials. The development of these ideas into polymer-based inorganic composites has been initiated although much further work is required particularly in establishing preferred crystallographic orientations and high inorganic volume fractions. One process yet to be explored is the ability to assembly inorganic materials into organised micro-architectures analogous to bone, teeth or shells. In biological systems, the assembly of bioinorganic minerals is regulated by flow processes whereby cells control the fluxes of ions and molecules to and from the mineralisation sites. Concepts such as self-assemble, dynamics, feedback and remodelling in inorganic materials synthesis are figments of our imagination. Their future realisation will revolutionise the design and fabrication of advanced inorganic materials.

5.6 REFERENCES

1. S Mann, *Nature*, **349**, 285 (1991).
2. P A Bianconi, J Lin and A R Strzelecki, *Nature*, **349**, 315 (1991).
3. J D Birchall, *Phil Trans R Soc Lond* A, **310**, 31 (1983).
4. W J Clegg, K Kendall, N Alford, T W Button and J D Birchall, *Nature*, **347**, 455 (1990).
5. S Mann, J Webb and R J P Williams, *Biomineralisation; Chemical and Biochemical Perspectives*, VCH Publishers, Weinheim (1989).
6. H A Lowenstam and S Weiner, *On Biomineralisation*, Oxford University Press, Oxford (1989).
7. M D Ross and K G Pote, *Phil Trans R Soc Lond B*, **304**, 445 (1984).
8. K M Towe, *Science*, **179**, 1007 (1973).
9. H A Lowenstam and D P Abbott, *Science*, **188**, 363 (1975).
10. H Setoguchi, M Okazaki and S Suga, in: *Origin, Evolution and Modern Aspects of Biomineralisation in Plants and Animals*, R E Crick, Ed, Plenum Press, New York, p 409 (1989).
11. M J Glimcher, *Phil Trans R Soc Lond B*, **304**, 479 (1984).
12. B R Heywood, N H C Sparks, R P Shellis, S Weiner and S Mann, *Connective Tiss Res*, **25**, 1 (1990).

13. W Traub, T Arad and S Weiner, *Proc Nat Acad Sci*, USA, **86**, 9822 (1989).
14. D B Spangenberg and C W Beck, *Auralia Trans Am Microsc Soc*, **87**, 329 (1968).
15. S Mann, N H C Sparks and R G Board, *Adv Microbial Phys*, **31**, 125 (1989).
16. S Mann, C C Perry, J Webb, B Luke and R J P Williams, *Proc R Soc Lond R*, **227**, 179–190 (1986).
17. K S Kim, D J Macey, J Webb and S Mann, *Proc R Soc Lond R*, **237**, 335 (1989).
18. M Freke and D J Tate, *Biochem Microbiol Technol Eng*, **3**, 29 (1961).
19. S Mann, N H C Sparks, R B Frankel, D A Bazylinski and H W Jannasch, *Nature*, **343**, 258 (1990).
20. B R Heywood, D A Bazylinski, A Garrett-Reed, S Mann and R B Frankel, *Naturwissenschaft*, **77**, 536 (1990).
21. B E Volcani and T L Simpson, *Silicon and Siliceous Structures in Biological Systems*, Springer Verlag, Berlin (1982).
22. S Mann and C C Perry, *CIBA Foundation Symposium*, **121**, 40 (1986).
23. C C Perry, in: *Biomineralisation; Chemical and Biochemical Perspectives*, S Mann, J Webb, R J P Williams Eds, VCH Publishers, Weinheim, p 223 (1989).
24. A Berman, L Addadi and S Weiner, *Nature*, **331**, 546 (1988).
25. A Berman, L Addadi, A Kvick, L Leiserowitz, M Nelson and S Weiner, *Science*, **250**, 664 (1990).
26. S Mann, *Structure and Bonding*, **54**, 125 (1983).
27. S Mann, in: *Biomineralisation: Chemical and Biochemical Perspectives* S Mann, J Webb and R J P Williams, Eds, VCH Publishers, Weinheim, p 35 (1989).
28. H A Lowenstam, *Science*, **211**, 1126 (1981).
29. K Simkiss and K M Wilbur, *Biomineralisation; Cell Biology and Mineral Deposition*, Academic Press (1989).
30. P M Harrison, P J Artymiuk, G C Ford, D M Lawson, J M A Smith, A Treffry and J L White, in *Biomineralisation; Chemical and Biochemical Perspectives*, S Mann, J Webb, R J P Williams, Eds, VCH Publishers, Weinheim, p 257 (1989).
31. A Treffry, P M Harrison, M I Cleton, W C de Bruijn and S Mann, *J Inorg Biochem*, **31**, 1 (1987).
32. P M Harrison, F A Fischback, T G Hoy and G H Haggis, *Nature*, **216**, 1188 (1967).
33. K M Towe and W P Bradley, *J Colloid Interface Sci*, **24**, 384 (1967).
34. W H Massover and J M Cowley, *Proc Nat Acad Sci*, **70**, 3847 (1973).
35. S Mann, J V Bannister and R J P Williams, *J Mol Biol*, **188**, 225 (1986).

292 *Stephen Mann*

36. T G St Pierre, S H Bell, D P E Dickson, S Mann, J Webb, G R Moore and R J P Williams, *Biochim Biophys Acta,* **870,** 127 (1986).
37. L Levi, A Luzzago, G Cesareni, A Cozzi, F Franceschinelli, A Albertini and P Arosio, *J Biol Chem,* **263,** 18086 (1988).
38. S Levi, J Salfeld, F Franceschinelli, A Cozzi, M Dorner and P Arosio, *Biochemistry,* **28,** 5179 (1989).
39. D M Lawson, P J Artymiuk, S J Yewdall, J M A Smith, J C Livingstone, A Treffry, A Luzzago, S Levi, P Arosio, G Cesareni, C D Thomas, W V Shaw and P M Harrison, *Nature,* **349,** 541 (1991).
40. S Mann, J M Williams, A Treffry and P M Harrison, *J Mol Biol,* **198,** 405 (1987).
41. V J Wade, S Levi, P Arosio, A Treffry, P M Harrison and S Mann, *J Mol Biol,* **221,** 1443 (1991).
42. J L Hutchison, S Mann, A J Skarnulis and R J P Williams, *J Chem Soc Chem Comm,* 634 (1980).
43. S Mann and, R J P Williams, *J Chem Soc Dalton Trans,* 311 (1983).
44. S Mann, A J Skarnulis and R J P Williams, *Israel J Chem,* **21,** 3 (1981).
45. B R Heywood and E D Eanes, *Calcif Tissue Int,* **41,** 192 (1987).
46. S Mann, J P Hannington and R J P Williams, *Nature,* **324,** 565 (1986).
47. S Mann and J P Hannington, *J Colloid Interface Sci,* **122,** 326 (1988).
48. S Bhandarkar and A Bose, *J Colloid Interface Sci,* **135,** 531 (1990).
49. C C Perry, in: *Proceedings of the 1st ANAIC Conference on Si and Sn,* Oxford University Press, Oxford, (1991).
50. S Mann, M J Kime, R G Ratcliffe and R J P Williams, *J Chem Soc Dalton Trans,* 771 (1983).
51. R P Blakemore, *Science,* **190,** 377 (1975).
52. S Mann, T T Moench and R J P Williams, *Proc R Soc Lond B,* **221,** 385 (1984).
53. N H C Sparks, S Mann, D A Bazylinski, D R Lovley, H W Janasch and R B Frankel, *Earth and Planetary Science Letters,* **98,** 14 (1990).
54. S Mann, R B Frankel and R P Blakemore, *Nature,* **310,** 405 (1984).
55. F F Torres de Arujo, M A Pires, R B Frankel and C E M Bicudo, *Biophysical Journal,* **50,** 375 (1985).
56. R B Frankel, G C Papaefthymiou, R P Blakemore and W D O'Brien, *Biochimica Biophysica Acta,* **753,** 147 (1983).
57. D A Bazylinski, R B Frankel and H W Jannasch, *Nature,* **334,** 518 (1988).
58. R B Frankel, R P Blakemore and R S Wolfe, *Science,* **203,** 1355 (1979).
59. N H C Sparks, L Courteaux, S Mann and R G Board, *FEMS Microbiological Letters,* **22,** 171 (1986).
60. T Matsuda, J Endo, N Osakabe and A Tonamura, *Nature,* **302,** 411 (1983).

61. S Mann, N H C Sparks and R P Blakemore, *Proc R Soc Lond B,* **231,** 469 (1987).
62. S Mann, N H C Sparks and R P Blakemore, *Proc R Soc Lond B,* **231,** 477 (1987).
63. L C Paoletti and R P Blakemore, *Journal of Bacteriology,* **167,** 73 (1986).
64. R M Garrels and C L Christ, in: *Solution, Minerals and Equilibria,* Harper and Row, New York (1965).
65. Y Tamura, K Ito and T Katsura, *J Chem Soc Dalton Trans,* 189 (1983).
66. Y A Gorby, T J Beveridge and R P Blakemore, *Journal of Bacteriology,* **170,** 834 (1988).
67. B S C Leadbeater, *Brit Phycol,* **7,** 195 (1972).
68. B S C Leadbeater, *Protoplasma,* **98,** 311 (1979).
69. S Mann and R J P Williams, *Proc R Soc Lond B,* **216,** 137 (1982).
70. B S C Leadbeater, *Phil Trans R Soc Lond B,* **304,** 529 (1984).
71. R H Iler, in: *Biochemistry of Silicon and Related Problems,* G Bendz, I Lundqvist, Eds, Plenum Press, New York, p 53 (1977).
72. N Watabe, *Calcif Tissue Res,* 114 (1967).
73. S Mann and N H C Sparks, *Proc R Soc Lond B,* **234,** 441 (1988).
74. K M Wilbur and N Watabe, *Ann N Y Acad Sci,* **82,** 109 (1963).
75. P Van der Wal, E W de Jong, P Westbroek and W C de Bruijn, *Protoplasma,* **118,** 157 (1983).
76. A M J Fitchinger-Schepmann, *PhD Thesis,* University of Utrecht (1980).
77. A H Borman, E W de Jong, M Huizinga, D J Kok, P Westbroek and L Bosch, *Eur J Biochem,* **129,** 179 (1982).
78. L Addadi and S Weiner, *Proc Natn Acad Sci,* **82,** 4110 (1985).
79. S Mann, J M Didymus, N P Sanderson, B R Heywood and E J A Samper, *J Chem Soc Faraday Trans,* **86,** 1873 (1990).
80. M D Ross and K G Pote, *Phil Trans R Soc Lond B,* **304,** 445 (1984).
81. J M Didymus, S Mann, N P Sanderson, B R Heywood and E J A Samper, in: *Proceedings of the Sixth International Conference on Biomineralisation,* S Suga, Ed, Springer Verlag, Berlin p 267 (1991).
82. R J P Williams, *Phil Trans R Soc Lond B,* **304,** 411 (1984).
83. S Mann, B R Heywood, S Rajam and V J Wade, in: *Proceedings of the Sixth International Conference on Biomineralisation,* S Suga, Ed, Springer Verlag, Berlin p 47 (1991).
84. S Weiner, *CRC Crit Rev Biochem,* **20,** 365 (1986).
85. S L Lee, A Veis and T Glonek, *Biochemistry,* **16,** 2971 (1977).
86. B Boyan-Salyers and A Boskey, *Calc Tiss Intern,* **30,** 167 (1980).
87. P K Wolber and G J Warren, *Trends Biol Sci,* **14,** 179 (1989).
88. A L DeVries, *Phil Trans R Soc Lond B,* **304,** 575 (1984).
89. S Weiner and W Traub, *FEBS Letts,* **111,** 311 (1980).
90. G J Warren and P K Wolber, *Cryo-Lett,* **8,** 204 (1987).

91.   S Weiner and W Traub, *Phil Trans R Soc Lond B,* **304**, 425 (1984).
92.   S Mann, *Nature,* **332**, 119 (1988).
93.   J P M de Vrind, E W de Vrind-de Jong, J W de Voogt, P Westbroek, F C Boogerd and R A An Rosson, *Appl Envir Microbiol,* **52**, 1096 (1986).
94.   W F Neuman and M W Neuman, *The Chemical Dynamics of Bone Mineralisation,* University of Chicago Press, Illinois (1958).
95.   B N Bachra in: *Biological Mineralisation,* I Zipkin, Ed, J Wiley & Sons, Chichester, p 845 (1973).
96.   L Addadi, J Moradian, E Shay, N G Maroudas and S Weiner, *Proc Natn Acad Sci* USA, **84**, 2732 (1987).
97.   L Addadi and S Weiner, in: *Biomineralisation: Chemical and Biochemical Perspectives,* S Mann, J Webb and R J P Williams, Eds, VCH Publishers, Weinheim, p 134 (1989).
98.   E M Landau, R Popovitz-Bior, M Levanon, L Leiserowitz, M Lahav and J Sagiv, *Molec Cryst Liq Cryst,* **134**, 323 (1986).
99.   E M Landau, S Grayer Wolf, M Levanon, L Leiserowitz, M Lahav and J Sagiv, *J Am Chem Soc,* **111**, 1436 (1989).
100. S Mann, B R Heywood, S Rajam and J D Birchall, *Nature,* **334**, 692 (1988).
101. S Mann, B R Heywood, S Rajam and J D Birchall, *Proc R Soc Lond A,* **423**, 457 (1989).
102. S Mann, B R Heywood, S Rajam, J B A Walker, R J Davey and J D Birchall, *Adv Mater,* **2**, 257 (1990).
103. N P Hughes, D Heard, C C Perry and R J P Williams, *J Appl Phys,* **24**, 146 (1991).

6 Clay Chemistry

Richard W McCabe

Inorganic Materials. Edited by Duncan W Bruce and Dermot O'Hare
© 1992 John Wiley & Sons Ltd

6.1   INTRODUCTION

Clay minerals have long been recognised as efficient materials for the promotion of many organic reactions.   They are highly versatile, affording both industrial (petroleum, etc.) and laboratory catalysts with excellent product-, regio- and stereo-selectivity [1, 2].   Clay catalysts have distinct advantages over homogeneous catalysts as the work-up of the reaction mixture is often very simple, i.e. the clay is removed by filtration.

Clays are powerful adsorbents with extremely high surface areas [1–4], especially when acid activated, and industrially they are used in such diverse applications as the co-reactant for carbonless copying paper development and the decolourising of vegetable oils.   Recent studies have shown that in suitable circumstances clays [5] and clay-supported chiral metal complexes [6, 7] may be used in asymmetric synthesis as they can, for example, either differentially adsorb one enantiomer from a racemate or adsorb enantiomers equally from a non-racemic mixture. They have also been used as selective supports for gas chromatography [8].

Early work with clays concentrated on either the cation exchange of the clay to increase Brønsted or Lewis acidity, or activation of the clay by treatment with strong acid [3, 4].   More recently, however, the repertoire of the clay chemist has expanded to include redox processes [3, 9], generation of reactive intermediates such as carbenes [10], and the use of clays as a support for metal salts [11–13] and complexes [14]. Pillaring of the clays gives a more clearly-defined reaction space and can produce catalysts which begin to rival the zeolites in stability and shape selectivity [e.g. 15–18].   Besides their use as catalysts and reactive supports, it has also been suggested that clays may have been involved in the formation of the original proteins and genetic material [19].

6.1.1   The Structure of Clay Minerals

Clay minerals are the most abundant sedimentary mineral group.   They predominate in the colloidal fractions of soils, sediments, rocks and waters and are classified as phyllosilicates (usually hydrous aluminosilicates).

In geology the word *clay* is used in two ways: firstly as a rock classification which generally implies an earthy, fine-grained material that develops plasticity on mixing with a limited amount of water. Secondly, it is used as a particle size term, which describes clays as materials which have particle sizes <4 μm, although the modern tendency is to define clays as having particle sizes <2 μm as fractions of this size generally give the very pure mineral.

Clay minerals have been shown by X-ray diffraction studies to be crystalline, even in their finest particles [20–27], although the presence of traces of amorphous materials has been verified in some clay samples [28]. Pauling [23] showed that micas, pyrophillite, chlorite and kaolinite had repeating layer structures, a fact which can be confirmed by electron microscopy [29].

Most of the important clay minerals are made up of combinations of two basic types of layers, the first consists of sheets of $[SiO_4]^{4-}$ tetrahedra arranged in hexagonal ring subunits (Figure 1). Each silica ring is combined with the two adjacent rings so as to form another 12-membered ring. The second type of layer is usually composed of either gibbsite $[Al_2(OH)_6]$ or brucite $[Mg_3(OH)_6]$ units which form octahedrally coordinated sheets, with the octahedra arranged so that each oxygen atom is shared with two neighbouring metal atoms (Figure 2).

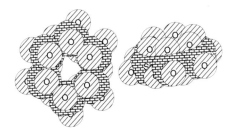

Figure 1 Silica subunits of the tetrahedral sheets

The analogous symmetry and almost identical dimensions of these tetrahedral and octahedral sheets allow them to be combined by sharing the apical oxygens of the silica layers with the successive octahedral layer(s). The silica sub-units are distorted from their ideally hexagonal symmetry to a ditrigonal symmetry, by an opposing rotation of alternate tetrahedra. This feature of the structure is generally accepted to be due to the slight misfit between the larger tetrahedral layer and the smaller octahedral layer. The strain thus generated is relieved by slight expansion of the octahedral layer and contraction of the tetrahedral layer. The remaining coordination sites of the octahedral metal are made up by bonding to hydroxyl groups rather than bridging oxygens.

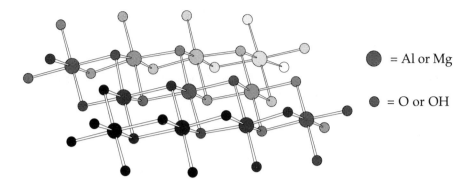

Figure 2 The arrangement of atoms in a dioctahedral layer

There are two dominant patterns of combining the layers. The first has alternating tetrahedral and octahedral sheets and is described as a 1:1 or T:O arrangement. This group includes kaolinite (Figure 3) $[Al_4Si_4O_{10}(OH)_8]$ — the main constituent of china clay — and crystolite $[Mg_4Si_4O_{10}(OH)_8]$. The main bonding forces between layers are hydrogen bonds between –OH on one layer and a bridging –O– on the next [30].

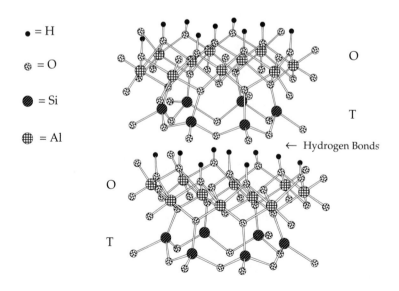

Figure 3 The idealised structure of kaolinite [30]

The second, and perhaps chemically more important type, has repeating units of tetrahedral:octahedral:tetrahedral layers and is described as a T:O:T or 2:1 arrangement. If the octahedral layer consists of M^{2+}-cations then all of the octahedral sites must be occupied to maintain electrical neutrality. However, if an M^{3+}-cation is used then only 2/3 of the octahedral sites need be occupied to maintain neutrality. The former mineral type is thus known as trioctahedral and the latter as dioctahedral. Some of the most common minerals in this class are:

Illite	$[K_{1.6}Si_{6.4}Al_{5.6}O_{20}(OH)_4] + (H_2O)_x$
Vermiculite	$[Mg_{6.6}Si_{6.8}Al_{1.2}O_{20}(OH)_4] + (H_2O)_x$
Biotite	$[K_2(Mg, Fe)_6Si_6Al_2O_{20}(OH)_4] + (H_2O)_x$
Montmorillonite	$[Na_{0.6}Al_{3.4}Mg_{0.6}Si_8O_{20}(OH)_4] + (H_2O)_x$
Hectorite	$[Li_{1.0}Mg_{5.5}Si_8O_{20}(OH)_4] + (H_2O)_x$
Beidellite	$[Ca_{0.35}Al_4(Si_{7.3}Al_{0.7})Si_8O_{20}(OH)_4] + (H_2O)_x$

The formulae given here are idealised general formulae and large variations in composition are encountered within each subgroup of minerals; for instance most montmorillonites have at least a few percent of Fe^{3+} cations replacing the octahedral Al^{3+} cations and Ca^{2+} cations often replace the interlayer Na^+.

One of the most useful of these clays is the smectite (swelling) clay, montmorillonite (Figure 4). Montmorillonite is the main constituent of bentonites and Fuller's Earth which are noted for their excellent swelling properties when wetted. However, due to its microcrystalline nature the structure of the mineral has had to be determined by analogy to similar minerals such as pyrophillite [31] and vermiculite [32]. An alternative arrangement of the silica layer has been suggested by Edelman and Favejee [36], in which alternate SiO_4 tetrahedra are inverted. However, it has been suggested [37] that each of these structures can be correct depending on the source of the montmorillonite.

The outstanding feature of the structure of these smectite clays is their fixed a (5.17 Å) and b (8.95 Å) crystallographic axes, but variable c-axis spacing [31]. Water and organic molecules [38] can enter the interlayer space and can alter the basal layer spacing, thus affecting the c-axis dimension. For a completely dry clay, the c-axis spacing depends upon the size and type of the interlayer cation present. The interlayer spacing, Δd, is an important parameter of the clay system as it gives a measure of the available reaction space during a clay/organic molecule interaction.

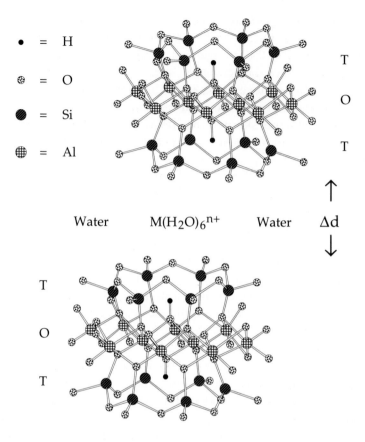

• = H

⊛ = O

◉ = Si

⊕ = Al

T

O

T

Water   M(H$_2$O)$_6^{n+}$ Water Δd

T

O

T

Figure 4 The idealised structure of a trioctahedral smectite clay showing the
interlayer aqueous metal cations [31–35]

6.1.2 Ion Exchange of Clays

6.1.2.1 Cation Exchange Clays — Cation Exchange Capacity (CEC)

To a greater or lesser degree, many clays have the ability to adsorb and
exchange cations from solution [39] and it is this cation 'storage' that
makes clays such an important component of many soils. A typical
montmorillonite can exchange over 100 millimoles of M$^+$ cations per 100 g
of clay, whilst a kaolinite only adsorbs a few tens of millimoles [40].

The ideal structures of the clay minerals depicted above are deviated
from in a number of ways which introduce charge imbalances into the

structure. The main causes of these charge imbalances are isomorphous exchange of Al^{3+} cations for Si^{4+} in the tetrahedral layers, and of Mg^{2+} and Al^{3+} cations in the octahedral layer [41] and crystal defects, usually at the crystallite edges [42]. The layers therefore have an overall negative charge which is balanced by adsorption of metal cations into the interlayer region of the clay mineral (Figure 4) [43].

A typical montmorillonite will have Na^+, Ca^{2+} or Mg^{2+} cations in the interlayer space. These cations are hydrated, usually hexa(aqua), and the remainder of the interlayer is filled by a variable amount of water [44] which can swell the clay. The interlayer cations are much less strongly bound than the layer cations and will thus easily exchange with cations from an aqueous solution [38].

Very small cations such as Li^+ initially occupy the interlayer region, but they can migrate on heating through the hexagonal cavities in the silica layer to occupy unfilled octahedral sites [45]. This decreases the ability of the mineral to adsorb cations.

The position in which these adsorbed cations reside depends upon the hydration of the clay, the size of the cation and whether edge or layer charges are involved. Na^+ [46] and Ca^{2+} [47] cations sit in the hexagonal cavities of the silica sheets in dry montmorillonite, but they move into the interlayer region as the clay is hydrated. In contrast, kaolinite, whose charge imbalance is due mainly to edge defects [30], adsorbs the metal ions at the edges of the particles. Excess water can separate the metal cation from the surface by imposition of extra aquation spheres [47].

When a solution of a metal cation is used to exchange the interlayer cations of a clay, it has been observed that the smaller the size of the exchange cation and the higher the charge on the cation, the more powerful that cation is at replacing the interlayer exchangeable cations [48, 49]. Similarly the ease of replacement of interlayer cations follows the reverse trend. Thus the series shown in Figure 5 can be constructed. The apparently anomalous positions of the H_3O^+, K^+ and NH_4^+ cations in this series is due to their almost perfect fit into the $(SiO_4)_6$-ring co-ordination sites between the clay layers [50].

The exact mechanism of cation exchange can be quite complex. If the metal cation is low in the series, e.g. Na^+, then some, but not all, of these cations may be displaced by H^+ from water even at a pH of 7 [51]. Thus in a typical cation exchange of Na^+ for Al^{3+}, e.g. by the method of Vaughan [15], using a 10-fold excess of say 0.3 mol dm^{-3} aluminium chloride (pH < 3), the first observation is that all of the Na^+ cations are displaced at the diffusion rate [51] by the very mobile H$^+$ cations; the much slower exchange of Al^{3+} cations for H$^+$ can then follow.

$$M^+ \ll M^{2+} < H_3O^+ \approx K^+ \approx NH_4^+ < M^{3+} < M^{4+}$$

Increasing Exchange Power
(Decreasing Ease of Exchange)

Figure 5 The exchange properties of cations with clays

The exact cationic species involved in the exchange reaction can depend very much on the pH (and hence concentration) of the exchange solution. Aluminium is particularly prone to pH effects as the hexa(aqua)-aluminium cation can aggregate into poly(oxoaluminium) species as the pH increases [52], so one must be very careful when exchanging a clay to control the pH carefully if a reproducible product is to be obtained.

6.1.2.2 Anion Exchange Clays

A series of clay minerals of lesser importance, whose charge imbalance or layer substitution pattern has given them positively charged layers, are known. Such minerals as hydrotalcite ($Mg_{4.5}Al_2O_{7.5}$) [53] and xonotlite ($Ca_6Si_6O_{17}(OH)_2$) [54] can be used as solid 'carriers' of hydroxy and *t*-butoxy anions, respectively.

6.1.3 Acidity of Clays

Several measurements of the acidity of clay surfaces have been carried out using surface conductivity [55], nuclear magnetic resonance [56] and Hammett indicator [57–61] methods. These experiments show that the effective pH of the water at the mineral surface is usually between 1.5 and 3, which is 4 to 5.5 units more acidic than that of bulk water.

Clay minerals show both Brønsted and Lewis acidity. Structural and environmental factors govern the degree to which each is present [62] and normally one type predominates for a given set of conditions.

Lewis acidity is normally associated with exposed Al^{3+} and Fe^{3+} at the 'broken' crystallite edges [4] and such acidity can be increased by heating the clay to >300 °C. However, this heat treatment not only removes the interlayer water and dehydrates the aquated metal cations in the interlayer but also leads to irreversible collapse of the clay layers.

The most important source of Brønsted acidity derives from dissociation of water molecules in the hydration sphere of the interlayer exchangeable cations [62–64] (Equation (1)).

$$[M(OH_2)_n]^{m+} \quad \rightleftharpoons \quad [M(OH_2)_{n-1}OH]^{(m-1)+} + H^+ \qquad (1)$$

The acidity depends upon the water content of the clay and to a lesser extent upon whether the layer charge arises mainly from substitution in the octahedral or tetrahedral sheet [57, 62, 65]. Acidity is maximised when the water content of the clay is low [57, 62, 65] (i.e. little dissipation of the M^{n+} charge by the excess water) and when highly polarising species such as M^{3+} cations are exchanged for the natural Na^+ and Ca^{2+} cations [63, 64]. Clays may be effectively dehydrated by heat or vacuum or by displacing the interlayer water with a non-polar organic solvent [66]. pK_As of -3 to -8 have been suggested for $[M(H_2O)_6]^{3+}$ cations in such an environment (cf. pK_As of *ca* 3.5 for these cations in aqueous solution) [67]. Care should be taken, however, not to overheat M^{3+} exchanged clays as this leads to *reduced* acidity [68] and collapse of the clay layers. As the acidity depends so critically upon the water content of the clay, they are best equilibrated to known relative humidity before use [69].

A further source of acidity is associated with the –OH groups of the octahedral layer which protrude into the interlayer region via the holes in the silica rings (Figures 3 and 4). The incidence of these protons may be increased by preparing a 'proton-exchanged' clay. This is achieved either by simply exchanging the clay with dilute acid [70], or less destructively, by exchanging the clay with ammonium ions and calcining at 200–300 °C to expel ammonia [71]. The exchanged protons can migrate into vacancies in the octahedral layer of dioctahedral clays where they protonate bridging oxygens (Equation (2)).

Alternatively the protons can attach to an oxygen of the tetrahedral sheet in a manner analogous to that suggested for X- and Y-zeolites [72].

These are thought to be minor contributors to the total acidity at low temperature as the hydrogens are tightly bound. However such acidity becomes much more important at high temperature or under anhydrous conditions, when exchangeable cation acidity becomes low [e.g. 73, 74].

6.1.4   Acid Activation of Clays

Smectite clays are often treated with strong mineral acid (acid activated) [75] to give materials of very high surface area (increasing from *ca* 60 $m^2\,g^{-1}$ to *ca* 300 $m^2\,g^{-1}$) which have excellent activity as adsorbents [76] and catalysts [77] (Figure 6). The application of acid-activated clays as a developer for carbonless copying paper [78], requires a high brightness of the material.   Acid activation improves the brightness mainly by removal of structural Fe^{3+} cations which cause the clay to be a grey or yellow colour.

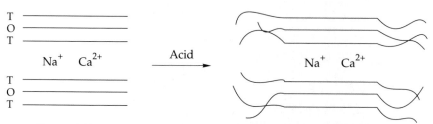

Figure 6 Diagrammatic representation of the effects of acid activation

The acid activation process is often quite severe (>5 mol dm^{-3} hot mineral acid for several hours) and destroys much of the bentonite layer structure as it removes iron, aluminium and magnesium from the octahedral layer [79]. Scanning electron microscopy [79] shows that the edges of the clay particles become very disordered and consist mostly of 'floppy' silica sheets (see Figures 7 and 8). The exchangeable cations are replaced mainly by Al^{3+} and H^+-cations.

6.1.5   Pillaring of Clays

When clays are compared as catalysts to the more rigid cage-like zeolites [80], they are found to be comparable at low temperatures but, due to a tendency to dehydrate and collapse their layers, clays are usually inferior at high temperatures (> *ca* 150 °C). One way to overcome this disadvantage is to use inorganic cations [15–18] to 'prop' the layers apart. Such a process is known as pillaring and in addition to the great improvements in structural integrity which are often obtained [81], these materials have the added advantage that the pillars themselves may be catalytically active [82].   Pillaring may also confer enhanced shape selectivity upon the clay [18].

Whilst certain organic molecules are capable of pillaring between the clay layers (e.g. 1,4-diaza-bicyclo[2.2.2]octane (1) [61, 83] and alkylammonium ions [84]), such pillars as expected are thermally unstable, thus, a variety of more stable inorganic pillars has been synthesised.

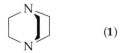

(1)

Numerous methods have appeared in the literature (e.g. [15, 81, 85–88]) for the preparation of alumina pillared clays with Δd's of *ca* 18 Å. Hydroxyaluminium pillars are readily produced either from alkaline solutions of aluminium ions or from a material called aluminium chlorhydrol (ACH, supplied by the Reheis Chemical Co, USA). The main polycationic species present in both of these solutions appears to be: $[Al_{13}O_4(OH)_{24}(H_2O)_{12}]^{7+}$ [89], which is first cation exchanged into the clay and the clay is washed with water. The polycation is then dehydrated by heating to give the alumina pillars. Literature procedures are rather vague as to the degree of washing and heating required to form the pillars, two factors which seem to be critical to the success of pillaring [90]. Table 1 shows the effect that the degree of washing can have on the interlayer spacing, Δd, of the clay before heating; clearly too much washing and the adsorbed polycation decomposes. Calcining at 200 °C for 15 minutes consistently gave a final Δd of 8.06 Å, whilst a temperature of 400 °C occasionally produced an interlayer distance of 8.06 Å, but usually the layers collapsed to 3.40 Å. This shows that the alumina pillared clays are not as stable as might be hoped and can be far less stable than zeolites.

The structure of the alumina pillars appears to be a column of 13 alumina octahedra arranged in three layers [16]. The top layer shares one oxygen atom with an inverted SiO_4 of the tetrahedral layer, whilst the bottom layer shares two oxygens with the lower tetrahedral layer.

Solutions of $[Nb_6Cl_{12}]^{2/3+}$ ions can be used to displace Na^+ ions from the interlayer of clays and form pillared clays with Δd spacings of *ca* 9 Å [16]. Hydrolysis of these pillars occurs when the clays are calcined *in vacuo* at 240 °C, the water present appears to oxidise the niobium to niobium(V) keeping the layers at a Δd of 9 Å. Tantalum solutions behave analogously.

Zirconia pillars are easily introduced by exchanging the clay with a solution of zirconium oxychloride and then calcining [91]. Such pillars appear to be highly stable.

Table 1 The effect of washing water on Al-pillared clay

Sample*	Proportion of washing water	Δd/ Å after calcining
Na$^+$-Tonsil 13	-	3.40
ACH-Tonsil 13	0	4.26
ACH-Tonsil 13	60%	7.21
ACH-Tonsil 13	600%	7.78
ACH-Tonsil 13	Until Cl$^-$ free	9.94 or 3.40

* Tonsil 13 is a German Montmorillonite from Moosburg, near Munich, and is supplied by Süd Chemie AG; ACH = aluminium chlorhydrol exchanged.

Hydroxy-chromium pillars [92] (Section 6.2.6.1) can be produced in various sizes by exchanging clays with solutions of Cr^{3+} cations which have been hydrolysed and polymerised by raising their pH with various proportions of sodium carbonate solution and heating for *ca* 17 hours at 95 °C [93, 94]. Each pillared clay has a different colour [90], a different sized pillar, and hence a different Δd [93]. Hydroxy-copper pillars have also been reported [95].

A second approach to pillaring clay has been introduced by Brindley *et al.* [96, 97]. The clay is exchanged with metal cations and then alkaline hydrolysis produces pillared clays with relatively low surface areas (Equation (3)).

$$\text{Ni}^{2+}\text{-Montmorillonite} \xrightarrow{\text{Ni}^{2+}, \text{OH}^-} [\text{Ni}_x(\text{OH})_{2x-y}]^{y+}\text{-Montmorillonite} \quad (3)$$

Vanadium-pillared clays have been used as regioselective oxidising agents [98, 99] (Section 6.2.4.1).

6.1.6 Clays as Supports for Reagents and Complexes

Metal salts may be deposited as a thin layer on the surface of an 'inert', high surface-area support such as an acid-activated clay. The usual method of deposition is to dissolve the metal salt in an *organic* solvent such as acetone, mix in the clay and evaporate *in vacuo* [e.g. 11–13]. Such thin layered metal salts often have very high and unusual reactivity due to the large number of exposed crystal edges.

Laszlo and Cornelis have pioneered the synthetic use of clay-supported inorganic reagents [e.g. 11–13]. One of their reagents, claycop (copper(II) nitrate supported on the acid activated clay K-10), has appeared in a recent Aldrich Chemical Catalogue (1990–91), where it is described as a powerful and selective nitrating agent for phenols [11]. Another of their materials, clayfen (iron(III) nitrate supported on K-10), has similar reactivity to claycop, but is less stable and is best freshly prepared.

A recent patent [13] shows that the Lewis acidity of poor Lewis acids such as copper, magnesium and zinc chlorides can be dramatically enhanced by support on an acid-activated clay.

Various transition metal complexes have been adsorbed into clays (e.g. tertiary arsine complexes of Ru(II) [100]) or synthesised from transition metal cations exchanged on to the inner surface of clays (e.g. tertiary phosphine complexes of Rh(I) [14]). These adsorbed complexes have been used as efficient heterogeneous catalysts for the hydrogenation of alkenes. Other (chiral) complexes have been used to induce asymmetry into organic syntheses [6, 7].

6.2 CLAY-MEDIATED ORGANIC REACTIONS

Clay-mediated organic reactions are of two general types; those carried out in a solvent at relatively low temperatures, i.e. <150 to 200 °C, and those carried out at higher temperatures, usually with the reactants in the gas phase. When non-acid-activated clays are used, the specificity of the solution reactions is often influenced by the interlayer spacing of the mineral, whilst at higher temperatures this becomes less important (unless the clay is pillared) due to the collapse of the structure. A further difference between the two temperatures is that as the clay dehydrates, it loses Brønsted acidity but gains Lewis acidity. Selected topics in the following survey will be treated in a little more depth so as to convey the versatility and limitations of clay-mediated organic reactions.

6.2.1 Clays as Brønsted Acids

For cation-exchanged clays, Brønsted sites essentially arise from dissociation of the water in the hydration sphere of the exchange cation (see Section 6.1.3). This acidity protonates adsorbed nitrogen bases such as amines and ammonia and, as expected, the equilibrium constant depends upon the basicity (pK_b) of the base [4]. These bases are not only protonated by Ca^{2+} or Mg^{2+} montmorillonite, but they also convert some of the aqua metal cations into non-exchangeable hydroxides [101–104]. Stronger bases (e.g. ammonia, pK_b 4.74) are adsorbed nearly to the cation exchange capacity of the clay, whilst weaker bases like pyridine (pK_b

8.85) are only adsorbed to about half of the exchange capacity [105]. Extremely weak bases (e.g. acetamide (pK_b 13.34) and urea (pK_b 13.82)) require M^{3+} cations for protonation [106]. When the amount of base exceeds the acidity available in the clay a hemi-salt $((B\cdots H\cdots B)^+)$ may form [102, 106–110].

Metalloporphyrins have been shown to become protonated and to be de-metallated by very dry montmorillonite and hectorite clays, re-metallation occurring as the acidity is lowered [111–113]. This shows that the clay must be at least as acidic as concentrated sulphuric acid as, for example, tin porphyrins do not protonate under less severe conditions [114].

Both cationic and zwitterionic forms of amino acids have been detected on a montmorillonite surface by IR spectroscopy [115]. Interestingly these amino acids were converted to peptides on heating to 180 °C.

6.2.2 Brønsted Acid Catalysis

The inherent Brønsted acidity of cation exchanged clays can be used to catalyse a very large number of organic reactions [e.g. 4, 9]. The catalytic reactions are often clean, high yielding and can offer routes to materials which are inaccessible or difficult to obtain by other more conventional means.

As mentioned previously (Section 6.1.3), the Brønsted acidity of clays usually arises due to the dissociation of water molecules in the primary hydration sphere of exchangeable metal cations in the clay interlayer region. Thus, due to the restricted environment in the clay interlayer, shape selectivity of the reaction products is often observed in cation exchanged clay catalysed reactions [e.g. 116].

Adams *et al.* [117] have produced a set of guidelines for assessing the Brønsted acid activity of cation exchanged clays:

1. (a) Cr^{3+} and Fe^{3+} are the most active interlayer cations.
   (b) Although Al^{3+} is also active, the exact procedures used for the ion-exchange and washing steps are critical for giving catalysts of reproducible activity.
2. (a) Below 100 °C, the reactions proceed provided they involve tertiary or allylic carbocation intermediates.
   (b) At 150–180 °C reactions involving primary and secondary carbocations are possible.
3. (a) Reactions of carbocations with unsaturated hydrocarbons take place overwhelmingly in the interlayer region of the clay, where the hydrocarbon double bond can be polarised.

(b) Reactions of carbocations with polar, oxygenated, species can take place on the surface of the clay particles as well as in the interlayer space.

4. (a) When acid-catalyzed reactions are performed in the liquid phase and involve tertiary carbocations, the most suitable solvents are those that provide miscibility; 1,4-dioxan is especially good.

(b) When more acid conditions are required for the formation of primary and secondary carbocations, a non-polar solvent is more efficacious.

Laszlo [118] has pointed out that the involvement of primary carbocations (rules 2(b) and 4(b)) is unlikely. Moreover, in rule 2(a) benzyl carbocations should also be included. Furthermore, as the acidity of the clay depends critically upon its water content [62, 63], it is usually advisable to equilibrate the clay to a known relative humidity (RH) to produce a reproducible acid strength [116].

As might be expected, the carbocations are most readily produced from tertiary alcohols (Equation (4)), e.g. *t*-butanol [e.g. 90, 119] and 1-phenyl-alkanols [120–122], or from substituted alkenes [e.g. 117]. The carbonium ion formed can subsequently react with nucleophiles such as water, alcohols, acids [e.g. 117], hydrogen sulphide [121, 122] and thiols [123, 124].

e.g.

$$\text{(4)}$$

Similarly, carbonium ions generated from primary amines by elimination of ammonia react with another molecule of primary amine to give a secondary amine [125]. Secondary and primary alcohols and terminal alkenes require higher temperatures [66] with, for example, 1-butanol only beginning to show products derived from butyl carbonium ions (N.B. the 1-butyl carbonium ion readily rearranges to the more stable 2-butyl carbonium ion) at 150 °C.

Aryl carbocations can be formed from diazonium salts in clays [126].

6.2.2.1 Deuterium Exchange

Acid-catalysed deuterium exchange can be carried out with fully deuterated (via ^2H_2O) clays [127]. Chloroform solutions of β-keto esters and β-diketones rapidly exchange their methylene protons whereas pyrrole exchanges all of its protons. Indoles exchange primarily at the 3-position, but prolonged exposure leads to further exchange.

6.2.2.2 Ether formation

M^{3+}-Exchanged montmorillonites can readily catalyse the addition of primary alcohols to alkenes to give ethers [66, 128]. Reactions work best if the acidity of the clay is increased by equilibration at low % RH (*ca* 12% is usually suitable) [116, 129]. M^{2+}-exchanged clays show some reactivity whilst M^+-exchanged clays are essentially unreactive [9, 83], thus confirming that the Brønsted acidity of the clay is mainly due to dissociation of water molecules in the hydration sphere of exchangeable interlayer cations. Further confirmation of this is given by the fact that if the clay catalyst is replaced by clays exchanged with diprotonated 1, 4-diazabicyclo[2.2.2]octane (1) [83] or heated for an extended period at a high temperature (*ca* 500 °C) to collapse the clay layers, the activity of the clay falls drastically [66]. For example, at 95 °C a collapsed M^{3+} clay has only *ca* 25% of the activity of a non-calcined clay [66]. Furthermore, at 150 °C the collapsed clay simply produces the same products as at 95 °C, whilst the normal clay begins to show products (e.g. alkene dimers and ethers) derived from reactions of the primary alcohol components of the reaction mixture. This suggests that there are two reaction sites, a surface site and a more reactive interlayer cationic site. Both sites can catalyse the reactions of carbocations derived from alkenes with alcohols or water, whilst the interlayer site can also catalyse the formation of alkenes from primary alcohols at high enough temperatures.

Terminal alkenes, such as 1-hexene, react slowly at 150 °C to give a mixture of 2-alkyl and 3-alkyl ethers (Equation (5)) via rearrangement of the intermediate secondary carbonium ion [66].

$$CH_2{=}CH{-}CH_2{-}R \xrightarrow[]{M^{3+}\text{-Clay}} CH_3{-}CH{\overset{CH_2{-}R}{\underset{O{-}R'}{<}}} + CH_3{\underset{CH_2{-}CH}{\overset{R}{<}}} \quad (5)$$

Terminal alkenes alone can react with the water present in M^{3+} cation-exchanged montmorillonites to produce 2-alkyl and 3-alkyl ether 'dimers' [116, 129]. The dialkyl ethers are produced in a reaction which proceeds by intermediate formation of an alcohol by hydration of the 1-alkene (Equation (6)). The reaction can also be achieved, at a slower rate, but in a more shape selective manner, with a Cu^{2+}-exchanged smectite, which produces the 2,2'-dialkyl ether almost exclusively [130, 131]. The reaction is not truly catalytic as the interlayer water cannot be effectively replenished without reducing the acidity of the clay too far.

$$
\begin{array}{c}
CH_2\!=\!CH\!-\!R' \\
\diagdown \; + \\
\diagdown \; H^+
\end{array}
\xrightarrow{Cu^{2+}\text{-Clay}}
\begin{array}{c}
\overset{+}{CH_3}\!-\!CH\!-\!R' \\
\diagup \; + \\
\diagdown \; :OH_2
\end{array}
\xrightarrow{-H^+}
\begin{array}{c}
CH_3\!-\!CH\!-\!R' \\
| \\
OH
\end{array}
\qquad (6)
$$

Substituted alkenes readily (<100 °C) protonate in clays to the more stable tertiary carbonium ions which can react with alcohols to give high yields of ethers [117]. 1,4-Butadiene can produce allylic carbonium ions, again below 100 °C [117], which react with methanol, for example, to give mainly 1-methoxy-2-butene. The ether is accompanied by a mixture of cyclic unsaturated C_6 hydrocarbons (formed by Brønsted processes) and the Diels-Alder product, 4-vinylcyclohexene (Section 6.2.6.1).

M^{3+}-Exchanged montmorillonites convert isobutene and methanol into the petroleum anti-knock additive, methyl *t*-butyl ether (Equation (7)) in high yield [83, 119, 132, 133]. Only small amounts of the by-products dimethyl ether, *t*-butanol, and what appears to be di(*t*-butyl ether), can be detected. Yields of ~60% after 50 minutes at 60 °C are achieved, an activity of *ca* 60% relative to the commercial ion exchanged resin used for this process [130]. Divalent cations give <10% yield under similar conditions [83]. *t*-Butanol can conveniently replace gaseous isobutene in preliminary experiments as it rapidly eliminates water in the acidic environment of the clay to give the alkene [90, 119], however, reactions are slower and the water formed eventually 'poisons' the clay catalyst.

$$
\begin{array}{c}
CH_3 \\
\diagdown \\
\quad\; C\!=\!CH_2 \\
\diagup \\
CH_3
\end{array}
\xrightarrow{H^+ \;\; H^+\text{-Clay}}
\begin{array}{c}
CH_3 \\
\diagdown \\
\quad\; C^+\!-\!CH_3 \\
\diagup \\
CH_3
\end{array}
\xrightarrow[-H^+]{MeOH}
\begin{array}{c}
\quad\quad CH_3 \\
\quad\quad \diagdown \\
CH_3\!-\!C\!-\!O \\
\diagup \quad\quad \diagdown \\
CH_3 \quad\quad CH_3
\end{array}
\qquad (7)
$$

Solvent effects for these reactions can be very great, with 1,4-dioxan producing a 6-fold rate increase over the solvent-free reaction [119]. Other solvents, whether polar, non-polar or coordinating, produce reaction rates which are less than 10% of that observed with 1,4-dioxan [83, 119]. The reason for the high activity of 1,4-dioxan is somewhat unclear as some of the other solvents used in the study (e.g. tetrahydrofuran) should produce similar miscibility and solubility effects, but in fact the reaction rate is about an order of magnitude lower than with 1,4-dioxan [83].

Primary alcohols give 1,1'-dialkylethers with Al^{3+} or Fe^{3+} clays [134], the best rates being observed with Al^{3+} cations. Addition of benzidine to the clay appears to enhance the rate of reaction, possibly due to benzidinium cations catalysing alkene formation [135].

Alkylene oxides have been used in the presence of bentonites [136] or group 11 or 12 cation-exchanged montmorillonites [137] to hydroxy-alkylate a wide range of alcohols.

Al^{3+} montmorillonite catalyses the cyclodehydration of alkanediols and dithiols to the corresponding cyclic ethers and thioethers [137, 138]. Experiments with S-(+)-pentane-1,4-diol [139] show competition for protonation at the primary and secondary alcohols followed by S_N2 dehydration to the heterocycle. Alumina-pillared clay is a poorer catalyst. The selective synthesis of O-alkylated phenols has been described [140].

Conversely, various clays and pillared clays can be used to cleave the Me–O bond of various anisoles [141]; pillared clays gave the best rates.

6.2.2.3 Ester and Lactone Formation

1-Phenylalkanols and a carboxylic acid readily form benzyl-carbonium-ion-derived esters (Equation (8)) on treatment with an acid-activated clay (K-10) [121, 122], e.g.

$$CH_2{=}CH_2 \ + \ HOOC{-}CH_3 \ \xrightarrow{\text{H}^+\text{-Montmorillonite}} \ C_2H_5O{-}\overset{\displaystyle O}{\underset{\displaystyle CH_3}{C}} \qquad (8)$$

Alkenes [142, 143], chloroalkanes [144] or alcohols [145] can esterify carboxylic acids (Equation (9)) over acid treated clays, either in the gas phase or in an autoclave, e.g.

$$C_3H_7{-}Cl \ + \ HOOC{-}CH_3 \ \xrightarrow[\text{200 °C, Autoclave}]{\text{H}^+\text{-Montmorillonite}} \ C_3H_7O{-}\overset{\displaystyle O}{\underset{\displaystyle CH_3}{C}} \qquad (9)$$

Cyclooctene-5-carboxylic acid forms a variety of lactone products, via a secondary carbonium ion intermediate, under reflux with a high boiling solvent (>140 °C) and M^{2+}- or M^{3+}- exchanged montmorillonite [117, 146].

A clay-catalysed ene reaction has been used to synthesise γ-lactones [147], e.g. isobutene and diethyl ketomalonate react over a cation exchanged or acid activated montmorillonite or a kaolin to give lactones.

Acid-doped clays were best, affording the highest reaction rates at low temperatures and demonstrating good regiospecificity.

Ketenes readily add to alcohols in the presence of acid-activated clays to give esters [148].

6.2.2.4 *Formation of Amides and Peptides*

Glycine, activated by intercalation into Cu^{2+} exchanged montmorillonite, gives low yields of hippuric acid with benzoic acid [149].

Amino acid adenylates give polypeptides on reaction with montmorillonites [150–152]. It has been suggested that the amino acid adenylate binds to the clay outer surface with the terminal amino group protruding into the activating interlayer region.

Al^{3+} Montmorillonite catalyses a Ritter type reaction between tertiary alcohols and nitriles to give substituted amides [153].

6.2.2.5 *Acetal, Ketal, Thioketal, Imine and Enamine Formation (Protection/Activation of Carbonyl Groups)*

Cyclic acetals and ketals can be synthesised from 1,2-diols [154] or epoxides [155] and carbonyl compounds in the presence of acid-activated or cation-exchanged clays. Acetic anhydride gives diacylate acetals under similar conditions [156].

Phenols and alcohols can be cleanly tetrahydropyranylated with dihydro-4*H*-pyran in the presence of an acid-activated clay (K-10) [157] (Equation (**10**)). Conversely, deprotection may be achieved by treatment of the tetrahydropyranyl ether with acetone and bentonite [158].

$$\text{(ring)O} + R{-}OH \xrightarrow[CH_2Cl_2]{K\text{-}10} \text{(ring)O-OR} \qquad (10)$$

A wide range of acetals and ketals have been synthesised, in moderate yield (40–70%), in the presence of a montmorillonite catalyst, from diols and enol ethers under reflux in benzene [159] (Equation (**11**)). It has been suggested that the reaction mechanism involves a protonated aldehyde or ketone [3], e.g.

$$\text{(diol)}\begin{matrix}OH\\OH\end{matrix} + \text{(ring)}{-}OCH_2CH_3 \xrightarrow[\text{Benzene}]{\Delta,\ \text{Clay}} \text{(product)} \qquad (11)$$

Trialkyl orthoformates react rapidly with carbonyl compounds in the presence of acid-activated clays to give acetals and ketals [160], whilst quaternary ammonium montmorillonite has been used as a phase transfer catalyst in the acetalisation of formaldehyde by alcohols in dichloromethane/aqueous sodium hydroxide mixtures [161]. Formaldehyde diethyl acetal can be used to protect alcohols by alcohol exchange in the presence of acid-activated clay [162] (Equation (12)), whilst a similar catalyst mediates the addition of diethyl acetals to enol ethers [163, 164], e.g.

$$(12)$$

Reasonable yields of dithianes, thioacetals, thioketals and thiochromanes have been synthesised from various ketones or aldehydes and the corresponding thiols in the presence of an acid-activated montmorillonite (KSF-Süd Chemie) [165] (Equation (13)), e.g.

$$HCHO \ + \ HS(CH_2)_3SH \xrightarrow[\text{Reflux}]{\text{KSF, Toluene}}$$

60% (13)

Montmorillonites and acid-activated clays have been used to form an imine in the initial stage of a synthesis of phenylalanine [166] (Equation (14)).

$$(14)$$

Acid-activated clays catalyse the formation of enamines from secondary amines (e.g. morpholine) and ketones in refluxing benzene [167].

6.2.2.6 Cleavage of 'Protected' Carbonyl Groups

Aldoximes and ketoximes can be deoxinated to their respective aldehydes or ketones by treatment with bentonite [168] (Eqns. (15) and (16)).

$$(15)$$

Where: R = OH, Me; R' = Me, H; and R = OMe, NO_2; R' = Me.

$$(16)$$

Where: R = C_3H_7; or R,R = $(CH_2)_5$.

Beckmann rearrangement accompanies the cleavage in certain cases (Equation (**17**)).

$$(17)$$

Where: R = OH, OMe; R' = Me.

Semicarbazones may be cleaved in up to quantitative yield by bentonite in ethyl acetate with a trace of hydrochloric acid [169] (Equation (**18**)), e.g.

$$100\% \quad (18)$$

Similar results to the above have been reported for clay-supported reagents such as clayfen and chromyl chloride (Section 6.2.7.1).

6.2.2.7 Cyclic Anhydrides

α,ω-Dicarboxylic acids generally require quite severe conditions to convert them to cyclic anhydrides [170], e.g. hot, concentrated sulphuric acid. Al^{3+}- [171], Fe^{3+}- or Cr^{3+}-exchanged [172] montmorillonite catalyses these reactions, usually with much higher yield (up to 100%) and specificity. Water is removed azeotropically from the reaction mixture (Equation (**19**)).

$$\text{diacid} \xrightarrow[\substack{\text{Solvent}\\\text{Reflux}}]{\text{M}^{3+}\text{-Clay}} \text{cyclic anhydride} + H_2O \qquad (19)$$

The mechanism of the reaction appears to involve initial protonation of one of the carbonyls of the diacid by the M^{3+}-exchanged clay [171].

Solvent effects are extremely important with the solvent controlling both the reaction temperature and the interlayer distance (Δd) of the clay (Table 2). Thus, toluene (b.p. 111 °C) is a good solvent for formation of 5-ring cyclic anhydrides, whilst longer chain diacids require higher boiling solvents. Large rings are formed slowly, probably due to poor entropy and ring-strain factors.

Table 2 Reaction times* and Δd for Al^{3+} montmorillonite in various solvents

Solvent	$\Delta d/\text{Å}$	B.p./ °C	% Conversion (with time/h)	
			Itaconic acid (2)	Glutaric acid
1,1,1-Trichloroethane	14.0	75	None	None
Iso-octane		99	100 (18)	None
Water (fresh clay)	3.2	100		
1,4-Dioxan	5.0	101	100 (18)	None (48)
Toluene	5.5	111	100 (2.5)	Trace (72)
Mixture of 1,1,1-Trichloroethane and 1,1,2,2-Tetrachloroethane (ratio adjusted to give b.p. 125 °C)		125	100 (7)	100 (12)
n-Octane	2.9	126	25 (24)	
Ethylbenzene		136	100 (4)	Little Product
o-Xylene		146		88 (22.5)
1,1,2,2-Tetrachloroethane	5.5	147	100 (3.5)	100 (7)

* All reactions were carried out at reflux using 0.5 g of clay, 2 g of diacid and 50 cm^3 of solvent

In contrast to similar cyclic anhydride formation reactions with zeolites (3A [171], 4A or 5A [172] molecular sieves), the reactions were very specific in that the double bond of itaconic acid does not migrate to give the more thermodynamically stable citraconic anhydride [173] (Equation (**20**)).

$$(20)$$

The method of exchange of the clay is very important if Z/E-isomerisation of exocyclic double bonds is to be avoided [172]. Butylidenesuccinic acid can isomerise with montmorillonite which has been exchanged at a relatively high pH and with a low excess of aluminium (Equation (**21**)), whilst 2-phenylethylidenesuccinic acid does not (Equation (**22**)).

$$(21)$$

$$(22)$$

It is supposed that a crude form of pillaring occurs at high pH (Section 6.1.5) which gives an interlayer spacing (Δd) sufficient to allow rotation only of the smaller butylidene group. Isomerisation of the double bond into the ring does not occur in either of these reactions, which suggests that the alkylidene bond may also be coordinated to the active site during the reaction (Equation (23)).

$$(23)$$

Facile Brønsted catalysis of the Z/E-isomerisation of fulgides (complex cyclic anhydrides) has been achieved with cation-exchanged (especially M^{3+}) bentonites [174]. Ring-closure of the fulgide is also found, which is surprising as such reactions are normally thermally-disallowed, photochemical processes [175] (Equation (24)). Large bathochromic shifts in the λ_{max} of the fulgide and ring-closed products relative to the compounds in free solution were observed. Protonation of the carbonyl group conjugated to the aryl moiety appears to be a necessary step in the reaction mechanism.

$$(24)$$

6.2.2.8 Synthesis of Some Heterocyclic Compounds

Phenols react with α,β-unsaturated carboxylic acids in the presence of Al^{3+}-exchanged montmorillonite to give chromanones in reasonable yield [176] (Equation (25)). However, the reaction gives several by-products and cleaner products are obtained with strong acid polymers, e.g.

$$(25)$$

Primary amines react with 1,4-diketones on an acid-activated clay (K-10) to give pyrroles in 90–98% yield [177] (Equation (**26**)).

$$\text{(diagram)} + R-NH_2 \xrightarrow[\underline{No}\ Solvent]{K\text{-}10} \text{(pyrrole)} \qquad (26)$$

Where: R = Et, Pr, Bz, Ph

Cognate reactions with hydrazines and 1,3-diketones give similar yields of pyrazoles (Equation (**27**)).

$$\text{(diketone)} + R''-NHNH_2 \xrightarrow[\underline{No}\ Solvent]{K\text{-}10} \text{(pyrazole)} \qquad (27)$$

Where: R, R' = Me, Ph; and R'' = Me, Ph

Microwave irradiation of ketones with phenyl hydrazine in the presence of acid-activated clay (KSF-Süd Chemie) gave high yields of 2-substituted indoles [178].

Dihydrothiazines, which are of interest in the synthesis of cephalosporin antibiotics, may be synthesised from thioamides and α,β-unsaturated ketones in the presence of acid catalysts [179]. Strong acid catalysts, usually anhydrous hydrogen chloride, mediate both the ring-closure reaction and dehydration of the intermediate tetrahydrothiazine. A similar reaction with Al^{3+}-, Cr^{3+}- or Fe^{3+}- exchanged montmorillonite as catalyst allows the tetrahydrothiazine to be isolated [180], such a product being unobtainable by other, more traditional means (Equation (**28**)).

$$ (28) $$

6.2.2.9 *Acid-catalysed Rearrangements and Ring Closures*

Clay catalysts rearrange polychlorocyclohexadienones to phenols [181]. Clay intercalates of 1,2-diols undergo pinacol rearrangement, on heating, to give 2-hydroxyketones [182, 183] in preference to simple dehydration. Selectivity increases with higher interlayer cation acidity and in the case of 2,3-diphenylbutan-2,3-diol both methyl and phenyl migration occur, the product ratio being different from that obtained under homogeneous conditions and on the surfaces of other solid acid catalysts (Equation (29)).

Montmorillonites catalyse the rearrangement of allyl phenyl ethers to *o*-allyl phenols [184] under very mild conditions. Extended reaction times result in ring-closure to give chromans or coumarins (Equation (30)), e.g.

α-Pinene rearranges to camphene and tricyclene, on treatment with 10% HCl-activated clays [185]. The rate of reaction appears to be related to the SiO_2/Al_2O_3 ratio of the clays used. In a process which mimics a proposed biogenetic pathway, the germacranolide (2) rearranged on bentonite to give a mixture of the two guaianolides (3) and (4) and the pseudoguaianolide (5) [186] (Equation (31)).

6.2.2.10 *Development of the Primary Dyes of Carbonless Copying Paper*

The Brønsted (and possibly Lewis) acidity of clay surfaces are used to develop carbonless copying paper leuco dyes. The colourless leuco dyes are contained in polymer microcapsules which are fractured under pressure, thus bringing the leuco dyes into contact with a so-called co-reactant [187], the major type of which is an acid-activated montmorillonite clay.

Two types of leuco dye are involved in the process, the primary and the secondary dyes [188, 189]. The primary dye develops immediately on adsorption onto the acid-activated clay, whilst the secondary dye develops more slowly, usually by a photoinitiated redox reaction (Section 6.2.4.3). The microcapsules are coated onto the back of the top sheet (CB) and the acid-activated clay is coated onto the front of the bottom sheet (CF). Intermediate sheets are coated front and back (CFB) (Figure 7).

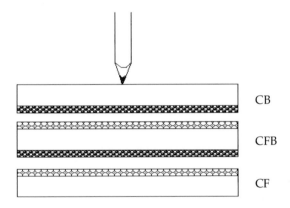

CB

CFB

CF

Figure 7 The operation of carbonless copying paper

A mixture of dyes is used to produce the colour black, but the commonest leuco dye used is crystal violet lactone (CVL), which produces the intensely-violet, propeller-shaped triphenylmethyl carbonium dye (really a zwitterion) on contact with the clay surface [190] (Equation (**32**)). Before use, a slurry of the clay is adjusted to a pH of 8–10, as too low a pH changes the colour of the dye by protonation of the $-NMe_2$ groups [61, 191].

(32)

The secondary dye is necessary as the primary dye has a low light fastness and fades on exposure to light, either by photochemical cleavage of the N–Me bonds [61, 192], or by Ph–C^+ cleavage to give a substituted benzophenone [61, 193, 194]. Singlet oxygen, produced via a photoexcited state of the dye itself [61], has also been implicated in the fading process.

6.2.3   Lewis-acid-catalysed Reactions

It has been shown, for example, by the thermal desorption of ammonia [195] and by the changes in the IR spectrum of pyridine adsorbed on clays [89, 196, 197], that as a clay is dehydrated with increasing temperature it loses its Brønsted acidity, and its Lewis acidity becomes more important. Many of the processes described below have complex mechanisms but are thought to be mainly Lewis-acid-catalysed.

6.2.3.1 Catalytic Cracking and Isomerisation Processes

The year 1930 showed the first industrial use of clays as catalysts for cracking petroleum [198]. At the high temperatures required (typically >300 °C), reactions appear to occur mainly by Lewis-acid-catalysed processes [199–201].

The efficiency of natural montmorillonites was considerably improved by acid activation [199, 202], which produced catalysts which were much more stable at the high temperatures required for the cracking reactions.

Although very many patents have appeared on the use of metal-doped and acid-activated clays as cracking catalysts, the major disadvantage with clays is that they are not fully stable to the hydrothermal (superheated steam) treatment required for oxidation of the coke build-up in the clay pores, during regeneration of the catalyst [203]. Thus, by the 1970's clays had been replaced by the more stable zeolites [203]. Recently however, pillaring of the clay layers by metal oxides (Section 6.1.5) has produced clays which rival the zeolites in stability [204, 205].

Besides simple cracking of long chain alkanes to shorter chain alkanes, clays doped with transition metals can also catalyse the isomerisation of *n*-alkanes to *iso*-alkanes [e.g. 206–209], probably via carbonium ion mechanisms [199–201]. An example [207] is given in Equation (33).

$$n\text{-}C_{4-7}\text{ alkanes} \xrightarrow[\text{H}_2,\ 250\ °\text{C}]{\text{Ni,Co-H}^+\text{ Montmorillonite}} iso\text{-}C_{4-7}\text{ alkanes} \quad 60\%\text{ yield} \quad (33)$$

In the absence of hydrogen alkenes are formed, thus alcohols such as 1-hexanol and cyclohexanol are dehydrated over clay catalysts at 250–450 °C, with significant amounts of isomerisation [210], to 2- and 3-hexenes and methylcyclopentenes respectively.

6.2.3.2 Methanol and 'Syn-Gas' Conversions

Alkenes (C_{2-4}) can be obtained from methanol (Equation (34)) by passage over either a cation-exchanged montmorillonite [211] or a tetramethylammonium montmorillonite into which silica has been deposited [212], or a zirconium pillared montmorillonite [213].

$$CH_3-OH \xrightarrow[350\ °\text{C}]{\text{Al}^{3+}\text{-Montmorillonite}} C_{2-4}\text{ alkenes} \quad (34)$$

Methyl formate has been obtained from methanol (Equation (35)) over a synthetic copper-exchanged fluorotetrasilicic mica (Cu-TSM) [214].

$$CH_3-OH \xrightarrow[300\ °\text{C}]{\text{Cu-TSM}} CH_3O-CHO \quad 68\% \quad (35)$$

Carbon monoxide and hydrogen mixtures (syn-gas) can be converted, at 175 °C, by transition-metal-exchanged- [215] or cobalt-substituted-montmorillonites [216] to a mixture of alkanes and alkenes (mainly *iso*-), in which the C_{5-12} hydrocarbons predominate. C_4–C_{10} hydrocarbons predominate with a Ru supported alumina pillared clay [217]. A zirconia-pillared clay/Cu/ZnO mixed catalyst gave methanol selectively [218].

6.2.3.3 Friedel-Crafts Alkylation and Acylation of Aromatics

Phenols can be readily C-alkylated by iodoalkanes [219] in the presence of Al^{3+} clay and can di-arylate formaldehyde [220] in the presence of acid-activated clay. Little *para*-selectivity was obtained with the products in the first case, but the latter process gave significant preference for the *para*-isomers. The mechanism of the latter reaction and those outlined below is almost certainly a mixture of Brønsted and Lewis processes.

A wide variety of alkenes (e.g. *n*-alkenes [221, 222], branched alkenes [223], substituted styrenes [224–226], cycloalkenes [227]) and polyalkenes [e.g. 221, 222] alkylate phenols with cation-exchanged or acid-activated clay catalysts. Polyalkylated [223] and *O*-alkylated [227] products were often obtained on modifying the reaction conditions. Similarly, alkenes plus anilines can give alkylanilines over clays [228, 229].

Unactivated, aromatic hydrocarbons can be alkylated by halides, alcohols or alkenes in the presence of cation-exchanged and pillared-clay catalysts [e.g. 230–234]. The catalyst efficiency appears to bear no relationship to that of the corresponding Lewis acids under homogeneous conditions. Toluene can be used to alkylate itself in the presence of bentonite and bromine [235] (Equation (**36**)). Aromatics with strong electron releasing groups give ring bromination only.

$$2C_6H_5\text{–}CH_3 + Br_2 \xrightarrow{\text{Bentonite}} \quad + \quad \tag{36}$$

33%

66%

Friedel-Crafts acylation of aromatic hydrocarbons with carboxylic acids can be catalysed by cation-exchanged montmorillonite [236] or Al^{3+}-exchanged tetrasilicic mica [237]. *para*-Substitution predominates and yields depend upon the exchange cation and the acid chain length [236].

6.2.3.4 Catalysis of Other Electrophilic Aromatic Substitution Reactions

Concentrated nitric acid plus montmorillonite nitrates toluene at 300 °C [238]; slight *para*-selectivity was noted. Improved reactivity (e.g. benzene, 195 °C) was obtained with cation-exchanged clays [239]. Mild nitrations were achieved with acyl nitrates (*in situ*) and clays [240–242], whilst toluene has been chlorinated in the presence of clay [243].

6.2.4   Redox Reactions

Redox reactions can be associated with transition metal sites in the clay, which reside either in the layers themselves, e.g. as Fe^{3+} replacement for octahedral Al^{3+}, or in the interlayer space as exchange cations, or as adsorbed metal complexes; the latter will be discussed with the clay supported reagents (Section 6.2.7.1).

6.2.4.1 Clay-mediated Oxidations

Aerobic oxidation of allylic alcohols can be achieved with Ni^{2+}-exchanged tetrasilicic mica [244], whilst Cr^{3+} mica can be used similarly to cleave oxidatively 2-propenyl aromatic compounds to acetophenones [245]. Clays exchanged with Co^{3+}, which has been generated *in situ*, selectively oxidise cyclohexane to a mixture of cyclohexanol and cyclohexanone [246].

Several patents have appeared for the acid-activated, clay-catalysed cleavage of cumene hydroperoxide to phenol and acetone [e.g. 247, 248].

Uranyl-exchanged, pillared-clays show higher selectivity than their non-pillared analogues for the photooxidation of ethanol and ether [82].

Yields in the clay-catalysed oxidative coupling of anisole to 4, 4'-dimethoxybiphenyl decreased with the reduction potential of the exchange-metal cation, i.e. $Fe^{3+}, Cu^{2+}, Co^{2+}, Al^{3+}$ and Ca^{2+} [249]. The same product can be obtained by a one-electron oxidation of 4-chloroanisole on Cu^{2+} smectite [250]; chloride ions were also detected. An analogous reaction occurs with anilines to give highly coloured (blue) benzidinium radical cations ($[p,p'\text{-}R_2NC_6H_4C_6H_4NR_2]^{+\cdot}$) [e.g. 251], which degrade further. The reaction does not require the presence of transition metals as the clay appears to force the benzidine into a planar conformation which lowers its redox potential [252].

Aluminium-pillared beidellite [253] and acid-activated montmorillonite [254] have been used to catalyse the hydroxylation of phenols and phenolic ethers with hydrogen peroxide. A vanadium-pillared clay showed markedly better yields with *para*-isomers compared to *ortho*-isomers in the hydrogen peroxide oxidation of ring-substituted benzyl alcohols [98], thus demonstrating the shape selectivity of the pillared clay. The same clay also showed unusual preference for epoxidation of internal allylic double bonds over those of terminal allylic alcohols [99].

High yields of epoxides are obtained from alkenes *gem*-disubstituted with electron-withdrawing groups (e.g. –CN and –COOR) and sodium hypochlorite [255]. The same conditions have been used to epoxidise alkenes generated *in situ* by a Knoevenagel reaction (Section 6.2.5.3).

6.2.4.2 Clay-mediated Reductions

Few reduction reactions mediated by clays alone have been reported, one example is the montmorillonite-(K-10) catalysed reduction of nitroarenes to anilines with hydrazine [256].

6.2.4.3 Carbonless Copying Paper Secondary Dye Development

As mentioned above (Section 6.2.2.10), carbonless copying paper uses two types of dye: primary and secondary. The primary dye develops by a clay surface-acid catalysed mechanism, whilst the secondary dye develops more slowly by a photoinitiated redox process [257]. Typical leuco dyes used are benzoyl leuco methylene blue (6) [257] and carbazolyl blue (S-RB) (7) [258] which develop to give very light-fast dyes (Equation (37)).

(6)

(37)

However, secondary dyes like S-RB (7) also develop in the dark to a variable extent simply on contact with the acid-activated clay [61]. There appear to be two mechanisms for this dark development: the first is catalysis by coordination to the Lewis acid sites on the clay edges [3, 61], whilst the second appears to be a redox process associated with the Fe^{3+} cations in the clay layers themselves [61]. The following facts support these propositions: dye development is low with iron-deficient clays; polyphosphates, which bind to Lewis sites, severely reduce development as does the reduction of the Fe^{3+} cations to Fe^{2+} with hydrazine [61].

6.2.5 Promotion of Reactions via Reactive Intermediates

6.2.5.1 Aldol Reactions Catalysed by Clays

Na^+ montmorillonite catalyses the condensation of glycolaldehyde to monosaccharides (mainly hexoses) apparently by an aldol process [259]. Al^{3+}-Exchanged montmorillonite catalyses the cross-aldol addition of silyl enol ethers [260, 261] to carbonyl compounds or acetals. Similarly, silyl ketene acetals and carbonyls give 3-silylether esters [262] (Equation (38)).

$$RCHO + R'_2C:C(OMe)OSiMe_3 \xrightarrow{Al^{3+}\text{-Clay}} RC(OSiMe_3)CR'_2COOMe \quad (38)$$

Acetals and carbonyls can be allylated with the acid-activated clay K-10 [263] (Equation (39)).

$$\begin{matrix} RCH(OMe)_2 \\ + \\ CH_2:CHCH_2SiMe_3 \end{matrix} \xrightarrow[CH_2Cl_2]{K\text{-}10} RCH(OMe)CH_2CH:CH_2 + Me_3SiOMe \quad (39)$$

6.2.5.2 Michael-Type Additions

Dual catalysis of the Michael addition of β-diketones to α,β-unsaturated ketones by $FeCl_3$ (homogeneous) and $NiBr_2$/clay has been reported [264].

Michael adducts can be obtained from silyl ketene acetals or silyl enol ethers and α,β-unsaturated esters or ketones with clay catalysts [265, 266]. The reactions are versatile, highly regiospecific and the silyl products readily cleaved.

3-Unsubstituted indoles add α,β-unsaturated ketones and esters at the 3-position [267]. 3-Methylindoles gave 2-substituted products, whilst 3-benzylindoles gave 2-substitution by 2,3-migration of the benzyl group.

6.2.5.3 The Knoevenagel Reaction Catalysed by Clays

Synthesis of alkenes via the Knoevenagel reaction from aldehydes and C-acids (e.g. Meldrum's acid [268] or malonate derivatives [255]) has been accomplished by clay catalysts.

6.2.5.4 Generation of Carbenes — Cyclopropanation

Certain transition metals and their cations, e.g. Cu and Rh, are noted for their ability to coordinate alkenes and to generate carbenes from diazoalkanes [269]. This can be used in the cyclopropanation of alkenes (Equation (40)).

$$\text{(structure)} + N_2CHCOOEt \longrightarrow \text{(structure)} \begin{smallmatrix} H \\ COOEt \end{smallmatrix} + N_2 \qquad (40)$$

A number of alkenes and dienes have been successfully cyclopropanated by ethyl diazoacetate at room temperature in the presence of a Cu^{2+}-exchanged montmorillonite catalyst [6]. So far, no enhancement of stereoselectivity relative to other similar catalysts has been observed.

6.2.6   Cycloaddition Reactions

6.2.6.1 *Catalysis of Diels-Alder Reactions*

Uncatalysed Diels-Alder reactions often require prolonged reaction times at elevated temperature and pressure, thus a great deal of effort has been expended on evaluating numerous catalyst systems with the aim of improving the rate and stereoselectivity of the reaction. Brønsted and Lewis acid [270] systems have proven effective, as has addition of radical cation sources [271] and the use of aqueous solvents [272].

An early example of Diels-Alder reactions associated with clay catalysts can be found in the industrially-important cyclodimerisation of oleic acid with acid-activated clays [273]. The reaction requires a high temperature (>230 °C) to effect the preliminary dehydrogenation of the oleic acid, thus there may be no actual catalysis of the Diels-Alder step (Equation (41)).

$$\text{(reaction scheme)} \qquad (41)$$

H $\xrightarrow{-H_2}$ $\xrightarrow{>230\,°C}$

Acid-
activated
COOH   Clay   COOH   COOH   COOH
                                                      COOH

The first true example of a Diels-Alder reaction catalysed by a clay was reported in 1978 by Downing *et al.* [274], who dimerised butadiene to vinyl cyclohexene over a Cu^+ montmorillonite at 100 °C (Equation (42)).

(42)

Transition-metal-cation-exchanged montmorillonites give Diels-Alder type dimers with butadiene and isoprene at lower temperatures (the reactions begin at <50 °C with TM^{3+} cations) [275, 276]. Diene dimer formation increases with cationic charge, however, the majority of the dimers are simply Brønsted acid catalysed ring-closure products.

Fe^{3+}-exchanged K-10 (an acid-activated clay, Süd Chemie) catalyses the Diels-Alder dimerisation of 1,3-cyclohexadiene in dichloromethane [277]. The reaction requires >200 °C for several days when uncatalysed, but goes to completion in just a few hours at 0 °C with the Fe^{3+} clay. Addition of *t*-butylphenol (*ca* 10%) as a radical source significantly improves the reaction rate [277], indicating a radical cation mechanism. The *endo/exo*-isomer ratio (4:1) was the same for catalysed and uncatalysed reactions (Equation (**43**)).

Endo   Exo   (43)

Fe^{3+} K-10 catalyses the Diels-Alder reactions of methyl vinyl ketone [278, 279] or acrolein [280] with cyclopentadiene, furan and 2,5-dimethylfuran (Equation (**44**)). These reactions gave comparable stereo-selectivities, but with a slower rate, in water [278, 279].

Where: X = CH_2, R = H   Endo   Exo   (44)
and X = O, R = H or CH_3

Non-acid-activated transition metal cation-exchanged clays give better control of the stereoselectivity [9, 90, 281]. Although the rate with M^{3+} clays is much greater than with M^{2+} clays, the former produce a greater proportion of the *endo*-isomer (Table 3).

The greater the layer charge of the mineral the smaller the interlayer spacing and the lower the rate of reaction. The *endo/exo* ratio also decreased suggesting that the less bulky transition state (Figure 8), which leads to the *exo*-isomer, is favoured as the interlayer spacing becomes more restrictive.

Endo-Transition State   *Exo*-Transition State

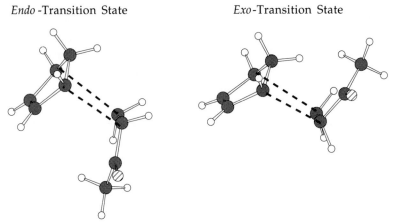

Figure 8 Comparison of the 'bulkiness' of the *Endo* and *Exo* transition states for the Diels-Alder reaction of cyclopentadiene with methyl vinyl ketone

Table 3 The clay-catalysed Diels-Alder reaction between cyclopentadiene and methyl vinyl ketone

M^{r+} Clay	% Yield in 20 min	Layer Charge[a]	Δd/ Å	*Endo/Exo*-Ratio
No Clay	17			19:1
Cr^{3+} Tonsil 13[b]	91	0.37	6.8	9:1
Fe^{3+} Tonsil 13	93	'		7:1
Al^{3+} Tonsil 13	20	'		11:1
Cu^{2+} Tonsil 13		'		5.5:1
Co^{2+} Tonsil 13		'		6:1
Ni^{2+} Tonsil 13		'		5.3:1
Fe^{3+} K-10	97	–	–	9:1
K-10				9:1
Cr^{3+} RLO1987[c]	80	0.60	3.2	6:1
Cr^{3+} Vermiculite[d]	43	0.65	3.1	2.5:1

[a] Charge per $(Si, Al)_4O_{10}(OH)_2$ unit; [b] A German montmorillonite; [c] A Brett Fuller's Earth; [d] A swelling vermiculite

Changing solvents had little effect on the *endo/exo* isomer ratio of the products, suggesting that the interlayer distance (Δd) is primarily controlled in this instance by the reactants and not the solvent [9, 90, 281].

A series of chromium-pillared clays (Table 4) were also good catalysts for these Diels-Alder reactions [90].

Table 4 Hydroxy-chromium-pillared clay catalysis of the Diels-Alder reaction between cyclopentadiene and methyl vinyl ketone

Ratio of Equivalents of Base : Cr	Colour of Clay	% Yield in 20 mins	Endo/Exo Ratio	Δd/ Å
0.0	Grey/Blue	90	8.5:1	4.08
0.5	Grey/Green	55	5.0:1	5.59
1.0	Light Khaki Green	42	4.5:1	6.18
1.5	Dark Green	59	5.5:1	16.69
2.0	Light Green	66	6.0:1	13.20

As expected, the *endo/exo* ratio decreased with lower reaction rate, however, the expected correlation with decreasing layer spacing, Δd, was not observed. This suggests that the 'width' of the pillars must be considered as well as their height, and that the formation of the less bulky *exo*-transition state becomes more favoured as the size of the interlayer pores becomes more restrictive. This pore-size-related selectivity is reminiscent of much of zeolite chemistry [282].

At low temperatures (*ca* −80 °C), when competition from the uncatalysed solution reaction becomes insignificant, a fairly delicate control of the stereoselectivity of the reaction between cyclopentadiene and methyl vinyl ketone can be achieved by selection of the water content of the clay [283] (the water content of clays can be controlled by equilibration in closed vessels over saturated salt solutions with various relative humidities (RH) [284]). Different profiles of isomer ratio versus RH are obtained with different metal cations and in the most favourable case (Cu^{2+} at 58% RH), an inversion of the normal *endo/exo* isomer ratio for this metal from *ca* 5.5 : 1 to *ca* 1:13 can be obtained.

Several other dienes, including 1,3-cyclohexadiene, isoprene, furan and pyrrole, but not the more aromatic thiophene, successfully react with methyl vinyl ketone in the presence of Cr^{3+}-exchanged montmorillonite [90, 281] (Equation (**45**)). With pyrrole the reaction soon came to a halt due to 'poisoning' by the highly-basic, secondary amine Diels-Alder product. Reactions of other dienophiles with cyclopentadiene showed limited success, e.g. methyl acrylate [9, 281] would only react with clay deferrated [285] prior to Cr^{3+} exchange, whilst methyl methacrylate simply polymerised onto the edges of the clay particle.

Where: $X = CH_2CH_2$, O or NH; $R = CH_3$ *Endo*   *Exo*
and $X = CH_2$; $R = CH_3$ or OCH_3

(45)

The mechanism of the catalysis is far from clear; both Brønsted and Lewis acid catalysis appear to be ruled out as the acidity of an Al^{3+} clay (which does not catalyse the reaction) would be expected to be greater than that of the transition metal M^{2+} clays, and comparable to that of the transition metal M^{3+} clays, both of which are catalysts. The necessity for a transition metal cation suggests a redox mechanism (one electron transfer to the dienophile [9]), but there appears to be no appreciable variation of rate with metals of quite different redox potential; e.g. $Cr^{3+} \rightarrow Cr^{2+}$, -0.41 V and $Fe^{3+} \rightarrow Fe^{2+}$, $+0.77$ V [286]. Whilst the presence of the clay will no doubt alter these solution redox potentials, it is difficult to imagine that they will become so similar as to give almost identical reaction rates. It is possible that the transition metal cations are acting as templates for the 'assembly' of the transition state, but there is no evidence for this and the question of the mechanism must remain open.

6.2.6.2 Other Cycloadditions

Stilbazolium cations adsorbed in clays undergo efficient and selective photodimerisation to the cyclobutane dimer [287]. *cis-trans* Isomerisation of the double bond was suppressed between the clay layers.

N-Benzylideneaniline and vinyl ethers give an azetidine by a regio- and stereo-specific clay-catalysed cycloaddition, together with a tetra-hydroquinoline, formed by *ortho*-insertion into the N-phenyl ring [288] (Equation (46)).

$$PhN = CHPh + R-CH=CH-OR' \longrightarrow$$

(46)

6.2.7 Clay-supported Reagents

There has recently been an upsurge in interest in the use of supported reagents as these heterogeneous reagents make the purification of the reaction products much simpler: the spent reagent is simply filtered off! The general characteristics of this type of chemistry are the low cost of the supported reagents, their often surprising reactivity and the ease with which reactions are carried out.

Much of the work to date uses acid-activated clays as a support, which provide a high surface dispersal of the reagent (see Section 6.1.6). Furthermore, the support can be used to stabilise an otherwise inaccessible species (e.g. anhydrous $Fe(NO_3)_3$ in clayfen) and even provide an active surface for further reaction (usually Brønsted-acid catalysed).

6.2.7.1 Stoichiometric Reactions with Clay-supported Reagents

Clayfen and claycop are acid-activated, clay- (K-10 from Süd Chemie) supported anhydrous iron(III) and copper(II) nitrate respectively. Due to its versatility and ease of use, the latter material has recently appeared in the Aldrich Chemical Co. catalogue of fine chemicals (reviewed [289]). Both materials are simply prepared by evaporating (rotary evaporator) acetone solutions of the nitrate salts onto the clay, where the salts exist as covalent, bidentate nitrato complexes [11, 12].

Iron(III) nitrate has not been isolated as the pure anhydrous compound and all attempts at its preparation result in decomposition. However, it does exist in solution in acetone [290] and can be stabilised by deposition onto the clay support. A similar effect is seen in the stabilisation of organolithium and Grignard reagents on dry clays [291].

Clayfen is reasonably stable under an inert atmosphere [2], but only survives a few hours of exposure to the atmosphere at room temperature. It is generally a very reactive material and has even been known to cause mild explosions in the dry state. In contrast to this, claycop is a much more stable material and can be left exposed to the atmosphere for several months [2].

Both reagents act as powerful nitrosating agents and as a source of nitrosonium ions compare very favourably with more expensive and less convenient reagents such as nitrosonium tetrafluoroborate, ytterbium nitrate and benzeneseleninic anhydride [2]. The following examples will serve to illustrate the versatility of these reagents.

Secondary alkyl alcohols are readily oxidised by clayfen and claycop to ketones under reflux in *n*-pentane or *n*-hexane (Equation (**47**)). Similar reactions with aromatic primary alcohols yield aldehydes only [292] (Equation (**48**)).

$$
\underset{\substack{|\\ \text{OH}}}{\text{R}-\text{CH}-\text{R}'} \xrightarrow[\substack{\text{Reflux}\\ n\text{-Alkane}}]{\substack{\text{Clayfen}\\ \text{or Claycop}}} \left[\underset{\substack{|\\ \text{O}-\text{NO}}}{\text{R}-\text{CH}-\text{R}'} \right] \xrightarrow[\text{Clay}]{\text{H}^+} \underset{\substack{\|\\ \text{O}}}{\text{R}-\text{C}-\text{R}'} \qquad (47)
$$

$$
\text{Ar}-\text{CH}_2\text{OH} \xrightarrow[\substack{\text{Reflux}\\ n\text{-Alkane}}]{\substack{\text{Clayfen}\\ \text{or Claycop}}} \text{Ar}-\text{CHO} \qquad (48)
$$

The fact that nitrous esters are formed as intermediates is confirmed by the fact that chiral alcohols and conformationally anchored cyclohexanols both retain their configuration after reaction [293]. The clay support is necessary to provide acidity for decomposition of the nitrite ester [294], as illustrated by Equations (**49**) and (**50**).

$$
\text{RR'CHONO} + \text{H}^+ \rightleftharpoons \underset{\substack{|\\ \text{H}}}{\text{RR'CH}-\text{O}^+-\text{NO}} \longrightarrow \text{RR'CHOH} + \text{NO}^+ \qquad (49)
$$

$$
\text{RR'CHONO} + \text{NO}^+ \longrightarrow \underset{\substack{|\\ \text{H}}}{\text{RR'C}}\!\!\overbrace{}^{\text{O}}\!\!\underset{\substack{\|\\ \text{O}}}{\text{N}^+}\!\!\overset{\overset{\text{O}}{\|}}{\text{N}} \longrightarrow \text{RR'C}{=}\text{O} + \text{H}^+ + 2\text{NO} \qquad (50)
$$

Benzoins may be readily oxidised to benzils in a very high yielding reaction [295] (Equation (**51**)).

$$
\underset{\substack{|\\ \text{OH}}}{\text{Ar}-\text{CH}-\text{CO}-\text{Ar}'} \xrightarrow[\substack{\text{Reflux}\\ n\text{-Alkane}}]{\substack{\text{Clayfen}\\ \text{or Claycop}}} \underset{\substack{\|\ \ \|\\ \text{O} \ \text{O}}}{\text{Ar}-\text{C}-\text{C}-\text{Ar}'} \qquad (51)
$$

Thiols are rapidly oxidised (minutes) to disulphides, the reaction proceeding via the highly-coloured alkyl thionitrite and the thiyl radical [296] (Equation (**52**)). It is possible that this process may be used as a convenient source of thiyl radicals for other reactions.

$$R-S-H \xrightarrow[\substack{\text{Room Temp.}\\\text{Solvent}}]{\substack{\text{Clayfen}\\\text{or Claycop}}} \underbrace{[R-S-NO]}_{\substack{\text{Intensely}\\\text{Coloured}}} \longrightarrow [R-S^\bullet] \longrightarrow R-S-S-R \quad (52)$$

Dithioketals are very useful protecting groups for carbonyl compounds, but their removal is often rather difficult. However, claycop and clayfen can cleave thioketals under very mild conditions and often with near quantitative yield [2, 12] (Equation (53)). When cyclic dithioketals are employed there is the added advantage that the polymeric disulphide by-product adsorbs onto the clay and can be removed by filtration. Selenoacetals are cleaved similarly [297].

$$\begin{array}{c} R \\ \diagdown \\ R' \end{array}\hspace{-0.3em}\begin{array}{c} S-R'' \\ \diagup \\ S-R'' \end{array} \xrightarrow[\substack{\text{Room Temp.}\\\text{Solvent}}]{\substack{\text{Clayfen}\\\text{or Claycop}}} \begin{array}{c} R \\ \diagdown \\ R' \end{array}\hspace{-0.3em}{=}O \quad (53)$$

Where:

$2\ \text{-S-R''} = \text{(-S-Et)}_2,\ \text{-S-(CH}_2)_2\text{-S-}\ \text{or}\ \text{-S-(CH}_2)_3\text{-S-}$

Thioketones may be converted to ketones by these reagents [298]. The intermediate appears to be the cyclic dioxo-disulphide represented in Equation (54).

$$\begin{array}{c} Ar \\ \diagdown \\ Ar \end{array}\hspace{-0.3em}{=}S \xrightarrow{\text{Clayfen}} \left[\begin{array}{c} Ar \\ | \\ O\!-\!O^{\diagup}Ar \\ Ar\!-\!\overset{\displaystyle |}{\underset{\displaystyle |}{}}S\!-\!S \\ Ar \end{array} \right] \longrightarrow \begin{array}{c} Ar \\ \diagdown \\ Ar \end{array}\hspace{-0.3em}{=}O \quad (54)$$

Reaction gives lower yield with Alkyl replacing Ar

Imino-type functional groups are often used in the derivatisation of aldehydes and ketones, but are rarely used as protecting groups due to difficulties in removing them after reaction. Clayfen, on the other hand, cleaves these imino-bonds rapidly and in good yield [299, 300] (Equation (55)).

$$\begin{array}{c} R \\ \diagdown \\ R' \end{array}\hspace{-0.3em}{=}N\text{-}X \xrightarrow{\text{Clayfen}} \begin{array}{c} R \\ \diagdown \\ R' \end{array}\hspace{-0.3em}{=}O \quad (55)$$

Where:

$X = \text{-NMe}_2,\ \text{-NH-CO-NH}_2,\ \text{-NH-Ph,}$
$\quad\quad \text{-NH-Ph-}p\text{-Me or -2,4-(NO}_2)_2\text{Ph}$

A similar cleavage of oximes (X = –OH) has been achieved with clay supported chromyl chloride [301], whilst clayfen has been used to oxidise an oxime to a nitro compound [302].

Clayfen can be used to convert hydrazides to the synthetically important azides in good yield [303] (Equation (**56**)).

$$R-NH-NH_2 \xrightarrow[CH_2Cl_2]{Clayfen} R-N_3 \qquad (56)$$

The mildness of the clayfen-mediated dehydrogenation of 1,4-dihydropyridines compares very favourably with the more vigorous traditional methods used for this reaction [304] (Equation (**57**)). Claycop is even more efficient, often giving yields in excess of 90% [305].

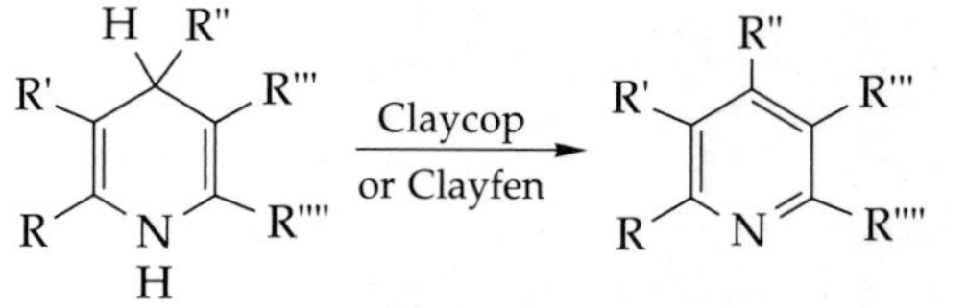

$$(57)$$

Nitrosonium ions have been implicated in the nitration of phenols, hence clayfen and claycop are found to be good reagents for the mild mono-nitration of phenols [11, 12, 305]. The reaction is regioselective in that no *meta*-products are formed (Equation (**58**)). For phenol itself, the *para*-isomer is favoured over a statistical distribution, whilst with *meta*-substituted phenols, one *ortho*-position is favoured over the other. Most other reactions follow the expected statistical distribution and normally labile groups such as aldehydes and nitriles are unaffected.

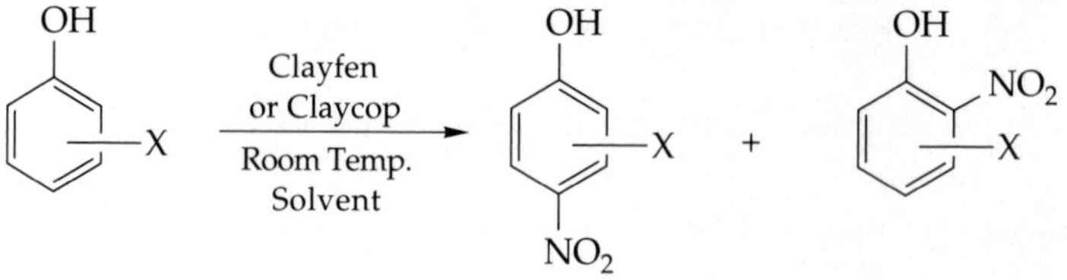

$$(58)$$

The mechanism in Scheme 1 has been proposed for these nitrations.

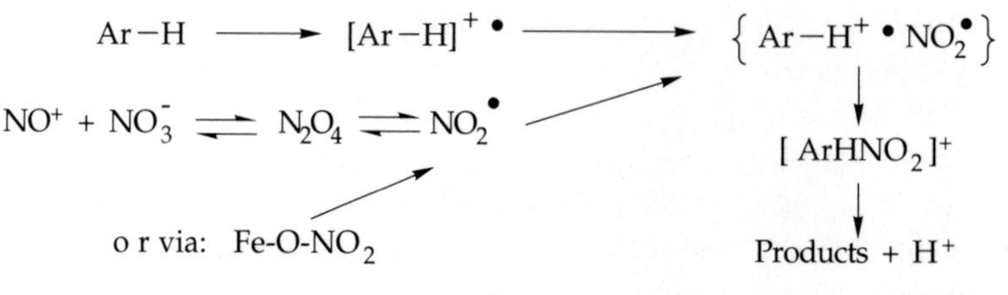

$$Ar-H \longrightarrow [Ar-H]^{+ \bullet} \longrightarrow \{ Ar-H^+ \bullet NO_2^\bullet \}$$

$$NO^+ + NO_3^- \rightleftharpoons N_2O_4 \rightleftharpoons NO_2^\bullet$$

$$[ArHNO_2]^+$$

$$o\,r\ via:\ Fe-O-NO_2$$

$$Products + H^+$$

Scheme 1 Proposed mechanism for the clay-catalysed nitration of phenols

Halobenzenes [306] and aromatic hydrocarbons [307–309] are also mononitrated by claycop in acetic anhydride, in reasonable to high yield and with good *para*-selectivity in many cases.

In the presence of a peracid, K-10 clay-supported $FeCl_3$ efficiently ring chlorinates toluene and anisole by a so-called biomimetic route [310]. The same supported reagent has been reported to chlorinate adamantane [2]. Mössbauer analysis showed that only Fe^{3+} remained after reaction.

Montmorillonite-supported sodium borohydride efficiently reduces ketones in dichloromethane [311].

Ketones may be oxidised to esters and styrene derivatives to acetals, by thallium(III) nitrate on K-10 [312]. This reagent has also been used to oxidise α-methylpyrroles to α-formylpyrroles [313].

6.2.7.2 *Clay-supported Catalysts*

Platinum group metal catalysts have been supported on clays and used for hydrogenation of alkenes and alkynes [e.g. 314, 315]. Shape selectivity was observed and catalytic activity varied with the Δd of the clay [315].

Traditional Friedel-Crafts procedures for alkylation or acylation of aromatic hydrocarbons like benzene require quite powerful Lewis acid catalysts such as aluminium chloride [316]. This can present great difficulties on an industrial scale as the aqueous work-up of the homogeneous reaction produces large volumes of acid for disposal. Supporting the Lewis acid on an aluminosilicate surface provides a catalyst which is recyclable, gives an advantageous work-up and often enhanced activity [13, 317, 318]. The catalyst is generally prepared by deposition of the metal chloride (Zn, Cu, Mg, Co, Ni, Cd or Al) onto the montmorillonite clay from methanolic solution and then activating at 50–300 °C. A wide range of aromatics have been formed with these catalysts by alkylation of benzene with, e.g. haloalkanes and paraformaldehyde.

Chiral *tris*-1,10-phenanthroline complexes of ruthenium and nickel adsorbed into montmorillonite have been used in the resolution of racemic mixtures [6, 7] and in the asymmetric synthesis of chiral sulphoxides (e.g. enantiomeric excesses >75 % [319]) by electrochemical [320] and periodate oxidation [321] of achiral sulphides. Optically-active *tris*(2,2'-bipyridyl)ruthenium(II) has also been used as a chiral photosensitiser in the aerobic photooxidation of achiral sulphides [320, 321].

Montmorillonite-bound palladium(II) bipyridyl [322, 323] and phosphine [324, 325] complexes are good stereo- and shape-selective catalysts for the hydrogenation of alkenes and alkynes, the latter giving *cis*-addition. The phosphine complexes also catalyse the selective hydrogenation of organic azides to amines [326] and the cross-coupling of vinyl acetate with iodobenzenes to give *trans*-stilbenes [327] (Equation (59)).

$$Me \overbigcirc{} -I \xrightarrow[\text{Montmorillonite / Bu}_3\text{N, 100 °C}]{(EtO)_3SiCH_2CH_2PPh_2PdCl_2} Me \overbigcirc{} \diagup\diagdown \overbigcirc{} -Me \quad (59)$$

$$+ \ CH_2{=}CHOAc$$

Several (silylamine)palladium(II) catalysts adsorbed on montmorillonite have been used for the quantitative, selective and sequential hydrogenation of nitroaromatics [328, 329]. Sensitive groups such as CHO were unaffected.

(*N,N*'-bis(Salicylideneaminato)ethylene)aquamanganese hexafluorophosphate supported on kaolin proved to be more selective in the epoxidation of cyclohexene with *t*-butyl hydroperoxide than the free complex [330]. However, cyclooctene showed no selectivity for epoxidation over formation of the *t*-butyl cyclooctenyl peroxide. A free radical process was demonstrated for the free complex whilst an oxometal mediated route was suggested for the kaolin-bound complex.

Esters or ethers were carbonylated, in the presence of montmorillonite-supported rhodium tri(butyl)phosphine [331] or nickel imidazole complexes [332], by carbon monoxide under medium pressure, to give anhydrides.

6.3   CONCLUSIONS

Originally clays were important as catalysts for the petroleum industry, but they were superceeded by the then more stable zeolites, however, pillaring of clays has produced clay based catalysts whose thermal stability competes with that of zeolites. Several workers are seeking to improve the activity of pillared clays and to utilise the advantage that clays can accommodate much larger organic molecules than zeolites.

Clays can be extremely shape and size selective, especially when intact, and the fact that clays have been used to control the stereochemical outcome of several organic reactions suggests that clays may prove to be of great use to the fine chemical and pharmaceutical industries.

The increasing interest in the use of supported reagents for producing 'environmentally-friendly' stoichiometric reagents and catalysts has already benefitted from the use of high surface area materials such as acid-activated clays. The high potential of intact clays for increased control of reactions offers numerous possibilities in this rapidly expanding field.

6.4 REFERENCES

1. P Laszlo, *Acc Chem Res*, **19**, 121 (1986).
2. A Cornelis and P Laszlo, in: *Chemical Reactions in Organic and Inorganic Constrained Systems*, R Setton, Ed, Reidel, New York, p 213 (1986).
3. B K G Theng, in: *Developments in Sedimentology No 35, Intern Clay Conf 1981*, H. Van Olphan and F. Veniale, Eds, Elsevier, Amsterdam, p 197 (1982).
4. B K G Theng, *The Chemistry of Clay-organic Reactions*, Adam Hilger, London (1974).
5. A Yamagishi, *Gendai Kagaku*, **191**, 35 (1987). CA107(7):58148y.
6. A Yamagishi, *Jpn Patent*, JP61204138 A2, (1986). CA106(9):66933v.
7. A Yamagishi and A Aramata, *Hyomen*, **24**, 442 (1986). CA105(21):190186x
8. See e.g. A I Zhukova, V S Kozlova, S V Bondarenko and Yu I Tarasevich, *Zh Praktil Khim (Leningrad)*, **62**, 1311 (1989). CA112(7):55101v
9. J M Adams, K Martin and R W McCabe, *J Inclusion Phenomena*, **5**, 663 (1987).
10. S Dyer, R W McCabe and B Osbourne, *in preparation*.
11. A Cornelis, P Laszlo and P Pennetreau, *Bull Soc Chim Belg*, **93**, 961 (1984).
12. M Balogh, A Cornelis and P Laszlo, *Tetrahedron Lett*, **25**, 3313 (1984).
13. C M Brown, S J Barlow, D J McQuarrie, J H Clark and A P Kybett, *Eur Patent*, EP 352, 878 A1 (1990). CA112:234978h.
14. E.g. T J Pinnavaia, R Raythatha, J Guo-Shu Lee, L J Halloran and J F Hoffman, *J Am Chem Soc*, **101**, 6891 (1979).
15. D E W Vaughan, R J Lusier and J S Magee, *US Patent* 4, 176, 090 (1979).
16. T J Pinnavaia, in: *Chemical Reactions in Organic and Inorganic Constrained Systems*, R Setton, Ed, Reidel, New York, p 151 (1986).
17. G Poncelet and A Schutz, in: *Chemical Reactions in Organic and Inorganic Constrained Systems*, R Setton, Ed, Reidel, New York, p 165 (1986).
18. E Kikuchi and T Matsuda, *Catal Today*, **2**, 297 (1988).
19. *Clay Minerals and the Origin of Life*, A G Cairns-Smith and H Hartman, Eds, Cambridge University Press, 1986.

20. A Hadding, Z *Krist*, **58**, 108 (1923).

21. F Rinne, Z *Krist*, **60**, 55 (1924).

22. S B Hendricks and W H Fry, *Soil Sci*, **29**, 457 (1930).

23. L Pauling, *Proc Nat Acad Sci*, **16**, 123 and 578 (1930).

24. W P Kelley, W H Dore and S M Brown, *Soil Sci*, **31**, 25 (1931).

25. G W Brindley, *X-Ray Identification and Crystal Structures of Clay Minerals*, The Mineralogical Society, London, (1951).

26. I I Ginsberg and R R Yashina, *Dokl K Sobraniyu Mezhdunar Komis po Izuch Glin Akad Nauk USSR*, 59 (1960).

27. H Beutelspacher and H W Van der Morel, *Acta Univ Cardinae Geol Supl*, **1**, 97 (1961).

28. R E Grim, *Second Int Congr Soil Mech*, **3**, 8 (1948).

29. H Vali and H M Koester, *Clay Miner*, **21**, 827 (1986).

30. J P Quirk, *Nature*, **188**, 253 (1961).

31. U Hoffman, K Endel and K Wiln, *Zeit Krist*, **86**, 340 (1933).

32. A McL Mathieson and G F Walker, *Am Mineralogist*, **39**, 231 (1954).

33. C E Marshells, *Zeit Kryst*, **91**, 433 (1935).

34. E Maegdefrau and U Hoffman, *Zeit Kryst*, **98**, 299 (1937).

35. S B Hendricks, *J Geol*, **50**, 276 (1942).

36. C H Edelman and J C L Favejee, *Zeit Kryst*, **102**, 417 (1940).

37. R E Grim and G Kulbicki, *Amer Mineralogist*, **46**, 1329 (1961).

38. R Fahn, *Kolloid-Zeitschrift*, **187**, 120 (1963).

39. G Foschamer, *Ann Roy Agr Coll Sweden*, **14**, 171 (1850).

40. Süd Chemie, Munich-Clay Data Sheets.

41. R W Grimshaw, *Chemistry and Physics of Clays*, Benn, London (1971).

42. U Hoffman, H P Boehm and W Gromes, *Z Anorg Allgem Chem*, **308**, 143 (1961).

43. G W Brindley and G Brown, *Crystal Structures of Clay Minerals and their X-Ray Identification*, Mineralogical Society, London (1980).

44. W F Bradley, R E Grim and G L Clark, *Zeit Kryst*, **97**, 216 (1937).

45. J L White, *Nat Acad Sci*, **Publ 456**, 133 (1956).

46. H Pezerat and J Mering, *Bull Groupe Franc Argiles*, **10**, 26 (1958).

47. Y Takeuchi and R Sadanaga, *Acta Cryst*, **12**, 945 (1959).

48. R E Grim and R H Bray, *J Am Ceram Soc*, **19**, 307 (1936).

49. A L Bar and H J Tenderloo, *Kolloid-Beihefter*, **44**, 97 (1936).

50. W P Kelley, *Cation Exchange in Soils*, Rheinhold, New York (1948).

51. R W McCabe, B Osbourne and P Patel, *in preparation*.

52. P L Heyden and A J Rubin, in: *Aqueous Environment Chemistry of Metals*, A J Rubin, Ed, Academic Press, New York, Chapter 9, (1974).

53. Y Nakajima and F Matsunaga, *Jpn Kokai Tokkyo Koho, Jpn Patents* 01304043–01304045 A2, (1989), CA112:234971a, CA112:234955y and CA112:197835m.

54. P Laszlo and P Pennetreau, *Tetrahedron Lett,* **26**, 2645 (1985).

55. J J Fripiat, A N Jelli, G Poncelet and J André, *J Phys Chem,* **69**, 2185 (1965).

56. R Touillaux, P Salvador, C Vandermeersche and J J Fripiat, *Israel J Chem,* **6**, 337 (1968).

57. D H Solomon, J D Swift and AJ Murphy, *J Macromol Sci Chem,* **A5**, 587 (1971).

58. H A Benesi, *J Am Chem Soc,* **78**, 5490 (1956).

59. F M Fowkes, H A Benesi, R B Ryland, W M Sawyer, K D Detling, E S Loeffler, F B Folckemer, M R Johnson and Y P Sun, *J Agr Food Chem,* **8**, 203 (1960).

60. T Henmi and K Wada, *Clay Miner,* **10**, 231 (1974).

61. M A Caine, *PhD Thesis,* Lancashire Polytechnic (1990).

62. M M Mortland and K V Raman, *Clays and Clay Miner,* **16**, 393 (1968).

63. J J Fripiat and M I Cruz-Crumplido, *Ann Rev Earth Planet Sci,* **2**, 239 (1974).

64. M M Mortland, *Trans 9th Int Congr Soil Sci,* 691 (1968).

65. M Frenkel, *Clays and Clay Miner,* **22**, 435 (1974).

66. J M Adams, D E Clement and S H Graham, *Clays and Clay Miner,* **31**, 129 (1983).

67. A Weiss, *Angew Chem Int Ed Engl,* **20**, 850 (1981).

68. C Breen, A Deane and J J Flynn, *Clay Miner,* **22**, 169 (1987).

69. J M Adams, J A Ballantine, S H Graham, R J Laub, J H Purnell, P I Reid, W Y M Shaman and J M Thomas, *J Catal,* **58**, 238 (1979).

70. I Barshad and A E Foscolo, *Soil Sci,* **110**, 52 (1970).

71. A C Wright, W T Grandquist and J V Kennedy, *J Catal,* **25**, 65 (1972).

72. J B Utterhoeven, L G Christener and W K Hall, *J Phys Chem,* **69**, 2117 (1965).

73. M W Tamele, *Disc Faraday Soc,* **No 8**, 270 (1950).

74. H E Swift and E R Black, *Ind Eng Chem, Prod Res Develop,* **13**, 106 (1974).

75. R Fahn, *SME-AIME Fall Meeting,* Tucson, Arizona, USA (1979).

76. S Branner, E Emmett and E Teller, *J Am Chem Soc,* **60**, 309 (1938).

77. W Franz, P Gunther and C E Hofstadt, *World Petrol Congr Proc 5th,* New York, **3**, 123 (1960).

78. B K G Theng, *Clays and Clay Miner,* **19**, 383 (1971).

79. R Kahn and K Fenderl, *Clay Miner,* **18**, 447 (1983).

80. J S Magee and J J Blazek, *Am Chem Soc Monograph,* **171**, (1976).

81. Anonymous, *Clays and Clay Miner,* **27**, 303 (1979).

82. Eg. S L Suib, J F Tanguay and M F Occelli, *J Am Chem Soc*, **108**, 6972 (1986).
83. J M Adams, D E Clement and S H Graham, *Clays and Clay Miner*, **30**, 129 (1982).
84. R M Barrer and D M McLeod, *Trans Faraday Soc*, **51**, 1290 (1955).
85. P A Diddams, J M Thomas, W Jones, J A Ballantine and J H Purnell, *J Chem Soc, Chem Commun*, 1340 (1984).
86. D E W Vaughan, R J Lusier and J S Magee, UK Patent GB 2,059,408A (1979).
87. G W Brindley and R E Sempels, *Clay Miner*, **12**, 229 (1977).
88. N Lahav and U Shani, *Clays and Clay Miner*, **26**, 116 (1978).
89. M L Occelli and R M Tindwa, *Clays and Clay Miner*, **31**, 22 (1983).
90. K Martin, *PhD Thesis*, University College of Wales Aberystwyth, (1986).
91. S Yamanaka and G W Brindley, *Clay Miner*, **27**, 119 (1979).
92. G W Brindley and S Yamanaka, *Am Mineralogist*, **64**, 830 (1979).
93. T J Pinnavaia, M S Tsou and S D Landau, *J Am Chem Soc*, **107**, 4783 (1985).
94. J A Laswick and R A Plane, *J Am Chem Soc*, **81**, 3564 (1959).
95. S Yamanaka, R Numata and M Hatori, *Int Clay Conf Abstracts*, Denver, 260 (1985).
96. S Yamanaka and G W Brindley, *Clays and Clay Miner*, **26**, 21 (1978).
97. G W Brindley and C C Kao, *Clays and Clay Miner*, **28**, 435 (1980).
98. B M Choudary, V L K Valli and A D Prasad, *J Chem Soc Chem Commun*, 721 (1990).
99. B M Choudary and V L K Valli, *J Chem Soc Chem Commun*, 1115 (1990).
100. M M Taqui Khan, S A Samad and M R H Siddiqui, *J Mol Cat*, **50**, 97 (1989).
101. M M Mortland, J J Fripiat, J Chaussidon and J B Utterhoeven, *J Phys Chem*, **78**, 994 (1963).
102. V C Farmer and M M Mortland, *J Chem Soc (A)*, 344 (1966).
103. S Yariv and L Heller, *Israel J Chem*, **8**, 935 (1970).
104. L Heller-Killai, S Yariv and M Riemer, *Proc Int Clay Conf Madrid*, 651 (1973).
105. M M Mortland, *Trans 9th Int Congr Soli Sci*, Adelaide, **1**, 691 (1968).
106. M M Mortland, *Clay Miner*, **6**, 143 (1966).
107. S Tahoun and M M Mortland, *Soil Sci*, **102**, 248 (1966).
108. S Yariv, L Heller, Z Sofer and W Bondheimer, *Israel J Chem*, **6**, 741 (1968).
109. V C Farmer and M M Mortland, *J Phys Chem*, **69**, 683 (1965).
110. P Cloos and R D Laura, *Clays and Clay Miner*, **20**, 259 (1972).

111. H Van Damme, M Crispin, F Obrecht, M I Cruz and J J Fripiat, *J Colloid Interface Sci*, **66**, 43 (1978).

112. P Cannesson, M I Cruz and H Van Damme, *Proc Int Clay Conf*, Oxford, 217 (1979).

113. S Abdo, M I Cruz and J J Fripiat, *Clays and Clay Miner*, **28**, 125 (1980).

114. J W Buchler, in: *Porphyrins and Metalloporphyrins*, KM Smith, Ed, Elsevier, New York, p 147 (1975).

115. V S Rak, *Fiz-Khim Mekh Liophil'nost Dispersnykh Sist*, **20**, 66 (1989). CA113(25):232006m.

116. J M Adams, J A Ballantine, S H Graham, R J Laub, J H Purnell, P I Reid, W Y M Shaman and J M Thomas, *J Catal*, **58**, 238 (1979).

117. J M Adams, T V Clapp and D E Clement, *Clay Miner*, **18**, 411 (1983).

118. P Laszlo, *personal communication*, (1986).

119. J M Adams, K Martin, R W McCabe and S Murray, *Clay Miner*, **34**, 587 (1986).

120. J J Fripiat, J Helsen and L Vielvoye, *Bull Groupe Fr Argiles*, **15**, 3 (1964).

121. D H Kubiek, *Belgian Patent*, BE 886261 (1982). CA96:51802c.

122. Phillips Petroleum, *US Patent*, US 096345 (1979).

123. Y Okuda, M Yoshihara, T Maeshima, M Fujii and T Aida, *Yakagaku*, **38**, 153 (1989). CA111:96739g.

124. Y Okuda, M Yoshihara, T Maeshima, M Fujii and T Aida, *Chem Express*, **4**, 17 (1989). CA111:232232r.

125. J A Ballantine, J H Purnell, M Rayanakorn, K J Williams and J M Thomas, *J Mol Catal*, **30**, 373 (1985).

126. W Bauer, M Langer and W Sperling, *Ger Patent*, DE 3524095 A1 (1987). CA106:119441s.

127. K R N Rao, R C Towill and A H Jackson, *J Indian Chem Soc*, **66**, 654 (1989).

128. J A Ballantine, M Davies, I Patel, J H Purnell, M Rayanakorn, K J Williams and J M Thomas, *J Mol Catal*, **26**, 37 (1984).

129. J M Adams, J A Ballantine, S H Graham, R J Laub, J H Purnell, P I Reid, W Y M Shaman and J M Thomas, *Angew Chem Int Edn Engl*, **17**, 282 (1978).

130. J M Adams, A Bylina and S H Graham, *J Catal*, **75**, 190 (1982).

131. J M Adams, A Bylina and S H Graham, *Clay Miner*, **16**, 325 (1981).

132. J M Adams, D E Clement and S H Graham, *J Chem Res S*, 254 (1981).

133. A Bylina, J M Adams, S H Graham and J M Thomas, *J Chem Soc Chem Commun*, 1003 (1980).

134. A M Habib, A A Saafan, A K Abou-Seif and M A Salem, *Colloids Surf*, **29**, 337 (1988).

135. A M Habib, M F Abd-El-Megeed, A Saafan and R M Issa, *J Inclusion Phenomena*, **4**, 185 (1986).

136.   A E Kapustin, V F Shvets M G Makarov, A V Simenido and F N Zeiberlikh, *Neftekhimiya*, **26**, 267 (1986). CA105:60263j.
137.   K Fujita, Y Ishida and J Suezawa, *Jpn Patent*, JP 62289537 A2 (1987). CA109:148882d.
138.   D Kotkar and P K Ghosh, *J Chem Soc Chem Commun*, 650 (1986).
139.   D Kotkar, S W Mahajan, A K Mandal and P K Ghosh, *J Chem Soc, Perkin Trans 1*, 1749 (1988).
140.   L G Lucatello and G E Smith, *UK Patent* 1265152 (1972). CA76:153330b.
141.   K A Carrado, R Hayatsu, R E Botto and R E Winans, *Clays and Clay Miner*, **38**, 250 (1990).
142.   J A Ballantine, M Davies, R M O'Neil, I Patel, J H Purnell, M Rayanakorn, K J Williams and J M Thomas, *J Mol Catal*, **26**, 57 (1984).
143.   R Gregory, D J H Smith and D J Westlake, *Clay Miner*, **18**, 431 (1983).
144.   M P Atkins, *UK Patent*, GB 2175300 A1, (1986). CA107:6522w.
145.   BASF AG, *Ger Patent*, 1211643 (1963).
146.   J M Adams, S E Davies, S H Graham and J M Thomas, *J Catal*, **78**, 197 (1983).
147.   J F Roudier and A Foucaud, in: *Chemical Reactions in Organic and Inorganic Constrained Systems*, R Setton, Ed, Reidel, New York, p 229 (1986).
148.   BASF AG, *Ger Patent*, DP 1643712 (1971).
149.   S Kessaissia, B Siffert and J B Donnet, *Clay Miner*, **15**, 383 (1980).
150.   A Katchalsky and G Ailam, *Biochem Biophys Acta*, **140**, 1 (1967)
151.   M Paecht-Horowitz, *Biosystems*, **9**, 93 (1977).
152.   M Paecht-Horowitz, J Berger and A Katchalsky, *Nature (London)*, **228**, 636 (1970).
153.   R M O'Neil, *UC Swansea Post-Doctoral Report*, (1982) (cf 203).
154.   J Y Conan, A Natat and D Privolet, *Bull Soc Chim Fr*, 1935 (1976).
155.   Süd Chemie AG, *Ger Patent*, DBP 1086241 (1961).
156.   Farbwerke Hoechst AG, *Ger Patent*, 1146871 (1963).
157.   S Hoyer, P Laszlo, M Orlovic and E Polla, *Synthesis*, 655 (1986).
158.   R Cruz-Almanaz, F J Perez-Flores and M Avila, *Synth Commun*, **20**, 1125 (1990).
159.   T Vu Moc, H Petit and P Maitte, *Bull Soc Chim Fr*, **15**, 264 (1979).
160.   E C Tailor and C Chaing, *Synthesis*, 467 (1977).
161.   A Cornelis and P Laszlo, *Synthesis*, 162 (1982).
162.   U Schafer, *Synthesis*, 794 (1981).
163.   D F Ishman, J T Klug and A Shani, *Synthesis*, 137 (1981).
164.   C Liu, R Huang and X Lin, *Cuihua Xuebao*, **8**, 177 (1987). CA108:37194r.
165.   B Labiad and D Villemin, *Synth Commun*, **19**, 31 (1989).

166. N Kametaka, N Nagato, K Hiromoto, S Soya and T Nozawa, *Ger Patent*, DE 3531084 A1, (1986). CA105:173046j.
167. S Hunig, K Hubner and E Benzing, *Chem Ber*, **95**, 931 (1962).
168. C Alvarez, A C Cano, V Rivera and C Marquez, *Synth Commun*, **17**, 279 (1987).
169. A C Cano, A A Cordoba, C Marquez and C Alvarez, *Synth Commun*, **18**, 2051 (1988).
170. A R Graham, B P McGrath and B Yeomans: *British Patent* 1, 359, 113 (1974). CA82:17726g.
171. R W McCabe, J M Adams and K Martin, *J Chem Res(S)*, 357 (1985).
172. R W McCabe, S Bhatia, D A Griffiths and P O'Toole, *submitted for publication*.
173. M C Galanti and A V Galanti, *J Org Chem*, **47**, 1572 (1982).
174. J M Adams and A Gabbut, *J Inclusion Phenom Mol Recognit Chem*, **9**, 63 (1990).
175. R B Woodward and R Hoffmann, *Angew Chem Int Edn Engl*, **8**, 781 (1969).
176. R W McCabe, *unpublished results*.
177. F Texier-Boullet, B Klein and J Hamelin, *Synthesis*, 409 (1986).
178. D Villemin, B Labiad and Y Ouhilal, *Chem Ind (London)*, 607 (1989).
179. S H Eggers, V V Kane and G Lowe, *J Chem Soc*, 1262 (1965).
180. D Buckley, S Dyer and R W McCabe, *in preparation*.
181. J Desmurs and S Ratton, *Eur Patents*, EP 262062 A1 and EP 262063 A2, (1988). CA109:170025y and CA109:210687z.
182. E Gutierrez and E Ruiz-Hitzky, *Mol Cryst Liq Cryst*, **161B**, 453 (1987).
183. E Gutierrez, A J Aznar and E Ruiz-Hitzky, *Stud Surf Sci Catal*, **41**, 211 (1988).
184. W G Dauben, J M Cogen and V Behar, *Tetrahedron Lett*, **31**, 3241 (1990).
185. S Kullaj, *Bul Shkencave Nat*, **43**, 81 (1989). CA112:21154z.
186. A Ortega and E Maldonado, *Heterocycles*, **29**, 635 (1989).
187. F Brunner, *Pulp and Paper International*, **54**, 24 (1982).
188. G Baxter, *Microencapsulation Processes and Applications*, J E Vandegaar, Ed, Plenum Press, New York (1974).
189. M E Rohmann and H Schoepke, *Wochenblatt für Papierfabrikation*, **110**, 767 (1982).
190. J C Petitpierre, *TAPPI, Prceedings of Coating Conference*, 157 (1983).
191. R Cigan, *Acta Chem Scand*, **12**, 1456 (1958).
192. P C Henriquez, *Recl Trav Chim Pays-Bas*, **52**, 991 (1953).
193. K Iwamoto, *Bull Chem Soc Jap*, **19**, 420 (1935).
194. R Bangert, W Aichelle, E Schollmeyer, B Weimann and H Herlinger, *Melliand Textilber*, **58**, 399 (1977).

346   *Richard W McCabe*

195. T Matsuda, M Asanuma and E Kikuchi, *Appl Catal*, **38**, 289 (1988).
196. D Plee, A Schutz, G Poncelet and J J Fripiat, in: *Catalysis by Acids and Bases*, B Imelik, C Naccache, G Coudurier, Y Ben Taarit and J C Vedrine, Eds, Elsevier, Amsterdam, p 343 (1985).
197. D Tichit, F Fajula, F Figueras, J Bosquet and C Gueguen, in: *Catalysis by Acids and Bases*, B Imelik, C Naccache, G Coudurier, Y Ben Taarit and J C Vedrine, Eds, Elsevier, Amsterdam, 351 (1985).
198. W Franz, P Gunter and C E Hofstadt, *Proc 5th World Petroleum Congress* , New York, (1959).
199. C L Thomas, J Hickey and G Strecker, *Ind Eng Chem*, **42**, 866 (1950).
200. C L Thomas, *Ind Eng Chem*, **41**, 2564 (1949).
201. A Grenall, *Ind Eng Chem*, **40**, 2148 (1948); **41**, 1485 (1949).
202. G A Mills, J Holmes and E B Cornelius, *J Phys Colloid Chem*, **54**, 1170 (1950).
203. J A Ballantine, in: *Chemical Reactions in Organic and Inorganic Constrained Systems*, R Setton, Ed, Reidel, New York, p 197 (1986).
204. N Lahav, U Shani and J Shabtai, *Clays and Clay Miner*, **26**, 107 (1978).
205. T Endo, M M Mortland and T J Pinnavaia, *Clays and Clay Miner* **28**, 105 (1980).
206. Chevron Research Co, *US Patents*, US 3655798 (1972) and US 3766292 (1973). CA76:156482p and CA80:14526q.
207. J J L Heinerman and M F M Post, *European Patent*, EP79091 (1983). CA99:125386f.
208. V N Parulekar and J W Hightower, *Appl Catal*, **35**, 249 (1987).
209. J F Brody, J W Johnson, G B McVicker and J J Ziemiak, *Solid State Ionics 1988*, **32–33**, 353 (1989).
210. M R Musaev, S R Mirzabekova, L I Alieva, A G Gasanov and Kh I Makhmudov, *Azer Neft Khoz*, **1**, 44 (1986). CA105:96988p.
211. Toyo Soda MFG KK, *Jpn Patent*, JP 8083635 (1983).
212. Agency of Ind Sci Tech, *Jpn Patent*, JP 8067340 (1983).
213. R Birch and C I Warburton, *J Catal*, **97**, 511 (1986).
214. Y Morikawa, T Goto, Y Moro-oka and T Ikawa, *Chem Lett*, 1667 (1982).
215. N L Industries Inc, *US Patent*, US 4217295 (1977).
216. Shell Int Res, *Belgian Patent*, BE 891410 (1980). CA97:112423z.
217. H Mori, T Okuhara and M Misono, *Chem Lett*, 2305 (1987).
218. G J J Bartley and R Burch, *Appl Catal*, **28**, 209 (1986).
219. Kh I Areshidze and G I Shetsiruli, *Izv Akad Nauk Gruz SSR, Ser Khim*, **12**, 237 (1986). CA106(21):175868y.
220. M Imanari, H Iwane, T Sugawara, S Ohtaka and N Suzuki, *European Patent*, EP 331173 A1, (1989). CA112:7160s.

221. V D Sukhoverkhov, F D Ovcharenko, A S Zhurba, I I Martsin, N V Alekseenko, V A Pistol'kors and T L Loichenko, *Ukr Khim Zh*, **52**, 594 (1986). CA107(1):6864j.
222. G O Chivadze and V V Khakhnelidze, *Heterog Catal*, **6**, 454 (1987).
223. M Imanari, H Iwane and T Sugawara, *Jpn Patent*, JP 62153235 A2, (1987). CA108:21490y.
224. K Takagi and Y Naruse, *Jpn Patent*, JP 01238549 A2, (1989). CA112:157851g.
225. M Imanari, H Iwane, T Sugawara and S Ohtaka, *Jpn Patent*, JP 63208545 A2, (1987). CA110:7846d.
226. K Taniguchi, Y Sigejo and Y Kurano, *Jpn Patent*, JP 62114942 A2, (1987). CA108:37376b.
227. G O Chivadze, V V Khakhnelidze, L Z Chkheidze and Ts I Naskidashvili, *Soobshch Akad Nauk Gruz SSR*, **121**, 105 (1986). CA106:66831k.
228. Y Shigeshiro H Oyoshi and T Fujita, *Jpn Patent*, JP 62155242 A2, (1987). CA109:149042s.
229. R Bacskai and H A Valerias, *Eur Patent*, EP 265932 A2, (1988). CA110:57288u.
230. P Laszlo and A Mathy, *Helv Chem Acta*, **70**, 577 (1987).
231. B E M Hassan, E A Sultan, F M Tawfik and S M Sappah, *Egypt J Chem*, **28**, 93 (1985).
232. T Matsuda, M Matsukata, E Kikuchi and Y Morita, *Appl Catal*, **21**, 297 (1986).
233. T Ishibashi, *Jpn Patent*, JP 01213241 A2 (1989). CA112:118421x.
234. T Minoe, N Shimitzu and T Tsubochi, JP 62294630 A2 (1987). CA110:153937a.
235. M Salmon, E Angeles and R Miranda, *J Chem Soc, Chem Commun*, 1188 (1990).
236. B Chiche, A Finiels, C Gauthier, P Geneste, J Graille and D Pioch, *J Mol Catal*, **42**, 229 (1987).
237. T Fujita and K Takahata, *Jpn Patent*, JP 61152636 A2 (1986). CA106:18104d.
238. J M Bakke and J Liaskan, *Ger Patent*, DP 2826433 (1979). CA90:121181b.
239. H Sato, K Hirose, K Nagai, H Yoshiokai and Y Nagaoka, *Eur Patent*, EP 343048 A1 (1989). CA112:197818h.
240. C Collet, A Delville and P Laszlo, *Angew Chem Int Edn Engl*, **29**, 535 (1990).
241. M Butters, K Smith, K Fry and N J Stewart, *Eur Patent*, EP 356091 A2 (1990). CA113:97183z.
242. A Cornelis, A Gerstmans and P Laszlo, *Chem Lett*, 1839 (1988).

348   *Richard W McCabe*

243. J M Bakke, J Liaskan and G B Loretzen, *J Prakt Chem*, **324**, 488 (1982).
244. T Fujita, K Mizuno and K Saeki, *Jpn Patent*, JP 62142134 A2, (1987). CA108:37278w.
245. M Kondo, M Tanaka and K Taniguchi, *Jpn Patent*, JP 02115140 A2, (1990). CA113:131744w.
246. K Hashimoto, Y Asahi and T Maki, *Jpn Patent*, JP 63303936 A2, (1988). CA111:6957z.
247. Phenolchemie GMBH, *Ger Patent*, Ger 1136713 (1962).
248. J F Knifton, *US Patents*, US 4870217 A (1989) and US 4898987 A (1990). CA112:118444g and CA112:234968e.
249. S Toma, K Cizmarikova and P Elecko, *Chem Pap*, **40**, 755 (1986).
250. N Govindaraj, M M Mortland and S A Boyd, *Environ Sci Technol*, **21**, 1119 (1987). CA107(19):175331p.
251. S Akyuz and T Akyuz, *Doga: Kim Ser*, **11**, 51 (1987).
252. M B McBride and M G Johnson, *Clays Clay Miner*, **34**, 686 (1986).
253. M Constantini and J M Popa, *Eur Patent*, EP 299893 A2 (1989). CA112:7155u.
254. M Constantini, J M Popa and M Gubelman, *Eur Patent*, EP 314583 A1 (1989). CA111:214219n.
255. A Foucaud and M Bakouetila, *Synthesis*, 854 (1987).
256. B H Han and D G Jang, *Tetrahedron Lett*, **31**, 1181 (1990).
257. E.g. W J Gensler, J R Jones, R Rein, J J Bruno and D M Bryan, *J Org Chem*, **31**, 2324 (1966).
258. Ciba-Geigy Corporation, *US Patent* 4,202,820 (1980).
259. N Evole Martil and F Aragon de la Cruz, *An Quim, Ser B*, **82**, 256 (1986). CA108(1):6297x
260. M Kawai, M Onaka and Y Izumi, *Chem Lett*, 1581 (1986).
261. M Kawai, M Onaka and Y Izumi, *Bull Chem Soc Jpn*, **61**, 1237 (1988).
262. M Onaka, R Ono, M Kawai and Y Izumi, *Bull Chem Soc Jpn*, **60**, 2689 (1987).
263. M Kawai, M Onaka and Y Izumi, *Chem Lett*, 381 (1986).
264. P Laszlo, M T Montaufier and S L Randriamahefa, *Tetrahedron Lett*, **31**, 4867 (1990).
265. M Kawai, M Onaka and Y Izumi, *J Chem Soc, Chem Commun*, 1203 (1987).
266. M Kawai, M Onaka and Y Izumi, *Bull Chem Soc Jpn*, **61**, 2157 (1988).
267. I Zafar, A H Jackson and K R N Rao, *Tetrahedron Lett*, **29**, 2577 (1988).
268. M T Thorat, M H Jagdale, R B Mane, M M Salunkhe and P P Wadagaonkar, *Curr Sci*, **56**, 771 (1987).
269. D S Wulfman, R S McDaniel Jr and B W Peace, *Tetrahedron*, **32**, 1241 (1976).
270. K N Houk and R W Strozier, *J Am Chem Soc*, **95**, 4094 (1973).

271. D J Bellville and N L Bauld, *J Am Chem Soc*, **103**, 718 (1981).
272. P Laszlo, J Lucchetti, *L'Actualité Chimique*, 42 (1984).
273. L S Newton, *Speciality Chemicals*, 17 (1984).
274. R S Downing, J van Amstel and A H Joustra, *US Patent* 4, 125, 483 (1978). CA90:71828v.
275. J M Adams, K Martin and R W McCabe, *Proc Int Clay Conf, Denver, 1985*, L G Schultz, H van Olphen and F A Mumpton, Eds, Clay Mineral Soc of Bloomington Indiana, p 324 (1987).
276. J M Adams and T V Clapp, *Clays Clay Miner*, **34**, 287 (1986).
277. P Laszlo and J Lucchetti, *Tetrahedron Lett*, **25**, 1567 (1984).
278. P Laszlo and J Lucchetti, *Tetrahedron Lett*, **25**, 2147 (1984).
279. P Laszlo and J Lucchetti, *Tetrahedron Lett*, **25**, 4387 (1984).
280. P Laszlo and H Moison, *Chem Lett*, **6**, 1031 (1989).
281. J M Adams, S, Dyer, K Martin and R W McCabe, *submitted for publication*.
282. S M Csicsery in: *Zeolite Chemistry and Catalysis*, Ed, J A Rabo, ACS Monograph **171**, Washington, 680 (1976)
283. S Dyer and R W McCabe, submitted for publication.
284. K Norrish, *Disc Farad Soc*, **18**, 120 (1954).
285. L van Leemput and M S Stuhl, *Clay Miner*, **17**, 209 (1982).
286. K M McKay and R A McKay, *Introduction to Modern Inorganic Chemistry*, 2nd Edn, p 79 (1972).
287. K Tagagaki, H Usami, H Fukaya and Y Sawaki, *J Chem Soc, Chem Commun*, 1174 (1989).
288. J Cabral, P Laszlo and M T Montaufier, *Tetrahedron Lett*, **29**, 547 (1988).
289. P Laszlo and A Cornelis, *Aldrichemica Acta*, **21**, 97 (1988).
290. A Naumann, *Ber Deutsch Chem Ges*, **37**, 4328 (1904).
291. B Penth, *Ger Patent*, DE 3637780 A1 (1988). CA108:167675j.
292. A Cornelis and P Laszlo, *Synthesis*, 849 (1980).
293. A Cornelis, P Y Herze and P Laszlo, *Tetrahedron Lett*, **23**, 5035, (1982).
294. D H R Barton, G C Ramsay and D Wege, *J Chem Soc*, 1915 (1967).
295. M Besemann, A Cornelis and P Laszlo, *C R Acad Sci Paris, Ser II*, **299**, 427 (1984).
296. A Cornelis, N Depaye, A Gerstmans and P Laszlo, *Tetrahedron Lett*, **24**, 3103 (1983).
297. P Laszlo, P Pennetreau and A Krief, *Tetrahedron Lett*, **27**, 3153 (1986).
298. S Chalais, A Cornelis, P Laszlo and A Mathy, *Tetrahedron Lett*, **19**, 2327 (1985).
299. P Laszlo and E Polla, *Tetrahedron Lett*, **25**, 3309 (1984).
300. P Laszlo and E Polla, *Tetrahedron Lett*, **25**, 4651 (1984).

301. M Salmon, R Miranda and E Angeles, *Synth Commun*, **16**, 1827 (1986).
302. P Pennetreau, M Balogh, I Hermecz and A Gerstmans, *J Org Chem*, **55**, 6198 (1990).
303. P Laszlo and E Polla, *Tetrahedron Lett*, **25**, 3701 (1984).
304. M Balogh, I Hermecz, Z Mészàros and P Laszlo, *Helv Chim Acta*, **67**, 227 (1984).
305. A Cornelis, P Laszlo and P Pennetreau, *J Org Chem*, **48**, 4771 (1983).
306. P Laszlo and P Pennetreau, *J Org Chem*, **52**, 2407 (1987).
307. A Cornelis, L Delaude, A Gerstmans and P Laszlo, *Tetrahedron Lett*, **29**, 5909 (1988).
308. A Cornelis, L Delaude, A Gerstmans and P Laszlo, *Tetrahedron Lett*, **29**, 5657 (1988).
309. P Laszlo and J Vandormael, *Chem Lett*, 1843 (1988).
310. L Delaude and P Laszlo, *Catal Lett*, **5**, 35 (1990).
311. A Sarkar, B R Rao and M M Konar, *Synth Commun*, **19**, 2313 (1989).
312. E C Taylor, C Chiang, A McKillop and J F White, *J Am Chem Soc*, **98**, 6750 (1976).
313. A H Jackson, K R N Rao and E Smeaton, *Tetrahedron Lett*, **30**, 2673 (1989).
314. M Nikles, D Bur and U Sequin, *Tetrahedron*, **46**, 1569 (1990).
315. S Shimatsu, T Hirano and T Uematsu, *Appl Catal*, **34**, 255 (1987).
316. G A Olah, *Friedel-Crafts Chemistry*, Wiley, New York (1973).
317. S J Barlow, J H Clark, M R Darby, A P Kybett, P Landon and K Martin, *J Chem Res (S)*, in press (1992).
318. J H Clark, A P Kybett, D J McQuarrie, S J Barlow and P Landon, *J Chem Soc, Chem Commun*, 1353, (1989).
319. A Yamagishi, *J Chem Soc, Chem Commun*, 290 (1986).
320. A Yamagishi, *Chem Aust*, **54**, 278 (1987).
321. T Hikita, K Tamaru, A Yamagishi and T Iwamoto, *Inorg Chem*, **28**, 2221 (1989).
322. B M Choudary and P Barathi, *J Chem Soc, Chem Commun*, 1505 (1987).
323. B M Choudary, G V M Sharma and P Barathi, *Angew Chem Int Edn Engl*, **28**, 465 (1989).
324. G V M Sharma, B M Choudary, M R Sarma and K K Rao, *J Org Chem*, **54**, 2997 (1989).
325. B M Choudary, K Mukkanti and Y V S Rao, *J Mol Catal*, **48**, 151 (1988).
326. G V M Sharma and S Chandrasekhar, *Synth Commun*, **19**, 3289 (1989).
327. B M Choudary and M R Sarma, *Tetrahedron Lett*, **31**, 1495 (1990).
328. K Mukkanti, Y V S Rao and B M Choudary, *Tetrahedron Lett*, **30**, 251 (1989).

329. Y V S Rao, K Mukkanti and B M Choudary, *J Mol Catal,* **49**, L47 (1989).
330. P S Dixit and K Srinivasan, *Inorg Chem,* **27**, 4507 (1988).
331. N Okada and O Takahashi, *Jpn Patent,* JP 61001628 A2, (1986). CA105:42338h.
332. N Okada and O Takahashi, *Jpn Patent,* JP 62135445 A2, (1987). CA108:111825r.

7 Polymeric Coordination Complexes: Bridging Molecular Metals and Conductive Polymers

Glen Eugene Kellogg and John G Gaudiello

Inorganic Materials. Edited by Duncan W Bruce and Dermot O'Hare
© 1992 John Wiley & Sons Ltd

7.1   INTRODUCTION

The constant search for materials with new and improved properties has led to a variety of investigations into inorganic and organometallic polymers [1–4]. The reasoning is that if these new materials combine some of the properties unique to either organic polymers or inorganic solids, they may provide materials suitable as engineering tools for important technological problems. The improved high-performance materials would extend applications to nearly every area of technology and manufacturing; the types of properties that could be improved span the range of electrical, magnetic, optical, chemical, etc. A wide variety of materials, broadly classified as inorganic and organometallic polymers, has been examined. Since an in-depth review in a single chapter is virtually impossible, we will introduce a few of the many classes of metal-containing polymers here, and direct the reader to pertinent literature. Polysiloxanes (or silicones) are perhaps the most commercially viable and familiar inorganic polymers. Research on these materials continues in both industrial and academic laboratories as new applications for high performance sealants arise [5–9]. Polysilanes and polycarbosilanes are receiving attention because of their potential as silicon carbide precursors or as photoresists in microelectronics [10–13]. Phosphorus-containing polymers, such as the poly(phosphazenes), have been studied since the early 1900s and have yielded important technological materials [14–16]. Boron-containing polymers have been synthesised as precursors to boron nitride [17, 18]. Another major class of metal-containing polymers are those constructed by polymerisation of organometallic monomers that contain the vinyl moiety [19, 20]. Also, condensation polymerisations of appropriately substituted organometallic complexes can yield metal-containing systems [21].

In this chapter, we focus on polymeric coordination compounds that have unique electronic (electrical, optical, and magnetic) properties. Reviews emphasizing other aspects of inorganic and organometallic polymers have appeared [1–4]. Developing an understanding of the significance of polymeric, metal-containing materials with enhanced electronic properties is best accomplished through examination of related materials: charge-transfer solids, conductive (organic) polymers, and (non-conductive) inorganic and organometallic polymers.

Organic and inorganic materials that possess the properties of semiconductors, traditional metals and superconductors have been an area of intensive study for over a decade [22–36] (see also Chapter 1). These low-dimensional (anisotropic), electrically conductive compounds are comprised of extended molecular arrays and are usually classified as belonging to one of two groups of materials: charge-transfer solids or organic/organometallic conductive polymers. Charge-transfer (CT) solids are often obtained by condensation of discrete molecular components, electron donors (reducing agents) and acceptors (oxidising agents) that have undergone partial charge-transfer. Organic conductive polymers are routinely obtained by chemically- or electrochemically-initiated polymerisation of the corresponding monomer. The resulting covalently-linked assemblies are then doped (partially oxidised or reduced) using chemical or electrochemical methodologies.

Both types of low-dimensional materials possess two important properties that give rise to their unique characteristics: a band-type electronic structure formed by closely-arranged, adjacent molecules which interact and a fractional, or partial, oxidation state. These materials possess the optical, magnetic and electrical properties usually associated with traditional metals. Most research effort in the development of charge-transfer solids has gone into maximising electrical conductivity in these materials. The unique combination of organic/organometallic materials with metallic characteristics results in a new and exciting class of materials. Since the creation of new molecular materials requires considerable synthetic skill, chemistry has been the science central to these breakthroughs.

The extended, low-dimensional arrays of CT solids and conductive polymers mandate that the electronic structure be described in terms of bands [37]. Figure 1 illustrates how discrete molecular orbitals interact to form this type of electronic state. The energy range of the individual states (bandwidth) is directly related to the degree of interaction between the orbitals from which the band is derived. For a simple tight-binding (Hückel) description the bandwidth is 4t, where t is the transfer integral (analogous to β, Hückel integral) [37–40]. Partial filling of these bands by either electron-removal from the valence band (oxidation) or electron injection into the conduction band (reduction) results in a metallic state.

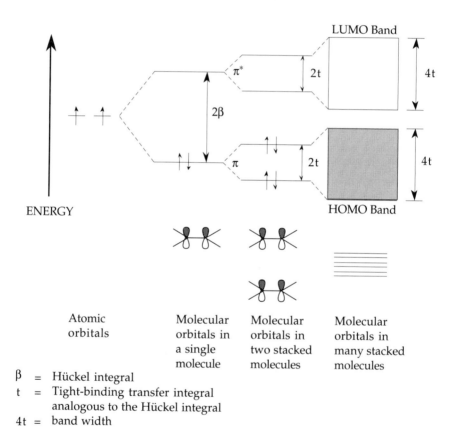

β = Hückel integral
t = Tight-binding transfer integral
     analogous to the Hückel integral
4t = band width

Figure 1 Schematic representation of band formation from discrete atomic and molecular orbitals. (Reprinted with permission from reference [24]. Copyright (1986) American Chemical Society)

7.1.1 Charge-transfer Solids: Molecular Conductors

7.1.1.1 *Historical Perspective*

Single-crystal CT solids (Figure 2) based on donors, such as tetrathiafulvalene (TTF) (**1**), bis(ethylenedithia)tetrathiafulvalene (BEDT-TTF or ET) (**2**), tetramethyltetraselenafulvalene (TMTSF) (**3**), tetraselenatetracene (TSeT) (**4**), and nickel phthalocyanine (NiPc) (**5**) and acceptors such as tetracyanoquinodimethane (TCNQ) (**6**), and I_2 have shown very high conductivities [22–26], and, in some cases, superconductivity when measured in the stacking (conductive) direction.

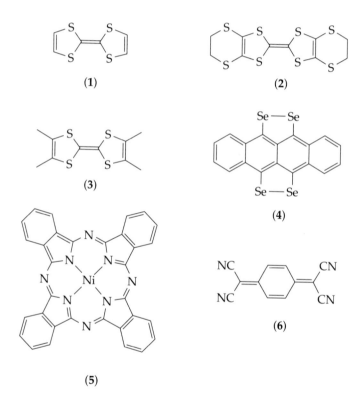

Figure 2 Typical donors and acceptors

In contrast to polymers, the major disadvantage of molecular metals is that there is very limited tunability: only unique combinations and stoichiometric ratios of donors and acceptors yield conductive materials. The key is stacking of the donors (or acceptors) in a manner allowing for band formation and thus partial charge-transfer. This is, of course, a delicate function of the components' shape and size, Madelung forces and crystallisation conditions, that is only met in a small proportion of cases [34, 35, 41–45]. This stacking requirement further leads to low-dimensionality in these molecular, metal-like materials. However, the most successful (i.e. superconducting) molecular conductors often have somewhat higher dimensionality to allow conduction between parallel chains [22–24].

7.1.1.2 *Thermodynamic/Electronic/Crystalline Considerations and Limitations*

The preparation and properties of CT solids is strongly dependent on redox potential, structure and polarisability of the discrete components [34, 35]. The formation of a fractional oxidation state is facilitated by small differences between the standard potentials (E^0) of the donors and acceptors. This can be estimated directly from the difference in E^0 values of the components or from the difference in ionisation potential (IP) of the donor and the electron affinity (EA) of the acceptor.

To facilitate band formation, the geometry of the molecular component(s) should be planar in both the neutral and ionic states. Donors and acceptors of similar size more easily allow for the formation of segregated stacks. Crystal packing and electronic considerations suggest a preference for symmetrical molecules. In addition, the electrostatic energy associated with the formation of the crystal lattice (Madelung energy) should be large. A polarisable and fluid electron density in the molecular components (usually associated with delocalisation) leads to enhanced orbital overlap between adjacent molecules in the stack.

7.1.1.3 *Examples: Materials and Properties*

Table 1 lists representative single crystal CT solids, their degree of partial charge-transfer and their room temperature electrical conductivities (σ_{RT}). These materials typically exhibit σ_{RT} in the range of 10^2 to 10^4 S cm^{-1}, halfway between the values for conventional metals and semiconductors. Their degree of fractional oxidation is usually around 0.5.

There are two general methods for preparing these solids. The simplest procedure involves the redox reaction between the neutral donor and acceptor under controlled conditions, which often results in crystal growth. Electrocrystallisation methods employ an inert electrode (typically Pt) which changes the oxidation state of one component, allowing for crystal growth at the electrode surface. This technique can be advantageous since it requires that only one component undergoes a change in oxidation state, thus permitting the introduction of redox-inert counterions.

Table 1 Typical CT solids and their electrical conductivities

CT Solid (X^+Y^-)	Oxidation State of X	σ_{RT} (S cm^{-1})	Ref
(TTF)(TCNQ)	0.59	500	[46, 47]
(TMTSF)(ClO$_4$)	0.50	650	[48]
(BEDT-TTF)$_2$(I)$_3$	0.50	30	[49]
(TSeT)$_2$I	0.50	1500	[50]
[NiPc][I]	0.33	550	[51]

7.1.2 Covalent-organic Arrays: Polymeric Conductors

7.1.2.1 Historical Perspective

Paralleling the growth and discovery of single crystal CT solids has been the field of electrically-conductive organic polymers [31, 52–55]. These systems are usually comprised of conjugated planar organic moieties that are normally 'pre-assembled' before 'doping' (partial oxidation or reduction of the polymer) occurs. Although a band-type electronic structure is formed upon polymerisation, the undoped materials are usually insulators (i.e. possess a fairly large band gap). Detailed understanding of the processes of conduction and the interpretation of physical measurements are hindered by poor structural data for the doped materials. Most are poorly crystalline or amorphous in nature. The mode of conduction is believed to be somewhat different from that of charge-transfer metals. Models based on the propagation of structural phase kinks (solitons) or the movement of spinless dications (bipolarons) have been proposed [56–60].

The first well-characterised and relatively highly-conductive, organic polymer was poly(acetylene), (CH)$_x$. MacDiarmid, Heeger, Shirakawa and co-workers obtained values for σ_{RT} of around 1000 S cm^{-1} by doping chemically polymerised (CH)$_x$ with I$_2$ and AsF$_6$ [61, 62]. A few years later, Diaz, Street and co-workers at IBM showed that pyrrole could be polymerised electrochemically and cycled between conducting and insulating states [63–65]. Following these initial reports, conductive polymers, based on a wide variety of organic moieties such as thiophene, aniline, and *p*-phenylene, have been reported. Representative examples are listed in Figure 3 and Table 2.

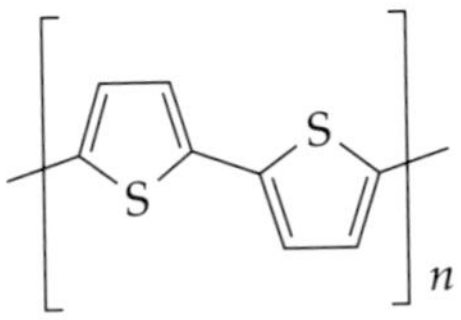

Poly(thiophene)

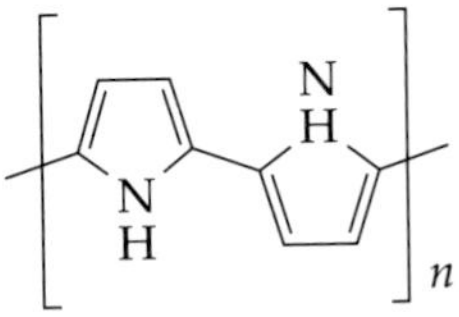

Poly(pyrrole)

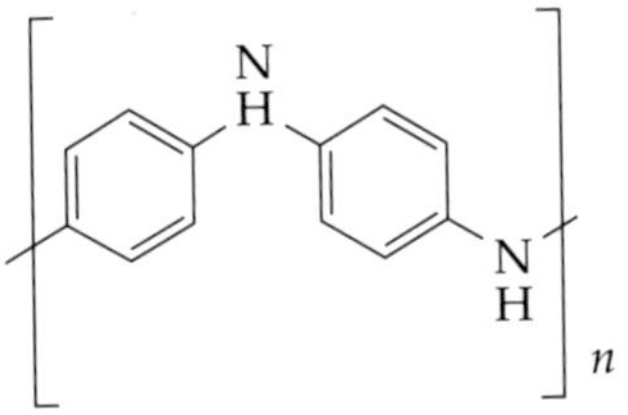

Poly(aniline)

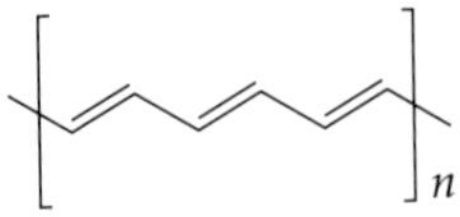

Poly(acetylene)

Figure 3   Conductive organic polymers

Table 2 Typical conducting polymers and their electrical conductivities

Polymer	degree of partial charge-transfer	σ_{RT} (S cm^{-1})	Ref
poly(acetylene)	0.1	1100	[66]
poly(pyrrole)	0.3	600	[67]
poly(thiophene)	0.24	10	[68]
poly(aniline)	0.5	5	[69]

7.1.2.2 *Polymerisation and Doping of Polymers*

Conductive organic polymers are routinely obtained by either chemically- or electrochemically-initiated polymerisation of the corresponding monomer. The resulting covalently-linked assemblies are then doped using chemical or electrochemical methodologies. They differ from CT solids in several key regards. These materials are physically and chemically more robust, are usually amorphous, and lend themselves more easily to processing and device fabrication. The extent of doping is more easily controlled and both soluble and insoluble systems can be prepared. The bandwidths of these materials are usually larger than typical CT solids.

7.1.3 Molecular Assemblies: Polymers Constructed with Metal-containing Macrocycles

7.1.3.1 *Electrically-conductive Metal-containing Polymers*

Building on the strengths of both classes of electrical conductors, rational methods have been developed to prepare modular molecular systems for exploiting the properties of both CT and polymeric materials. These systems are comprised of CT components (building blocks) held together by chemical linkages [44, 70]. The extended arrays thus created have the chemical stability and enforced stacking of polymers and the charge-transport properties of CT solids. Many of the limitations and uncertainties associated with the crystal growth of segregated stacks in CT solids (Madelung forces) are either eliminated or reduced. In addition, this strategy allows for unprecedented examination of the physical properties of metal-like molecular assemblies in response to variations in the interplanar spacing, off-axis counter-ion identity and degree of partial charge-transfer. In effect, this architecture creates a designed polymeric array of molecular metal donors that can be represented as $[MLZ]_n$ (where M = metal, L = macrocycle, Z = bridging group).

Molecular heteroatom macrocycles such as phthalocyanine (Pc), tetrabenzoporphyrin (TBP), and glyoximates, have the requisite features to form metal-like CT solids under proper crystallisation conditions. CT solids comprised from these macrocycles with and without a central metal atom have been demonstrated [71–76]. The metal can act as an obvious bridgehead for chemical linkage, allowing for the assembly of low-dimensional metallomacrocyclic arrays.

7.1.3.2 *Strategies for Assembly*

The nature of the bridgehead metal and bridging group dictates the type of linkage leading to polymer formation (Figure 4). Hanack describes three classes of polymeric metallomacrocycles based on the bonding and coordinating properties of the Z bridging groups [70]. Class A systems are those where the bridging group bonds by coordination at each end. An example is $[Fe(Pc)(pyrazine)]_n$. Other Class A bridging groups (ligands) are bipyridine and tetrazine. Class B systems are those where the bridging group covalently bonds at one end and bonds coordinatively at the other, such as in $[Fe(Pc)(CN)]_n$, $[Co(Pc)(SCN)]_n$, or $[Al(Pc)F]_n$. If the bridging group can covalently bond at both ends, it is considered Class C. $[Si(Pc)(C_2)]_n$ and $[Si(Pc)(NCN)]_n$ are members of this class. This motif provides the required template for investigating the importance of several key parameters, such as macrocycle spacing, identity of off-axis counterions, and degree of partial charge-transfer. With the $[MLZ]_n$ architecture the effect of varying the spacing between adjacent macrocycles can be examined by changing the central metal atom (M) and/or the linkage groups (Z). The synthetic procedures and physical properties of these systems are described in more detail in Section 7.2.

A   coordinative-coordinative $\longleftarrow$ Z $\longrightarrow$ M(L) $\longleftarrow$ Z $\longrightarrow$

B   coordinative-covalent $\longrightarrow$ Z $\longrightarrow$ M(L) $\longrightarrow$ Z $\longrightarrow$

C   covalent-covalent $\longrightarrow$ Z $\longrightarrow$ M(L) $\longrightarrow$ Z $\longrightarrow$

Figure 4 Classes of polymeric metallomacrocycles

7.1.3.3 Considerations for Assembly

The theoretical importance of ring-ring spacing in designing a polymeric molecular metal would appear to be qualitatively clear; interplanar spacing controls the overlap between orbitals on adjacent macrocycles and, hence, the magnitude of the transfer (tight-binding) integral or bandwidth. The polymeric $[MLZ]_n$ system has two features that can be chemically adjusted to affect interplanar spacing. The first is choice of the bridging group, Z, while the second, somewhat more subtle, is the choice of the central metal.

   The nature and composition of the macrocycle also greatly influences the inherent properties of the system. The accessible redox states and bandwidths of the extended array are dependent on the character of the frontier orbitals of the component macrocycles. The structure of these orbitals is a function of their electronic environment including the presence of electron-donating or electron-withdrawing substituents. Furthermore, the substituents can be used to regulate the solubility of the array and thus the methodology required for doping (i.e. homogeneous or heterogeneous).

Also, because these planar macrocyclic components are prone to slipped-stack crystallographic packing in the undoped state [77, 78], the rôle of the backbone is to hold the array components in a co-facial arrangement. This enforces the structure as well as enhances the orbital overlap and thus charge-transport.

7.1.4 Techniques for the Characterisation of Polymeric and Molecular Conductors

The understanding and development of these materials is facilitated by the use of an array of characterisation techniques. This section will summarise these methods and some of the experimental/theoretical considerations implied by the techniques. Later, specific data for the $[\{M(R_4Pc)O\}(X)_y]_n$-doped polymer systems (M= Si, Ge; R= H, *t*-butyl) will be presented. Foremost, and perhaps best recognised of the characterisation methods for synthetic conductors, are conductivity experiments. These give a quantitative value for electrical conduction but can often be misleading due to sample preparation/form and measurement considerations [45, 71, 79, 80]. Thermoelectric power, which exploits the conversion of thermal to electrical energy (i.e. as in a thermocouple), is much less dependent on sample constitution than conductivity measurements and thus gives a rather comprehensive view of the electrical and electronic properties of materials [81, 82]. Magnetism probes the electronic structure of a conductor by characterising the collective properties of interacting charge carriers [83–85]. Two limiting case descriptions of collective properties (Curie-like or Pauli-like) can be invoked and experimentally observed with magnetic susceptibility and electron spin resonance (ESR) techniques. Optical reflectance measurements reveal the frequency of conduction electrons in the metal-like state [83, 86–88]. Perhaps the most important characterisation tools for understanding and quantifying the results of the other measurements are the techniques of X-ray crystallography. Precisely-known molecular structures, either through single-crystal studies of molecular conductors or powder diffraction studies of polymers, provide an experimental foundation for theoretical investigations into the mechanisms of conductivity [89–92]. Finally, as described in Section 7.3, voltammetric (electrochemical) techniques can be used not just to dope these materials but also to probe quantitatively the electronic structure and chemical stability of soluble systems.

7.1.4.1 Conductivity: DC and AC at Variable Temperature

Room temperature, DC conductivity studies are the most simple and straightforward characterisation of a molecular or polymeric material. The typical measurement geometry uses four probes, two for current application and two for voltage measurement. This technique gives a numerical value (usually reported in Siemens/cm; $S\ cm^{-1}$) for the sample, but tells little about the electronic structure of the material. Single crystal samples often have vastly superior (2–3 orders of magnitude) conductivity relative to polycrystalline or powder samples because of reduced or eliminated interparticle contact resistance and anisotropy [45, 71, 83, 93]. Tables 1 and 2, which set out some typical values of room temperature conductivity for molecular and polymeric conductors, demonstrate the wide range of properties observed.

On the other hand, variable temperature DC conductivity is much more informative. For metals and metal-like samples, increasing conductivity with decreasing temperature is commonly observed. The limit is, of course, superconduction, which is infinite conductivity at or below a critical temperature (T_c). For single crystal CT solids, negative slopes $(d\sigma/dT)$ are observed, but the variable-temperature conductivity of many materials is discontinuous due to disorder, Peierls distortions, etc. The variable-temperature conductivity of polymers is a more complicated picture. Because of random isotropy and lack of crystallinity in most polymer samples, DC conductivity measurements fail to record accurately the conduction down a single polymer chain. Instead, the measured conductivity reflects the effects of a large number of contacts between adjacent chains and particles, and typically exhibits the temperature dependence of a semiconductor. A number of conductivity models have been proposed to reconcile these effects. There are two primary classes of models: hopping, where charge carriers 'hop' over the electrical barriers erected by contact resistance between chains or crystallites [94]; and tunnelling, where charge carriers 'tunnel' through the barriers [95]. Usually, hopping conductivity models are applied to materials with low doping levels, while tunnelling models are applied to more highly-doped materials. Detailed analysis of variable temperature DC conductivity measurements in terms of these models for doped polycrystalline metallomacrocycle polymers suggests that these materials are inherently as conductive as the corresponding CT material.

Additional information concerning the transport mechanism for conductive systems can be obtained by frequency-dependent (AC) conductivity measurements. Unlike DC measurements where only the net charge that crosses the entire sample is measured, AC studies yield information about the motion of charge carriers in a particular state. The movement of carriers across band gaps, mobility edges, or through extended metal states can be differentiated. The frequency range studied is typically between 10^{-4} and 10^{13} Hz [96–98].

7.1.4.2 Thermoelectric Power

The thermoelectric power (TEP) experiment measures the Seebeck coefficient (S, V/K) as a function of temperature by placing a temperature gradient across a (crystal) sample and measuring the induced voltage. Several features of thermopower data are noteworthy. First, the sign of the thermopower (S) is indicative of the nature of the charge carriers. Positive thermopower indicates 'hole' (radical cation) carriers, while negative thermopower indicates electron (radical anion) carriers. Second, the slope, dS/dT, further defines the nature of the sample. If dS/dT is negative, the sample is best described as an insulator; if dS/dT is near zero, the sample is best described as a semiconductor; if dS/dT is positive, the sample is best described as being metal-like. Within the metal regime, smaller S values indicate greater metal-like character. Also, the value of dS/dT is related to the tight-binding bandwidth for metal-like materials [83]. Thermopower results are much less dependent upon sample form than conductivity measurements. Hence, single-crystal and polycrystalline studies give comparable results [82, 83].

7.1.4.3 Magnetism

The magnetic properties of molecular and polymeric metals can be studied by three different methods, all of which are appropriate for both crystals and polycrystalline powders. The first, static magnetic susceptibility, measures the concentration of spin carriers as a function of temperature. Analysis of the susceptibility (χ) versus T profile can reveal the electronic character of the material, i.e. whether the carriers are Curie-like or Pauli-like. Curie-like magnetic susceptibility exhibits an inverse temperature dependence and follows the Curie–Weiss law. For most conductive materials the Curie-like behaviour is small and generally can be ascribed to defects, impurities, and possible disorder [45, 50, 71, 83, 99, 100]. High magnetic susceptibility at very low temperature is indicative of Curie-like, random, localised free spins. Pauli-like susceptibilty describes the magnetic behaviour of totally delocalised (i.e. metal) free spins. In some solid-state physical descriptions, metals are depicted as nuclei surrounded by a 'sea' of electrons. In general, magnetically-derived tight-binding bandwidths from Pauli susceptibility values are smaller than those determined by thermopower and optical measurements [40, 45, 50, 71, 83, 101, 102].

Complementary information can be obtained with electron spin resonance (ESR) spectroscopy. Here the g values indicate the character of the carrier (e.g. free electron), while measured line widths suggest the nature of conduction and the degree of electron delocalisation [83]. Hyperfine splittings (if observed) can be fingerprints for the atomic identity of a localised charge. In addition, cross-polarisation magic angle spinning (CPMAS) nuclear magnetic resonance is a straightforwardly applied and powerful structural/dynamic probe for molecular and polymeric conductors [103].

7.1.4.4 *Optical Reflectance*

Optical reflectance spectroscopy provides another invaluable probe of electronic structure and electrical properties of low-dimensional solids and polymers. Interpretation of optical reflectance spectra within the framework of simple free-electron models, by application of the Drude model of metals, has often provided useful electronic information [86, 104]. The key observable in an optical reflectance spectrum is the plasma edge that corresponds to the plasma frequency of conduction electrons in the 'metal'. From this, the optical conductivity and optical tight-binding bandwidth can be derived [83]. The optical reflectance experiments are particularly well suited to polycrystalline samples (i.e. polymer) because the effects of interparticle contacts are minimised [83] and corrections for anisotropy are possible [71]. However, the isotropy of single crystal molecular metal samples can be examined by polarisation experiments where the sample/beam geometry is parallel or perpendicular with respect to the crystal (stacking) axis.

7.1.4.5 *X-Ray Crystallography*

Molecular CT solids rely on crystallisation techniques to achieve the metal-like state. Single crystal, X-ray crystallographic experiments on these materials have revealed some fascinating details of structure that have greatly aided the interpretation of the other physical data. Some studies have actually captured the essence of dynamic processes, such as the Peierls distortion or anion ordering as functions of temperature [105].

Perhaps the major problem in understanding the electrical, electronic, magnetic, and optical properties of polymeric conductors is the scarcity of complete structural information. Most polymers have poor crystallinity, leading to broad, ill-defined diffractions in powder/pellet crystallographic experiments [106–108]. Thus, even though the molecular structure of the monomer units and the general nature of the covalent linkages between monomers are known, little can be implied about chain structure and interactions. In some ways, however, this lack of crystallinity is attractive as a physical property for fabrication of devices based on electroactive polymers.

7.2 EXAMPLES OF POLYMERIC SYSTEMS

7.2.1 Polymers Constructed with Non-Covalent Bridging Ligands

Materials constructed with coordinative-coordinative and coordinative-covalent ligands (Class A and B, respectively) have been prepared and studied by a number of groups [109–138]. Hanack and co-workers have pioneered the development of systematic methods for the synthesis of a wide variety of materials based on a number of different macrocycles (porphyrin, hemiporphyrazine, phthalocyanine). The majority of Class A polymers contain a transition metal with octahedral coordination (Ru, Fe, Co, Mn, Cr, Rh), allowing for linear bridges and thus stacked systems. Typical bidentate bridging groups are unsubstituted and substituted pyrazine (pyz) [112–117], 4,4'-bipyridine (bipy) [118, 119], 4,4'-bipyridylacetylene (bipyac) [120], 1,4-diisocyanobenzene (dib) [112–115, 116,119–124, 126, 127], 1,4-diazabicyclo[2.2.2]octane [119, 129], or tetrazine (tz) [112, 123, 130]. Using these bridges results in an interplanar spacing between the macrocycles that is too large to maintain the π–π overlap normally required for high conductivity. However, a conduction pathway can occur along the central axis of the polymer via the transition metal and delocalised π-bridge. Increases in conductivity upon doping are comparable to systems having a π–π macrocycle pathway.

The preparation of these polymers is usually accomplished by reacting the solid metallomacrocycle with either neat ligand or with dissolved ligand in a refluxing solution. The resulting polymer is typically insoluble, allowing for easy isolation. Purification is typically done by washing the precipitate or by high vacuum sublimation to remove excess ligand. Oxidation of these polymers yields conductive materials. For example, $[PcFe(pyz)]_n$ has been doped using I_2, resulting in oxidations up to 0.51, i.e. $[\{PcFe(pyz)\}(I_5)_{0.51}]_n$ and a σ_{RT} of 0.2 S cm^{-1} [135]. Other Class A bridge polymers (dib, bpy, tz, bpyac, etc.) have been chemically or electrochemically doped yielding conductive materials.

Class B polymers usually contain either a transition or main-group metal (Al, Ga) with typical bridges being CN [118, 123, 126, 129, 131–133, 135–137], SCN [118, 123, 138] or F [139–144]. These materials can be prepared by axial anion displacement in a coordinatively-unsaturated monomer ($nM(L)X + nZ \longrightarrow [M(L)Z]_n$) or by reacting the dihalogenated metallomacrocycle with the bridging ligand ($nM(L)X_2 + nZ \longrightarrow [M(L)Z]_n$). Chemical doping using I_2 or nitrosonium salts yields respectable conductivites. For example, $[(PcAlF)I_{3.3}]_n$ exhibits $\sigma_{RT} = 5$ S cm^{-1} [142] while typical cyanide complexes such as $[(PcCoCN)I_{1.6}]_n$ have $\sigma_{RT} = 0.6$ S cm^{-1} [145].

7.2.2 Polymers Constructed with Covalent Linkages

There are significant reasons to focus on this group of co-facially-joined metallomacrocyclic polymers as opposed to the ligand-bridged polymers. Perhaps the most compelling reason is that these materials can be extremely robust and have rather long chain lengths [106]. It is important that these systems have the shortest known ring-ring spacings. Maybe the most investigated class of co-facially-linked polymers are those based on silylphthalocyanine as the macrocycle, i.e. $[Si(Pc)O]_n$ (Figure 5).

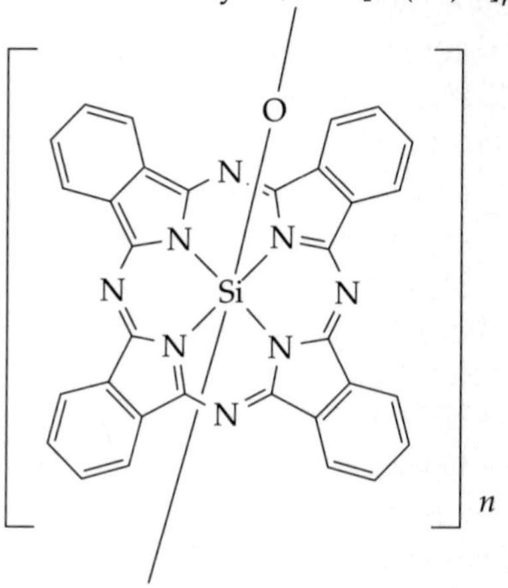

Figure 5 Polymeric silylphthalocyanines

The research efforts of the Northwestern University group led by Tobin Marks have revealed a wealth of information about this unique class of material [44, 45, 71, 83, 89, 103, 106, 145]. It is likely that other polymers constructed from metallomacrocycles via covalent linkages will have similar, or perhaps enhanced, properties. However, this section summarises the results from [Si(Pc)O]$_n$ studies because of the clear picture of electrical, electronic, magnetic, and optical evolution in response to doping afforded by these materials.

7.2.2.1 Outline of Synthetic Approach

The poly(phthalocyaninatometalloxane) materials [M(Pc)O]$_n$ can be prepared by adaptations of the synthetic methodology pioneered by Kenney and co-workers [146, 147]. The polymerisation proceeds from dihydroxysilylphthalocyanine by high temperature and vacuum (400 °C, 10^{-4} τ) condensation. Chain lengths of 120 ± 30 monomer repeat units are typically achieved with silicon phthalocyanines (70 ± 40 with germanium, 100 ± 40 with tin) [145]. These polymers are extraordinarily robust due to the thermodynamic stability of the phthalocyanine moiety and the strength of metal-oxygen bonds.

7.2.2.2 Chemical/Phase-doped Polymers: Doping Methodology

Virgin (undoped) samples of [M(Pc)O]$_n$ (M = Si, Ge) are essentially insulators, with only a small residual conductivity attributable to impurities and defects. As expected, chemical doping of [M(Pc)O]$_n$ polymers yields materials with metal-like physical properties, but understanding the doping process itself has some rather significant consequences. There are several issues of central importance with respect to how the [M(Pc)O]$_n$ polymers interact with chemical oxidants, and how this chemistry impacts upon collective properties. The first is whether macrocycle oxidation actually occurs and the nature of the products formed. Second is the question of whether the doping is homogeneous or heterogeneous (i.e. phase doped). The third question concerns the importance of choice in oxidant/counterion in determining the oxidation level of the doped material.

The [M(Pc)O]$_n$ polymers can be chemically doped with dissolved or gaseous halogens (most experiments have been conducted with dissolved I$_2$). Although stoichiometries for [{M(Pc)O}I)$_y$]$_n$ as high as y = 2–3 could be achieved, vacuum drying invariably yields a sample with y = 1.1. Resonance Raman spectroscopy showed that the form of iodine present in the doped species was I_3^-, suggesting that the appropriate formulation for the fully-doped polymer is $[\{M(Pc)O\}(I_3)_{0.36}]_n$. [71, 148].

Another revealing aspect of doping $[M(Pc)O]_n$ is that there is a clearly-observed structural change concurrent with the macrocyclic oxidation. While the undoped polymer crystallises in the orthorhombic space group *Ibam*, the fully-doped material crystallises in the same tetragonal space group ($P4$) as [Ni(Pc)][I] (the analogous CT analogue [71]). This permits an unprecedented chance to monitor structural changes in response to incremental changes in polymer oxidation.   As the oxidant is incrementally added, the starting (orthorhombic) phase is observed to diminish and the doped (tetragonal) phase grows in. At intermediate doping levels ($0 < y < 0.36$), the starting and doped phase co-exist. This important observation strongly suggests that the doping process is *heterogeneous*.

Chemically doping $[Si(Pc)O]_n$ with nitrosonium salts (NO^+X^-, $X =$ BF_4, PF_6, SbF_6, ClO_4) yields the same doping chemistry and structural features as halogen doping [45].   By varying the NO^+X^-:$[Si(Pc)O]_n$ stoichiometry and reaction time, a wide range in y is achieved, but in no case was $y > 0.36$.   Again, X-ray studies suggested the lower-doped materials were heterogeneous. For $[Si(Pc)O]_n + I_2$, Br_2, or ClO_4^- all data exist to construct Born-Haber cycles to determine U, the crystal binding energy of the lattice. The difference $U - \Delta H^0{}_{CT}$, where $\Delta H^0{}_{CT}$ is the standard enthalpy of charge-transfer, is -91 kcal mol^{-1} for I_3^-, -91 kcal mol^{-1} for Br_3^-, and -84 kcal mol^{-1} for ClO_4^- [45].   This suggests that the crystal binding energies for complexes of a wide variety of oxidant/polymer products are very similar, and there may be a thermodynamic stability for the $y = 0.36$ products.

7.2.2.3 *Chemical/Phase-doped Polymers: Physical Characterisation*

The extensive studies on the chemically-doped siloxyphthalocyanine polymers, in particular $[\{Si(Pc)O\}(I_3)_{0.36}]_n$, indicate that these materials are metal-like in the doped state.   Furthermore, all available evidence suggests a heterogeneous doping process in that most physical properties can be modeled by assuming a physical mixture of doped and undoped phases in the intermediately ($y < 0.36$) oxidised samples. Section 7.2.5, gives a detailed discussion of the physical properties for this system, including a comparison and contrasting of the results.

7.2.2.4 *Electrochemically-doped Polymer: Doping Methodologies*

Electrochemical (oxidative) doping and subsequent undoping of the insoluble $[Si(Pc)O]_n$ polymer reveals further information about the unique physicochemical relationship between the undoped and doped structural phases. In general, these materials are easily oxidised as slurries or microcompaction/electrode composites with a wide variety of electrolytes. The choice of electrolyte (counterion) is governed by its electrochemical stability and relative size. Electrochemical Potential Spectroscopy (ECPS) of the polymer/electrolyte system is a particularly attractive electroanalytical/electrosynthetic method because each data point is obtained under equilibrium conditions and accurately maps the doping process [89, 149, 150]. Figure 6, which is the ECPS curve for slurry doping of $[Si(Pc)O]_n$ in $[Bu_4N][BF_4]$/acetonitrile, shows that the primary doping process occurs initially near 1.1 V versus SSCE (saturated sodium calomel electrode) and is marked by a sharp discontinuity in y at that potential. Near 1.4 V, y reaches a plateau of *ca* 0.50 electrons per Si(Pc)O, i.e. $[\{Si(Pc)O\}(BF_4)_{0.50}]_n$. Reversed cycling to less positive potentials (undoping) returns the material to the neutral, undoped state ($y = 0$).

Sharp discontinuities in y, as seen in electrochemical doping of $[Si(Pc)O]_n$, are usually associated with phase changes (most commonly structural reorganisation) [89, 149–151]. In this case, the reorganisation is a change from an orthorhombic to a tetragonal structure. X-ray diffraction results demonstrate that the electrochemically-doped $[\{Si(Pc)O\}-(BF_4)_{0.50}]_n$ polymeric metals are nearly isostructural with the chemically-doped analogues. Most interesting, however, is that the electrochemically-undoped (cycled) material has *retained* the tetragonal structure of the doped materials, albeit with no off-axis counterions.

The ECPS behaviour of once-cycled, tetragonal $[Si(Pc)O]_n$ is considerably different from that of the orthorhombic phase (Figure 6). Here, the onset of doping occurs at 0.3–0.4 V versus SSCE and y as a function of E_{appl} is a smooth, continuous function over a nearly 1.0 V range up to $y = 0.50$ (at *ca* 1.20 V). That the tetragonal $[Si(Pc)O]_n$ phase can be oxidised and counterions inserted at a much lower potential than for the orthorhombic phase is an example of the 'break-in' phenomenon often observed in the electrochemistry of conductive polymers [89, 152]. Clearly in the $[Si(Pc)O]_n$ case the cause of the 'break-in' is structural in origin, having to do with the transformation of polymer chain packing in the crystal lattice. The smooth, monotonic transformation from the undoped ($y = 0$) tetragonal polymer to the fully-doped ($y = 0.50$) $[\{Si(Pc)O\}X_y]_n$ polymer suggests that y is a continuous function for this system, and that the system is at the maximum y for each point along the curve. (Each point on the ECPS curve is at equilibrium).   Thus, materials at intermediate potential are homogeneously doped, and they can be 'tuned' to a desired level of oxidation or 'band-filling'.   This last point will become clear with results from the array of physical measurements applied to these unique polymers.

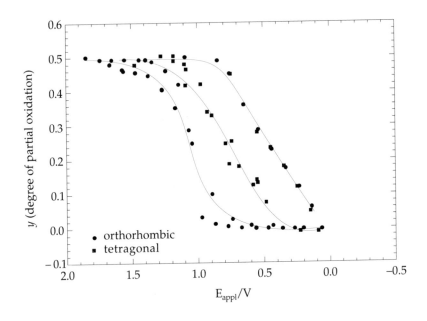

Figure 6 ECPS curve for 'slurry' doping of $[Si(Pc)O]_n$ in $[Bu_4N][BF_4]/CH_3CN$. Two cycles are shown: doping/undoping orthorhombic $[Si(Pc)O]_n$ and doping/undoping tetragonal $[Si(Pc)O]_n$. (Reprinted with permission from reference [89]. Copyright (1989) American Chemical Society)

7.2.2.5 *Electrochemically-doped Polymers: Physical Characterisation*

X-ray diffraction data support the concept of homogeneous doping for the oxidation of the tetragonal $[Si(Pc)O]_n$ polymer. No significant broadening of the diffraction peaks occurs, and a smooth expansion of the lattice parameter, a, and a slight smooth contraction of the lattice parameter, c, are observed with increasing y for the $[\{Si(Pc)O\}(BF_4)_y]_n$ samples [89].

Counterion-dependent structural and compositional effects were not seen with chemical doping, but with the wider variety of counterions and larger dynamic range afforded by the electrochemical doping methodology, some unique counterion effects are observed. First, the maximum doping level achievable strongly correlates with the size of the inserted counterion, with small counterions yielding higher oxidation levels [89]. Secondly, very large counterions (such as 1-pyrenesulfonate or $[Mo_8O_{26}]^{4-}$ [153]) require higher applied potentials for insertion due to a significant kinetic overpotential.

Room temperature conductivity data as a function of y are plotted in Figure 7 for $X = BF_4^-$ (chemical- and electrochemical-doped) and I_3^- (chemical-doped). The general behaviour is similar to that observed for numerous other conductive polymer systems. At a given y, chemically- and electrochemically- doped materials have roughly the same conductivities but the inherent uncertainty of compacted powder measurements allows few detailed comparisons to be made.

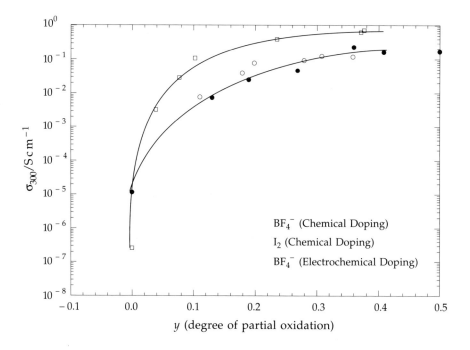

Figure 7 Room-temperature conductivity for $[\{Si(Pc)O\}X_y]_n$, $X = I_3, BF_4$

Variable-temperature conductivity data for two families of doped materials, I_3^- (chemically-oxidised) and BF_4^- (electrochemically-oxidised), are presented (log σ versus 1/T format) in Figure 8A and 8B, respectively. The behaviour is similar in both cases, generally with falling apparent activation energies (slopes) and increasing metallic character at progressively higher doping levels. The functional form in which polymer conductivity depends on temperature has been a subject of extensive discussion for poly(acetylene), poly(*p*-phenylene) and several other systems [32, 33, 81, 154, 155]. Likewise, considerable effort has been expended to understand the functional form of the σ versus T behaviour of the $[\{Si(Pc)O\}(X)_y]_n$ polymers [45, 71, 83].

For the chemically(inhomogeneously)-doped materials the σ(T) data can be best fitted (for *all y*) to a fluctuation-induced tunnelling (FIT) model originally developed for composites [95, 155]. The tunnel junction parameters obtained through the FIT analysis are fairly invariant over the range of *y* [45, 71]. This is certainly consistent with the notion that these samples are best represented as physical mixtures of doped and undoped phases with conductive regions separated by insulating barriers. For the electrochemically-doped polymers the tunnel junction parameters evolve with increasing *y* and are in fact physically unrealistic for lightly-doped samples [83]. For these materials hopping-type models are applicable, suggesting a different mechanism of conduction.

Variable-temperature, thermoelectric-power (TEP) data for homogeneous (electrochemically-prepared) $[\{Si(Pc)O\}(BF_4)_y]_n$ polymers are shown in Figure 9. At low doping levels (*y* < 0.19), the magnitude and temperature dependence of the TEP in these materials is characteristic of a *p*-type molecular semiconductor, whereas between the approximate limits $0.19 \leq y \leq 0.27$, the TEP data evolve toward that typical of a *p*-type low-dimensional metal. The transition to negative S at lower temperatures and high *y* and noticeable curvature of these data may be purely an artifact of interparticle contact resistance. An evolution in electronic structure is evidenced by the TEP data in response to the variable doping level. This implied transition appears to be related to the movement of the Fermi level across a mobility edge (arising from disorder, see below) as valence band depletion occurs.

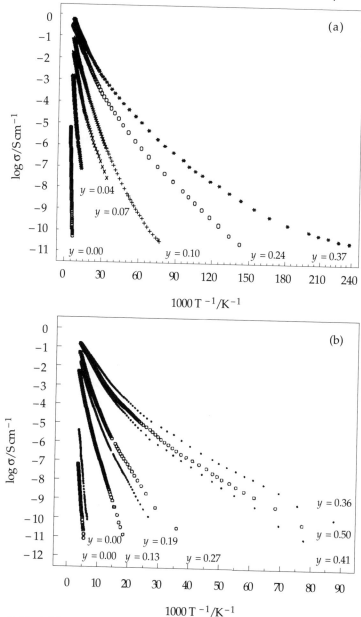

Figure 8 Variable-temperature conductivity (log σ versus 1/T format) for $[\{Si(Pc)O\}X_y]_n$, (A) X = I_3, (B) X = BF_4. (Figures 8A and 8B Reprinted with permission from references [71] and [83], Copyright (1983) and (1989) American Chemical Society)

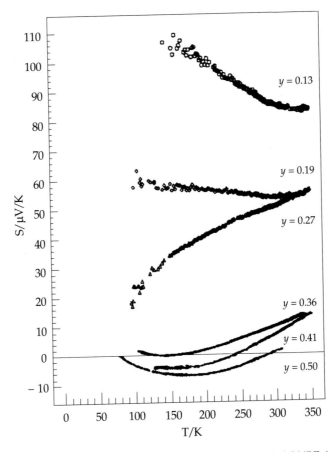

Figure 9 Variable-temperature thermopower for $[\{Si(Pc)O\}(BF_4)_y]_n$. Reprinted with permission from reference [83], Copyright (1989) American Chemical Society)

For doped $[\{Si(Pc)O\}(X)_y]_n$ polymers, compressed polycrystalline samples were examined with optical reflectance spectroscopy over frequency ranges as large as 100–50,000 cm^{-1}. All spectral data were analyzed using a Drude-like model [104]. For the chemically-doped (heterogeneously-doped) polymers, e.g., $[\{Si(Pc)O\}(I_3)_y]_n$, progressive oxidation yields a plasma-like edge (ω_p) at *ca* 3600 cm^{-1} independent of doping level (Table 3) [45, 71]. The Drude analysis (and fit) of the plasma edge yields optical parameters that can be used to calculate 'optical conductivities' that are quite similar to estimated *on-axis* electrical conductivities.

Table 3 Optical reflectance results for $[\{M(Pc)O\}X_y]_n$, $X = I_3$ BF_4, Tos

Compound (ω_p), c m^{-1}	Plasma Freq (4t), eV	Bandwidth (σ_{opt}), S cm^{-1}	Optical Conductivity	Ref
$[\{Si(Pc)O\}[I_3]_{0.37}]_n$	4540	0.60	400	[71]
$[\{Si(Pc)O\}[BF_4]_{0.36}]_n$	4710	0.64	355	[83]
$[\{Si(Pc)O\}[BF_4]_{0.50}]_n$	5390	0.63	450	[83]
$[\{Ge(Pc)O\}[I_3]_{0.37}]_n$	4210	0.48	200	[71]
$[\{Si(Pc)O\}[Tos]_{0.23}]_n$	4020	0.71	160	[83]
$[\{Si(Pc)O\}[Tos]_{0.37}]_n$	4350	0.55	210	[83]
$[\{Si(Pc)O\}[Tos]_{0.52}]_n$	4660	0.49	190	[83]
$[\{Si(Pc)O\}[Tos]_{0.67}]_n$	5000	0.48	275	[83]

A much more intriguing trend in optical reflectance was observed with the electrochemically-doped $[\{Si(Pc)O\}(X)_y]_n$ polymers. Specular reflectance data for polycrystalline compaction of the X = tosylate compounds are shown in Figure 10. Visual analysis revealed few differences between these and the chemically-oxidised polymer data, but Drude-model fitting of the plasma edge for the metal-like ($y > 0.20$), electrochemically-oxidised polymer samples indicates that the edge *shifts* to higher frequencies with increasing oxidation (e.g. Table 3, $[\{Si(Pc)O\}[Tos]_y]_n$ data). This observation, unprecedented for polymeric metals, confirms the picture of homogeneous doping. Parameters derived from the optical analysis are set out in Table 3. It will be shown below that the shift of ω_p with increasing y conforms very well to the theoretical expectations of tight-binding theory. Also significant is that the optical reflectance spectroscopy indicates that materials of low y have localised carriers (as evidenced by semiconductor-like or insulator-like electronic spectra of these species) [86], in agreement with the observations of charge transport experiments.

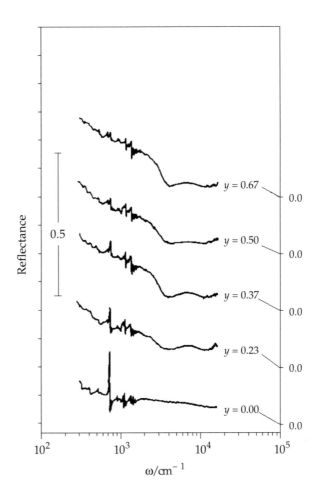

Figure 10 Optical reflectance spectra for $[\{Si(Pc)O\}Tos_y]_n$. (Reprinted with permission from reference [83]. Copyright (1989) American Chemical Society)

The static magnetic susceptibilities of doped $[\{M(Pc)O\}X_y]_n$ polymers were collected with a SQUID susceptometer. These data are analyzed in terms of two contributions: Curie-like and Pauli-like spin densities (Equation (1)).

$$\chi = \chi_P + A\,T^{-\alpha} \tag{1}$$

Figure 11 displays the deconvoluted data for (A) chemically ($I_3{}^-$)-doped and (B) electrochemically ($BF_4{}^-$)-doped polymers. The Curie contribution ($AT^{-\alpha}$) is uniformly small for the $[\{Si(Pc)O\}(I_3)_y]_n$ materials (less than 0.025 spins/Pc) and probably arises from impurities and defects.

The Pauli contribution linearly correlates with y and is a much more significant contribution to the total susceptibility. This is, of course, consistent with other evidence that the chemical doping (halogen, nitrosonium salts, etc.) is heterogeneous.

More significant, however, is the y-dependence of susceptibility for the electrochemically-doped $[\{Si(Pc)O\}(BF_4)_y]_n$ materials (Figure 11B). Two rather different types of behaviour are evident upon incremental doping (increasing y). First, low y samples have a large Curie contribution, peaking at a maximum of *ca* 0.13 spins/Pc at $y = 0.20$. Secondly, high y samples have predominantly Pauli-like behaviour with a noticeable plateau of *ca* 0.17 spins/Pc for $y > 0.25$. This behaviour is again consistent with localised charge carriers at low oxidation levels and delocalised charge carriers (metal-like behaviour) at high oxidation levels. Also evident from these data is the presence of a sudden change in electronic structure near $y = 0.20$.

Complementary information regarding electronic structure and conduction electron spin dynamics is available from electron spin resonance (ESR) spectroscopy. Free-electron g values are observed for all samples, i.e. for both chemically- and electrochemically-doped materials. Rather narrow, isotropic line widths can be associated with the absence of heavy atoms in extensive communication with the carriers.

In general, the electrical, electronic, optical, and magnetic properties of $[\{Si(Pc)O\}X_y]_n$, particularly for the electrochemically-prepared samples, present a coherent picture of evolving electronic structure with increasing incremental doping. The next section will present a simple model, based on tight-binding theory, to explore this evolution. It is actually quite rare to have a polymeric metal with (homogeneous) tunable band-filling. The additional advantage of the extensive array of available physical data allows for unprecedented insight into the electronic structure of molecular metals.

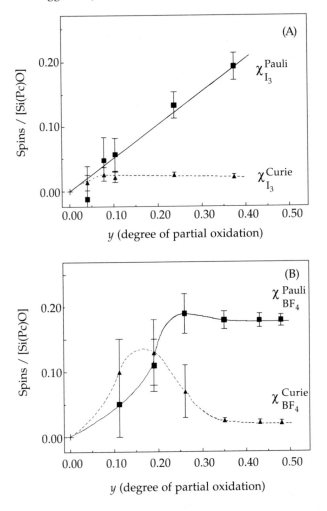

Figure 11 Magnetic spin density for $[\{Si(Pc)O\}X_y]_n$, (A) $y = I_3^-$, (B) $y = BF_4^-$

7.2.2.6 Tunable Band Filling and Band Structure

In this section, the seemingly diverse information derived from physical techniques (conductivity, thermopower, optical reflectance, and magnetic susceptibility) applied to the $[\{Si(Pc)O\}X_y]_n$ polymer system will be reconciled in terms of a single model. The key to this model and to the interpretation of the physical data is the unique structural rigidity of the polymer.   The most straightforward and simple theories for interpretation of physical properties of one-dimensional metals are based on the tight-binding model that was introduced in Figure 1. If we can treat our polymer as an infinite chain that has no external interactions, the mathematics describing the observed properties are simplified. The Density of States (DOS) for such a polymer is shown in Figure 12A. The DOS is a profile of the band structure as a function of energy. If the band is occupied by electrons the DOS corresponds to what would be experimentally measured with photoelectron spectroscopy. The width of this band, the tight-binding bandwidth, is 4t.

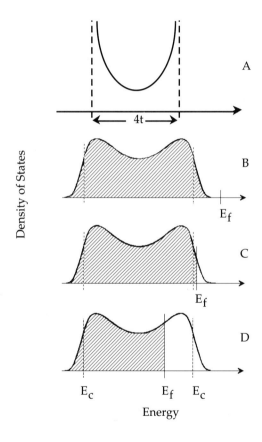

Figure 12 Band structure in response to doping. (Reprinted with permission from reference [83]. Copyright (1989) American Chemical Society)

The models derived for understanding thermopower, optical reflectance and magnetic susceptibility are based on this idealised one-dimensional band. In reality, however, there are contacts between adjacent chains, impurities, structural defects, disorder, and limited chain lengths, and the density of states for a polymer is more similar to that shown in Figures 12B–12D. E_c is a 'mobility edge' separating the localised states near the edge of the band from the metal-like states in the heart of the band [79, 94, 156]. E_f is the Fermi level which separates occupied states (orbitals) from unoccupied states.

The low-doped polymers behave as would be expected for localised charge carriers: the magnetic susceptibilty is Curie-like, the optical reflectance and thermoelectric power are (p-type) semiconductor-like, and the variable temperature conductivity is best described in terms of 'hopping' (between localised states) models. This is consistent with oxidation of the tail region of the DOS (Figure 12C).

At higher y (ca $\geq$ 0.20), all of the physical measurements show significant change: the magnetic susceptibilty is now dominated by Pauli-like spins, the optical reflectance shows a clearly observable plasma edge, the thermoelectric power suggests metal-like properties for the materials, and the variable temperature conductivity is best described in terms of a 'tunnelling' (through interparticle junctions) model that is appropriate for metals. This transformation to a metal-like state with changing oxidation level is known as an Anderson transition [156, 157], and in terms of the model of Figure 12, is the point at which the moving (with oxidation) Fermi level crosses the mobility edge.

Further oxidation improves the metallic properties of the polymers, but at a slower pace than in the y < 0.20 regime: the room temperature conductivity and Pauli magnetic susceptibility are virtually constant, while the thermoelectric power shows continued evolution to more metal-like materials and the plasma edge *shifts* to higher energy with increasing y in the metallic regime. This data is consistent with the description of Figure 12D, i.e. the oxidation is occurring over the delocalised states of the band where the DOS is nearly constant with respect to energy.

More detailed information about the band structure of the polymer can be obtained through closer examination of these physical results, and their relationship to theoretical and experimental electronic structure investigations. Marks, Fragala, *et al.* [145, 158] performed detailed photoelectron spectroscopic studies of monomer and dimer analogues to the $[M(Pc)O]_n$ polymers. The key observation and interpretation is that the observed dimer splitting in the first ionisation of the dimer spectra (0.29 eV for M = Si, 0.19 eV for M = Ge), corresponds to 2t (Figure 1). Thus, this is an *exsitu* probe of the tight-binding bandwidths for these two polymers (0.58 for $[Si(Pc)O]_n$ and 0.38 for $[Ge(Pc)O]_n$). DVM Xα calculations imply a bandwidth of 0.76 eV for the silicon polymer and 0.50 for the germanium polymer [40].

The tight-binding bandwidth is a key parameter that indicates, in a concise manner, the effect of designed tunability of interplanar spacing and degree of partial oxidation on the band structure of the 'metal'. While experimental methods for determining t (or 2t) are rare (cf. photoelectron spectroscopy), the bandwidth (4t) can be derived from some of the physical measurements described in this chapter. First, let us explore the behaviour of the tight-binding bandwidth with interplanar (molecule–molecule) spacing. While $[Ni(Pc)][I_3]_{0.33}$ is not a polymer, it is isostructural with the $[\{M(Pc)O\}(I_3)_{0.36}]_n$ polymers with the exception of the interplanar spacing (3.24 Å versus 3.30 Å for M = Si and 3.48 Å for M = Ge) [71]. Figure 13A shows derived bandwidths for these species from optical and magnetic susceptibility data plotted as a function of c, the interplanar spacing. Essentially, all collective properties are sensitive to this structural parameter. The observed dependence of increasing bandwidth with decreasing spacing is consistent with greater overlap between molecular orbitals on adjacent monomers leading to a richer band structure and a correspondingly larger bandwidth.

Figure 13B displays the 4t versus y behaviour for the $[\{Si(Pc)O\}(BF_4)_y]_n$ polymeric metals. The most notable feature is the nearly flat response for $y > 0.3$ evident in derived bandwidths for optical and magnetic measurements. This is as expected, but never observed, for a molecular system with only minor structural variation with increasing band-filling. The thermopower results are more sensitive to y, perhaps due to the effects of residual interparticle resistance. At lower y (ca 0.2, not shown in Figure 13B), the 4t values become unrealistically large as the tight-binding model is not valid outside the metallic regime. In terms of the model of Figure 12, recall that the equations for calculating 4t are based on the ideal band structure (Figure 12A) and are most unreliable in the band tails where charge carriers are localised [83]. In the metallic regime the density of states is essentially flat for both the ideal (12A) and non-ideal (12B–D) bands, and the tight-binding model is most accurate.

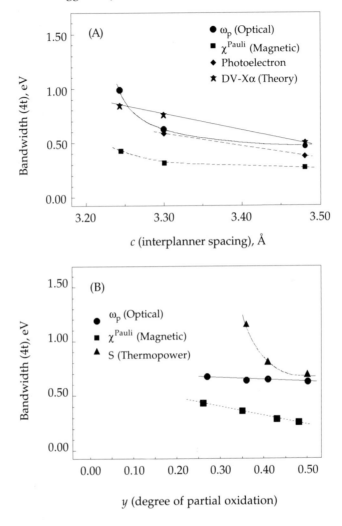

Figure 13 Dependence of bandwidth (4t) on (A) interplanar spacing, (B) degree of partial oxidation (y)

7.3 ELECTROCHEMICAL CHARACTERISATION OF SOLUBLE POLYMERIC COORDINATION COMPLEXES

Although most polymeric systems are insoluble in conventional solvents, and the vast majority of the applications envisaged for the materials involve solid state devices, the study of the inherent doping process, chemical stability, and band tunability is more easily accomplished by investigating soluble systems. Doping can again be done via chemical or electrochemical means. The use of electrochemical methods is particularly attractive in that conventional solution (homogeneous) voltammetry can be employed to probe the redox properties of these materials. Homogeneous, electrochemical doping does not suffer from many of the problems or complications typically encountered with the analogous heterogeneous processes. For example, the structural changes and space-filling limitations associated with counterion insertion [89], and the inseparability of Faradaic and capacitive current components are absent [159]. As illustrated below by the voltammetric characterisation of the soluble phthalocyanine polymers ($[\{M(t\text{-}Bu)_4Pc\}O]_n$, M = Si, Ge) [160–162], the percentage charge-transfer, chemical stability as a function of band-filling, and electronic structure of these systems can be determined using conventional electrochemical means [163–165].

7.3.1 Homogeneous Electrochemical Doping and Voltammetric Characterisation of Soluble Polymers

In a fashion analogous to the ECPS heterogeneous doping of the insoluble materials [89] (Figure 6), conventional sequential controlled potential coulometry (CPC) cycles [166, 167] can be used to determine the potential/band-filling profile of these soluble systems [163, 164]. Figure 14 shows the percentage charge-transfer versus potential profiles for the silicon and germanium polymers. The band filling for both oxidation and reduction can be easily tuned from 0 to 100% ($y = 0$ to 1) using conventional electrolytic techniques. The redox response for an integral change in oxidation state occurs over a 1–1.75 V range, compared to the several hundred millivolt range observed for conventional redox couples and polymers [168–170] and is indicative of a material with a band-type electronic structure. As discussed below, the potential range over which this transition occurs can be related to the bandwidth of the material when corrected for the effects of cumulative charging.

Each point represents the average percent charge-transfer determined from a CPC cycle [170]. The error bars depict the difference between the initial doping process and the conversion back to neutral material. Their small value suggests that no redox activity is lost or gained upon cycling and that no 'break-in' phenomenon is observed. These results, when compared with those obtained for the electrochemical oxidation of the insoluble analogue ($[Si(Pc)O]_n$), provide a basis for comparison between homogeneous (solution) and heterogeneous (solid state) doping of a conductive system. ECPS results with a variety of anions illustrate that, except for the extremely bulky anions, the maximum doping stoichiometries achievable are dictated mainly by anion size [89, 153].

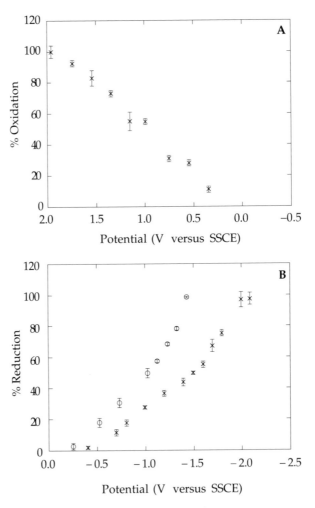

Figure 14 Degree of oxidation (A) and reduction (B) versus potential determined by sequential controlled potential coulometry (CPC) cycles of $[Si\{(t\text{-}Bu)_4Pc\}O]_n$ (x) and $[Ge\{(t\text{-}Bu)_4Pc\}O]_n$ (o). Each point represents the mean value for a complete oxidation/reduction or reduction/oxidation cycle. The error bars correspond to the difference between the reduction and oxidation values of the cycle. (Adapted from references [163] and [165]

Until the maximum doping level is approached (67% for the most highly-doped system), the percent oxidation versus potential profile is independent of the nature of the anion. The degree of oxidation at a given potential for the solid state doping is nearly identical to that observed for the soluble polymer.

This suggests that the data obtained from solid state studies can accurately represent the inherent redox properties of conductive systems. These results also imply that the choice of counterion for solid state doping is important and may limit the degree to which the material can be doped but not necessarily the potential at which doping occurs.

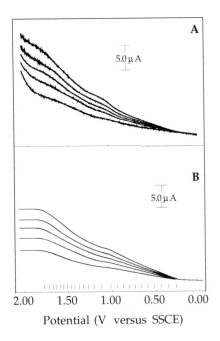

Figure 15 (A) Rotating disk electrode (RDE) voltammograms for the oxidation of a 32.1 μM (0.627 mg cm^{-3}) solution of [Si{(t-Bu)$_4$Pc}O]$_n$ in 1,1,2,2-tetrachloroethane. The increase in current near 1.85 V is a background process arising from the oxidation of the solvent. (B) Simulated voltammograms. The standard potentials used to model the system are represented by vertical lines on the potential axis. (Reprinted with permission from reference [163]. Copyright (1991) American Chemical Society)

Shown in Figures 15A–17A are the rotating disk (RDE) [172], differential pulse (DPV) [173] and cyclic voltammograms (CV) [174] for the doping of the soluble systems. As described below, the responses are characteristic of a chemically-stable Nernstian system. Thus, the voltammograms represent the thermodynamic potentials required to initiate incremental changes in band-filling. The RDE voltammograms show a Faradaic response that again occurs over a 1.75 V range and reaches a limiting current near 1.80 V. Their shape is independent of angular frequency (ω), and current increases in magnitude with $\omega^{1/2}$ at all potentials.

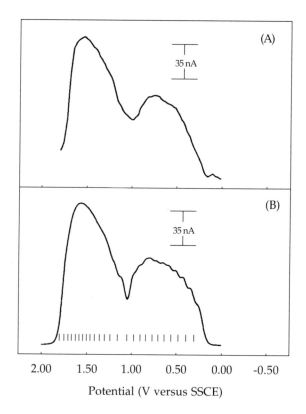

Figure 16 (A) Differential pulse voltammogram (DPV) for the oxidation of a 32.1 μM (0.627 mg cm^{-3}) solution of [Si{(t-Bu)$_4$Pc}O]$_n$ in 1,1,2,2-tetrachloroethane. (B) Simulated voltammogram. The standard potentials used to model the system are represented by vertical lines on the potential axis. (Reprinted with permission from reference [163]. Copyright (1991) American Chemical Society)

This behaviour is expected only for totally reversible (Nernstian) systems and suggests that there are no kinetic limitations for the oxidation of the polymer anywhere along the entire voltammogram [172]. The limiting current near 1.82 V implies that the polymer has reached its maximum degree of oxidation and correlates well with the CPC data (Figure 14A), which also reaches a limiting value near 1.80 V.

In contrast to the rather featureless response for the percentage charge-transfer versus potential profile obtained from the CPC and RDE data, the derivative output of DPV makes it extremely sensitive to subtle changes in redox chemistry and is an ideal technique for the characterisation of soluble conductive systems [173]. A DPV is essentially a derivative of an RDE voltammogram. Figure 16 gives the DPV for the oxidation of the $[Si\{(t\text{-}Bu)_4Pc\}O]_n$. The response shows that the oxidation is not a smooth, continuous process over the entire potential range. The voltammogram appears to contain two distinct regions of electroactivity. The potential range of electroactivity is similar to that observed in the CPC and RDE experiments. As expected, the potential for the decrease in current at 1.85 V correlates with the potential for the limiting current observed in the RDE voltammograms.

As illustrated by the oxidative cyclic voltammetric characterisation of $[Si\{(t\text{-}Bu)_4Pc\}O]_n$ (Figure 17), CV provides little, if any, information concerning the redox profile of chemically stable soluble conductive systems. The current-potential response is similar to the featureless capacitive charging current observed for a solution containing no electroactive material. However, as described below, homogeneous kinetics can easily and accurately be followed using this technique.

7.3.2 Digital Simulations of Voltammetric Results

Although the voltammograms are unlike the typical responses observed for conventional Nernstian systems, their properties can be understood and quantified when analysed via digital simulations. These types of numerical methods have proven invaluable in elucidating the mechanism of complicated redox events [175–178]. Since the polymers undergo reversible oxidation and reduction and lose or gain one electron per site at the maximum degree of charge-transfer, they were modelled as an array of redox couples undergoing sequential electron-transfer. The sequential Nernstian redox of the polymer can be represented as shown in Scheme 1.

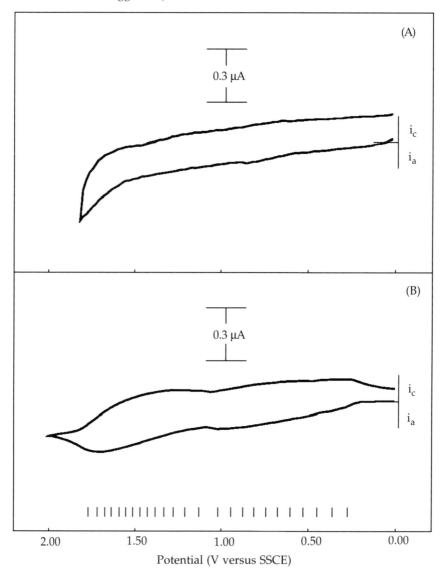

Figure 17 (A) Cyclic voltammogram (CV) for the oxidation of a 32.1 μM (0.627 mg cm^{-3}) solution of [Si{(t-Bu)$_4$Pc}O]$_n$ in 1,1,2,2-tetrachloroethane. The increase in current near 1.85 V is a background process arising from the oxidation of the solvent. (B) Simulated voltammogram. The standard potentials used to model the system are represented by vertical lines on the potential axis. (Reprinted with permission from reference [163]. Copyright (1991) American Chemical Society)

$$A_m \pm n_1 e^- \rightleftharpoons A_{m-1}B_1 \qquad E_1^0$$

$$A_{m-1}B_1 \pm n_2 e^- \rightleftharpoons A_{m-2}B_2 \qquad E_2^0$$

$$\bullet$$
$$\bullet$$
$$\bullet$$

$$A_{m-(m-1)}B_{m-1} \pm n_m e^- \rightleftharpoons B_m \qquad E_m^0$$

Scheme 1 Representation of a Nernstian redox process of a polymer

Each reaction has its own unique standard potential (E^0) and is in equilibrium with the other sites in the assembly. Shown in Figures 15B–17B are the corresponding simulations based on this model [179]. The standard potentials used to generate the voltammograms were carefully chosen to produce observed experimental results. For all the voltammetric techniques both the current magnitude and shape agree well with the experimental results and the difference in E^0 values allows for an estimate of the degree of interaction between the sites.

7.3.3 Following Chemical Stability as a Function of the Degree of Band-Filling by Voltammetric Techniques

The chemical stability of the assemblies as a function of band-filling is also more easily studied using standard voltammetry. These methodologies have a timescale ranging from 10^{-4} to 1 s [167], allowing for a wide range of coupled homogeneous reactions to be followed and quantified. Figures 18A and 18B show the CVs for the reduction of [{Ge(t-Bu)$_4$Pc}O]$_n$ when cycled between 0 and −1.45, and −1.95 V, respectively [164]. Curve A corresponds to cycling the polymer between 0 and 100% reduction (Figure 14).

The voltammogram remains constant upon repeated cycling, suggesting that the material is chemically stable at all degrees of reduction between 0 and 100%. At potentials greater than − 1.45 V (curve B) a large increase in current is observed, and two new conventional oxidation/reduction waves are observed at − 0.35 and − 0.83 V.

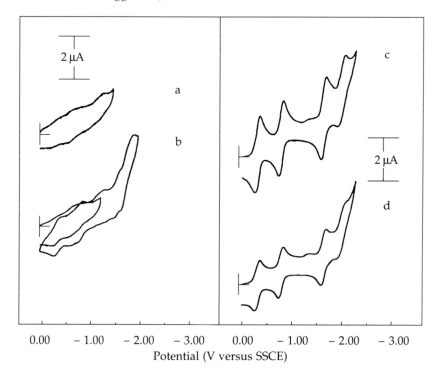

Figure 18 Cyclic voltammograms (CV) for the reduction of a 33.4 µM (0.413 mg cm^{-3}) solution of $[Ge\{(t\text{-}Bu)_4Pc\}O]_n$ in THF. (A) Virgin polymer; (B) two successive scans of a virgin polymer; (C) response after CPC cycle at –1.65/0.00 V; (D) cyclic voltammogram for the reduction of a 881 µM (0.743 mg cm^{-3}) solution of $[Ge\{(t\text{-}Bu)_4Pc\}(OH)_2]$ in THF. (Adapted from reference [165])

Upon a CPC cycle at – 1.65/0.00 V, the polymer fragments into monomeric units that exhibit a voltammetric response (Figure 18c) identical to that of the dihydroxy{tetra(t-butyl)phthalocyaninato}-germanium monomer (repeat unit) (Figure 18D). Quantitative analysis of the voltammetric results [174, 180, 181] suggest an E_nCEE/EEE-type mechanism (Scheme 2), a chemical reaction coupled between distinct series of electron-transfer reactions [167]. The first order decomposition reaction has a rate constant of 0.2 s^{-1}.

Reduction:

'E_n' $[Ge\{(t\text{-}Bu)_4Pc\}O]_n + ne^- \rightleftharpoons [Ge\{(t\text{-}Bu)_4Pc\}O]_n^-$ $E_{1/2(n)} = -1.35$ V

'C' $[Ge\{(t\text{-}Bu)_4Pc\}O]_n^- \xrightarrow{K_c = 0.2\,s^{-1}} n[Ge\{(t\text{-}Bu)_4Pc\}O]^-$

'E' $[Ge\{(t\text{-}Bu)_4Pc\}O]^- + e^- \rightleftharpoons [Ge\{(t\text{-}Bu)_4Pc\}O]^{2-}$ $E_{1/2} = -0.83$ V

'E' $[Ge\{(t\text{-}Bu)_4Pc\}O]^{2-} + e^- \rightleftharpoons [Ge\{(t\text{-}Bu)_4Pc\}O]^{3-}$ $E_{1/2} = -1.68$ V

Oxidation:

'E' $[Ge\{(t\text{-}Bu)_4Pc\}O]^{3-} \rightleftharpoons [Ge\{(t\text{-}Bu)_4Pc\}O]^{2-} + e^-$ $E_{1/2} = -1.68$ V

'E' $[Ge\{(t\text{-}Bu)_4Pc\}O]^{2-} \rightleftharpoons [Ge\{(t\text{-}Bu)_4Pc\}O]^- + e^-$ $E_{1/2} = -0.83$ V

'E' $[Ge\{(t\text{-}Bu)_4Pc\}O]^- \rightleftharpoons [Ge\{(t\text{-}Bu)_4Pc\}O] + e^-$ $E_{1/2} = -0.35$ V

Scheme 2   Representation of the redox behaviour of $[Ge\{(t\text{-}Bu)_4Pc\}O]_n$

7.3.4   Correlations Between Electronic Structure and Voltammetric Response

The voltammetric responses can also yield information concerning the electronic structure of conductive systems. Correlations between voltammetric response and band structure and energy as a function of band-filling are relatively rare [182, 183]. Bandwidths estimated from tight-binding band calculations (extended Hückel type) and other more sophisticated methods [145, 184–191] agree well with the potential range of electroactivity when corrected for incremental changes in oxidation state (cumulative charging) [165]. The change in band energy as a function of charging may be approximated by:

$$E = \varepsilon - A \text{ charge} \qquad (2)$$

where E is the redox potential of the polymer at a specific oxidation state, ε is the corresponding energy in the band calculation (adjusted relative to a reference electrode), and A is a proportionality constant [165]. Though A is very difficult to determine *a priori*, it may be estimated from the redox potentials of the corresponding monomer. The difference in redox potential between the 0/+1 and +1/+2 oxidation states of the bis(trimethylsiloxy)[tetra(t-butyl)phthalocyaninato]silicon is about 0.90 V [163]. Since oxidation involves removing electrons sequentially from the HOMO (Figure 1), the difference between the two states can be used to estimate the change in orbital energy due to charging. The net charge at a given ε may be found by integrating the DOS from E_f to that value. For the oxidation process [192]:

$$\text{charge} = -\frac{1}{2} \int_{E_f}^{\varepsilon} \text{DOS (e)} \, de \qquad (3)$$

This treatment is valid only if charging does not alter the band structure, except for shifting the zero of energy. Shown in Table 4 are the calculated 'charge-corrected' and 'redox-derived' bandwidths for $[\{Si(t\text{-}Bu)_4Pc\}O]_n$.

In addition, the structure of the calculated charge-corrected DOS agrees well with the DPV for the oxidation and reduction of the silicon polymer [165, 193]. These results are shown in Figure 19 and suggest that the redox response of conductive systems exhibiting Nernstian behaviour is governed predominantly by the band structure of the material.

Table 4 Comparison between band calculations modified for charging and differential pulse voltammograms of $[\{M(t\text{-}Bu)_4Pc\}O]_n$

	Calculation, eV	Calculation corrected for charging, eV	'Redox-derived values', eV
LUMO Band	0.62	1.5	1.6
HOMO Band	0.32	1.7	1.7
Band Gap	0.28	0.30	0.55

Oxidation

Reduction

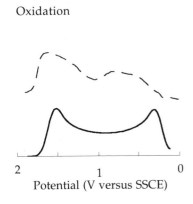
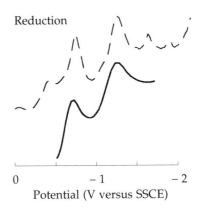

2   1   0   0   – 1   – 2
Potential (V versus SSCE)   Potential (V versus SSCE)

Figure 19 Modified DOS calculations (solid line) and experimental differential pulse voltammograms (dashed line) for oxidation (left) and reduction (right) of [Si{(t-Bu)$_4$Pc}O]$_n$

7.4   SUMMARY AND OUTLOOK

Over the past decade extensive procedures for the preparation and characterisation of a wide range of conductive systems have been developed [1–4, 9, 14, 22, 31]. Interest is now turning away from developing general methods for the synthesis and doping of these compounds and toward the discovery of rational and facile approaches for the preparation of materials with specific and predictable properties. Building on this foundation, custom-designed materials, synthesised and doped for a specific application are now possible.

In this chapter we have discussed and described the synthesis, 'doping', and characterisation of a very unique class of polymeric coordination complexes. Because these polymeric metallophthalocyanine materials can be considered as hybrids between molecular metals and conductive polymers, a wealth of information concerning both can be obtained. Particularly exciting is the breadth of the understanding of electronic structure obtained from the electrochemical measurements described in Section 7.3.

However, real progress in designing and synthesising new inorganic and organometallic polymeric materials that take advantage of the unique electronic and structural properties of d and f metals will apparently be reserved for the future. This is an exciting frontier of synthetic, inorganic chemistry with intrinsically, metal-containing polymers [194, 195]. Many desirable properties could be envisaged from metal-containing polymers, and a wide range of unprecedented 'synthetic devices' are possible. The availability of polymeric materials with useful magnetic, optical (including non-linear), as well as electrical properties, will certainly influence future microelectronic fabrication. This will clearly be one of the important arenas of materials chemistry research into the next century.

Also important is the development of an understanding of the electronic mechanism of conduction and related phenomena. As yet, no unified theory has emerged that explains these effects in much more than a handful of materials classes, let alone across the broad spectrum of synthetic conductors. This is a challenging and formidable goal for solid state theorists. We believe, however, that multidisciplinary studies, utilising the skills of chemists, physicists, material scientists and others, and application of a broad range of physical and analytical techniques, may reveal the keys to unlocking this understanding.

7.5   ACKNOWLEDGEMENTS

The authors became interested in metal-containing polymers while postdoctoral fellows at Northwestern University (NU). We wish to thank our NU, Michigan State University, Virginia Commonwealth University and IBM collaborators, co-authors and associates: Donald Abraham, Manuel Almeida, Robert Burton, John Butler, Stephen Cain, Steve Carr, Thomas Chapaton, Patricia DePra, Richard Freer, David Gale, Tamotsu Inabe, Mercouri Kanatidis, Carl Kannewurf, Line LeBlevenec, Joseph Lyding, WenBin Liang, Joseph Lomax, Henry Marcy, Tobin Marks, William McCarthy, Michael Moguel, John Schlueter, Simon Semus, Stephen Tetrick, Lauren Tonge, Paul Toscano, and Fred Wireko. We also wish to acknowledge the support and editorial help of our wives, Diane S Kellogg and Debora J Gaudiello.

7.6   REFERENCES

1.   C E Carraher Jr, J E Sheats and C U Pittman Jr, *Advances in Organometallic and Inorganic Polymer Science*, Eds, Marcel Dekker, New York (1982).

2. J E Sheats, C E Carraher Jr, C U Pittman Jr, M Zeldin and B Currell, *Inorganic and Metal-Containing Polymeric Materials*, Eds, Plenum, New York (1990).
3. M Zeldin, K J Wynne and H R Allcock, *Inorganic and Organometallic Polymers*, Eds, ACS Symposium Series Vol 360, Washington (1988).
4. H R Allcock, *Chem Eng News*, **63**, 22 (1985).
5. I Yilgor, W P Stickle Jr, D Tyagi, G L Wilkes and J E McGrath, *Polymer, London*, **25**, 1800 (1984).
6. J D Summers, C A Arnold, R H Bott, L T Taylor, T C Ward and J E McGrath, *Polym Prept*, **27**, 403 (1986).
7. M Zeldin and J-M Xu, *J Organomet Chem*, **320**, 267 (1987).
8. K Nate, M Ishikawa, N Imamura and Y Murakami, *J Polym Sci Polym Chem Ed*, **24**, 1551 (1986).
9. G B Goodwin and M E Kenney, in: *Inorganic and Organometallic Polymers*, M Zeldin, K J Wynne and H R Allcock, Eds, ACS Symposium Series: Washington Vol 360, p 238 (1988).
10. R West, *J Organomet Chem*, **300**, 327 (1986).
11. A R Wolff and R West, *Applied Organomet Chem*, **1**, 7 (1987).
12. D Seyferth, in: *Inorganic and Organometallic Polymers*, M Zeldin, K J Wynne and H R Allcock, Eds, ACS Symposium Series Vol 360, Washington, p 21 (1988).
13. R D Miller, B L Farmer, W Fleming, R Sooriyakumaran and J Rabolt, *J Am Chem Soc*, **109**, 2509 (1987).
14. H R Allcock, in: *Inorganic and Organometallic Polymers*, M Zeldin, K J Wynne and H R Allcock, Eds, ACS Symposium Series Vol 360, Washington, p 250 (1988).
15. H R Allcock, J L Desorcie and G H Riding, *Polyhedron*, **6**, 119 (1987).
16. P Wisian-Neilson, R R Ford, R H Neilson and A K Roy, *Macromolecules*, **19**, 2089 (1986).
17. C K Narula, R T Paine and R Schaeffer, *Polym Prepr (Am Chem Soc Div Polym Chem)*, **28**, 454 (1987).
18. K J L Paciorek, D H Harris and R H Kratzer, *J Polym Sci, Polym Chem Ed*, **24**, 173 (1986) .
19. D W Macomber, W P Hart, M D Rausch, R D Priester Jr and C U Pittman Jr, *J Am Chem Soc*, **104**, 884 (1982) .
20. C U Pittman Jr and A Hirao, *J Polym Sci, Polym Chem*, **15**, 1677 (1977).
21. K Gonsalves, R W Lenz and M D Rausch, *Appl Organomet Chem*, **1**, 81 (1987).
22. Proceedings of the International Conference on the Science and Technology of Synthetic Metals, *Synth Metals*, **17-19** (1987).
23. D Jerome and L G Caron, Eds, *Low-dimensional Conductors and Superconductors*, Plenum, New York (1987).

24. D O Cowan and F M Wiygul, *Chem Eng News*, **64**, 28 (1986).
25. K Bechgaard and D Jerome, *Sci American*, **247**, 52 (1982).
26. P M Chaikin and R L Greene, *Physics Today*, **39**, 24 (1986).
27. J M Williams, *Prog Inorg Chem*, **33**, 183 (1985).
28. J R Ferraro and J M Williams, *Introduction to Synthetic Electrical Conductors*, Academic Press, New York (1987).
29. J S Miller, Ed, *Extended Linear Chain Compounds*, Vols 1–3, Plenum Press, New York (1982).
30. S Roth and H Bleier, *Adv Phys*, **36**, 385 (1987).
31. T A Skotheim, Ed, *Handbook of Conducting Polymers*, Vols 1–2, Marcel Dekker, New York (1986).
32. R H Baughman, J L Bredas, R R Chance, R L Elsenbaumer and L W Shacklette, *Chem Rev*, **82**, 209 (1982).
33. G Wegner, *Angew Chem Int Ed Engl*, **20**, 361 (1981).
34. J S Miller and A J Epstein, *Angew Chem Int Ed Engl*, **26**, 287 (1987).
35. G Saito and J P Ferraris, *Bull Chem Soc Jpn*, **53**, 2141 (1980).
36. C B Duke, *Synth Metals*, **21**, 5 (1987).
37. R Hoffmann, *Angew Chem Int Ed Engl*, **26**, 846 (1987).
38. M-H Whangbo and K R Stewart, *Isr J Chem*, **23**, 573 (1983).
39. M C Böhm, *Chem Phys*, **86**, 17 (1984).
40. W J Pietro, M A Ratner and T J Marks, *J Am Chem Soc*, **107**, 5387 (1985).
41. J B Torrance, *Acc Chem Res*, **12**, 79 (1979).
42. F Wudl, in: *The Physics and Chemistry of Low Dimensional Solids*, L Alcacer, Ed, D Reidel, Dordrecht, p 265 (1980).
43. F M Wiygul, R M Metzger and T J Kistenmacher, *Mol Cryst. Liq Cryst*, **107**, 115 (1984).
44. T J Marks, *Science*, **227**, 881 (1985).
45. T Inabe, J G Gaudiello, M K Moguel, J W Lyding, R L Burton, W J McCarthy, C R Kannewurf and T J Marks, *J Am Chem Soc*, **108**, 7595 (1986).
46. L B Coleman, M J Cohen, D J Sandman, F G Yamagishi, A F Garito and A J Heeger, *Solid State Commun*, **12**, 1125 (1973) .
47. D Jerome and H J Schulz, *Adv in Physics*, **31**, 299 (1982).
48. K Bechgaard, K Carneiro, F B Rasmussen, M Olsen, G Rindorf, C S Jacobsen, H J Pedersen and J C Scott, *J Am Chem Soc*, **103**, 2440 (1981).
49. G W Crabtree, K D Carlson, L N Hall, P T Copps, H H Wang, T J Emge, M A Beno and J M Williams, *Phys Rev B*, **30**, 2958 (1984).
50. B Hilti, C W Mayer and G Rihs, *Helv Chim Acta*, **61**, 1462 (1978).
51. C S Schramm, R P Scaringe, D R Stojakovic, B M Hoffman, J A Ibers and T J Marks, *J Am Chem Soc*, **102**, 6702 (1980).
52. J Heine, *Top Curr Chem*, **152**, 2 (1990).
53. N C Billingham and D Calvert, *Adv Polym Sci*, **90**, 3 (1989).

54. J E Frommer and R R Chance, *Encyclopedia of Polymer Science and Engineering*, Vol 5, p 462 (1986).

55. R H Baughman, J L Bredas, R R Chance, R L Elsenbaumer and L W Shacklette, *Chem Rev*, **82**, 209 (1982).

56. W P Su, J R Schrieffer and A J Heeger, *Phys Rev Lett*, **42**, 1698 (1979).

57. W P Su, J R Schrieffer and A J Heeger, *Phys Rev B*, **22**, 2099 (1980).

58. A R Bishop, D K Campbell and K Fesser, *Mol Cryst, Liq Cryst*, **77**, 253 (1981).

59. J Tinka Gammel and J A Krumhansl, *Phys Rev B*, **24**, 1035 (1981).

60. J L Bredas, R R Chance and R Silbey, *Phys Rev B*, **26**, 58431 (1982).

61. H Shirakawa, E J Louis, A G MacDiarmid, C K Chiang and A J Heeger, *J Chem Soc, Chem Commun*, 578 (1977).

62. C K Chiang, M A Druy, S C Gau, A J Heeger, E J Louis, A G MacDiarmid, Y W Park and H Shirakawa, *J Am Chem Soc*, **100**, 1013 (1978).

63. K K Kanazawa, A F Dias, R H Geiss, W D Gill, J F Kwak, J A Logan, J F Rabolt and G B Street, *J Chem Soc, Chem Commun*, 854 (1979).

64. A F Diaz, J I Castillo, J A Logan and W Y Lee, *J Electroanal Chem*, **129**, 115 (1981).

65. A F Diaz and K K Kanazawa, in *Extended Linear Chain Compounds*, J S Miller, Ed, Plenum Press, New York, p 417 (1982).

66. S C Gau, J Milliken, A Pron, A G MacDiarmid and A J Heeger, *J Chem Soc, Chem Commun*, 662 (1977).

67. G Street, T Clarke, M Kroumbi, K Kanazawa, V Lee, P Fluger, J Scoot and G Weiser, *Mol Cryst, Liq Cryst*, **83**, 253 (1982).

68. M Kobayashi, J Chen, T C Chung, F Moraes, A J Heeger and F Wudl, *Synth Metals*, **9**, 77 (1984).

69. W R Salaneck, I Lundstrom, W S Haung and A G MacDiarmid, *Synth Metals*, **13**, 291 (1986).

70. M Hanack, *GIT Fachz Lab*, **2**, 75 (1987).

71. B N Diel, T Inabe, J W Lyding, K F Schoch Jr, C R Kannewurf and T J Marks, *J Am Chem Soc*, **105**, 1551 (1983).

72. M Hanack, S Deger, U Keppeler, A Lange, A Leverenz and M Rein, *Synth Metals*, **19**, 739 (1987).

73. J P Collman, J T McDevitt, C R Leidner, G T Yee, J B Torrance and W A Little, *J Am Chem Soc*, **109**, 4606 (1987).

74. K J Wynne, *Inorg Chem*, **24**, 1339 (1985).

75. R S Nohr, P M Kuznesof, K J Wynne, M E Kenney and P G Siebenman, *J Am Chem Soc*, **103**, 4371 (1981).

76. E A Orthmann, V Enkelmann and G Wegner, *Makromol Chem Rapid Commun*, **4**, 687 (1983).

77. C J Brown, *J Chem Soc A*, 2494 (1968).

78. C J Brown, *J Chem Soc A*, 2488 (1968).
79. C C Ku and R Liepins, *Electrical Properties of Polymers: Chemical Principles*, Hanser Publishers, Munich (1987).
80. D E Schaefer, F Wudl, G A Thomas, J P Ferraris and D O Cowan, *Solid State Commun*, **14**, 347 (1974).
81. Y-W Park, A J Heeger, M A Druy and A G MacDiarmid, *J Chem Phys*, **73**, 946 (1980).
82. J F Kwak, G Beni and P M Chaikin, *Phys Rev B*, **13**, 641 (1976).
83. M Almeida, J G Gaudiello, G E Kellogg, S M Tetrick, H O Marcy, W J McCarthy, J C Butler, C R Kannewurf and T J Marks, *J Am Chem Soc*, **111**, 5271 (1989).
84. J C Scott, A F Garito and A J Heeger, *Phys Rev B*, **10**, 3131 (1974).
85. L C Tippie and W G Clark, *Phys Rev B*, **23**, 5486 (1981).
86. C S Jacobsen, in *Low-Dimensional Conductors and Superconductors*, D Jerome and L G Caron, Eds, Plenum Press, New York, p 253 (1987).
87. M R Madison, L B Coleman and R B Somoano, *Solid State Commun*, **40**, 979 (1981).
88. C S Jacobsen, K Mortensen, J R Andersen and K Bechgaard, *Phys Rev B*, **18**, 905 (1978).
89. J G Gaudiello, G E Kellogg, S M Tetrick and T J Marks, *J Am Chem Soc*, **111**, 5259 (1989).
90. J P Pouget, J C Pouxviel, P Robin, R Comes, D Begin, D Billaud, A Feldblum, H W Gibson and A J Epstein, *Mol Cryst, Liq Cryst*, **117**, 75 (1985).
91. R H Baughman, N S Murthy and G G Miller, *J Chem Phys*, **79**, 515 (1983).
92. M Maxfield, T R Jow, S Gould, M G Sewchok and L G Shacklette, *J Electrochem Soc*, **135**, 299 (1988).
93. C K Chiang, M A Druy, S C Gau, A J Heeger, E J Louis, A G MacDiarmid, Y W Park and H Shirakawa, *J Am Chem Soc*, **100**, 1013 (1978).
94. N F Mott and E A Davis, *Electronic Processes in Non-Crystalline Materials*, 2nd edn, Clarendon Press, Oxford (1979).
95. P Sheng, *Phys Rev B*, **21**, 2180 (1980).
96. T A Skotheim, Ed, *Handbook of Conducting Polymers*, Chapter 29, Marcel Dekker, New York (1986).
97. M Abkowitz, D F Blossey and A I Lakatos, *Phys Rev B*, **8**, 3400 (1975).
98. G Guillaud and N Rosenberg, *J Phys E*, **13**, 1287 (1980).
99. M Takahashi, T Sugano and M Kinoshita, *Bull Chem Soc Jpn*, **57**, 26 (1984).
100. L C Isett, *Phys Rev B*, **18**, 439 (1978).
101. S Mazumdar and S N Dixit, *Phys Rev B*, **34**, 3683 (1986).

102. H Gutfreund, O Entin-Wohlman and M Weger, *Mol Cryst, Liq Cryst*, **119**, 457 (1986).
103. P J Toscano and T J Marks, *J Am Chem Soc*, **108**, 437 (1986).
104. F Wooten, *Optical Properties of Solids*, Academic Press, New York (1972).
105. M Almeida, M G Kanatzidis, L M Tonge, T J Marks, H O Marcy, W J McCarthy and C R Kannewurf, *Solid State Commun*, **63**, 457 (1987).
106. C W Dirk, T Inabe, K F Schoch Jr and T J Marks, *J Am Chem Soc*, **105**, 1539 (1983).
107. D C Bott and J N Winte, *Polym J*, **28**, 601 (1987).
108. K J Wynne and G B Street, *Macromolecules*, **18**, 3261 (1985).
109. J P Collman, J T McDevitt, C R Leidner, G T Yee, J B Torrance and W A Little, *J Am Chem Soc*, **109**, 4606 (1987).
110. J P Collman, J T McDevitt, G T Yee, M B Zisk, J B Torrance and W A Little, *Synth Met*, **15**, 129 (1986).
111. R S Nohr, P M Kuznosof, K J Wynne, M E Kenney and P G Siebenman, *J Am Chem Soc*, **103**, 4371 (1981).
112. M Hanack, A Lange, M Rein, R Behnisch, G Renz and A Leverenz, *Synth Met*, **29**, F1 (1989).
113. M Hanack and A Leverenz, *Springer Ser Solid-State Sci*, **76**, 318 (1987).
114. M Hanack and A Leverenz, *Synth Met*, **22**, 9 (1987).
115. M Hanack, *Isr J Chem*, **25**, 205 (1985).
116. M Hanack, *Mol Cryst, Liq Cryst*, **105**, 133 (1984).
117. B N Diel, T Inabe, N K Jaggi, J W Lyding, O Schneider, M Hanack, C R Kannewurf, T J Marks and L H Schwarz, *J Am Chem Soc*, **106**, 3207 (1984).
118. M Hanack, S Deger, A Lange and T Zipplies, *Synth Met*, **15**, 207 (1986).
119. O Schneider and M Hanack, *Chem Ber*, **116**, 2088 (1983).
120. M Hanack, U Keppeler and J H Schulze, *Synth Met*, **20**, 347 (1987).
121. M Hanack and G Renz, *Chem Ber* **123**, 1105 (1990).
122. M Hanack and R Grosshans, *Chem Ber*, **122**, 1665 (1989).
123. M Hanack, *Mol Cryst, Liq Cryst*, **160**, 133 (1988) and references therein.
124. U Keppeler and M Hanack, *Chem Ber*, **119**, 3363 (1986).
125. S Deger and M Hanack, *Synth Met*, **13**, 319 (1986).
126. M Hanack, A Datz, R Fay, K Fischer, C Hedtmann-Rein, U Keppeler, J Koch and W Stoeffler, *Polym Prepr*, **25**, 232 (1984).
127. U Keppeler, O Schneider, W Stoeffler and M Hanack, *Tetrahedron Lett*, **25**, 3679 (1984).
128. S Deger and M Hanack, *Springer Ser Solid-State Sci*, **63**, 327 (1985).
129. M Hanack, A Datz, W Kobel, J Koch, J Metz, M Mezger, O Schneider and H J Schulze, *J Phys*, 633 (1983).

130. M Hanack, S Deger, U Keppeler, A Lange, A Leverenz and M Rein, *Conduct Polym, Proc Workshop*, L Alcacer, Ed, Reidel, Dordrecht, p 173 (1986).

131. N Fahmy and M Hanack, *Recl Trav Chim Pays Bas*, **109**, 235 (1990).

132. M Hanack, S Deger, U Keppeler, A Lange, A Leverenz and M Rein, *Synth Met*, **19**, 739 (1987).

133. M Hanack and R Fay, *Recl Trav Chim Pays Bas*, **105**, 427 (1986).

134. M Hanack and C Hedtmann-Rein, *Z Naturforsch, B*, **40**, 1087 (1985).

135. M Hanack and X Muenz, *Synth Met*, **10**, 357 (1985).

136. A Datz, J Metz, O Schneider and M Hanack, *Synth Met*, **9**, 31 (1984).

137. O Schneider and M Hanack, *Z Naturforsch, B*, **39**, 265 (1984).

138. M Hanack, C Hedtmann-Rein, A Datz, U Keppeler and X Muenz, *Synth Met*, **19**, 787 (1987).

139. J P Linsky, T R Paul, R S Nohr and M E Kenney, *Inorg Chem*, **19**, 3131 (1980).

140. P M Kuznesof, R S Nohr, K J Wynne and M E Kenney, *J Macromol Sci Chem*, **A16**, 299 (1981).

141. R S Nohr, P M Kuznesof, K J Wynne, M E Kenney and P G Siebenman, *J Am Chem Soc*, **103**, 4371 (1981).

142. K J Wynne and R S Nohr, *Mol Cryst, Liq Cryst*, **81**, 234 (1981).

143. R S Nohr and K J Wynne, *J Chem Soc, Chem Commun*, 1210 (1981).

144. P Brant, R S Nohr, K J Wynne and D C Weber, *Mol Cryst, Liq Cryst*, **81**, 255 (1982).

145. E Ciliberto, K A Doris, W J Pietro, G M Reisner, D E Ellis, I Fragala, F H Herbstein, M A Ratner and T J Marks, *J Am Chem Soc*, **106**, 7748 (1984).

146. M K Lowery, A J Starshak, J N Esposito, P C Krueger and M E Kenney, *Inorg Chem*, **4**, 128 (1965).

147. R D Joyner and M E Kenney, *Inorg Chem*, **1**, 717 (1962).

148. R C Teitelbaum, S L Ruby and T J Marks, *J Am Chem Soc*, **102**, 3322 (1980).

149. A H Thompson, *Rev Sci Instrum*, **54**, 229 (1983).

150. J H Kaufman, T-C Chung, A J Heeger, *J Electrochem Soc*, **131**, 2847 (1984).

151. M S Whittingham, *Prog Solid State Chem*, **12**, 41 (1978).

152. P G Pickup and R A Osteryoung, *J Am Chem Soc*, **106**, 2294 (1984).

153. G E Kellogg, J G Gaudiello, J A Schlueter, S M Tetrick, T J Marks, H O Marcy, W J McCarthy and C R Kannewurf, *Synth Metals*, **29**, F15 (1989).

154. M Audenaert, G Gusman and R Deltour, *Phys Rev B*, **24**, 7380 (1981).

155. E K Sichel, J I Gittleman and P Sheng, *Phys Rev B*, **18**, 5712 (1978).

156. P W Anderson, *Phys Rev*, **109**, 1492 (1958).

157. N F Mott, *Metal-Insulator Transitions*, Taylor and Francis, London (1974).

158. K A Doris, E Ciliberto, I Fragala, M A Ratner and T J Marks, *Isr J Chem*, **27**, 337 (1986).

159. S W Feldberg, *J Am Chem Soc*, **106**, 4671 (1982) and references therein.

160. O Schneider, J Metz and M Hanack, *Mol Cryst, Liq Cryst*, **81**, 273 (1982).

161. J Metz, G Pawlowski and M Hanack, *Z Naturforsch B*, **38**, 378 (1983).

162. For preparing the precursors to these polymers see: M Hanack, J Metz and G Pawlowski, *Chem Ber*, **115**, 2836 (1982).

163. D C Gale and J G Gaudiello, *J Am Chem Soc*, **113**, 1610 (1991).

164. L F LeBlevenec and J G Gaudiello, *J Electroanal Chem*, **312**, 97 (1991).

165. S R Cain, D C Gale and J G Gaudiello, *J Phys Chem*, **95**, 9584 (1991).

166. A J Bard and K S V Santhanam, in: *Electroanalytical Chemistry*, A J Bard, Ed, Marcel Dekker, New York, Vol 4, p 215 (1970).

167. AJ Bard and L R Faulkner, *Electrochemical Methods*, Wiley, New York (1980).

168. D E Brennam and W E Geiger, *J Am Chem Soc*, **101**, 3399 (1979).

169. G H Brown, T J Meyer, D O Cowan, C LeVanda, F Kaufman, P V Roling, and M D Rausch, *Inorg Chem*, **104**, 506 (1975).

170. J B Flanagan, S Margel, A J Bard and F C Anson, *J Am Chem Soc*, **100**, 4248 (1979).

171. Although Faraday's Law implies that the degree of polymerisation must be known to calculate the degree of charge-transfer (to determine the number of electrons transferred per mole of polymer), only the number of electrons *transferred per redox site* and the *concentration of redox sites* is required. The degree of partial charge-transfer can be determined by relating the amount of Coulombs passed to the number of redox sites present.

172. V G Levich, *Physicochemical Hydrodynamics*, Prentice-Hall, Englewood Cliffs, N J, (1962).

173. E P Parry and R A Osteryoung, *Anal Chem*, **37**, 1635 (165).

174. R S Nicholson I Shain, *Anal Chem*, **36**, 706 (1964) .

175. S W Feldberg, in: *Electroanalytical Chemistry*, A J Bard, Ed, Marcel Dekker, New York, Vol 3, p 199 (1969).

176. K B Prater and A J Bard, *J Electrochem Soc*, **117**, 207 (1970).

177. D Britz, *Digital Simulations in Electrochemistry, Lecture notes in Chemistry*, Springer Verlag, Heidelberg (1981).

178. S W Feldberg, *Computers in Chemistry and Instrumentation*, J S Mattson, H B Mark and H C MacDonald Eds, Marcel Dekker, New York, Vol 2, p 185 (1972).

179. T C Chapaton, D C Gale and J G Gaudiello *J Electrochem Soc*, (submitted).

180. R S Nicholson and I Shain, *Anal Chem,* **37,** 175 (1965).
181. J M Seveant, *Electrochim Acta,* **12,** 753 (1967).
182. R B Kaner, S J Porter, D P Nairns and A G MacDiarmid, *J Chem Phys,* **90,** 5102 (1989) and references therein.
183. Wrighton and co-workers have related changes in conductivity for conductive polymer-based transistors to the potential range of electroactivity and thus degree of band-filling. D Ofer, R M Crooks and M S Wrighton, *J Am Chem Soc,* **112,** 7869 (1990).
184. F M-H Whangbo and K R Stewart, *Isr J Chem,* **23,** 133 (1983).
185. E Canadell and S Alvarez, *Inorg Chem,* **23,** 573 (1984).
186. W J Pietro, T J Marks and M A Ratner, *J Am Chem Soc,* **107,** 5387 (1985).
187. A M Schaffer, M Gouterman and E R Davidson, *Theor Chim Acta,* **30,** 9 (1973).
188. E Orti, J L Bredas and C Clarisse, *J Chem Phys,* **92,** 1228 (1990).
189. F W Kutzler and D E Ellis, *J Chem Phys,* **84,** 1033 (1986).
190. P D Hale, W J Pietro, M A Ratner, D E Ellis and T J Marks, *J Am Chem Soc,* **109,** 5943 (1987).
191. E Orti and J L Bredas, *Synth Met,* **29,** F115 (1989).
192. Note that the integration runs from the Fermi level, E_f, *down* to energy, ε. The factor of $1/2$ is introduced to account for the fact that only spin-up (or spin-down) electrons are removed. For reduction, the charge also may be calculated with the same expression except that the integration runs upward from E_f to ε.
193. The excitation waveform for DPV consists of a series of short, small-amplitude pulses (dE_p) on a slowly-stepped potential (E). The current is measured before and near the end of the pulse and the difference is plotted versus potential (E). For an adiabatic (reversible or Nernstain) process, the magnitude of the current is proportional to the number of accessible redox states at the given potential. Thus, the current at any potential for a DPV is a measure of the number of states in the range E to $E \pm dE_p$. This is analogous to the DOS from a band calculation.
194. J C Bailar Jr, in: *Organometallic Polymers,* C E Carraher Jr, J E Sheats and C U Pittman Jr, Eds, Academic Press, New York, p 313 (1978).
195. J M Manriquez, G T Yee, R S McLean, A J Epstein and J S Miller, *Science,* **252,** 1415 (1991).

8 Metal-containing Liquid Crystals

Duncan W Bruce

Inorganic Materials. Edited by Duncan W Bruce and Dermot O'Hare
© 1992 John Wiley & Sons Ltd

8.1 INTRODUCTION

While we are used to thinking of matter as existing in one of three forms, solid, liquid and gas, things are never quite so simple as they seem and our understanding so far tells us that in addition to these more familiar states of matter, there are two others. The first is the plasma state which is normally found in quite hostile environments, while the second is the liquid crystal state.

The liquid crystal state was recognised in the middle of the 19th century [1] in nerve myelin, although its discovery is often ascribed to Reinitzer and his discovery of liquid crystallinity in cholesteryl benzoate (Figure 1; **1**) and cholesteryl acetate (Figure 1; **2**) [2].

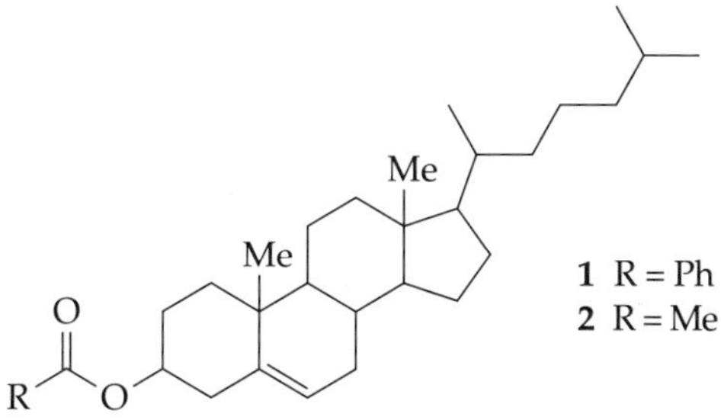

1 R = Ph
2 R = Me

Figure 1   Reinitzer's cholesteryl liquid-crystalline esters

Since these initial discoveries, liquid crystals have become a major, multidisciplinary field of research. Following the work of Gray [3] and co-workers in the early 1970s who synthesised materials with the correct properties and sufficient stability to be used widely in commercial applications, a vigorous industry devoted to displays has resulted.

Research into liquid crystals covers a very wide range of scientific disciplines, from the biological sciences (cell membranes are liquid crystalline), through chemistry to physics, mathematics and electronic engineering. It is a constantly-expanding field with new applications continually being developed and new materials being synthesised to meet these needs. One of the newer developments has been the effort directed to the synthesis of liquid crystal molecules containing metal atoms and many of us who work with them feel that this is an area with tremendous potential. This chapter aims to try to justify this confidence, but in order so to do, it will first be necessary to introduce some of the basic concepts and facets of the liquid crystal state.

8.2   THE THERMOTROPIC LIQUID CRYSTAL STATE

As the name might suggest, a liquid crystal has properties which are reminiscent of both the solid and liquid state and it may be helpful to regard them as either *disordered solids* or as *ordered liquids*. Liquid crystals are grouped into two broad classifications depending on the method used to destroy the order associated with the solid state. The first, and the one which will principally concern us, are the *thermotropic* liquid crystals where heat is applied and causes transitions between the various states. The second are the *lyotropic* liquid crystals where the order of the solid state is disrupted by the action of a solvent, usually water; heat may also effect phase changes but only in combination with the action of the solvent.

Before proceeding further, it will be of use to introduce some of the terminology commonly associated with liquid crystals. A material which has liquid crystal properties is referred to as a *mesogen* and is said to exhibit *mesomorphism*; something which is 'liquid-crystal-like' is known as *mesogenic*, although something mesogenic is not necessarily mesomorphic. To avoid ambiguity, the (conventional) liquid state is referred to as the *isotropic* state. The temperature at which a material passes from the solid state into a mesophase is referred to as the *melting point*, while the temperature at which the mesophase transforms into an isotropic fluid is called the *clearing point* as this is the temperature at which the opaque liquid crystal becomes a clear liquid.

8.2.1   Calamitic Mesophases

The mesophases of thermotropic liquid crystals are divided into two principal types. Liquid crystals derived from rod-like molecules are said in general to form *calamitic* phases while those derived from disc-like molecules are said to be generally *discotic*. Chemical features of the different types of molecule will be discussed in Section 8.2.3 and so for the purposes of the discussion below, simple rods and discs will be used to describe calamitic and discotic materials, respectively.

8.2.1.1 The Nematic Phase

The nematic phase has the simplest structure of all of the mesophases, is very fluid and is also the most disordered mesophase – it is abbreviated, N. It is characterised by the long axes of the molecules pointing on average in the same direction, but with no positional correlation; note that the representation of the phase (Figure 2) does not distinguish one end of the molecule from the other, i.e. it is the long axes which correlate and not the molecules themselves.

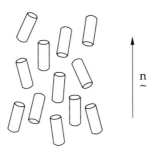

Figure 2 Schematic representation of the molecular arrangement in the nematic phase

The parameter known as the *director* ($\underset{\sim}{n}$) defines the net orientation of the molecules within the phase.

8.2.1.2 The Chiral Nematic Phase

If a molecule shows a nematic phase as a racemic modification, then in the enantiomerically pure material (or in a mixture containing an excess of one enantiomer), the nematic phase will be replaced by the chiral nematic phase (N*) whose structure is represented in Figure 3. Note that N* phases may also be generated by doping mesomorphic hosts with chiral (and not necessarily mesomorphic) materials.

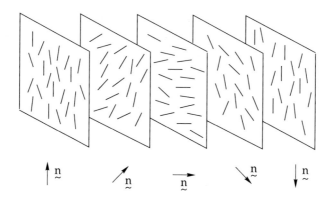

Figure 3 Schematic diagram of the chiral nematic phase

This phase is also widely known as the cholesteric phase (Chl*) due to its initial discovery in materials derived from a cholesterol core, although the term chiral nematic is more accurate. Due to the packing constraints imposed by the materials being chiral, the molecules cannot simply align side-by-side in the phase and the long axis of one molecule will be slightly offset with respect to that of its neighbour. The net effect is for the director to precess through the phase, describing a helix which may be left- or right-handed depending on both the sense of the chirality and its position in the molecule [4]; two enantiomers of the same material will describe opposite twist senses for the helix. The pitch (p) of the helical twist is often of the order of the wavelength of visible light and is very sensitive to temperature. Coupled with the fact that the N* phase exhibits selective Bragg reflection of light of the wavelength equal to np (where n is the average refractive index of the material), then N* materials appear to change colour with temperature, making them useful as thermometers and in aircraft testing [5]. A more extensive description of the properties and applications of N* materials may be found elsewhere [6].

8.2.1.3 *The True Smectic Phases [7]*

The smectic phases are more highly ordered than the nematic phase and are characterised by some positional correlation of the molecules into layers, in addition to orientational correlations. The simplest smectic phase is the smectic A (SmA) phase which is schematically represented in Figure 4a (note that the label 'A' and the other following labels carry no significance and are simply historic). As in the nematic phase, the long axes of the molecules are oriented on average in the same direction but in addition, the molecules are loosely associated into layers, with the orientational direction perpendicular to the layer normal; diffusion between the layers occurs readily and the phase is fluid. In fact, this is a very idealised scheme and the layers in this (and the following phases) are much less well defined; a more precise description of the structure of the smectic phases may be found elsewhere [7, 8]. If the SmA is modified slightly by tilting the molecules within the layer plane, then another smectic phase, the smectic C (SmC) is obtained (Figure 4b).

Alternatively, the SmA may be modified by retaining the orthogonality of the molecules with respect to the layer normal and introducing hexagonal symmetry into the layer so that the molecules sit at sites which describe a hexagonal net; this is the smectic B (SmB) phase (Figure 5). As with all smectic phases, the SmB is fluid and interlayer diffusion of the molecules is facile, although rotation about the molecular long axes is concerted.

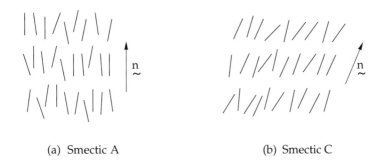

(a) Smectic A   (b) Smectic C

Figure 4 'Conventional' (schematic) representation of (a) the SmA and (b) SmC phases

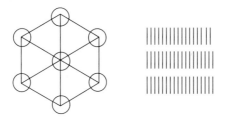

Figure 5 Schematic structure of the SmB phase showing the hexagonal ordering within the layers and the orthogonal nature of the phase

Two other smectic phases are obtained as tilted variations of the SmB. Thus, the smectic I (SmI) phase may be regarded as an SmB phase which is tilted towards a *vertex* of the hexagonal net, while the smectic F (SmF) phase may be regarded as an SmB phase which is tilted towards the *edge* of the hexagonal net (Figure 6).

These five phases are true smectic phases and in normal phase sequences would be expected to be found in the order shown in Scheme 1 [9]; the nematic and isotropic (I) phases are included for completeness.

In addition, the SmC, SmI and the SmF may exist as chiral modifications (SmC*, SmI* and SmF*) either by doping with a chiral additive or by resolving a racemic material which shows one or more of the phases. Because of the low (C_2) symmetry in these phases, the molecular dipoles align within the layers which are then ferroelectric.

However, the chirality also requires that the direction of the ferroelectricity precesses through space from one layer to the next and so in the bulk sample, the ferroelectricity is lost.

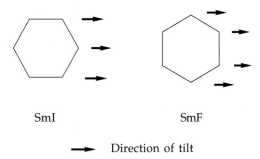

SmI   SmF

➤ Direction of tilt

Figure 6   Schematic representation of the SmF and SmI mesophases

$$I \rightarrow N \rightarrow SmA \rightarrow SmC \rightarrow SmI \rightarrow SmF \rightarrow SmB$$

Increasing order (decreasing temperature)

Scheme 1   Normal thermodynamic ordering of liquid crystal mesophases

In most cases, mesophases are found on both heating and cooling a material, so that Sequence 1 in Scheme 2 is fully reversible; such mesophases are termed *enantiotropic*. However, in some cases a particular mesophase may only appear on cooling a material and is therefore metastable (Sequence 2, Scheme 2); these phases are termed *monotropic*.

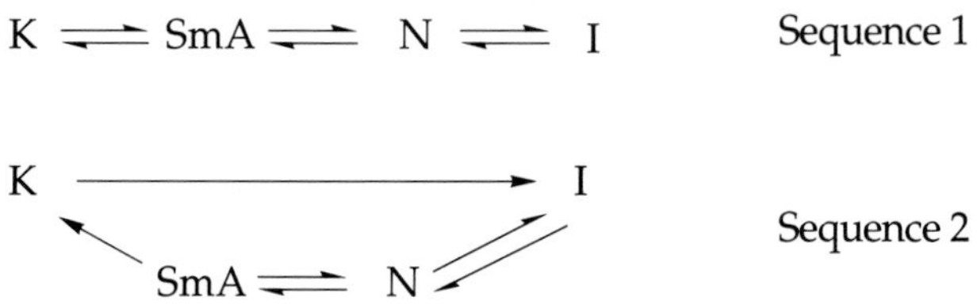

K ⇌ SmA ⇌ N ⇌ I   Sequence 1

K ⟶ I
SmA ⇌ N   Sequence 2

Scheme 2   Phase sequences illustrating enantiotropic and monotropic mesomorphism

8.2.1.4 The Crystal Smectic Phase [7]

In addition to the true smectic phases, there is another class of mesophases, formerly also called smectic phases, which do possess extra positional order and which are derived from the true smectics. These crystal smectic phases are characterised by the appearance of inter-layer correlations and, in some cases, by the loss of molecular rotational freedom. Thus, the (crystal) B, G and J phases are SmB, SmF and SmI phases respectively with inter-layer correlations, while the E, H and K phases are B, G and J phases which have lost rotational freedom (note that there are both smectic and crystal B phases). These phases possess considerable disorder and are therefore properly intermediate between the solid and liquid states.

The nature and number of any mesophases which are formed by a given material must be determined experimentally, although some predictions are sometimes possible.

8.2.2 Discotic Mesophases

Discotic phases are formed (as the name suggests) by disc-like molecules and their discovery is often ascribed to Chandrasekhar in 1977 [10] who described the mesomorphism of some hexaalkanoates of benzene (Figure 7), although Skoulios had recognised the existence of columnar phases (also discotic) of metal soaps many years earlier [11].

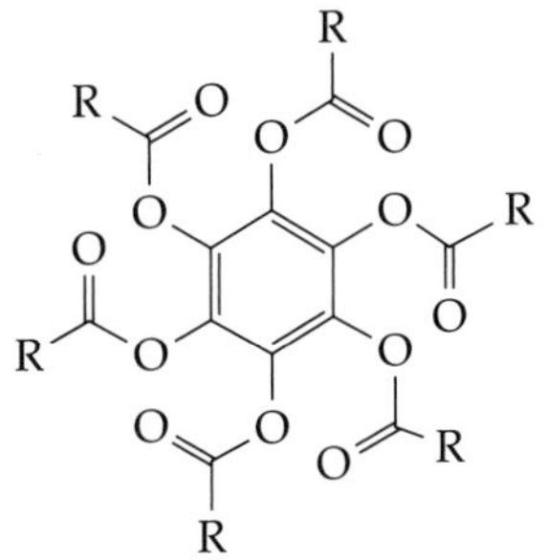

Figure 7   Chandrasekhar's discotic hexaalkanoates of benzene

In contrast to calamitic phases, discotics correlate their short axes and several phase variations are know. The nematic discotic (N_D) is directly analogous to the calamitic nematic, being a very fluid phase possessing only orientational order of the (short) molecular axes (Figure 8); materials showing the N_D phase are quite rare.

Figure 8   Schematic structure of the N_D phase

More common are the various columnar phases which are characterised by the symmetry of the side-to-side molecular arrangement and the presence or absence of order within the columns (Figure 9).   In the disordered hexagonal columnar phase (D_{hd}), the molecules are arranged in columns which are further organised into an hexagonal array; within the columns there is liquid-like order; the ordered analogue differs only in that there is order within the columns and might be considered as the discotic equivalents of the crystal smectic phases.   In addition, rectangular (D_{ro} and D_{rd}), oblique (D_o) and tilted (D_t) modifications are known.   Recently, the first examples of columnar nematic phases have been described by Praefke [12] and by Ringsdorf [13]; these are nematic arrays of columnar stacks.

Figure 9   Schematic structure of the D_h phase

8.2.3 The Chemical Nature of Thermotropic Mesogens

8.2.3.1 Calamitic Materials

The general structure of calamitic mesogens is often given as shown in Figure 10 (after Toyne [14]):

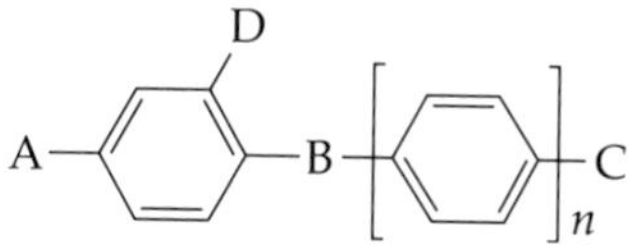

Figure 10   A general molecular structure for calamitic mesogens

Such a model describes materials possessing a rigid aromatic core with two terminal functions (A and C), a linking group (B) and possibly a lateral substituent (D) and reflects the fact that the bulk of the known liquid crystal systems are aromatic. In such a model, at least one of the terminal groups is likely to be an alkyl or alkoxy chain, while the other might be a small, polarisable group which extends the length of the molecule without extending its breadth (e.g. $-CN$, $-NO_2$, $-OMe$). The linking group (B) would also preserve the overall molecular linearity and would serve to extend the conjugation of the system (e.g. $-CH=CH-$, $-C\equiv C-$, $-CH=N-$, $-N=N-$, $-CO_2-$). There are many thousands of liquid crystal molecules known and examples may be sought in the two volumes of *Flüssige Kristalle in Tabellen* [15].

In general, the molecular requirements expressed in the above diagram can be summarised as:

i)   they (the molecules) must be structurally anisotropic;
ii)   they must possess (in most cases) a permanent dipole and
iii) they must possess a high anisotropy of polarisability.

Because of the structural anisotropy of the molecules, the intermolecular forces between them are also anisotropic and it is these weak dispersion forces together with the shape anisotropy which are responsible for stabilising the mesophases. In general, the more rod-like the molecule, the stronger will be the anisotropic dispersion forces and hence the more stable the mesophase. In addition, the Maier-Saupe theory [16] attempts to relate both molecular length and the anisotropy of electronic polarisability (Δα) to the clearing temperature of the nematic phase, asserting that the clearing point is proportional to both.

While such a general model is quite adequate to describe many of the known liquid crystal species, developments in synthetic chemistry mean that new mesomorphic materials exist which do not conform to the model [17]. For example, alicyclic rings such as cyclohexane, bicyclooctane and cubane have been successfully incorporated, and flexible spacers (e.g. –CH$_2$–CH$_2$–) are commonplace. More recently, Griffin [18] and Luckhurst [19] have described dimeric systems linked by alkyl chains, while Eidenschink [20] has described a mesomorphic material centred on a tetrahedral carbon!

In any given series of materials conforming more or less to the model in Figure 10, some broad generalisations concerning structure/mesophase relationships may be made. In general, nematic phases are favoured at shorter alkyl(oxy) chain lengths, as the effect of the core predominates and the molecules are more like rigid rods. As the chain length increases, the nematic phase will first be further stabilised (Maier-Saupe theory) until smectic phases appear on account of the further association promoted by mutual interaction of the chains thus stabilising layers. As the chains get longer and the smectic phases stabilise, the nematic phase will disappear. The replacement of alkyl chains by alkoxy chains will tend to stabilise both the crystal phase and the mesophases. Readers are referred elsewhere for more detailed discussions of structure/property relationships [14].

8.2.3.2 Discotic Materials

The structure of discotic materials is somewhat simpler than that of calamitic systems and is shown in Figure 11. The disc-like core is often aromatic and is in general surrounded by six or eight alkyl chains.

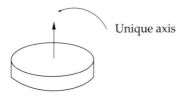

Figure 11   Schematic representation of molecular structure of discotic materials

Figure 12 shows some representative examples of discotic materials and it is interesting to note the hexaaza-18-crown-6 derivative [21] which shows that absolute rigidity is not actually required in the material core.

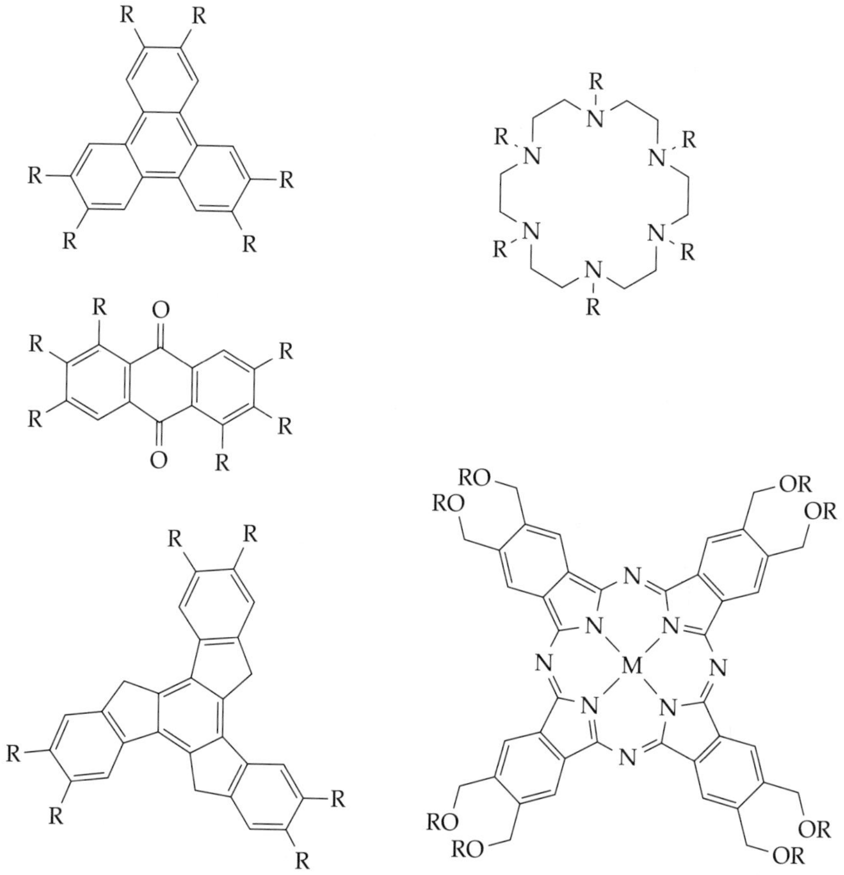

Figure 12 Representative examples of discotic materials

8.2.4 Physical Properties of Thermotropic Liquid Crystals

Probably the single most important thing about liquid crystal phases is that as fluids they are anisotropic; this means that their physical properties are likewise anisotropic and it is this feature which is the basis for the widespread application of the materials. Consider, for example, the refractive indices of a nematic phase (Figure 13).

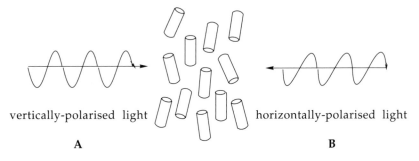

vertically-polarised light   horizontally-polarised light

A   B

Figure 13   The interaction of polarised light with a nematic phase

The figure assumes that the molecules have a greater polarisability along their long axis than along either of their two short axes. In case A, the electric vector of the light is coincident with the direction of greatest polarisability and so the light is retarded. However, in case B, the electric vector of the light is coincident with the direction of least polarisability and so the light is not retarded. The consequence is that the material has two refractive indices, $n_{\parallel}$ (case A) and $n_{\perp}$ (case B), and the difference between these ($n_{\parallel} - n_{\perp}$) is termed the birefringence and is given the symbol Δn. (Because the nematic phase has $D_{\infty h}$ symmetry, the two short axes are effectively equivalent and so the two possible $n_{\perp}$ reduce to only one). Materials which have $n_{\parallel}$ greater than $n_{\perp}$ are said to have a positive birefringence ($\Delta n > 0$) while materials which have $n_{\perp}$ greater than $n_{\parallel}$ are said to have a negative birefringence ($\Delta n < 0$).

Of course, this anisotropy of properties is not confined to refractive index and it is immediately apparent that the electronic polarisability in the mesophases is anisotropic ($\Delta\alpha = \alpha_{\parallel} - \alpha_{\perp}$). Similarly, dielectric permittivity ($\Delta\varepsilon$) and (dia)magnetic susceptibility ($\Delta\chi$) are anisotropic. Measurement of these quantities gives important information about the organisation and properties of the materials in the mesophase.

Such properties are determined primarily by molecular features and as such can be tuned at the molecular level. The values for the birefringence and dielectric anisotropy of the two-ringed cyano nematogens shown in Figure 14 illustrate this well, both quantities decreasing as the polarisability of the materials decrease (data from reference [22]).

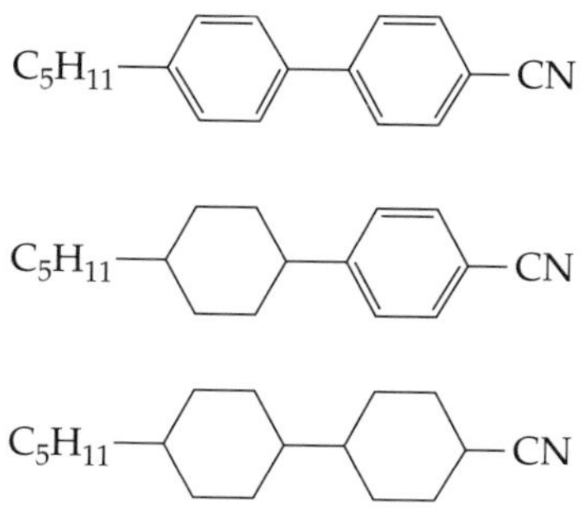

	Δn	$\Delta \varepsilon$

C_5H_{11}—⬡—⬡—CN   0.18   11.5

C_5H_{11}—⬡—⬡—CN   0.1   9.7

C_5H_{11}—⬡—⬡—CN   0.06   4.4

Figure 14 Birefringence and dielectric anisotropy as a function of molecular composition

8.3 LYOTROPIC LIQUID CRYSTALS [23]

If a surfactant molecule, such as cetyltrimethylammonium bromide (CTAB), is dissolved in water, at some specific concentration (known as the *critical micelle concentration*) the molecules will organise and form micelles. This formation of micelles is driven by the hydrophobic effect [24] (which is entropically driven) which acts to minimise the unfavourable interactions between the solvent and the long alkyl chains. If more CTAB is then added, the concentration of micelles increases until their concentration becomes so high that they themselves organise to form ordered arrays. These ordered arrays constitute a lyotropic liquid crystal phase. There are several well-characterised lyotropic liquid crystal phases and a host of so-called intermediate phases whose characterisation is not unequivocal. The micelles may be formed by cationic (e.g. CTAB), anionic (e.g. $C_{12}H_{25}OSO_3Na$) or non-ionic (e.g. $C_{12}H_{25}(OCH_2CH_2)_6OH$) surfactants.

As it is the aggregation of micelles which leads to the formation of lyotropic mesophases, then it is not surprising to learn that different micelle types usually give rise to different mesophases. Spherical micelles in general give rise to cubic mesophases (termed I_1 and V_1 phases) in which the spheres can be considered as being sited at the lattice points of a cubic close-packed structure. Rod micelles give rise to hexagonal mesophases (termed H_1) which consist of a hexagonal array of the rods, while disc micelles give rise to a lamellar phase (L_α) which is a solvent-separated bilayer phase. Schematic diagrams of each of these phases may be found in references [23 and 25]. In addition, there are the so-called 'reversed-phase' analogues of the hexagonal (H_2) and cubic phases (I_2 and V_2) in which the micelles are reversed so that the polar head groups are found at the centre of the micelle along with a small quantity of water.

Of course, individual compounds do not necessarily form only one mesophase and a series is often found whose existence is determined by a particular ratio of amphiphile to solvent at a given temperature; a typical phase diagram for an amphiphile in water is given in Figure 15. Readers are referred elsewhere [23] for a more detailed consideration of lyotropic systems.

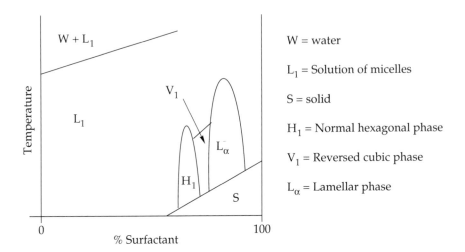

Figure 15   'Typical' two-component phase diagram for a lyotropic system

In addition to the lyotropic mesophases formed by surfactant amphiphiles, two other classifications are generally recognised. The first of these comprises rigid polymers which can form mesophases in both aqueous and non-aqueous solvents [26], while the second comprises aggregates in water of rigid, polarisable polyaromatics; these are the so-called 'chromonic' mesophases [27]. In these phases, the polyaromatics stack (driven by the π–π-interactions) and these stacks constitute the mesogenic unit which is known to organise into one of two well-characterised mesophases. In the nematic (N) phase, the stacks are well-separated by water and possess the one-dimensional orientational order characteristic of nematics, while in the hexagonal (or M) phase (found at lower water content) the stacks are arranged on a hexagonal net.

Much interest in these phases arises because several drugs (e.g. the anti-asthmatics, disodium chromoglycate and sodium 5-hexyl-7-(5-methylsulphoniumidoyl)xanthone-2-carboxylate; Figure 16) and some dyes display this type of mesomorphism. Many of the known metal-containing lyotropics show chromonic phases (Section 8.10).

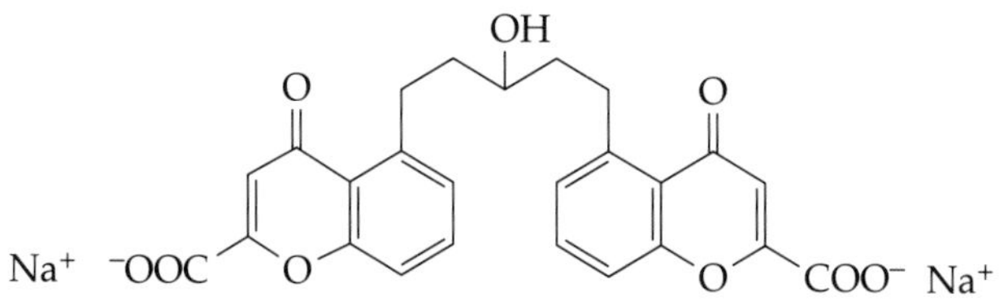

Disodium chromoglycate

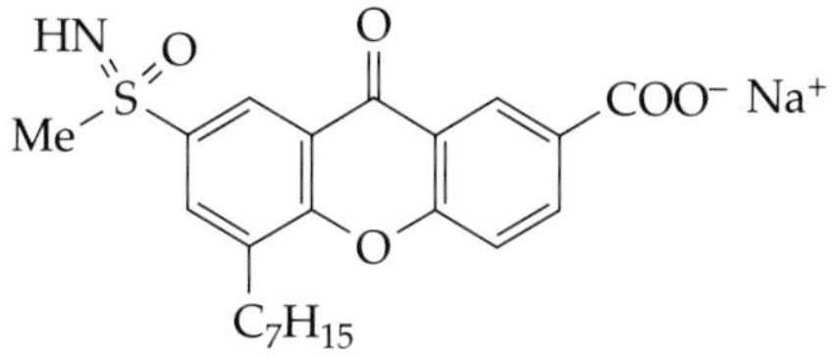

Sodium 5-hexyl-7-(5-methylsulphoniumidoyl)xanthone-2-carboxylate

Figure 16 Chromonic anti-asthmatic drugs

8.4 MESOPHASE CHARACTERISATION

Once a material is synthesised, how do we find out which (if any) mesophases it possesses? There are three techniques which are widely-used, namely polarising hot-stage optical microscopy, differential scanning calorimetry and small angle X-ray scattering.

8.4.1 Polarised Optical Microscopy [8, 28]

This is usually the first technique used to characterise the mesomorphism of a compound and, in skilled hands, can lead to complete characterisation of a material in most cases. Typically, the powdered sample (< 1 mg) is sandwiched between two microscope cover slips and then placed on a suitably-controlled hot-stage through which there is an optical path. The hot-stage (working optimally between − 50 °C and +400 °C) is mounted on the working stage of the microscope. The technique makes use of the birefringence of mesophases in that the light incident on the sample is first plane polarised, and there is a second polariser at 90° to the first between the sample and the objectives (shown schematically in Figure 17).

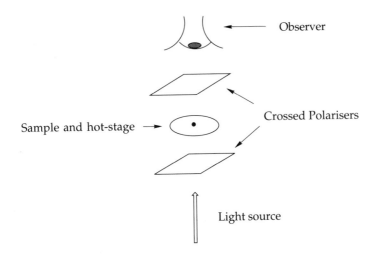

Figure 17   Schematic view of a polarised, hot-stage microscope

Thus, in the mesophase two refracted rays result and these interfere with one another to give characteristic interference patterns (strictly, the light becomes eliptically polarised); this contrasts with the situation where the material is in the isotropic state when the sample appears black between the crossed polars as only one refractive index is seen. These textures are diagnostic of a particular mesophase and are most useful and best developed when they are obtained as part of a cooling sequence.

A very useful method often used in conjunction with microscopy is miscibility. The simplest use of the technique is to bring two materials together on the cover slip in their mesophases in a contact preparation; the identity of the mesophase of one of the materials is already known. If the two materials are co-miscible, then both have the same mesophase (at the temperature in question) and this can then be a useful method of phase identification. Unfortunately, if the two materials are immiscible then no information is obtained as two materials in the same phase are not necessarily miscible (e.g. water and chloroform which are both isotropic!).

8.4.2 Differential Scanning Calorimetry (DSC)

In the DSC experiment, the change in the heat capacity of the sample is recorded as a function of temperature leading to a measure of the enthalpy change accompanying a phase transition. From the thermodynamic point of view, melting transitions are strongly first order while liquid crystal-liquid crystal transitions are weakly first order or may be second order (e.g. SmA–N and SmC–SmA can be second order). Information about the phase transition may therefore be derived from the relative magnitudes of the transition enthalpies, so that melting enthalpies are (obviously) much larger than those found for an N–I transition. While such information is quite useful, it does not allow generalisations to be made and the corresponding entropy changes are more useful.

The technique is therefore strictly complementary to optical microscopy, as all changes in optical texture do not necessarily correspond to a change in mesophase type, while all phase changes do not always lead to an easily identifiable change in texture. Thus DSC traces should always be compared with the results of the optical study to be sure of proper correspondence

8.4.3 Small-angle X-ray Scattering [29]

X-ray scattering is a very powerful method for mesophase identification and can provide much information; it is often the only unequivocal means of phase identification. The experiment is rather simple in principle and relies on the fact that mesomorphic structures are periodic and can therefore diffract; a schematic cartoon of the experiment is shown in Figure 18.

Thus in a smectic phase, diffraction lines corresponding to both the layer periodicity and side-to-side periodicity can be observed; comparison of the observed layer periodicity with the calculated molecular length can give information about tilt angles and interdigitation. Similarly, if there is symmetry in the smectic layer this is also observed as a series of signals in the wide-angle region. If samples are aligned (possible with fields of around 0.6 T for a nematic), then additional orientational information is available.

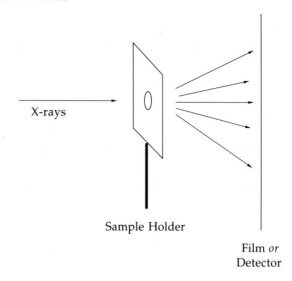

Sample Holder

Film *or*
Detector

X-rays

Figure 18   Schematic diagram of the X-ray scattering experiment

8.5   METAL COMPLEXES AS LIQUID CRYSTALS

Metal-containing liquid crystals are in general terms no different from purely organic materials in that they form the same types of mesophases. For the purposes of this chapter, the complexes will be organised by general ligand type (rather then mesophase type) which actually turns up one or two interesting observations concerning the relationship between molecular shape and mesophase type. The only exceptions are lyotropic phases of amphiphilic complexes which are treated separately (Section 8.10).

At this point it is worth a few lines to correct a very common misnomer which seems to perpetuate in the literature. Metal-containing liquid crystals (also known as *metallomesogens* [30]) are commonly and incorrectly all grouped together as *organometallic liquid crystals*, without any thought about the type of complex being described. While there are several examples of organometallic systems, the majority are in fact *coordination compounds*. Thus, organometallic complexes are defined as those which possess a direct bond (σ- or π-) between a metal and an *organic* carbon centre; those which possess metal–carbon monoxide bonds are also included by convention.

8.5.1 Historical Perspective

The first liquid crystals to contain a metal were the alkali metal salts of aliphatic carboxylic acids, reported by Vorländer in 1910 [31]. On heating, these ionic materials pass through a series of solid phases (later characterised in detail by Skoulios [32]) before melting into a classical lamellar phase. In 1923, Vorländer also reported some diarylmercury(II) complexes (Section 8.9) which he reported as showing smectic phases [33]. For the next 50 years, reports of metal-containing systems were scarce; Green, Haller and Young [34] reported some trimethyl -silyl, -germyl and -stannyl compounds in 1971 and Malthête and Billard reported a ferrocene derivative in 1976 [35].

However in 1977, Giroud and Mueller-Westerhoff [36] reported some mesomorphic dithiolene complexes of Ni(II) and this work is widely accepted as being the beginning of systematic research into metal-containing systems. Thus the subject is young with the bulk of the work having been carried out since the mid 1980s.

The vast majority of the calamitic materials so far synthesised have used metals which readily exist in a d^8–d^{10} configuration, thus exhibiting a geometry which is either linear or based on a square plane. This means that Rh(I), Ir(I), Ni(II), Pd(II), Pt(II), Cu(II), Ag(I), Au(III) and Zn(II) will feature heavily. The reason for this choice is not difficult to appreciate, given that materials containing such a metal core will most easily satisfy the basic structural requirements for a calamitic mesogen. While systems based on these metals will and should continue to attract attention, reports of mesogens based on iron(III) and oxovanadium(IV) represent a break from these planar geometries and point to exciting new avenues into which the subject must develop. What follows is an overview of the area which is not intended to be comprehensive, but which aims to give an overview of the work which has been carried out.

8.6 MESOMORPHIC COORDINATION COMPLEXES OF MONODENTATE LIGANDS

8.6.1 Complexes of Organonitriles

After the work of Vorländer with diarylmercury(II) systems, the next work to be published on mesogens with monodentate ligands was our own work [37, 38] with Pd(II) and Pt(II) complexes of the alkyl- and alkoxy-cyanobiphenyls (abbreviated as nCB and nOCB respectively, where n indicates the number of carbon atoms in the chain). These cyanobiphenyls were synthesised by Gray and co-workers [3] in Hull in the early 1970s and were responsible for the commercialisation of liquid crystal display technology [39]. These materials show enantiotropic nematic phases at chain lengths $5 < n < 9$ and SmA phases for $8 < n < 12$; clearing points (T) are in the range $30 < T < 60$ °C for nCB and $68 < T < 90$ °C for nOCB.

The *trans*-geometry (Figure 19) of the complexes ($[MCl_2(n(O)CB)_2]$) was established by single crystal X-ray structure determinations of $[PdCl_2(5CB)_2]$, $[PtCl_2(5CB)_2]$ [40], $[PtCl_2(6CB)_2]$ [41] and $[PtCl_2(8CB)_2]$. Four series of materials resulted: $[PdCl_2(nCB)_2]$ (**3a**), $[PdCl_2(nOCB)_2]$ (**3b**), $[PtCl_2(nCB)_2]$ (**4a**) and $[PtCl_2(nOCB)_2]$ (**4b**).

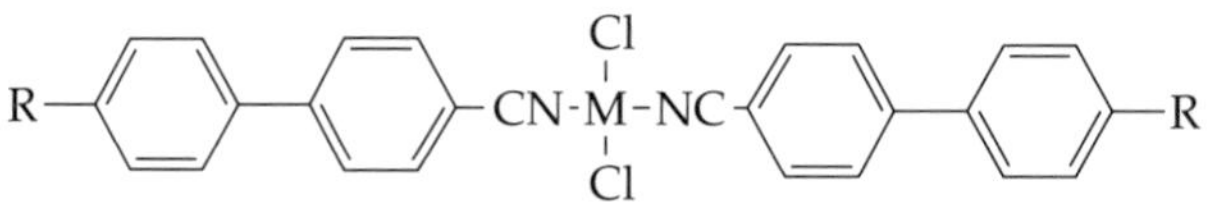

$$M = Pd \ (3)$$
$$M = Pt \ (4)$$
$$R = alkyl \ (a)$$
$$R = alkyloxy \ (b)$$

Figure 19   Palladium and platinum cyanobiphenyl complexes

In each series, the predominant phase was nematic (in series (**3a**) this was always monotropic), although at longer chain lengths SmA and SmC phases were found; enantiotropic phases emerged at much shorter chain lengths in the complexes compared to the ligands. The melting points of the palladium complexes were typically in the range 100–120 °C, while for platinum that range was 160–190 °C. Clearing points were also quite different so that while most of the platinum complexes cleared in the range 190–230 °C (series **4b** higher than **4a**), series (**3b**) cleared between 90–140 °C and series (**3a**) between 70–100 °C [42]. The rewarding result from this study was the appearance of a SmC phase, not possessed by any of the ligands, showing that mesomorphism was modified on complexation.

In addition to these cyanobiphenyl complexes, some Pd(II) complexes of related ligands (Figure 20) were made, namely the 4-(4'-alkylcyclo-hexyl)phenylcarbonitriles (PCHn) and *trans,trans*-4'-alkylbicyclohexyl-4-carbonitriles (CCHn).

The complexes of the PCH ligands all showed monotropic nematic phases, while for the CCH complexes, an enantiotropic nematic phase was stabilised by $n = 7$ [43]. These are mentioned here as their physical properties will be discussed in Section 8.6.4.

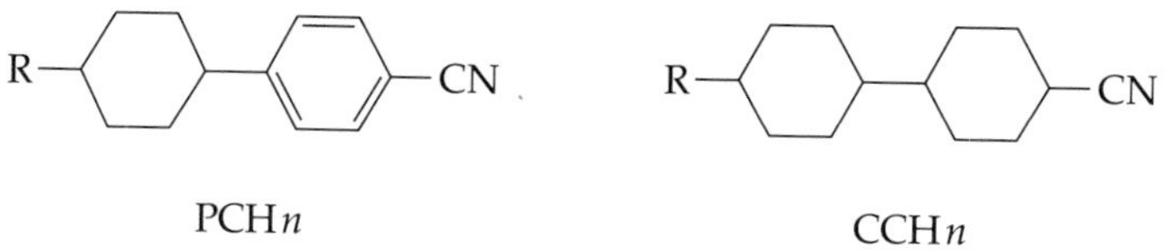

PCH*n*   CCH*n*

Figure 20   Nitrile ligands containing cyclohexyl rings

8.6.2   Complexes of Substituted Pyridines

One of the main problems associated with the nitrile complexes was the lability of the metal–ligand bond, especially for M = Pd, and so more strongly-binding ligands were sought. Various substituted pyridines have been used in this context and the greatest amount of work has been carried out with the *trans*-4-alkoxy-4'-stilbazoles (Figure 21) [44], synthesised by a palladium-catalysed Heck coupling between 4-vinylpyridine and an alkoxyiodobenzene.

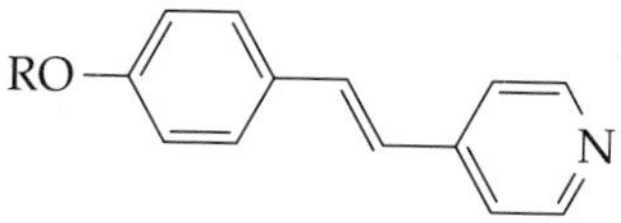

Figure 21   4-Alkoxystilbazoles

Study of the stilbazoles themselves revealed that they had a limited mesomorphism, showing a narrow-range SmB phase above a wider range crystal SmE; the compounds cleared in the range 80–90 °C.

8.6.2.1   *Palladium and Platinum Complexes of Substituted Pyridines*

Complexes of Pd(II) and Pt(II) analogous to those described above for the cyanobiphenyls were then synthesised, but it was found that few of these complexes were mesomorphic so that only the longest-chain derivatives (R > C_9H_{19}–) showed a SmC phase at elevated temperatures (>200 °C). However, two modifications of these materials produced materials with much lower melting (and clearing) points. In the first, the two chloride ligands in the Pd(II) complexes were replaced by aliphatic carboxylic acids to give the complexes shown in Figure 22 [45].

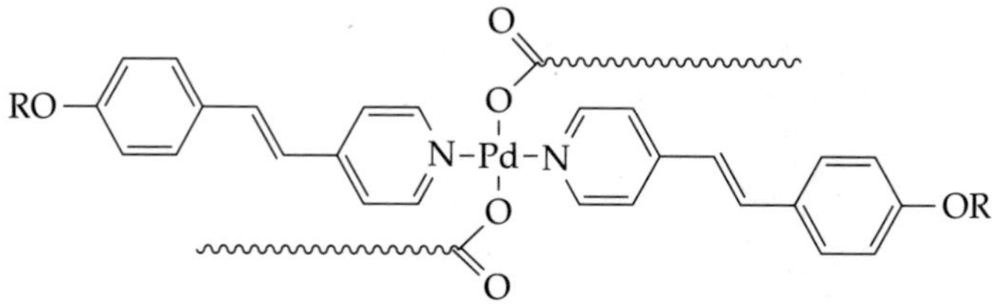

Figure 22   Mesomorphic palladium carboxylate complexes

As might have been expected by comparison with studies of laterally-substituted organic compounds, melting and clearing temperatures were dramatically reduced and nematic phases were preferred over smectic phases (Scheme 3).

$R = C_8H_{17}; R' = C_6H_{13}$ $\qquad$ K $\xrightarrow{156}$ N $\xrightarrow{164}$ I

$R = C_{12}H_{25}; R' = C_6H_{13}$ $\qquad$ K $\xrightarrow{144}$ N $\xrightarrow{164}$ I

Scheme 3   Phase behaviour of some palladium carboxylates

In all of these complexes it was assumed (again by comparison with related organic systems) that the lateral chains lie parallel to the long axis of the molecule as suggested in Figure 22.

The second modification (Figure 23) was to lower the symmetry of the system by replacement of one of the stilbazoles in the Pt complexes with an alkene [45, 46].

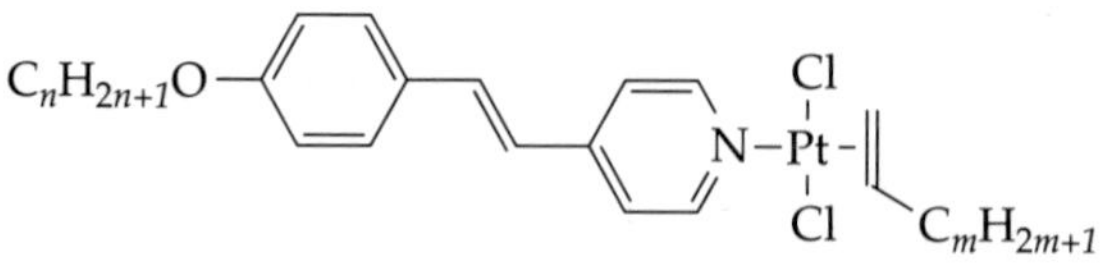

$n = 12; m = 3$ $\qquad$ K $\xrightarrow{61}$ SmA $\xrightarrow{90}$ I

$n = 12; m = 6$ $\qquad$ K $\xrightarrow{70}$ SmA $\xrightarrow{94}$ I

Figure 23   Structure of the mesomorphic Pt-alkene complexes and examples of their transition temperatures

These materials now showed much reduced melting and clearing points and SmA phases were seen for most derivatives synthesised (Figure 23). It was also noted that in general, the materials were non-mesomorphic when $n + m < 8$, showed monotropic mesomorphism when $8 < n + m < 13$ and gave enantiotropic mesophases for $n + m < 13$ [46].

8.6.2.2 Rhodium and Iridium Complexes of Substituted Pyridines

Reaction [47] of the alkoxystilbazoles with $[M_2(\mu\text{-}Cl)_2(COD)_2]$ (M = Rh, Ir; COD = 1,5-cyclooctadiene) in solution under an atmosphere of CO led to a series of complexes (Figure 24) analogous to those described by Serrano and co-workers [48, 49] using the related pyridine ligands (Figure 26).

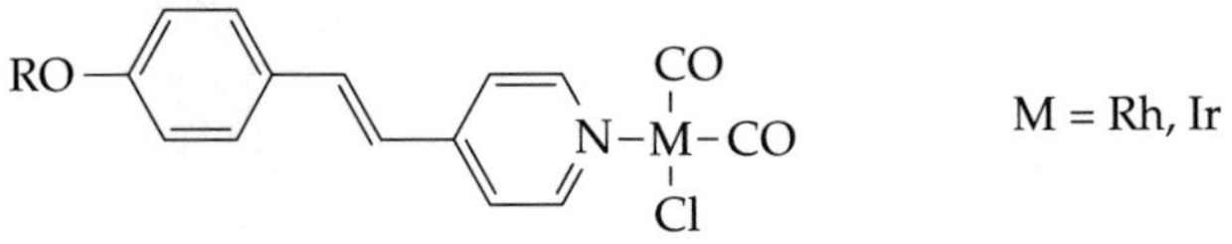

Figure 24   Structure of the mesomorphic stilbazole complexes of Rh(I) and Ir(I)

The rhodium complexes were yellow/orange in the solid state, characteristic of mononuclear Rh(I), while the iridium complexes were burgundy in the solid state and yellow in isotropic solution or in the melt. Examination of the v_{CO} absorptions in the infrared spectra implied that the iridium complexes were associated in some way in the solid state, possibly in a stacking arrangement as found in $[IrCl(CO)(py)]$ [50].

Both series of materials showed very similar mesomorphism; the phase diagram for the iridium complexes is shown in Figure 25.

The behaviour was quite typical of relatively simple dipolar materials where a nematic phase was found at short chain lengths, giving way to a SmA phase in higher homologues. While the iridium complexes could be cycled in and out of the isotropic phase with few problems of decomposition, the rhodium complexes began to decompose rapidly on melting in air.

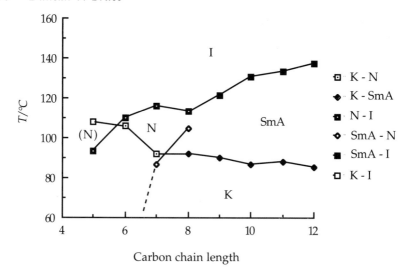

Figure 25   Phase diagram for the complexes *cis*-[IrCl(CO)$_2$(*n*-OPhVPy)]
(Reproduced by kind permission of the Royal Society of Chemistry)

Complexes related to these had previously been synthesised by Serrano (Figure 26) [48, 49].

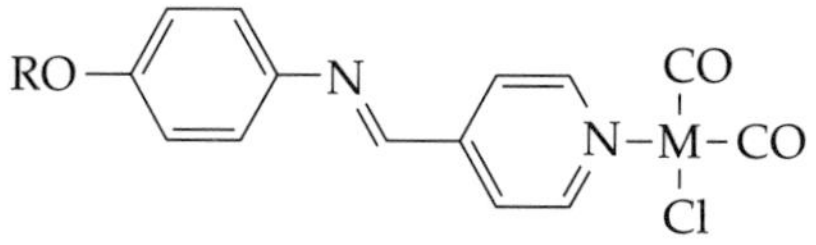

RO—⟨⟩—N ... N—M—CO   M = Rh, Ir

Figure 26   Structure of the mesomorphic alkoxypyridylbenzylideneaniline complexes of Rh(I) and Ir(I)

The mesomorphism of these complexes ([MCl(CO)$_2$(*n*-OPhIPy)]) was less extensive than in the related complexes with stilbazole ligands, and enantiotropic mesophases (N and SmA) were only established at the octyloxy homologue for both metals. The phase diagram for the iridium derivatives is shown in Figure 27 and clearly shows the much lower clearing temperatures and melting points compared to the stilbazole derivatives; the phase diagram of the rhodium congeners is very similar. The differences between the mesomorphism of the complexes of the two ligands is almost certainly due to the ligand geometry in the imine species, where the Ph–C=N–Ph torsion angle is typically 90° leading to a pronounced lowering in anisotropy.

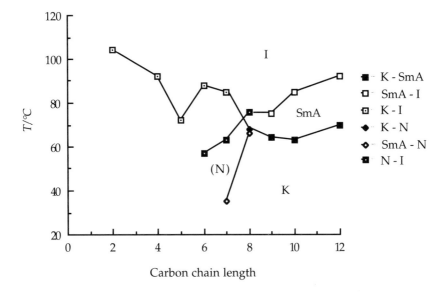

Figure 27   Phase diagram for the complexes *cis*-[IrCl(CO)$_2$(*n*-OPhIPy)]

Reaction of the rhodium derivatives with Me$_3$NO and an equivalent of ligand led to the *trans*-disubstituted derivatives, *trans*-[MCl(CO)(*n*-OPhIPy)$_2$], which curiously were non-mesomorphic, decomposing above 300 °C.

An interesting point arises when the phase behaviour of these rhodium and iridium complexes is compared with that of the parent pyridine ligands. Thus, while the stilbazoles showed a crystal smectic E phase and a narrow-range SmB phase with melting and clearing temperatures in the range 80–85 °C, their complexes with rhodium and iridium melted at around the same temperature (despite the large increase in molecular weight) and showed N and SmA phases which persisted to between 110–140 °C. For the imine ligands, the contrast was perhaps even sharper as the ligands themselves were non-mesomorphic. This pointed to the fact that the group *cis*-[MCl(CO)$_2$] (M = Rh, Ir) was rather efficient in the promotion of mesomorphism and preliminary attempts to quantify the effect suggested that it was as least as good as –CN in organic systems [14].

8.6.2.3 Silver Complexes of Substituted Pyridines

Reaction of substituted pyridines with silver salts led to linear silver(I) complexes according to Figure 28.

$$2L + AgX \longrightarrow [L-Ag-L][X]$$

$$X = BF_4, NO_3, CF_3SO_3, C_{12}H_{25}OSO_3, C_8H_{17}OSO_3$$

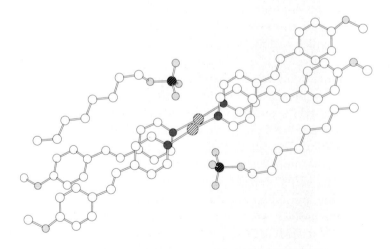

$$L = \quad N \bigcirc\!\!-X-\bigcirc\!\!-OC_nH_{2n+1}$$

X = CH=CH	*n*-OPhVPy
X = CH=N	*n*-OPhIPy
X = CO$_2$	*n*-OPhEPy

Figure 28   Some of the substituted pyridines used to synthesise mesomorphic Ag(I) complexes

The X-ray single crystal structure of one derivative, [Ag(1-OPhVPy)$_2$] [C$_8$H$_{17}$OSO$_3$], has been determined [51] and is shown in Figure 29 with water of crystallisation omitted.

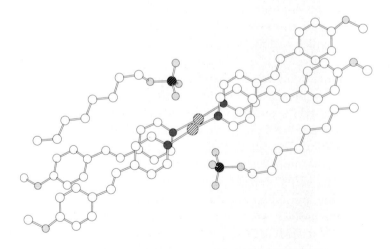

Figure 29   Structure of one of the crystallographically-independent dimers of [Ag(1-OPhVPy)$_2$][C$_8$H$_{17}$OSO$_3$] (Reproduced from reference [51] by permission of the Royal Society of Chemistry)

Two silver-containing cations were held into a dinuclear arrangement by the alkylsulphate anions which formed a bridge, although the Ag–O distances of 2.8–2.9 Å meant that the interactions were ionic rather than covalent. The two silver ions were separated by 3.19 Å which ruled out metal–metal bonding but did not necessarily preclude some intermetallic interaction. The unit cell contained two independent dimers of this type linked via a water molecule, hydrogen bonded to an octylsulphate from each dimer.

Initial work on these systems concentrated on complexes of alkoxystilbazoles with $AgBF_4$, which produced hygroscopic, light-sensitive materials forming SmC and SmA phases at long chain lengths and at very high temperatures; clearing points were typically around 300 °C with extensive decomposition [52]. Similarly, high-melting and high-clearing materials were obtained with X = NO_3 and CF_3SO_3 (OTf) with all three series of ligands [53, 54]; the phase diagram of the triflate salts of the alkoxystilbazoles is shown in Figure 30 [55].

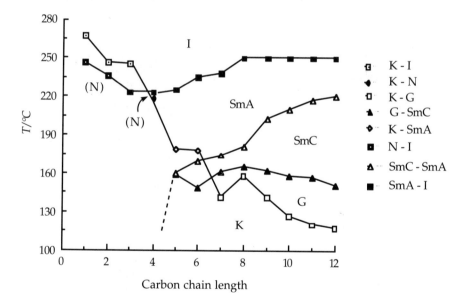

Figure 30   Phase diagram for the complexes [Ag(n-OPhVPy)$_2$][OTf] (reproduced from reference [55] by permission of Gordon and Breach)

This phase diagram is quite typical of the NO_3 and OTf salts of many of the silver(I) complexes of pyridine ligands and shows SmA and SmC mesophases at the longest chain lengths, while as the chain length shortens, the phase behaviour changes and for *n* < 5, the mesomorphism either disappears (L = *n*-OPhVPy, X = NO₃) or changes to show a nematic phase (L = *n*-OPhVPy, X = OTf; L = *n*-OPhIPy, X = BF₄). Such materials were the first examples of anhydrous, ionic liquid crystals to show a nematic phase [56].

The complexes became much more easily studied when the dodecylsulphate anion was incorporated. This was analogous to the introduction of a lateral chain in an organic system and the effects were very similar so that melting points were reduced (the molecules do not pack so well in the solid state) and clearing points fell from around 300 °C in the BF₄ salts to around 180 °C; nematic phases were also seen (Figure 31).

A very curious feature of the stilbazole complexes [Ag(*n*-OPhVPy)₂][DOS] was the appearance of a cubic mesophase (cub) between the SmC and SmA phase for *n* ≥ 6 (Figure 31).

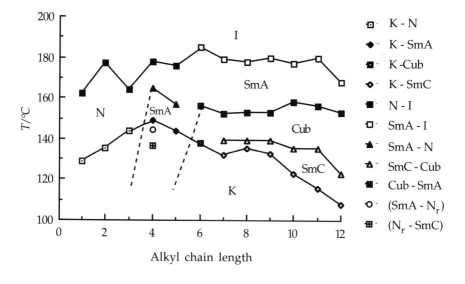

Figure 31   Phase diagram for the complexes [Ag(*n*-OPhVPy)₂][DOS] (reproduced from reference [52] by permission of Gordon and Breach)

Cubic phases, which are viscous, optically-isotropic mesophases usually bounded (in calamitic systems) by very fluid phases (SmC, SmA, N, I), are still rather uncommon [57–59] and are not well understood. Small changes in the nature of a material can cause the cubic phase to disappear, so that replacement of the DOS anion by octylsulphate causes the SmA phase to extend downwards in temperature to occupy the space in the phase diagram previously occupied by the cubic phase [51].

8.6.3 Acetylide Complexes

In 1978 Takahashi [60] reported that polymeric acetylide complexes of Pt(II) (Figure 32) could be formed by the reaction of $[PtCl_2(PBu_3)_2]$ with 1,4-butadiyne or 1,4-diethynylbenzene in the presence of CuI and base. Molecular weight average $(\overline{M}_w)$ degrees of polymerisation of up to 7×10^4 (corresponding to a number average $(\overline{M}_w)$ of 108) were obtained. For polymers with $(\overline{M}_w) \approx 10^5$ and at weight percentages (in trichloroethene) of around 36%, lyotropic nematic phases were observed [61].

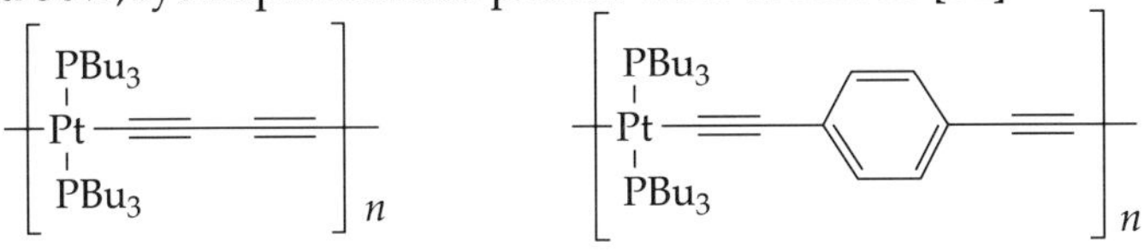

Figure 32 Polymeric Pt acetylide complexes described by Takahashi

Studies of the homometallic (Figure 33; M = M') and heterometallic (Figure 33; M = Pt, M' = Pd or Ni; M = Pd, M' = Ni) polymers of 1,4-butadiyne by ^{31}P NMR techniques showed that all possessed a negative diamagnetic anisotropy ($\Delta\chi < 0$), that is, they were shown to align with their long molecular axes perpendicular to the applied magnetic field. This was interpreted largely in terms of the negative diamagnetic anisotropy of the carbon–carbon triple bond [62].

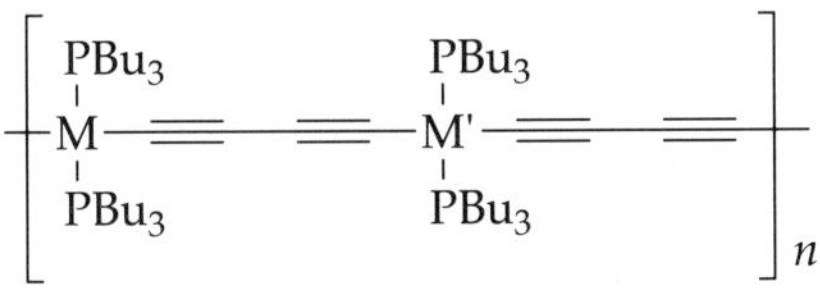

Figure 33 Structure of the mixed-metal butadiyne polymers

However, subsequent studies [63] of related polymeric materials showed that in fact the magnetic anisotropy was a more complex balance of ligand and metal. Thus, while the butadiyne homopolymer of Pd(II) (Figure 33, M = M' = Pd) had $\Delta\chi < 0$, the block copolymer (Figure 34) had

$\Delta\chi > 0$ due to the fine balance of the opposite magnetic anisotropies of benzene and carbon–carbon triple bonds in the complexes, which may well be enhanced by conjugation.

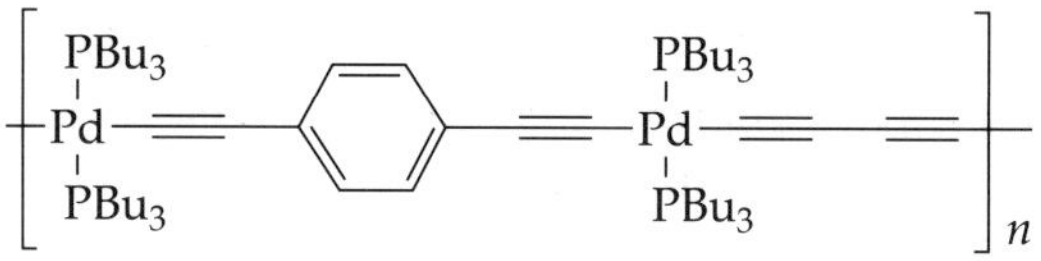

Figure 34   Homometallic block copolymer with $\Delta\chi > 0$

This work with polymeric acetylides them prompted studies of related low molar mass systems, and the Pt(II) complexes in Figure 35 were found to show N and SmA phases [64].

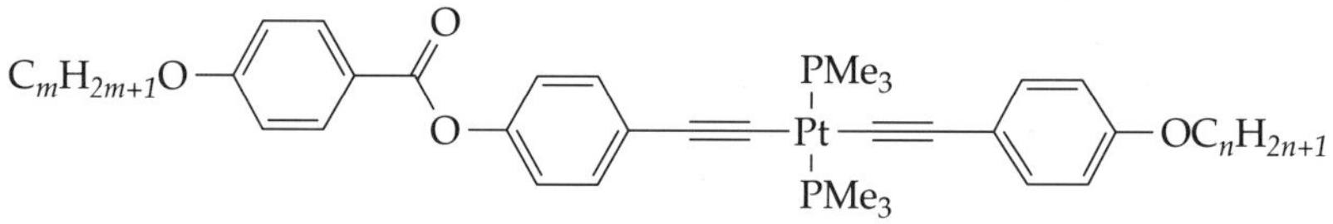

$m = 12;\ n = 12$ $\qquad$ K $\xrightarrow{174}$ SmA $\xrightarrow{177}$ N $\xrightarrow{180}$ I

$m = 8;\ n = 8$ $\qquad$ K $\xrightarrow{173}$ N $\xrightarrow{206}$ I

Figure 35   Structure of the mesomorphic Pt-acetylide complexes and examples of their transition temperatures

This shows that the relative bulk of the two trimethylphosphine groups in the centre of the complex does not have an adverse effect on the overall structural anisotropy of the complex, and indeed the high melting and clearing points suggest that the system may in fact tolerate slightly larger groups.

8.6.4   Physical Properties of Complexes of Monodentate Ligands

In order to try to understand the effect which a metal may have in a liquid crystalline system, it is necessary to study the physical properties of the complexes in addition to their phase behaviour.   One of the properties likely to be affected by the inclusion of a metal centre is the mean polarisability, $\bar{\alpha}$, due to the presence of a centre of highly-polarisable electron density. This is in fact the case and Table 1 shows some results [65].

In each case, the increase is large and is probably best exemplified by the iridium complexes where the addition of the group *cis*-[IrCl(CO)$_2$] virtually doubles the linear polarisability.

Table 1 Polarisability data for some mesomorphic metal complexes

Compound	$\bar{\alpha}/10^{-40}$ J^{-1} C^2 m^2	$\Delta\alpha/10^{-40}$ J^{-1} C^2 m^2	Ref
5CB	37.5	19.4	[22]
PCH5	36.2	16.0	[22]
CCH5	35.5	12.6	[22]
trans-[PdCl$_2$(5CB)$_2$]	82 ± 8	45 ± 9	[65]
trans-[PtCl$_2$(5CB)$_2$]	90 ± 10	—	[65]
trans-[PdCl$_2$(PCH7)$_2$]	86 ± 9	39 ± 8	[65]
trans-[PdCl$_2$(CCH5)$_2$]	79 ± 8	30 ± 6	[65]
n-OPhVPy	35 ± 4	28 ± 6	[65]
cis-[IrCl(CO)$_2$(5-OPhVPy)]	59 ± 6	54 ± 10	[65]
cis-[IrCl(CO)$_2$(7-OPhVPy)]	62 ± 6	59 ± 12	[65]
cis-[IrCl(CO)$_2$(9-OPhVPy)]	60 ± 6	49 ± 10	[65]
[Ag(6-OPhVPy)$_2$][OTf]	82 ± 8	–	[66]
[Ag(8-OPhVPy)$_2$][OTf]	92 ± 9	–	[66]
[Ag(10-OPhVPy)$_2$][DOS]	114 ± 14	–	[66]

If the polarisability is enhanced, then it is also likely that some effect on the birefringence will be seen as these two parameters are related. Measurements were carried out in the nematic phases of [IrCl(CO)$_2$ (8-OPhVPy)], [PdCl$_2$(n-OCB)$_2$] and [Ag(n-OPhVPy)$_2$][DOS] and were compared against the ligand where possible. These data are summarised in Table 2.

Further, cooling the mixture [PdCl$_2$(xOCB)$_2$] to T/T$_{NI}$ = 0.84 ($\approx$ 113 °C) results in a birefringence of 0.33 which can be extrapolated to *ca* 0.4 at room temperature.

These data (Tables 1 and 2) clearly demonstrate that the high electron density in metals does contribute to the properties of metal-based liquid-crystalline complexes.

Table 2 Birefringence data for some mesomorphic metal complexes

Compound	Δn^*	Ref
5OCB	0.1308	[67]
6OCB	0.1433	[67]
[PdCl$_2$(5OCB)$_2$]	0.1608	[67]
[PdCl$_2$(6OCB)$_2$]	0.1636	[67]
[PdCl$_2$(xOCB)$_2$]**	0.1730	[67]
[IrCl(CO)$_2$(8-OPhVPy)]	0.1659	[67]
[Ag(1-OPhVPy)$_2$][DOS]	0.1814	[67]
[Ag(2-OPhVPy)$_2$][DOS]	0.1968	[67]

* Measured directly at a reduced temperature of 0.985 (i.e. T/T_{NI} = 0.985); this gives a more accurate comparison of the data. ** This represents a preparation using equimolar amounts of 2OCB, 4OCB and 6OCB; the mixture contains all eight possible complexes.

8.7 MESOMORPHIC COORDINATION COMPLEXES OF BIDENTATE LIGANDS

8.7.1 Dithiolene Complexes with Two Chains

The first complexes of this type were those of Giroud-Godquin and Mueller-Westerhoff (Figure 36) and their synthesis marked the beginning of recent interest in metal-containing systems [36, 68–71].

In designing the complexes, it had been proposed that the two fused five-membered M–S–C–C–S rings would act in total like a phenyl ring, making the complexes analogous to 4,4″-dialkylterphenyls. This however turned out to be something of an oversimplification, given the profound effect that the metal had on the mesomorphism of these complexes; thus while the nickel and platinum derivatives were mesomorphic, those of palladium were not; the explanation advanced for this anomaly related to metal–metal bonding in the solid state. Thus in the solid state of (unsubstituted) bis(dithiolato)palladium(II), the Pd–Pd distance is 2.79 Å [72] which is shorter than the Pd–Pd distance in metallic palladium.

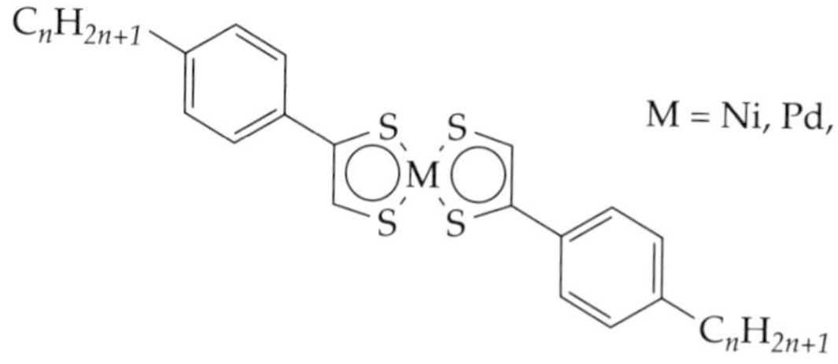

C_nH_{2n+1}

M = Ni, Pd, Pt

C_nH_{2n+1}

M = Ni	$n = 4$	K $\xrightarrow{117}$ N $\xrightarrow{175}$ I
M = Ni	$n = 8$	K $\xrightarrow{121}$ SmC $\xrightarrow{191}$ I
M = Pt	$n = 4$	K $\xrightarrow{158}$ N $\xrightarrow{202}$ I
M = Pt	$n = 8$	K $\xrightarrow{150}$ SmC $\xrightarrow{209}$ I

Figure 36 Structure of the mesomorphic dithiolene complexes and examples of their transition temperatures

It was argued therefore that in the alkylphenyldithiolene complexes, the same arrangement existed and it was the stability of the metal–metal interaction which precluded melting into the mesophase prior to decomposition. However, consideration of the structure of the related unsubstituted platinum complex [72] reveals a Pt–Pt separation of 2.77 Å. Given the greater extent of the $5d_{z^2}$ orbital (in Pt) relative to the $4d_{z^2}$ (in Pd), it would be expected that any Pt–Pt interaction would be stronger and so mesomorphism would be suppressed here too. Of course the vital pieces of information are missing, namely the M–M distances in the mesogenic Pd and Pt complexes. Until these numbers are known it is difficult to assess the potential contribution of any solid state M–M interactions to the mesomorphism (or lack of it) in these systems. If single crystals of the materials cannot be obtained, then EXAFS would provide the necessary data.

The identity of the SmC and N mesophases in these complexes was established by miscibility with the organic mesogen 4-octyloxy-benzoyloxy-4″-pentylbiphenyl (Figure 37).

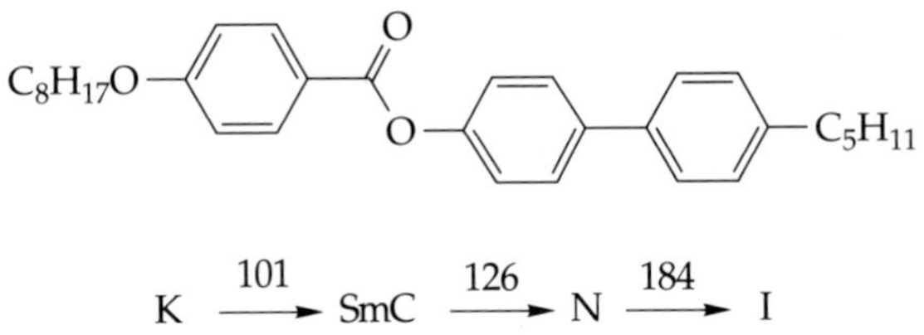

$$ K \xrightarrow{101} SmC \xrightarrow{126} N \xrightarrow{184} I $$

Figure 37   Structure and mesomorphism of 4-octyloxybenzoyloxy-4"-pentylbiphenyl

An interesting point here is that the phase diagrams showed no untypical features. This is important as it has been argued [30] that intermolecular metal–ligand or metal–metal interactions are important in stabilising the mesophases of mesomorphic metal complexes. If this were true, then as the concentration of the organic mesogen increased across the diagram, such interactions would necessarily be modified and this might be expected to be manifest in some way in the binary phase diagram. Similar binary phase diagrams (Sections 8.7.5 and 8.7.8) and EXAFS studies of dithiobenzoates (Section 8.7.3.1) further contribute to this debate.

8.7.2   Dithiolene Complexes with Four and Eight Chains

The original work in this area was carried out by Veber *et al.* who synthesised the tetrakis(alkoxyphenyl)dithiolenes shown in Figure 38 [73].

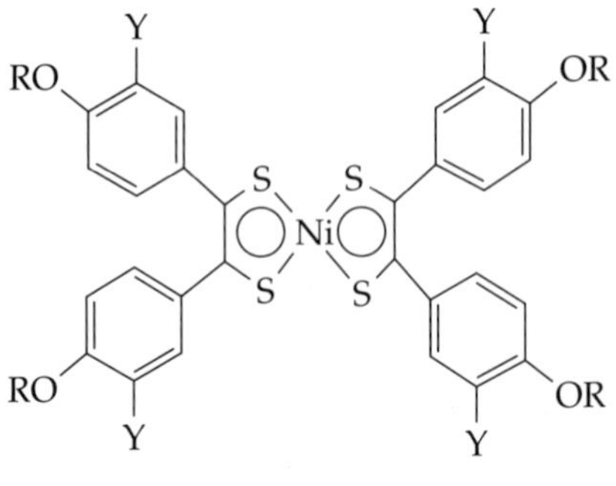

Figure 38   Polysubstituted nickel dithiolenes

For the complex with Y = H and R = $C_{12}H_{25}$, they described a material with a discotic mesophase which existed between 124 and 166 °C. Three years later, Ohta [74, 75] published the same materials, obtained by an improved route, with R = C_9H_{19} and $C_{11}H_{23}$ and again claimed discotic mesomorphism. Both groups identified the materials as electron-acceptors on the basis of cyclic voltammetry measurements. However, all of the complexes made by Veber and Ohta were investigated by Levelut using X-ray scattering techniques [76]. She concluded that the so-called mesophases of the materials were not in fact discotic but crystalline, with some small amount of disorder along the crystallographic *b*-axis.

Discotic liquid crystals of this type were eventually reported by Ohta [77] who used the strategy previously employed by Giroud-Godquin [78] for the generation of discotic mesophases in β-diketonate complexes (see Section 8.7.8), namely to attach eight peripheral chains to the flat core (Figure 38; Y = $C_{12}H_{25}O$, R = $C_{12}H_{25}$). The materials so produced were reported to have a D_{hd} phase between 84 and 112 °C and these claims were supported by X-ray scattering data which gave lines representing hexagonal symmetry and showed a diffuse band at 4.4 Å corresponding to the melted alkyl chains.

8.7.3 Dithiobenzoates

Most of the work with calamitic dithiocarboxylates has centred on studies on metal complexes of 4-alkoxydithiobenzoic acids (abbreviated to *n*-odtbH) shown in Figure 39.

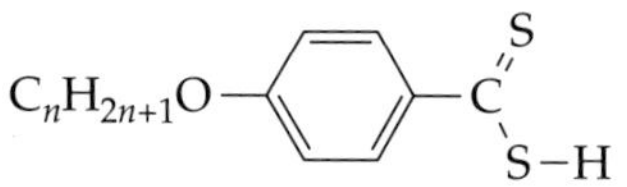

Figure 39 Structure of the 4-alkoxydithiobenzoic acids

Complexes of the empirical formula [M(*n*-odtb)$_2$] have been synthesised for M = Ni, Pd and Zn and some square planar gold(III) complexes showing SmA phases ([AuX$_2$(*n*-otdb)]; X = Cl, Br, Me) were also described [79]. The geometry of some of the complexes was established by single crystal X-ray studies and one palladium and two zinc complexes were characterised in this way.

8.7.3.1 Nickel and Palladium Dithiobenzoates

The palladium complex, [Pd(8-odtb)$_2$], was shown [79] by X-ray crystallography to have the expected square planar structure about the metal with local D$_{2h}$ symmetry (Figure 40). The molecules associated in pairs with intermolecular Pd–S contacts of 3.38 Å.

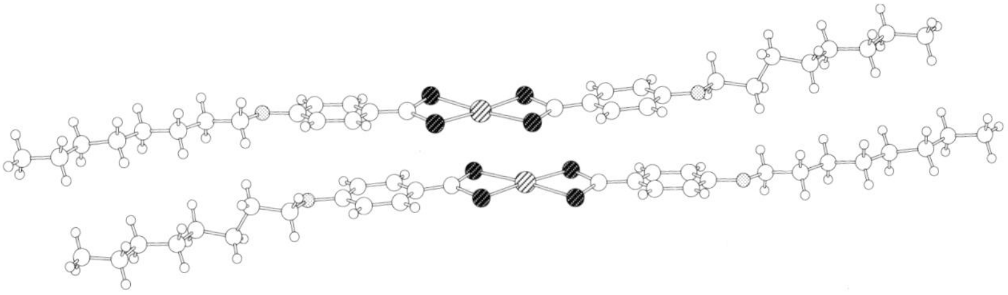

Figure 40   Molecular structure of [Pd(8-odtb)$_2$] (reproduced from ref [79] by permission of the Royal Society of Chemistry)

These complexes showed a nematic phase at short chain length which was replaced by a smectic C phase (established by X-ray methods) at longer chain lengths; below the SmC phase was a crystalline smectic phase. The transition temperatures of the materials were high and melting occurred typically at around 210–230 °C, while clearing (accompanied by extensive decomposition) was around 320 °C. While no structure determination was made on the nickel derivatives, they were assumed also to be square planar, D$_{2h}$ systems as ^{1}H NMR showed them to be diamagnetic. The mesomorphism of these materials was very similar to that of the palladium complexes, except that melting and 'clearing' points (see below) were reduced by some 30 °C and 80 °C, respectively. These materials also possessed the same ordered smectic phase found in the palladium complexes and this was proposed by Ohta [80] (on the basis of X-ray evidence) to be a crystal smectic H phase. Further studies by Richardson [79] on the same materials did not permit the unequivocal identification of the phase.

The palladium complexes were further studied using EXAFS [81]. In the solid state at room temperature, the data were readily fitted to the known crystal structure. This arrangement persisted through several crystalline modifications until the ordered smectic phase was entered. In this phase, the Pd–S distance lengthened to 3.8 Å where interaction must be precluded. Further heating into the SmC phase showed only four short

Pd–S vectors, confirming that the all significant intermolecular interactions had now disappeared. This would also tend to contradict the idea that intermolecular M–M or M–L interactions are of fundamental importance in metal-containing systems.

The major difference in the nickel series was that at temperatures around 230 °C, the blue bis(dithiobenzoates) rearranged (via some intermolecular process which also led to some decomposition) to form the red, mixed-ligand (alkoxydithiobenzoato)(alkoxytrithiobenzoato)-nickel(II) species (Figure 41A) which were also mesomorphic. These species were identified by Ohta [80] by extracting and purifying a heated mixture, while they were identified by ourselves via independent synthesis [79] using routes described by Fackler [82].

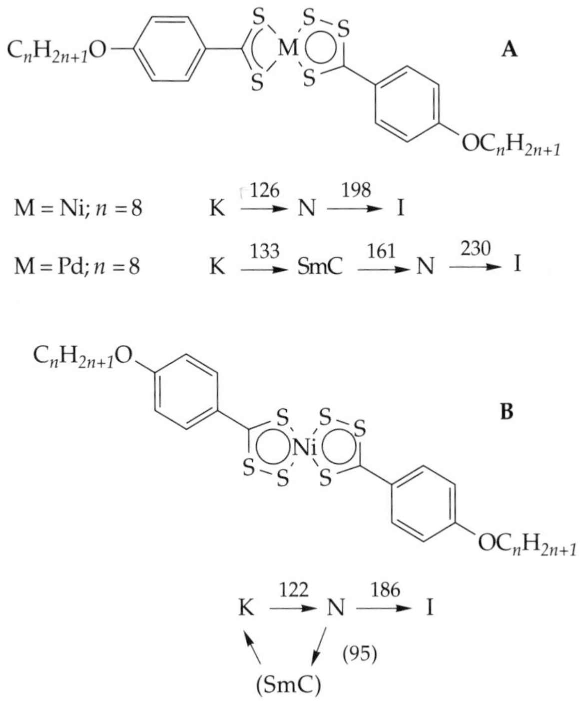

$M = Ni; n = 8$ $K \xrightarrow{126} N \xrightarrow{198} I$

$M = Pd; n = 8$ $K \xrightarrow{133} SmC \xrightarrow{161} N \xrightarrow{230} I$

$K \xrightarrow{122} N \xrightarrow{186} I$

(95)

(SmC)

Figure 41   Mesomorphic trithiobenzoate complexes of Ni(II) and Pd(II)

As might be expected from the much less linear shape of the mixed-ligand species, both the melting and clearing points were much reduced when compared with the symmetric bis(dithiobenzoate) parent compounds. While the symmetric bis(trithiobenzoate) complexes (Figure 41B) were not observed as a product of the thermal degradation, they could be independently synthesised, albeit in low yield, by reaction of the appropriate bis(alkoxytrithiobenzoato)zinc(II) complex with $NiCl_2.6H_2O$ [79]. The melting and clearing points of these materials were also lower than those of the parent bis(dithiobenzoate) complex and were in fact very similar to those found in the skeletally-isostructural dithiolenes described by Giroud-Godquin (Section 8.7.1).

8.7.3.2 Zinc Dithiobenzoates

Reaction of a sodium alkoxydithiobenzoate with zinc acetate in dilute acetic acid led to complexes of the empirical formula $[Zn(n\text{-odtb})_2]$ [79]. The as-obtained materials were analytically pure orange powders which showed a complex mesomorphism, apparently involving more than one species. However crystallisation yielded red crystals which had a well-defined mesomorphism, giving mainly nematic and some smectic C mesophases (Figure 42).

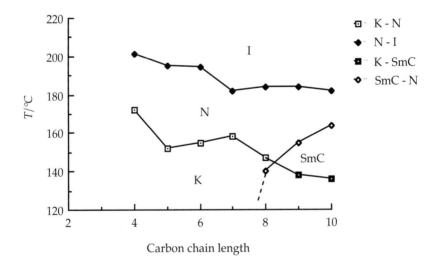

Figure 42   Phase diagram for the zinc dithiobenzoate complexes (reproduced in modified form by permission of the Royal Society of Chemistry)

Single crystal X-ray studies of the butoxy and octyloxy homologues [79] showed that these crystalline materials were in fact dimers (Figure 43) containing an eight-membered Zn–S–C–S–Zn–S–C–S ring. Such structures had previously been observed in dialkyldithiocarbamates of Zn(II) [83], although the unsubstituted dithiobenzoate of Zn(II) was in fact monomeric [84].

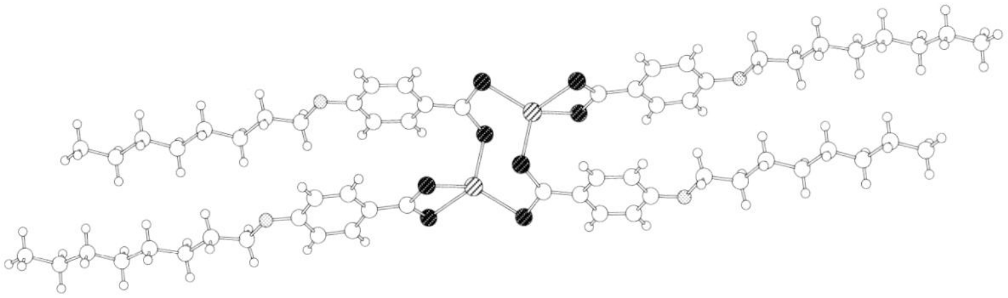

Figure 43   Single crystal structure of [Zn$_2$(8-odtb)$_4$] (reproduced by permission of the Royal Society of Chemistry)

The interesting feature of these materials was that while they were dimeric in the solid state, in solution in chloroform or toluene they were found to be monomeric [79].   What then was the species found in the nematic phase at elevated temperatures?   The answer was provided by EXAFS studies which clearly showed that in the nematic phase of [Zn$_2$(8-odtb)$_4$] at 160 °C, the dimer held together [82].

8.7.3.3   Other Dithio Ligands

Other dithio systems which have been studied are the alkyldithioacetates and alkylxanthates of Ni(II) described by Ohta [85] (Figure 44).

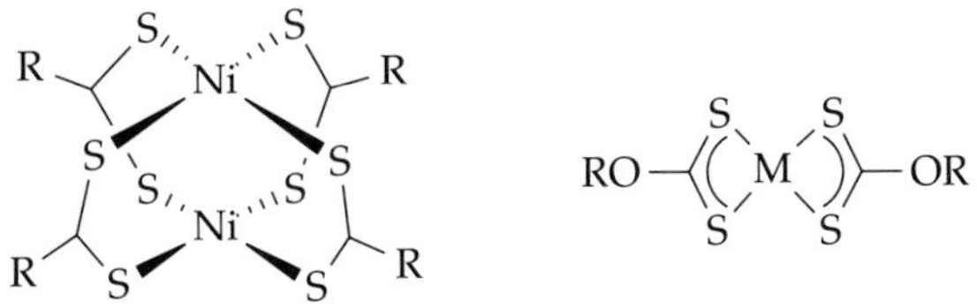

Figure 44   The alkyldithioacetates and alkylxanthates described by Ohta

Dithioacetates with R ranging from ethyl to dodecyl were prepared and all those with chains longer than butyl were reported to be mesomorphic, showing a monotropic mesophase which was described as being 'liquid-crystal-like'. On the basis of X-ray scattering and infrared data, a non-crystalline lamellar structure was proposed. However, examination of the published X-ray data (little variation in the strong, low-angle peaks through all the phases) and photomicrographs (sharp, well-defined edges) leads to the thought that this mesophase is not liquid crystalline and may well be of the crystal smectic type.

The related xanthate complexes showed a much more complex behaviour. Thus, only the longest chain xanthate with an even parity (dodecyl) showed any mesophase. However, odd-parity chains from pentyl and longer exhibited a complex thermal behaviour which involved the melting of the 'as-obtained' crystals into an isotropic state followed by, on heating, crystallisation to another crystal state which then melted again to another isotropic state. Thus, two isotropic states were observed for pentyl, heptyl and nonyl derivatives while for the undecyl homologue, three isotropic states were found; in addition, the nonyl and undecyl derivatives showed a monotropic mesophase probably of the crystal smectic type. Such multiple melting behaviour is curious and may possibly be attributed to some structural transformation between species with different melting points; similar transformations can account for the appearance of re-entrant mesophase phenomena [86].

8.7.4   Physical Properties of Mesogens with Bidentate Sulphur Ligands

The mesogenic nickel dithiolenes were found [87] to have reduction potentials (against calomel) of +0.06 V and − 0.76 V and as such were expected to form charge-transfer complexes with the mesomorphic, substituted tetrathiafulvalene (TTF) derivatives (Figure 45), which had oxidation potentials (also against calomel) of +0.43 V and +0.88 V.

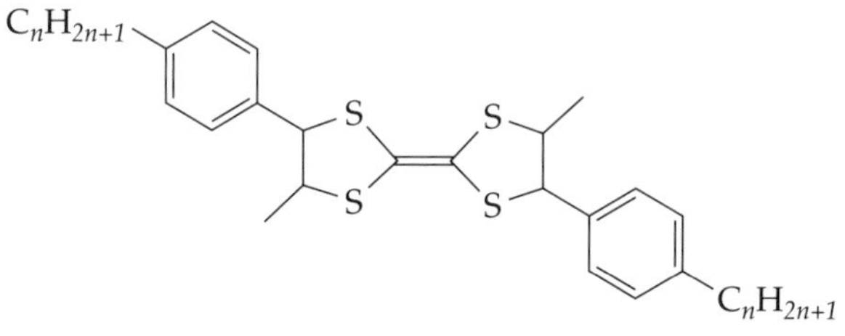

Figure 45   Structure of the mesomorphic TTF derivatives

Charge-transfer complexes were indeed formed and in a variety of stoichiometries; mixtures containing the decyl derivative of the nickel dithiolenes (Figure 36; R = $C_{10}H_{21}$) were always smectic.

The butyl derivative of these nickel complexes was investigated as a dye in a guest-host system using 4-pentyl-4'-cyanobiphenyl (5CB) as the host [88]. The complex was found to have a large extinction coefficient of absorption ($\varepsilon = 28,000 \text{ dm}^3 \text{ cm}^{-1} \text{ mol}^{-1}$ at 860 nm) and dissolved in 5CB at concentrations up to 10% (w/w). At 20 °C, such solutions showed a dichroic ratio (R_D, defined as $A_{||}/A_{\perp}$; A = absorbance) of 4.97 and a dye order parameter of 0.57.

Linear dichroism measurements were also made for the mesomorphic dithiobenzoates of Ni(II), Pd(II) and Zn(II) [89]. They were made difficult by the very low solubility of the complexes in most solvents and data were typically obtained at dye concentrations of < 0.5 wt. %; the solvents used were the commercial eutectic mixture of cyanobiphenyls and cyanoterphenyls, E7 (Merck) for the palladium complexes and the commercial mixture of cyanobicyclohexanes, ZLI2830 (Merck) for the nickel and zinc complexes. The nickel and palladium complexes both had two absorptions (Ni: λ_{max} = 595 nm, MLCT, $\varepsilon = 17,970 \text{ dm}^3 \text{ cm}^{-1} \text{ mol}^{-1}$ and λ_{max} = 375 nm, ligand-based, $\varepsilon = 127,220 \text{ dm}^3 \text{ cm}^{-1} \text{ mol}^{-1}$; Pd: λ_{max} = 432 nm, MLCT, $\varepsilon = 52,950 \text{ dm}^3 \text{ cm}^{-1} \text{ mol}^{-1}$ and λ_{max} = 335 nm, ligand-based, $\varepsilon = 55,530 \text{ dm}^3 \text{ cm}^{-1} \text{ mol}^{-1}$) while the zinc complex showed only the ligand-based absorption(λ_{max} = 372 nm, $66,020 \text{ dm}^3 \text{ cm}^{-1} \text{ mol}^{-1}$). However, because of the aromatic nature of the E7 host for Pd, dichroism data could only be obtained for the longer-wavelength absorption.

It was found that R_D increased according to Zn (4.2–4.7) < Ni (5.1–6.0) for the ligand-based absorption and according to Ni (7.0–10.12) < Pd (8.7–12.8) for the charge-transfer absorption; indeed this latter absorption for palladium gave dye order parameters of up to 0.8.

8.7.5 Complexes with N, O Donor Sets and Four Terminal Chains

A number of salicylaldimate complexes based on the ligand structures shown in Figure 46 have been synthesised using the metals copper, nickel, palladium, vanadium and iron.

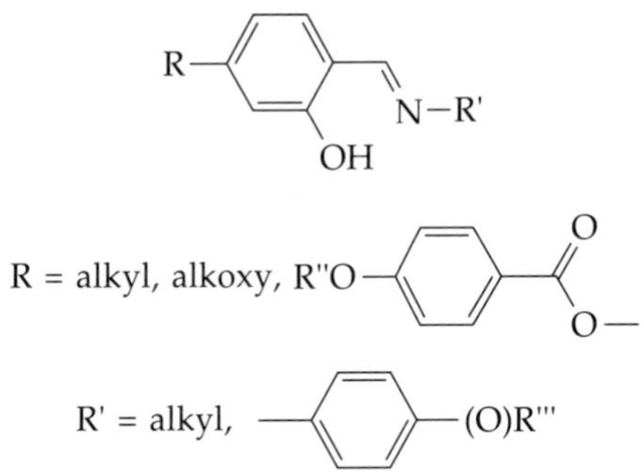

R = alkyl, alkoxy, R"O—⟨benzene⟩—C(=O)—O—

R' = alkyl, —⟨benzene⟩—(O)R'''

Figure 46 Structure of the salicyladimate ligands used to generate mesomorphic complexes

The original work in this area was carried out by Ovchinnikov and co-workers who synthesised copper complexes of the type shown in Figure 47 [90, 91].

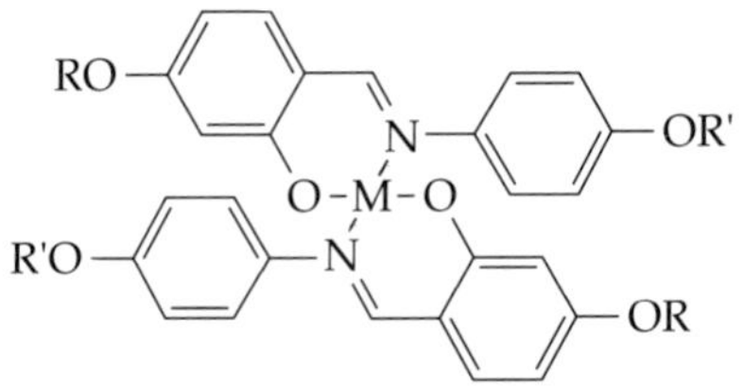

Figure 47 The copper salicylaldime complexes first synthesised by Ovchinnikov

These complexes showed both smectic and nematic mesophases and as such, were the first examples of materials with a paramagnetic nematic phase. Later, the same group reported an analogous (paramagnetic) complex of V(IV) (Figure 47; M = V=O) [92], although the exact nature of the mesophase was not reported by them until a few years later [93]. This paper was largely overlooked and a very important point was missed, namely that a mesomorphic metal-containing system containing a square pyramidal centre had been identified in a calamitic molecule. This idea was further re-emphasised by the synthesis of an Fe(III) complex (Figure 48) which showed a SmA phase [94].

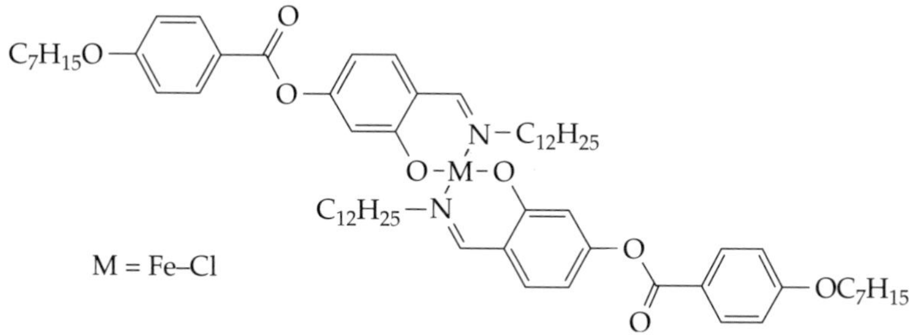

Figure 48 Mesomorphic Fe(III) complex

Following these initial studies by Ovchinnikov, the study of mesomorphic salicylaldimate complexes became very popular and at the present time, some seven or more groups internationally are active in this area, leading to some duplication of effort.

Marcos and co-workers [95, 96] examined several series of Cu(II) and Ni(II) complexes with the structures shown in Figures 49A and 49B, while independently, Hoshino and co-workers [97] reported the mesomorphism of some copper derivatives of the type shown in Figure 49B.

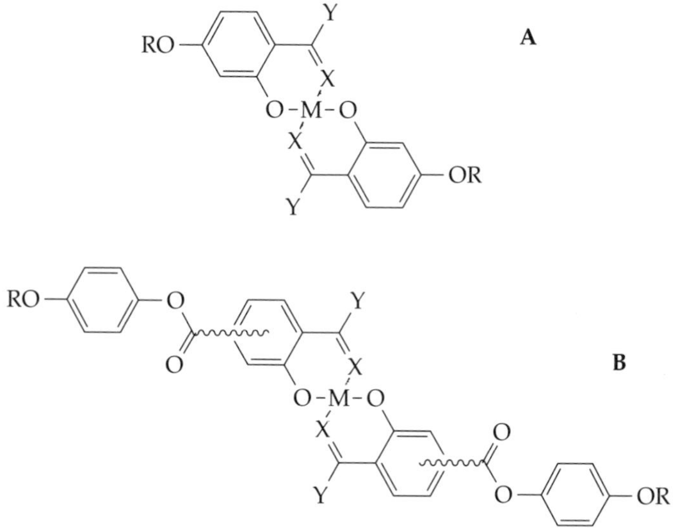

Figure 49 Marcos' and Hoshino's salicylalimate complexes

None of the complexes (Figure 49A) with X = O, Y = H, Me and M = Cu, was mesomorphic [95] and all melted straight to an isotropic fluid at elevated temperatures. However, for the related copper complexes with X = p-N–C_6H_4–R' and Y = H, smectic mesophases were observed which were reported to have an optical texture corresponding to SmA and SmC; however, on the basis of the viscosity of the mesophases, they were assigned as *discotic smectics* following the same classification used by Ovchinnikov in his earlier work. The complexes (Figure 49B) were synthesised for Y = H, M = Cu, R = C_mH_{2m+1} (alkoxybenzoate group *para* to the ring position of the imine carbon) and X = p-N–C_6H_4–OC_nH_{2n+1} by both Marcos [96] and Hoshino [97], with Marcos varying n for m = 10, while Hoshino varied m for n = 2; interestingly, both papers were submitted (to different journals) within three days of one other! Marcos reported that the complexes were mesomorphic for $1 \leq n \leq 14$ (m = 10), with a nematic phase observed from n = 1–10 and a SmC phase seen for n = 3–14; melting points were reported to be around 155–180 °C with clearing points between 232–267 °C. For $6 \leq m \leq 14$ (n = 2), Hoshino reported a nematic phase with a SmC phase for $11 \leq m \leq 18$; melting points were in the range 156–222 °C with clearing points between 251–275 °C. Curiously, the one compound which was made by both groups (m = 10, n = 2) was assigned the same phase behaviour but at different temperatures (Marcos: K • 161 • N • 255 • I; Hoshino: K • 189 • N • 271 • I). Marcos also synthesised the related Ni(II) complexes and reported that at low values of n a nematic phase was seen, whereas at larger n only a crystal smectic phase was observed.

The smectic phase of these materials was again classified as discotic smectic by Marcos, although curiously the nematic phase was not considered to be a discotic nematic. However, in a later publication [98] which considered complexes (Figure 49B) with X = H, Me, R = C_mH_{2m+1}, M = Cu, X = –NC_nH_{2n+1} and with the 4'-alkoxybenzoate group attached in either the 4- or 5- position relative to the imine carbon, the SmC and nematic phases found were investigated by X-ray scattering techniques and were reported to show, "no appreciable differences if compared with those of conventional mesophases." That the mesophases were 'conventional' was also show by Sirigu and co-workers who constructed a phase diagram for complexes of the type shown in Figure 47 and dibutyl-p-terphenyl-4,4''-dicarboxylate (which has a SmA phase) and found the mesophases of the two to be miscible in all proportions [99].

Hoshino [100, 101] and Sirigu [102] also reported examples of complexes (Figure 49B) with R = C_mH_{2m+1}, Y = H and X = NC_nH_{2n+1} for M = Cu [100, 102], Ni and V=O [101], while in addition, Marcos reported vanadyl derivatives of this type as well as those with $R = C_mH_{2m+1}$, $Y = H$ and $X = N-C_6H_4-C_nH_{2n+1}$ [103] . When one or both of n and m were short, only the nematic phase was seen and indeed for low values of n, only the nematic phase was see up to $m = 18$. A significant difference is noted in comparing the clearing temperatures of the nickel and oxovanadium derivatives where $n = 3$ and m was varied, namely that the clearing temperatures of the vanadyl complexes are significantly lower by around 25 °C. This was undoubtedly due to the lower symmetry (local D_{2h} for nickel and C_{2v} for vanadyl) which leads to a lower structural anisotropy and hence a lower clearing point. In fact, for the same ligand derivatives, Hoshino reports a general decrease in mesophase stability according to Ni > Cu > V=O, the lower position of copper being accounted for [101] by its greater susceptibility to tetrahedral distortion which would grossly reduce the structural anisotropy.

Investigations of the vanadyl and copper complexes by EPR techniques [101, 103, 104] showed that the magnetic anisotropy was positive for the vanadyl and nickel complexes and negative for those of copper. Some of the vanadyl derivatives were shown to be easily aligned in the presence of a magnetic field [103]. ^1H NMR studies [101] of some nickel derivatives in isotropic solution showed evidence for aggregation phenomena as evidenced by line broadening which it was assumed originated from the partial paramagnetism, which in turn resulted from some axial Ni$\cdots$O interaction; it was suggested that these interactions may persist in the mesophase. Given the very high preference of Ni(II) to form octahedrally-coordinated complexes, such conclusions are not unreasonable although the fact that the nematic phase is so prevalent for these complexes would suggest that any intermolecular interactions in the mesophase are relatively unimportant. Evidence for such interactions might be found by X-ray scattering studies where cybotactic nematic features (nematic phases containing smectic domains) may well be observed. However, a study of similar Ni(II) complexes [104] by EPR techniques showed unequivocally that the Ni(II) complexes were diamagnetic in the mesophase, suggesting that the line broadening in Hoshino's NMR experiments may have some other origin.

Ghedini and co-workers [105] investigated a series of salicylaldimate complexes of the structure shown in Figure 50, where $R = CH_3$, C_2H_5, C_3H_7, C_4H_9 and $R' = C_{12}H_{25}$ and C_7H_{15}.

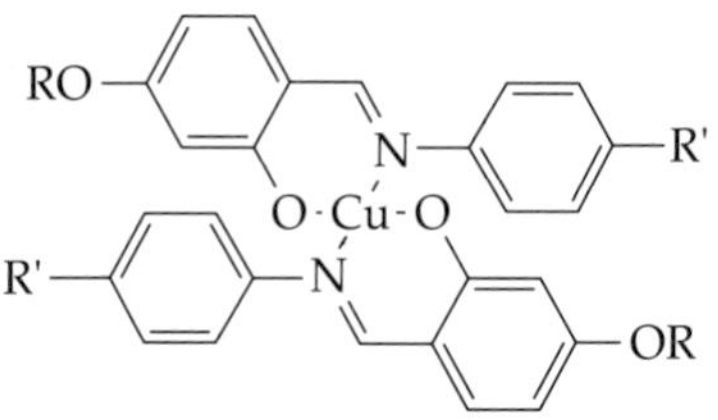

Figure 50 Ghedini's salicylaldimate copper complexes

While the ligands showed SmC, SmA and N phases, all of the copper complexes showed only a SmA phase with the exception of one which showed a SmB phase as well. Melting points for the ligands were typically between 50–70 °C, while clearing points were between 80–90 °C; for the corresponding complexes these temperature ranges were 110–120 °C and 100–150 °C respectively.

Low-angle X-ray scattering experiments were carried out in the SmA mesophase of two of these copper complexes, namely [Cu(DOBBA)$_2$] (Figure 50; R' = C$_{12}$H$_{25}$, R = C$_4$H$_9$) and [Cu(HOBBA)$_2$] (Figure 50; R' = C$_7$H$_{15}$, R = C$_4$H$_9$) and in the SmB phase of [Cu(DOBBA)$_2$] [106] (mesomorphism of [Cu(DOBBA)$_2$]: K • 90 • E • 98 • SmB • 114 • SmA • 140 • I). Results obtained in the SmA phase showed that the complexes had a negative diamagnetic anisotropy and that there was a weak, diffuse feature pointing to an in-layer Cu–Cu correlation at 8.5 Å corresponding to a side-to-side arrangement of the molecules. A ribbon-like structure was proposed for the SmB phase on the basis of the X-ray results and again a side-to-side correlation, this time of 8.6 Å, was found. Further studies of the SmB phase by EXAFS [107] revealed a Cu–Cu correlation at 3.85 Å showing the phase to have solid-like translational order and therefore to be intrinsically biaxial, although macroscopic biaxiality has not yet been demonstrated in these systems.

Sirigu [108, 109] and later Cortieu [110] used the same basic framework to synthesise a number of Cu, Ni and Pd complexes with the structure shown in Figure 51, as well as further derivatives of the compounds in Figures 47 and 50.

Sirigu showed that the copper complexes (Figure 51; R, R' = C$_n$H$_{2n+1}$CO$_2$) had SmA and SmC phases typically in the range 180–210 °C while the related complexes (Figure 51; R, R' = C$_n$H$_{2n+1}$O) had the same phases between 140–170 °C. However, the palladium derivatives of the same ligands showed the same mesophases and while the melting point increased by between 20–60 °C, the mesophase range was greatly extended, being typically 20 °C wider. Given that the copper and palladium complexes are largely isostructural in the solid state, it may be that at such elevated temperatures, the copper complexes may start to distort towards a tetrahedral arrangement, thus effectively shortening the molecule and reducing the clearing point.

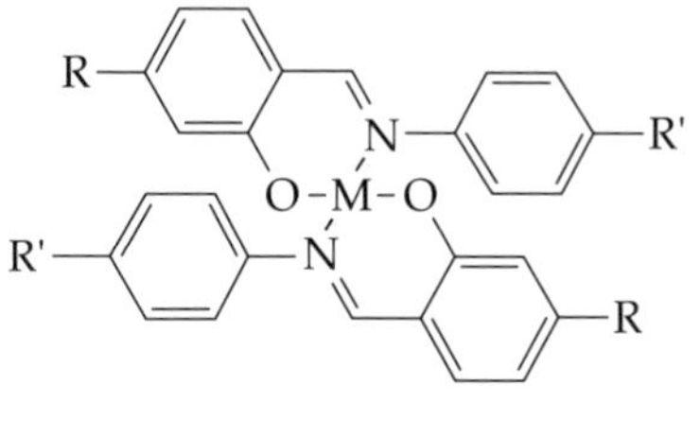

Figure 51 General structure of the metal salicylaldimine complexes synthesised by Sirigu and Cortieu

Cortieu synthesised the related complexes with R = $C_nH_{2n+1}OC_6H_4CO_2$ and R′ = C_mH_{2m+1} or $C_mH_{2m+1}O$ with Cu and Ni; one example with Pd was reported. While none of the Ni complexes showed any mesomorphism, the Cu complexes with $1 \leq n \leq 16$ and R′ = C_4H_9 all showed only the nematic phase, with melting and clearing points in the range 220–158 °C and 233–264 °C respectively. If, however, n was fixed at 16 and R′ was $C_mH_{2m+1}O$ with m varying from 4–16, then an additional (unidentified) smectic phase was observed. Initial melting points were in the range 117–145 °C while clearing points varied from 215–252 °C. In a related piece of work [111], Cortieu studied halogenated derivatives of some of the Cu complexes (Figure 52).

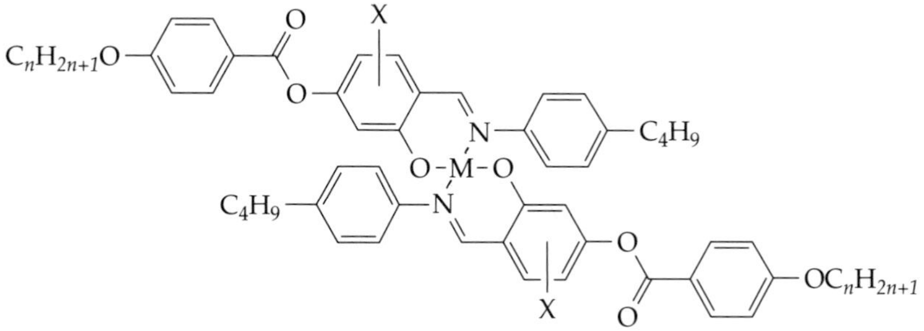

Figure 52 Structure of the halogenated copper complexes synthesised by Cortieu

The halogen (X) was either Cl or Br and was situated in either the 3- or 5-position relative to the ring position bearing the imino carbon. In all cases, the clearing temperatures of the nematic phase were reduced when compared to the non-halogenated analogues and these reductions were always greater for X = Br. For X = Cl, the reduction in clearing temperature was greatest for 5-substitution whereas for X = Br, no simple relationship was found.

Polymeric derivatives of these complexes have also been made [99] (Figure 53) where the rigid copper-based cores are linked by a dodeca(methylene) group or a tri(ethyleneoxide) group to give co-polymers which show a monotropic SmA phase.

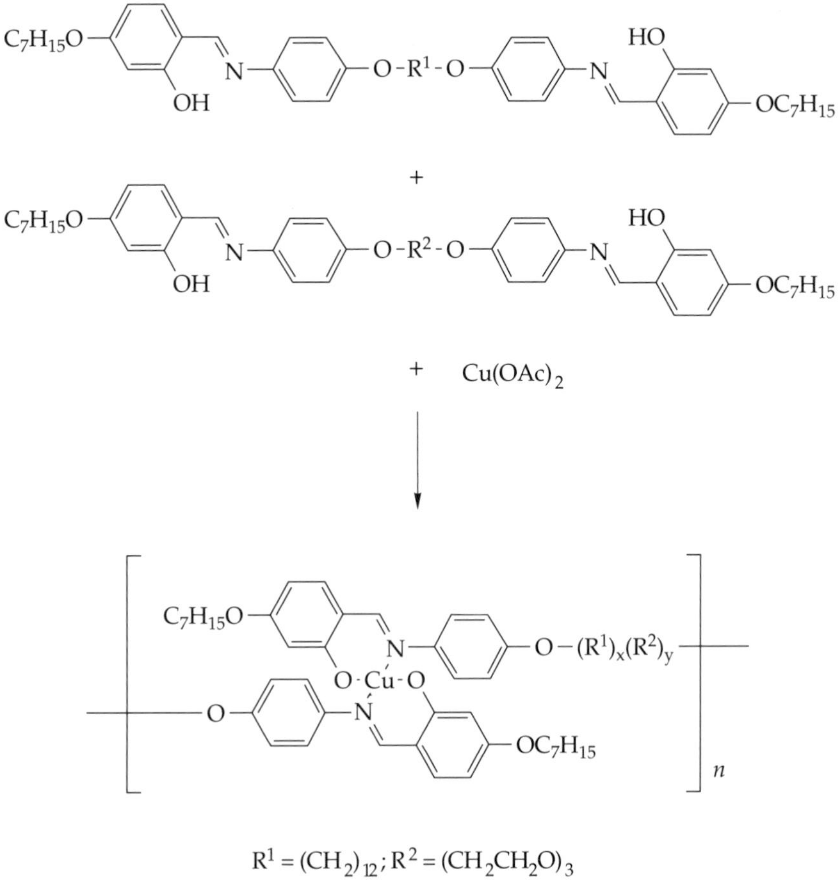

$$R^1 = (CH_2)_{12}; R^2 = (CH_2CH_2O)_3$$

Figure 53 Synthesis and structure of the copper salicylaldimine co-polymers

Finally in this section, there were two reports [112, 113] of the same bis(alkoxysalicylidene)ethylenediamine complexes of Ni(II) and Cu(II) (Figure 54). In the earlier paper [112], the materials were reported to show SmA mesophases at elevated temperatures, while in the later report [113], the mesophase was not identified beyond being smectic and the transition temperatures were quite different.

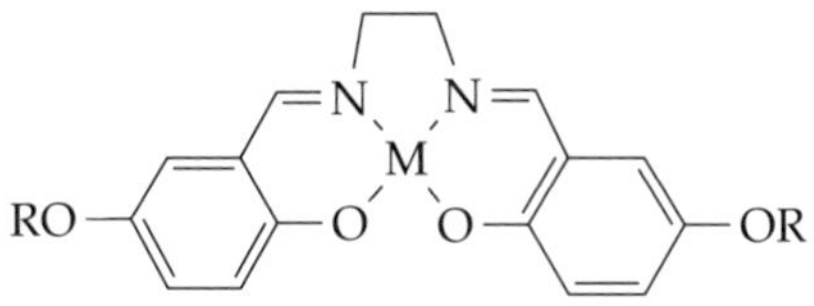

Figure 54 Structure of the bis(alkoxysalicylidene)ethylenediamine complexes of Ni(II) and Cu(II)

The earlier paper also reported that while the melting point of the copper complexes ($\approx$ 250 °C) was up to 60 °C higher than that of the nickel complexes, the latter cleared (with decomposition) some 30 °C higher at around 300 °C and as such had a much wider mesophase range. My money (for what it's worth) goes with the earlier paper!

8.7.6 *Ortho*-metallated Palladium Complexes with Four or More Terminal Chains

8.7.6.1 Ortho-*palladated Azo Complexes*

One of the earlier contributions to the revival of metal-containing liquid crystal systems was the synthesis of the *ortho*-metallated palladium complexes of mesogenic azobenzenes (Figure 55) by Ghedini and co-workers, which represented the first systematic attempt to coordinate metals to known liquid crystal systems.

Initial studies [114] investigated the complexes for which X = Cl, R = OEt and R' = $C_4H_9CO_2$, $C_6H_{13}CO_2$ and $CH_2=CH(CH_2)_8CO_2$ and the related complexes with L = PPh$_3$, py, quinoline and aniline [115]. All of the dinuclear complexes showed an enantiotropic nematic phase with melting points in the range 165–212 °C and clearing points between 185–215 °C; comparison with the behaviour of the free ligand is interesting in these complexes. The free ligands melt typically in the range 64–78°C and clear between 107 and 126 °C showing a typical mesophase range of around

50 °C. However, while a marked increase in melting point might have been expected (and was indeed observed) on complexation to form a dinuclear complex, the corresponding increase in clearing temperature is a little more puzzling as it would have been expected that the structural anisotropy would have decreased. This was, however, not the case and very stable nematic phases resulted, all be they with a much reduced range. The fact that complexation of such systems enhances mesophase stability was further demonstrated in the complexes (Figure 55) for which $R = C_{11}H_{23}$ and $R' = $ Me or Et [116]. Here it was found that the monotropic nematic phase of the parent ligands (T_{NI} 76 and 67 °C respectively) was stabilised much more than the crystal phase to give materials with enantiotropic nematic phases ($T_{NI} \approx 170$ °C).

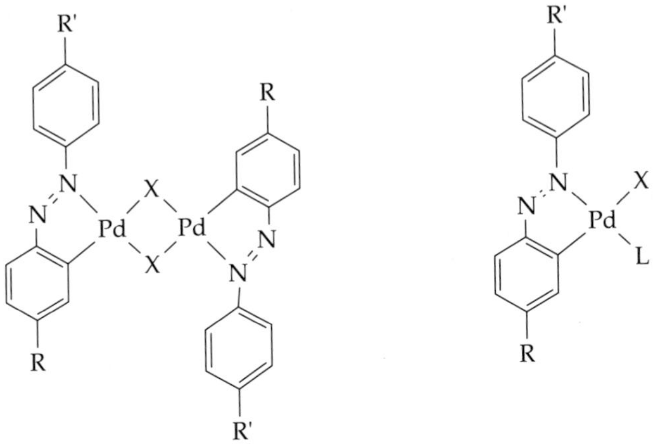

Figure 55   Ghedini's *ortho*-palladated Pd(II) complexes

The rôle of the bridging halogen in these systems was also investigated [117] and it was found that the melting point increased according to Cl(K-N) < Br(K-SmA) < I(K-SmA) and that the temperature at which the nematic phase first appeared increased according to the same order. Clearing points, however, followed a different trend, namely Br > I > Cl. The reasons for these various changes are unclear, although the introduction of a new phase (SmA) for Br and I possibly suggests some effect on the geometry of the complex, dependent on the nature of the bridging halogen. Molecular *and* macroscopic biaxiality have been proposed in these materials on the basis of order parameter measurements made by infrared techniques and on the basis of calorimetry and optical microscopy [118].

While none of the related mononuclear complexes with triphenylphosphine or aniline was mesomorphic, those with L = py and quinoline gave materials with nematic and smectic phases. In the case of the PPh$_3$ complex the lack of mesomorphism was probably due simply to the bulk of the ligand, whereas intermolecular hydrogen bonding may have been responsible in the case of aniline. For L = py, the nematic phase stability was similar to that found in the parent dinuclear systems (T$_{NI}$ = 235 °C), while for L = quinoline, melting points and clearing points were somewhat reduced (K • 136 • SmA • 151 • N • 180 • I), almost certainly due to the increased bulk of the quinoline over the pyridine.

Other *ortho*-metallated azobenzenes were reported by Hoshino [119] (Figure 56). In common with the parent ligands, the complexes were nematic, although melting and clearing points were raised by 60 and 102 °C, respectively, on complexation giving the complexes a much wider nematic range. Variable temperature ^{1}H NMR experiments of the $n = 4$ derivative in isotropic solution in CDCl$_3$ confirmed the presence of dynamic processes which were interpreted on the basis of a model assuming that the lateral chain was in fact sitting in the central cleft of the molecule as depicted in Figure 56.

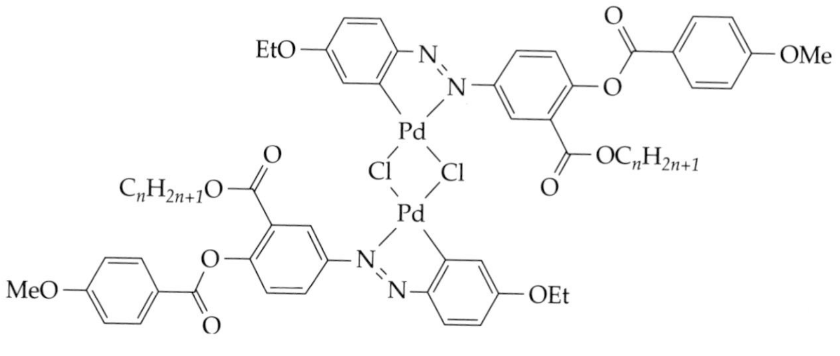

Figure 56   Hoshino's *ortho*-palladated azobenzenes, showing the conformation accounting for the observed dynamic behaviour

Related complexes based on imines have been described by Espinet and co-workers [120] (Figure 57).

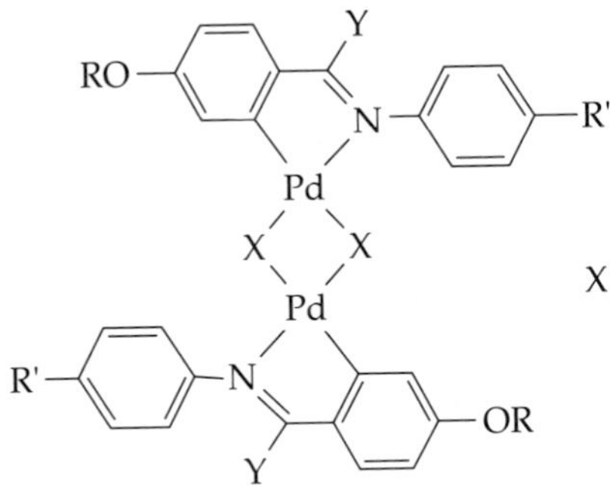

458 *Duncan W Bruce*

Figure 57 Espinet's *ortho*-palladated imine complexes

Complexes based on three ligands were initially investigated, namely $R = R' = C_{10}H_{21}$ and $Y = H$, $R = C_{10}H_{21}$, $R' = C_{10}H_{21}O$ and $Y = H$ and $R = R'$ $= C_{10}H_{21}$ and $Y = Me$. In general, OAc was found to be an ineffective bridging ligand in these systems and the complex of only one ligand showed any mesomorphism (monotropic SmA). For the other bridging ligands, SmA phases were commonly observed along with a SmC phase for two examples where $X = Cl$. Mesomorphic ranges were typically 80–100 °C for $X = Cl$ or Br, while for $X = SCN$, 30 °C was the maximum range for the SmA phase. Interestingly, the lateral methyl group did not lead to any significant effects in the complexes even though it had suppressed mesomorphism in the parent ligand.

8.7.6.2 Ortho-*palladated Azine Complexes*

These studies were then extended to the synthesis [121] of palladium complexes of symmetric azines (Figure 58).

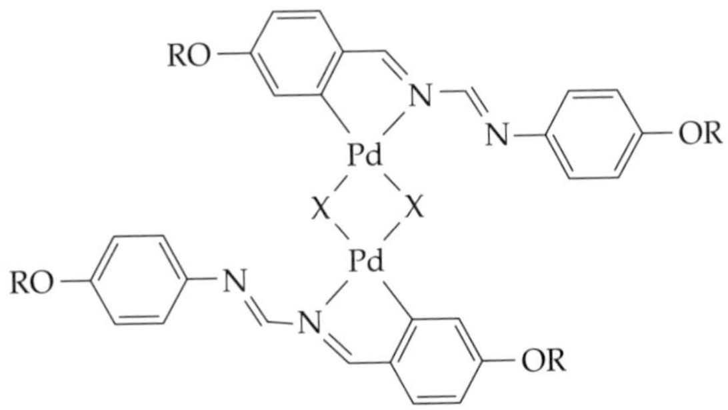

Figure 58 Dinuclear palladium complexes of azine systems

In the complexes where X = Cl and Br, only the *trans* isomer was observed (by ^{1}H NMR), while for X = SCN, two isomers were observed in a 60:40 ratio and it was assumed that these were *cis* and *trans*. In each of these examples, the complex was assumed to be planar by comparison with related structures in the literature [122]. However, where X = OAc, the situation was more complex and *trans* and *cis* isomers in the ratio 3:1 were consistently produced.

^{1}H NMR studies went on to show that the *trans* isomer was optically active and hence the structure had to be that of an 'open book' (Figure 59, R = Me), although in the synthesis a racemic mixture was produced.

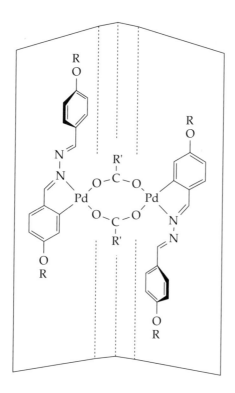

Figure 59   Schematic structure of the chiral *trans*-isomer of the μ-carboxylato palladium azines (from reference [123])

For the non-acetate-bridged dimers, the only mesophase seen was SmC which was typically in the range 102–249 °C (for R = C_9H_{19}) while for the acetate-bridged complexes, a nematic phase was seen for $C_6H_{13} \leq R \leq$

C_8H_{17} and for $R \geq C_7H_{15}$, a SmC phase was seen in addition. Mesophase ranges were very much larger in the planar materials.

Using the fact that the *trans*-isomer was chiral, a derivative was synthesised where the bridging carboxylate group was the optically-pure (R)-2-chloropropionate [123]. Synthesised from the μ-Cl_2 species by reaction with the sodium salt of the acid, a mixture was produced which was shown by ^1H NMR to have the following composition: *trans*-$\Lambda R,R$ (34%), *trans*-$\Delta R,R$ (34%), *cis*-R,R (32%). Thus, while the *trans*-components described a pair of diastereoisomers, the *cis* isomer was optically pure by virtue of the chiral acid groups. The mixture so produced had the phase sequence K • 102 • SmC* • 119 • SmA • 149 • I and physical measurements showed that the SmC* phase was ferroelectric, with a rise time of around 330 ms at a square wave voltage of ± 17 V and 0.5 Hz and a cell thickness of 11 μm. Such response times were some three orders of magnitude longer than those found in calamitic SmC* phases, something which was probably due to the greater viscosity of the palladium SmC* phase which in turn resulted from the molecular shape.

8.7.6.3 Ortho-*palladated Pyrimidine Complexes*

Other work in this area by Ghedini considered the products of the cyclopalladation of the well-known liquid crystals, the 2-phenyl-pyrimidines [124]. Initially, dimeric products were obtained and then they were further reacted with species able to cleave the $(\mu$-$Cl)_2$ bridge to give a series of mononuclear derivatives (Figure 60).

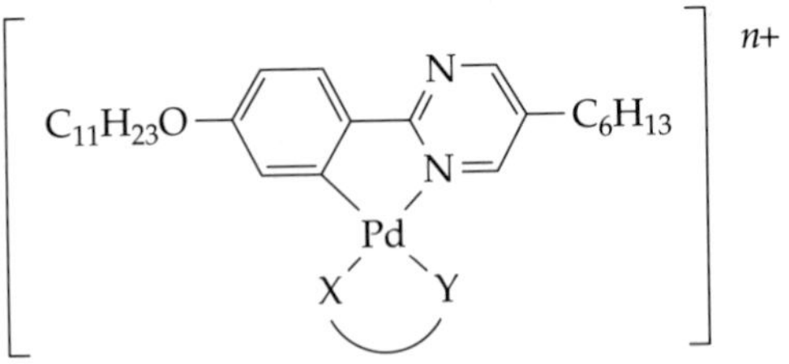

Figure 60 Mononuclear palladium complexes of 2-phenylpyrimidine derivatives

For X-Y = 8-hydroxyquinolate (n = 0) and X–Y = 1,10-phenanthroline (n = 1), the complexes were non-mesomorphic. However, when X–Y = acac (n = 0), a material with a monotropic SmA phase resulted (K • 83 • I ; I • 68 • SmA) and when X–Y = 2,2'-bipyridine (n = 1 with BF_4 counterion), a material with an enantiotropic nematic phase was produced (K • 146 • N •

158 • I); related complexes with PF_6 or SbF_6 anions were non-mesomorphic. These are further rare examples of ionic materials showing a thermotropic nematic phase (see Section 8.6.2.3).

Further studies of the same system [125] looked at the dimeric precursors to such monomeric species and considered the complexes shown in Figure 61.

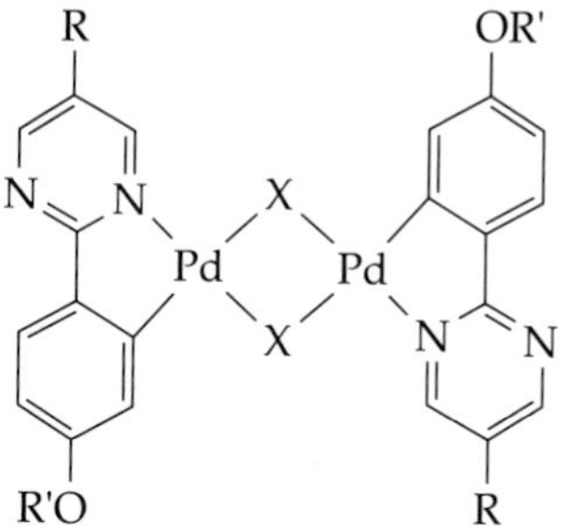

X = Cl, Br, I, OAc

Figure 61   Dinuclear 2-phenylpyrimidine derivatives

Four different ligands were studied: HL^1 (R = C_6H_{13}, R' = Me), HL^2 (R = C_9H_{19}, R' = Me), HL^3 (R = C_6H_{13}, R' = $C_{11}H_{23}$) and HL^4 (R = C_9H_{19}, R' = C_9H_{19}). None of the complexes with X = OAc and no derivatives of HL^1 were mesomorphic; other non-mesomorphic combinations were HL^2 with X = I, and HL^4 with X = Cl. Of the remaining materials, all had the SmA phase (the parent ligands showed nematic phases) typically between 105 and 202 °C, while two materials (HL^2 with X = Cl (5) and HL^4 with X = I (6)) were reported to have another smectic phase (SmX) above the SmA – both phases exhibited a focal conic texture. In the normal scheme of phases, there is, of course, no smectic phase above SmA and no comment as to the nature of this new phase was offered. The enthalpies of transition between SmA and SmX (J g^{-1}) were reported as 25.1 (5) and 5.95 (6) and the former might suggest that the SmA phase was misidentified and that perhaps SmX is in fact SmA while 'SmA' is some more ordered phase. In the latter compound (6), the 'SmA' phase could be fluid with this smaller value of ΔH, but miscibility and/or X-ray studies are needed to characterise fully these phase sequences.

8.7.7   Complexes of β-Diketonates with Two or Four Terminal Chains

The ease of synthesis of β-diketonate ligands has made this area quite fertile and has led to many interesting (and sometimes controversial) findings. The first people to suggest that β-diketonates may be useful as ligands to generate mesomorphic complexes were Bulkin, Rose and Santoro who in 1977 synthesised the palladium complex shown in Figure 62 (M = Pd, R = $C_8H_{17}O$) and suggested that it may be mesomorphic [126].

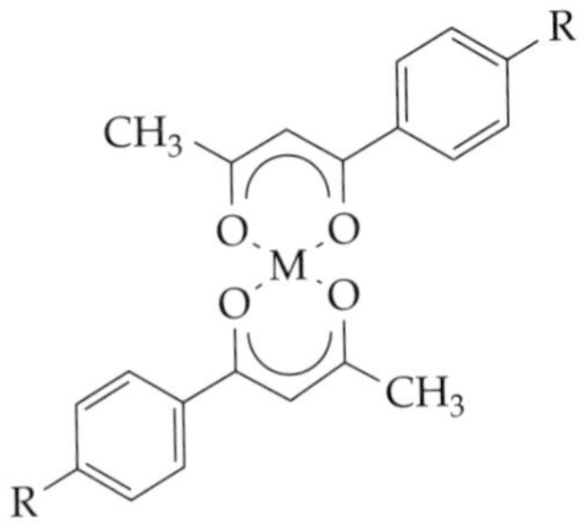

Figure 62   Structure of early mesogenic β-diketonate complexes

The monomeric nature of this yellow complex was established by osmometric methods, but observations of any potential mesomorphism were hampered by the severe darkening on heating above 170 °C due to decomposition. Curiously, no-one appears to have been tempted to go back and re-examine these materials or at least if they have, they have chosen not to publish their findings, possibly suggesting that the materials are not in fact mesomorphic.

In 1980 and 1981, there followed a series of papers from Ohta [127] on the copper derivatives (Figure 63; M = Cu, R = R' = alkyl) in which complex solid state polymorphism was described and studied by X-ray and infrared techniques.

Then, also in 1981, Giroud-Godquin and Billard [128] described the copper complex (Figure 63; M = Cu and R = R' = $C_{10}H_{21}$) which they reported to have a undefined discotic mesophase between 46 and 68.5 °C; in common with many discotic materials the clearing enthalpy (35.5 kJ mol^{-1}) was greater than the melting enthalpy (28.8 kJ mol^{-1}). Using X-ray techniques, Levelut [129] suggested that the discotic phase was in fact disordered lamellar crystal and indeed, Martins and Giroud-Godquin

[130] later suggested that the mesophase was lamellar as a result of NMR studies.

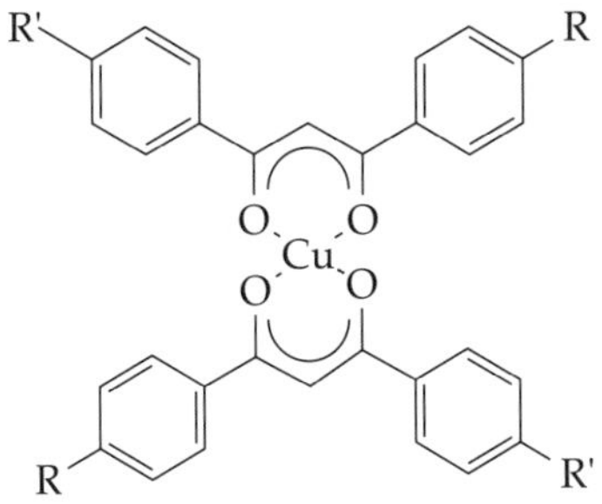

Figure 63   The β-diketonate complexes studied mainly by Giroud-Godquin and Ohta

Ohta later reported [131] the related complexes (Fig, 63; M = Cu, R = R' = C_8H_{17}; M = Cu, R = R' = $C_8H_{17}O$) in which he showed the 'D$_1$' phase of the octyl derivative to be miscible with the 'discotic' phase of Giroud-Godquin's compounds and the formation of a eutectic point at a particular composition, which he took as evidence that the phase was in fact liquid crystalline. That the phases were co-miscible, indeed, indicated that they are of the same type, but the fact that there was a particular eutectic composition does not prove liquid-crystallinity.

In general, the β-diketonates with alkoxyphenyl groups showed wider ranges for the mesophase than those with alkylphenyl groups, and in the case of the alkylphenyl derivatives, when R ≠ R', melting and clearing temperatures were reduced [127, 132].

Further X-ray studies [133] were later carried out on alkoxyphenyl derivatives of the complexes in Figure 63 which further established the lamellar nature of these mesophases and showed that there was no columnar structure (Figure 64).

Other studies by Ohta of such materials, many of which showed complex melting patterns, are collected in reference [134]. The X-ray single crystal structure of the compound shown in Figure 63 (R = R' = $C_8H_{17}O$) was reported by Usha and Vijayan [135].

A most interesting development in this field was made by Chandrasekhar and co-workers when in 1986 they published a copper β-diketonate which, in common with the copper salicylaldimine published by Ovchinnikov [91], was both nematic and paramagnetic [136]

(Figure 65; R = Me, Et, OMe, OEt, OPr – all nematic phases were monotropic).

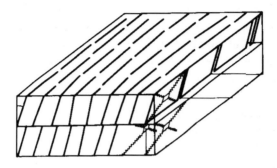

Figure 64   Representation of the lamellar structure of the mesophase of the copper β-diketonates (from ref [133])

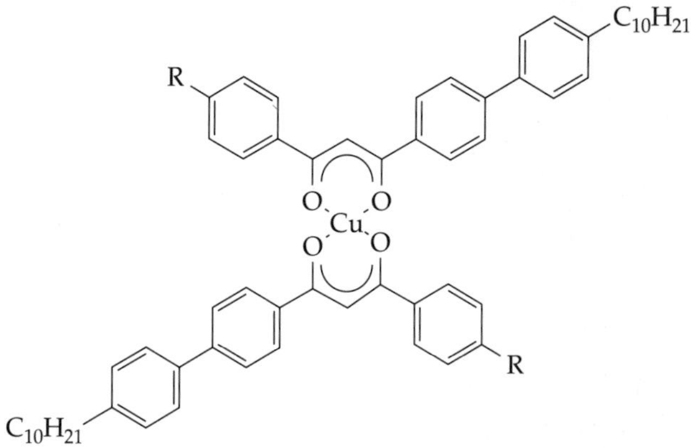

Figure 65   Chandrasekhar's paramagnetic, nematic copper β-diketonates

The calamitic nature of the nematic phase was confirmed in a later paper [137] which reported miscibility studies with the well-known nematogen 4''-pentyl-4-cyanoterphenyl (5CT). In the same paper, it was reported that the materials in the nematic phase were correlated in an anti-parallel fashion, although given the symmetric, non-dipolar nature of the complexes, it is difficult to see precisely what was meant by such a concept.

However, more than a year later, two further papers appeared [138, 139] in which it was claimed that the nematic phase was in fact biaxial and these claims were backed up by conoscopic [140] observations on samples which were held in the correct (homeotropic) alignment by surface coatings and by the application of an external AC field (3 kHz). Further evidence was apparently furnished by the observation that the phase sequence I $\rightarrow$ N$_u$ $\rightarrow$ N$_b$ (N$_b$ is the biaxial nematic phase and N$_u$ the uniaxial nematic phase) was observed in mixtures of the complex (99.6%) and 5CT (0.4%). While the evidence would certainly seem to support the claims being made, concerns have been expressed as to whether the biaxiality observed by conoscopy was induced by the applied field and other experienced workers have been unable to reproduce the observations of biaxiality [141]. The issue is, however, very important as biaxial nematic materials can be aligned about two (and hence all three) axes, meaning that the molecules can be switched about their low inertial axis, a much faster process than the switching in a uniaxial nematic which is about the high inertial axis (Figure 66).

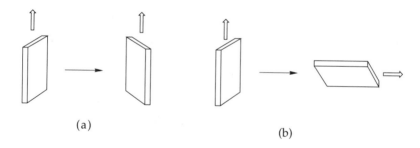

(a)

(b)

Figure 66 Switching about the low inertial axis in a biaxial nematic (a) and about the high inertial axis in a uniaxial nematic (b)

Understandably, these reports by Chandrasekhar produced some interest and three groups in particular pursued the molecular design on which these copper complexes were based. Thus, Mühlberger and Haase [142] synthesised the unsaturated derivatives (Figure 67) which also showed monotropic nematic phases; one derivative was successfully crystallised and showed the *trans* disposition of the ligands about the copper centre and the planar nature of the complex.

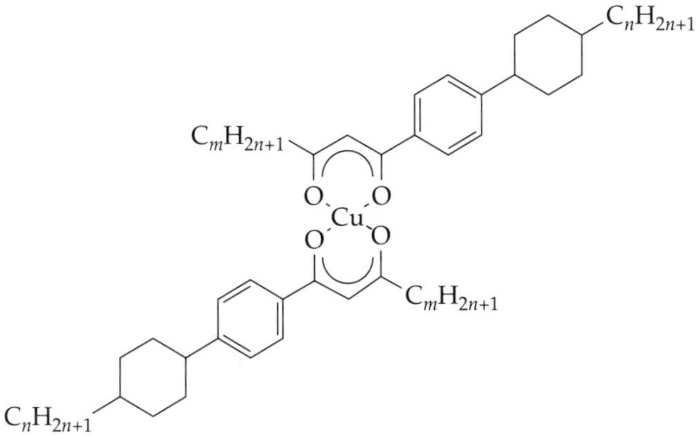

Figure 67   Mühlberger and Haase's copper β-diketonate complexes

Ohta [143] also followed this initiative and synthesised the biphenyl equivalents of Mühlberger and Haase's materials with $m = 1$. The materials in general showed enantiotropic mesophases but curiously showed an ordered discotic phase (D_{ro}) despite being more rod-like in shape than either of the previous examples. Toyne *et al.*, [144] carried out a more systematic investigation of such structures and found that complexes with the structure shown in Figure 68A only ever showed monotropic nematic mesophases while complexes with the structure shown in Figure 68B gave exclusively enantiotropic phases of the calamitic type.

An interesting and important conclusion from their work, and one which is equally applicable to the *ortho*-metallated complexes (Section 8.7.6) and the salicylaldime materials (Section 8.7.5) described above, is that "the terms 'rod-like' and 'disc-like' (discotic) are misleading terms to use to describe the extremes of molecular shape". There is then a large 'grey area' between rods and discs.

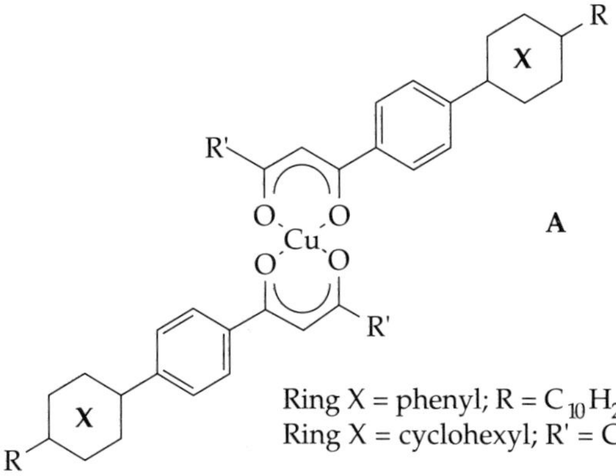

Ring X = phenyl; R = $C_{10}H_{21}$ R' = CH_3, H
Ring X = cyclohexyl; R' = CH_3, C_7H_{15}

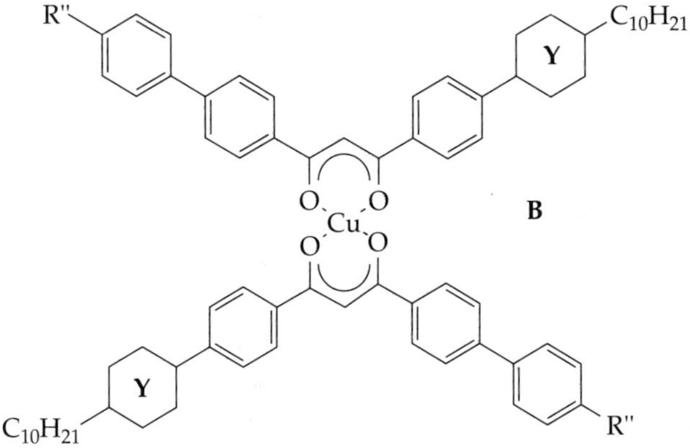

Ring Y = phenyl; R" = $C_{10}H_{21}$, CH_3, C_2H_5, F
Ring Y = cyclohexyl; R" = OCH_3

Figure 68 Toyne's β-diketonates

Another class of β-diketonates with promise are those containing the oxovanadium(IV) core, as such materials are paramagnetic and have a large off-axis dipole. Styring [145] has described a vanadium complex (Figure 69) which was reported to have a D_{ho} mesophase, confirmed by miscibility studies with tetrakis(tetradecanoato)dicopper(II) (Section 8.7.9).

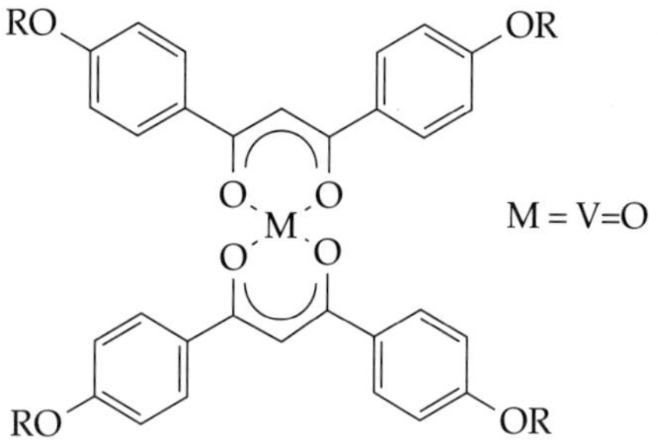

Figure 69 Styring's vanadyl β-diketonate

Given the very wide range of β-diketonates which are now available, this and related systems will obviously be fertile areas for future study.

8.7.8 Complexes of β-Diketonates with Eight Terminal Chains

Given that most of the β-diketonate complexes of copper with four peripheral chains had been shown to give rather ordered mesophases, the ligands were modified by Giroud-Godquin so that each contained four peripheral chains giving a resultant complex with eight chains [79, 146] (Figure 70).

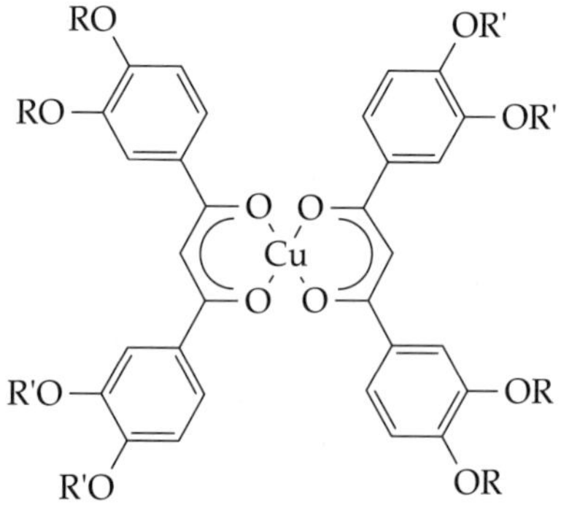

Figure 70 Giroud-Godquin's eight-chained copper β-diketonates

These complexes were shown to have a hexagonal discotic mesophase by its miscibility with the known discogen, triphenylenehexa-decylalkanoate (Figure 71). Although the exact nature of the phase was not unequivocally established, arguments based on enthalpy changes, miscibility and optical texture appeared to suggest that the phase was in fact D_{hd}, that is, a disordered hexagonal discotic [79].

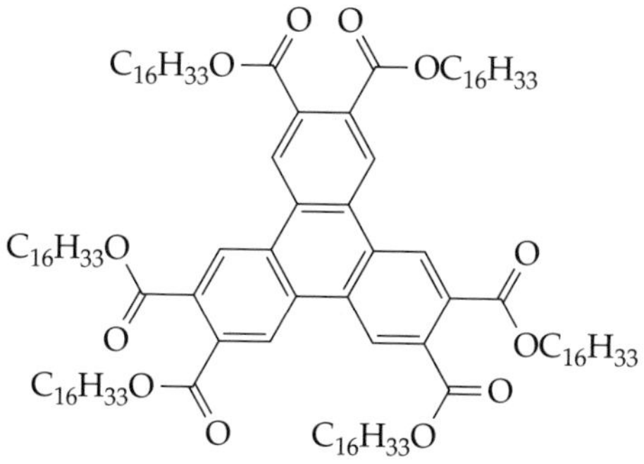

Figure 71   The known discogen, triphenylenehexedecylalkanoate (**x**)

What is particularly interesting about these results is that the binary phase diagram shows complete miscibility and implies nothing 'untoward' about the nature of the D_h mesophase of the copper complexes. This would appear to provide quite strong evidence that intermolecular metal/ligand or metal/metal interactions can play no significant part in the stabilisation of the mesophase of the copper complexes as such interactions would obviously be disrupted in a mixture with the triphenylene derivative.

8.7.9   Dinuclear Complexes of Alkanoic Acids with Four Peripheral Chains

The dinuclear carboxylates of Cu(II) have been known for some time and while it has been appreciated for some time that they are liquid crystalline [147, 148], it was not until 1984 that they were characterised as columnar discotics by X-ray techniques [149]. Thus, while the complexes adopt a lamellar structure in the solid state, the mesophase structure is that of a hexagonal array of columns, and Cu-K edge EXAFS studies showed [150] that the dimeric copper(II) complex forms the core of the columns which are stabilised in the columnar direction by intermolecular Cu–O interactions (Figure 72), giving the columns a polymeric nature.

The phase transition in these materials has been investigated by a variety of techniques and is therefore very well characterised. Dilatometric studies [151] showed that there was a sharp increase in molar volume at the phase transition and that the change was totally due to the molten nature of the alkyl chains.

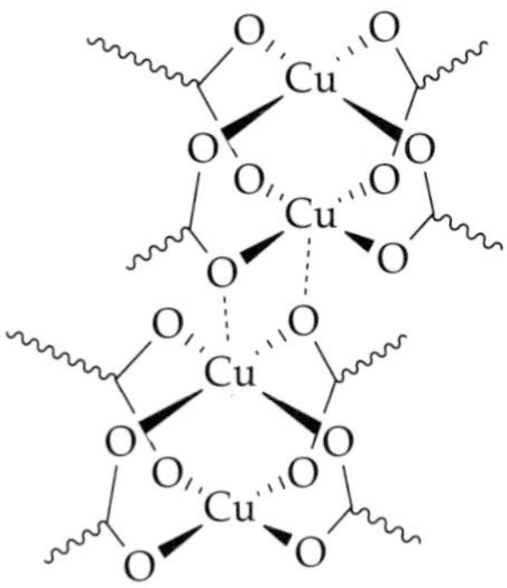

Figure 72   Structure of the copper alkanoates with the column in the mesophase

That the dinuclear arrangement of the core was preserved on passing into the mesophase was confirmed by the EXAFS experiments [150] which further showed that the intermolecular interactions persisted, although the X-ray [149] data (which showed a shorter columnar repeat distance in the mesophase (4.7 Å) than the solid state (5.2 Å)) and magnetic susceptibility [152] measurements (which showed a reduction in the magnetic moment of 0.04 μ_B at the transition to the mesophase) implied that there was some change which was probably attributable to a modification of the relative disposition of the dimers within the columns. The structure of the mesophases of these materials is very similar to that of the related compounds of calcium which were studied some 20 years earlier [11].

Similar studies were then carried out on the related tetrakis(alkanoato)dirhodium(II) complexes and a combination of X-ray and EXAFS techniques showed a similar picture of effectively no change in the core structure or the polymeric nature of the columns on passing from the solid state to the mesophase [153]. Mesomorphic dinuclear tetrakis(alkanoato)dimolybdenum(II) species have also been described [154], (although characterisation did not benefit from X-ray techniques), as have tetrakis(alkanoato)diruthenium(II,II) [155] and tetrakis-(alkanoato)diruthenium(II,III) [156] species. Both species show columnar mesophases and in the (II,II) materials, sharp changes in the magnetic moment at the solid–mesophase transition were ascribed to a core distortion in contrast to the earlier cases of the Cu(II) and Rh(II) systems.

The magnetic study of the Ru(II,II) system also provided valuable information on the electronic singlet ground state and thermally-accessible triplet excited state of these materials.

8.8 MESOMORPHIC COORDINATION COMPLEXES OF POLYDENTATE LIGANDS

8.8.1 Metallophthalocyanines

Much work on metallophthalocyanines has been carried out, notably by Simon and co-workers, and several interesting effects have been observed.

The first report (in 1982) described the synthesis of the octasubstituted phthalocyanines and their copper complexes [157] (Figure 73).

The copper complex (Figure 73; M = Cu, R = $CH_2OC_{12}H_{25}$) was reported to show a mesophase between 53 and 300 °C (where decomposition occurred) and was characterised by X-ray scattering techniques which showed lines with reciprocal spacings of 1: $\sqrt{3}$: $\sqrt{4}$: $\sqrt{7}$ implying a hexagonal arrangement of discotic columns separated by 34 Å with a 3.8 Å inter-ring distance within the column. Although the phase was not classified, the fact that the X-ray line corresponding to 3.8 Å was broad suggests that the phase was disordered and was therefore probably D_{hd}.

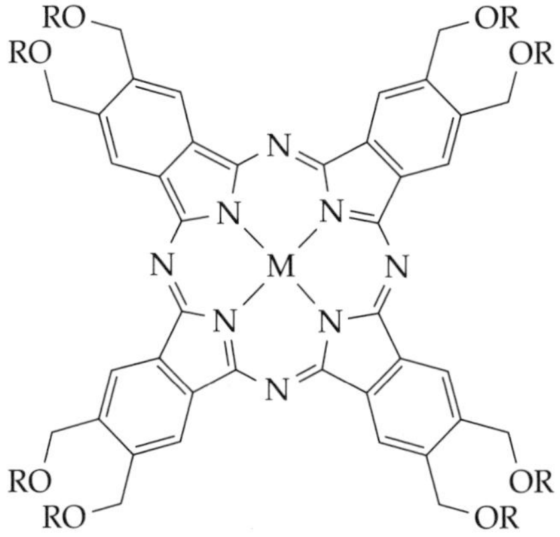

Figure 73 Structure of the octasubstituted metallophthalocyanines

The related metal-free phthalocyanine [158] possessed a slightly narrower mesomorphic range (between 78 and 264 °C) and this time with an inter-ring distance of 4.8 Å; in this case it was suggested that the mesophase was more ordered and hence might be classified as D_{ho}. Other metallated derivatives were synthesised and again, wide mesomorphic ranges were observed (Figure 73; M = Zn 72–>300 °C; M = Mn 44–280 °C) [159]. Of these various phthalocyanine systems [160], some are worthy of further discussion.

One elegant variation on the phthalocyanine structure was to attach crown ether moieties to the periphery of the main ring to give the materials shown in Figure 74 [161].

These most interesting compounds were reported as having a monotropic mesophase at 150 °C (although melting points were not given, nor was it at all clear whether 150 °C represented a transition temperature) and X-ray scattering studies showed the mesophase to be based on a two-dimensional square lattice (lattice parameter 20.8 Å) which stacked to give ion channels described by the crown ether rings, with a ring-ring spacing of 4.2 Å.

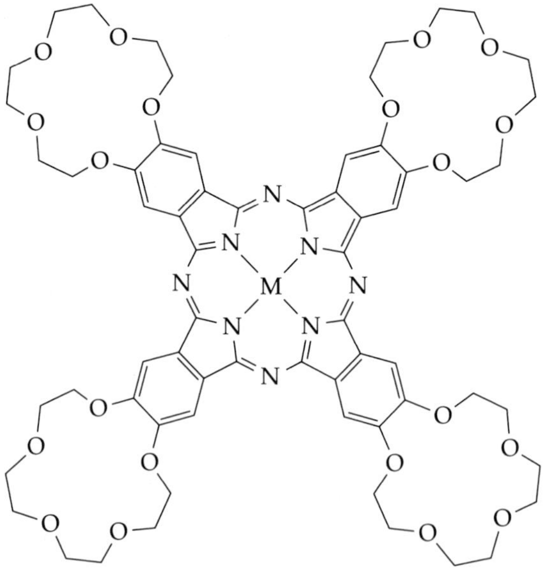

Figure 74 Crown-ether modified phthalocyanine

Another series of materials derived from a dihydroxy-silylphthalocyanine (Figure 73; M = *trans*-Si(OH)$_2$) [162] showed hexagonal mesophases from −7–300 °C and it was found that if held in the mesophase at around 180 °C for several hours, the compounds experienced a polycondensation reaction which eliminated water and led to the formation of polymeric materials with a polysiloxane 'spine' (Figure 75, M = Si; non-mesomorphic derivatives of these compounds are the major subject of Chapter 7). Related polymeric materials based on tin have also been described (Figure 75, M = Sn) [163].

The phase behaviour of the silicon-containing polymers was quite different from that of the monomer and showed a lamellar periodicity of 31 Å from room temperature to 60 °C where clearing occurred. In this lamellar phase, X-ray scattering showed that the rings were separated by 3.4 Å within a column and that the alkyl chains were molten.

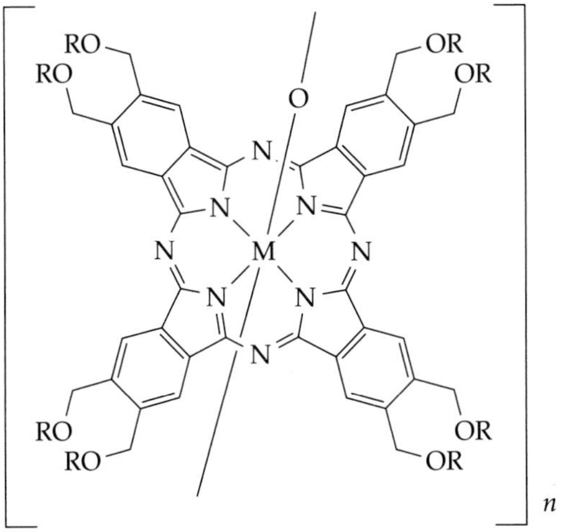

Figure 75 Structure of polymeric phthalocyanines

The cavity size of a phthalocyanine is 1.6 Å and metals with larger diameters are therefore required to sit out of the plane; such is the case with Pb(II) (ionic radius 2.4 Å). Thus, the compound (Figure 73; M = Pb, R = C_8H_{17}) was found to exhibit a discotic mesophase between – 45 and 158 °C and X-ray scattering studies showed that the mesophase was of the columnar hexagonal type with the intra-column repeat distance being 7.4 Å, i.e. twice the thickness of one lead phthalocyanine molecule. This led to the suggestion that the molecules were stacked antiferroelectrically in a tilted stack (Figure 76) [164].

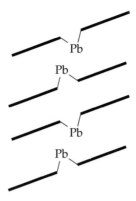

Figure 76   Proposed antiferroelectric stacking in the mesomorphic lead
phthalocyanine

Such a double structure was also found in the case of lutetium
phthalocyanines, except that in this case two phthalocyanines were
associated with each metal ion in a dimeric arrangement in which the
two rings were mutually parallel and rotationally displaced by 45° [165].
These materials demonstrated an anisotropy of conductivity of the order
of 10^7 and a conductivity of $3.9 \times 10^{-5}\,\mathrm{s\,cm^{-1}}$ at 10 GHz [166].

Phthalocyanines with a different substitution pattern (Figure 77) were
described by Cook and co-workers [167] and again, stable discotic
mesophases were obtained whose range was enhanced by complexation to
copper. These and related materials were shown to be far red or near
infrared absorbers [168].

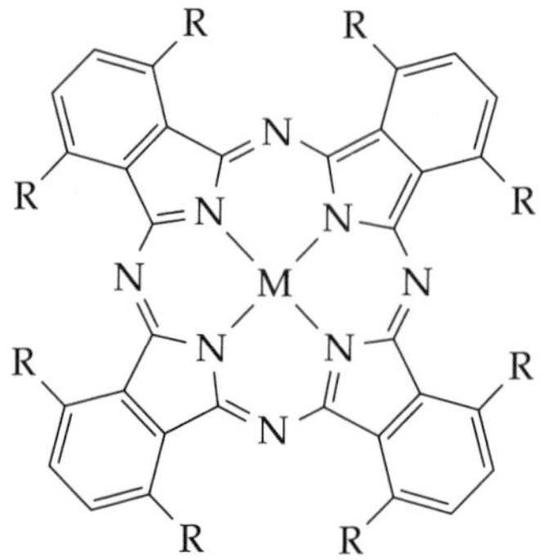

Figure 77   Mesomorphic 1,4,8,11,15,18,22,25-octasubstituted phthalocyanines

8.8.2   Metalloporphyrins

While there have been relatively extensive studies of mesomorphic phthalocyanines, there has been very much less work carried out on the related porphyrin systems, quite probably due to the much greater synthetic effort which has to be expended. The first report of a mesogenic porphyrin was by Goodby [169] who synthesised substituted uroporphyrins as their bis(hydrochlorides) and found a monotropic discotic phase, stable over 0.1 °C (no metals here!). Later, some 5,15-disubstituted copper(II) octamethylporphyrin derivatives were reported [170] (Figure 78; R' = Me, M = Cu) and while none of the pure materials was mesomorphic, a monotropic schlieren texture was reported for mixtures of certain porphyrins with long chain alkyl chlorides; the 'phase' was never identified.

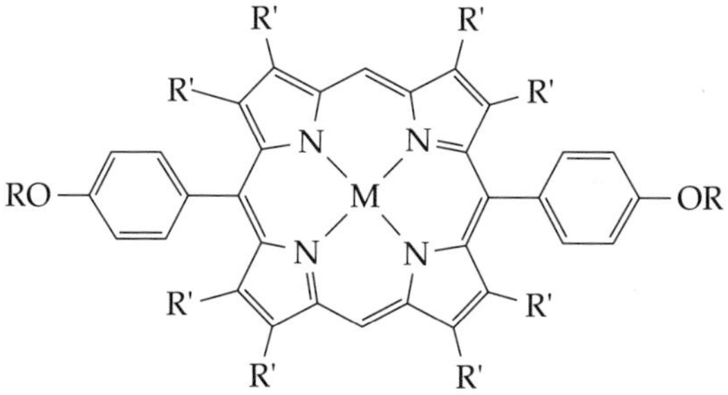

Figure 78   5,15-Disubstituted metalloporphyrins

Related systems (Figure 78; R' = H, M = Zn) were reported from our laboratories [171] and we were able to show in these materials the first examples of porphyrins with a calamitic mesophase, namely the crystal SmB.

The most systematic work in this area, however, was carried out by Gregg, Fox and Bard who described the synthesis and physical properties of some 2,3,7,8,12,13,17,18-octaethanol-substituted [172] (Figure 79) and octaacetic-acid-substituted porphyrins [173] containing a variety of metals. While the free-base octaethanol porphyrins showed very narrow-range discotic phases (maximum of a few °C), the metallated materials (M = Zn, Cd, Cu, Pd) showed mesophase ranges of the order of 50 °C.

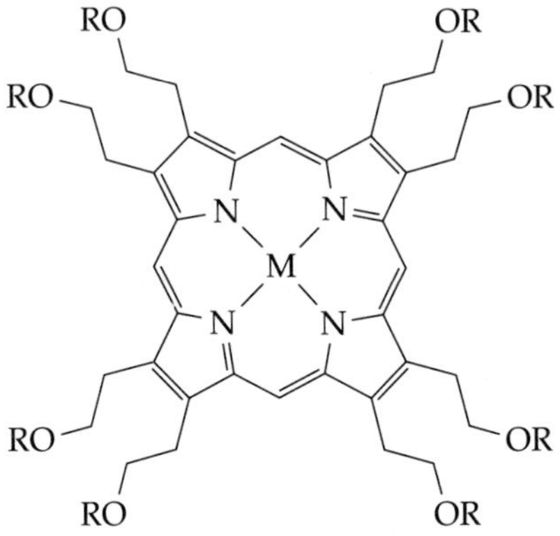

Figure 79   Mesogenic octasubstituted porphyrins

8.9   MISCELLANEOUS ORGANOMETALLICS

The largest group of materials in this classification is that containing a ferrocenyl group in the structure. The first such example was reported in 1976 by Malthête and Billard [35] and showed smectic mesophases (Figure 80).

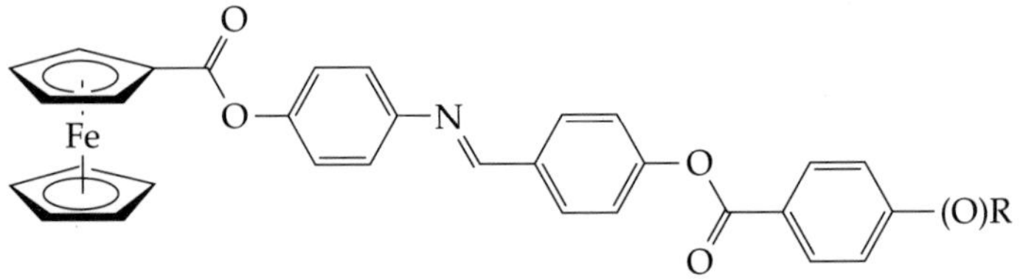

Figure 80   Smectogenic ferrocene complex

Other systems examined have been derived from 1,1'-disubstituted ferrocenes (Figure 81). The 1,1'-bis(alkoxybiphenyl) derivatives (Figure 81; X = Y = p-C$_6$H$_4$–OR) showed monotropic smectic phases which were either SmA or SmC, and single crystal X-ray diffraction studies (at 160 and 295 K) showed that the molecules existed in the extended 'S'-conformation (as opposed to the 'U'-shape which would result if the two alkoxybiphenyl groups were both on the same side of the molecule) and preliminary X-ray scattering results in the mesophase showed a lamellar spacing of around 47 Å which supports the existence of the 'S'-conformation in the mesophase [174].

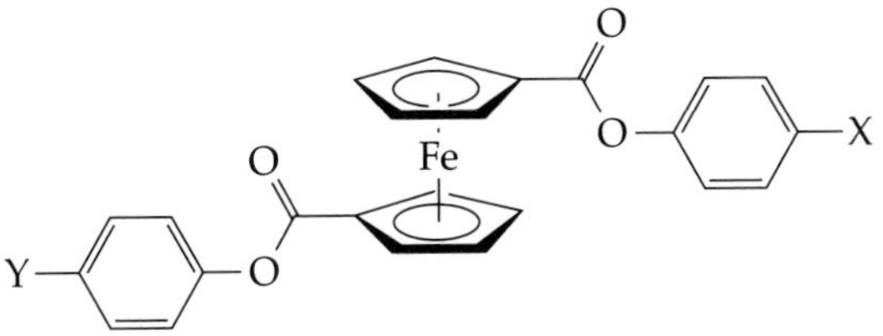

Figure 81   Core structure of 1,1'-disubstituted ferrocenes

In another study, symmetric (Figure 81; X = Y = OR) and unsymmetric (Figure 81; X = OR, Y = OH) 1,1'-disubstituted materials were examined, but in the pure materials, mesophases were either unidentified or irreproducibly obtained [175]. The only unequivocal identification of a mesophase appeared to be a wide-range nematic obtained in a 4:1 mixture of two complexes (1 part X = Y = $OC_{10}H_{21}$ and 4 parts X = $OC_{10}H_{21}$, Y = OH).

A most interesting group of complexes were the mesogenic butadiene iron tricarbonyl materials reported by Ziminski and Malthête [176] (Figure 82).

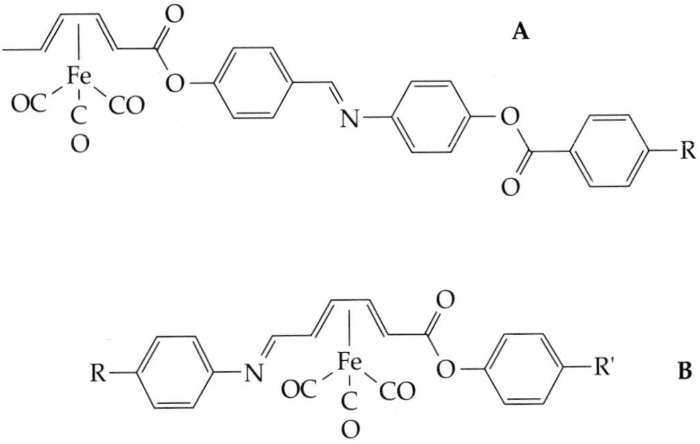

Figure 82   Mesogenic iron-tricarbonyl derivatives

All of the terminally-substituted complexes (Figure 82A) showed wide-range nematic phases (range > 100 °C), while the disubstituted complexes (Figure 82B; R = R') showed nematic phases at short-to-medium chain lengths (R = R' = OC_6H_{13} to $OC_{10}H_{21}$) and SmA phases as the chains grew longer (R = R' = OC_9H_{19} to $OC_{12}H_{24}$).

Finally in this section, it is worth noting the very first organometallic liquid crystals reported in 1923 by Vorländer and showing smectic phases (Figure 83) [33].

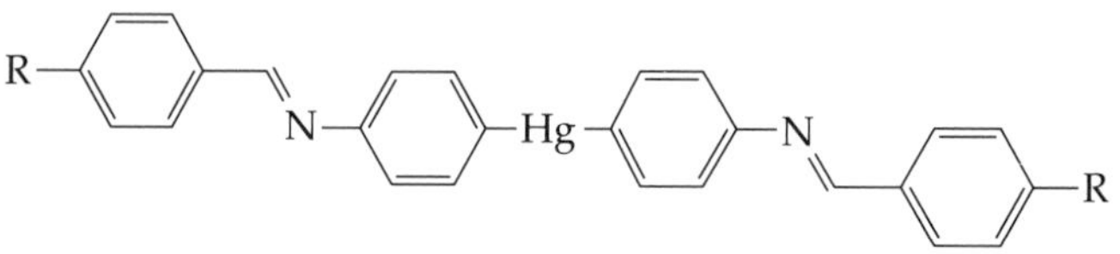

Figure 83   Vorländer's diarylmercury mesogens

8.10 LYOTROPIC LIQUID CRYSTALS FROM AMPHIPHILIC COMPLEXES

These materials will be discussed together as they are relatively few; simple salts of metals (e.g. metal carboxylates) are excluded.

While some groups had been aware of the possibilities of creating surfactant metal complexes [177], few had (or indeed have) taken the next step and investigated their behaviour in concentrated solution, searching for lyotropic behaviour. The first reports in this area were Gaspard's peripherally-carboxylated copper(II) phthalocyanines [178] whose phase behaviour was never properly defined (Figure 84, X, Z = H; Y = CO_2^-, M = Cu). Related complexes with M = Co, Ni, Pd and Pt were also synthesised but were reported not to show lyotropic organisation [179]. Later, Usoltceva [180] synthesised several peripherally-carboxylated phthalocyanines and successfully characterised their phase behaviour as chromonic (Section 8.3). Thus, all of the complexes shown in Figure 84 for which M = 2H, Cu, Zn or Co(II) and X = H, Y = COOH and Z = H or COOH were shown to possess chromonic nematic and hexagonal phases in aqueous ammonia. However, lyotropic mesomorphism was suppressed when the phthalocyanine ring was substituted according to X = COOH and Y = Z = H or when the central metal ion was Al(III). In this latter case, suppression of the mesomorphism was due to the formation of μ-oxo dimers ([PcAl–O–AlPc]).

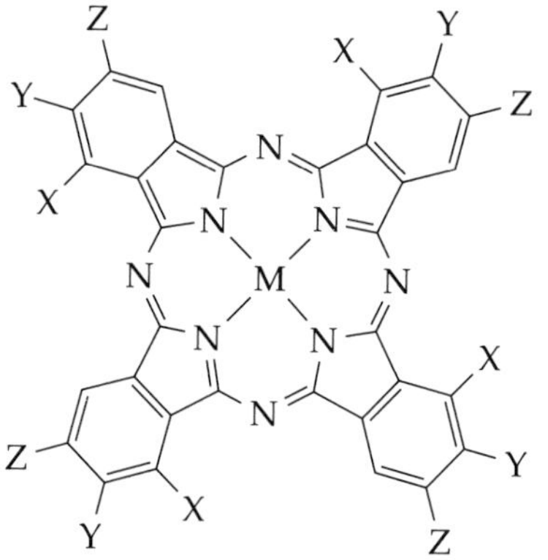

Figure 84   Anionic phthalocyanine liquid crystals

Following the initial work of Le Moigne and Simon [181] in defining the *annelides*, they went on to report a cobalt(III) annelide which showed a room-temperature lamellar phase (Figure 85).

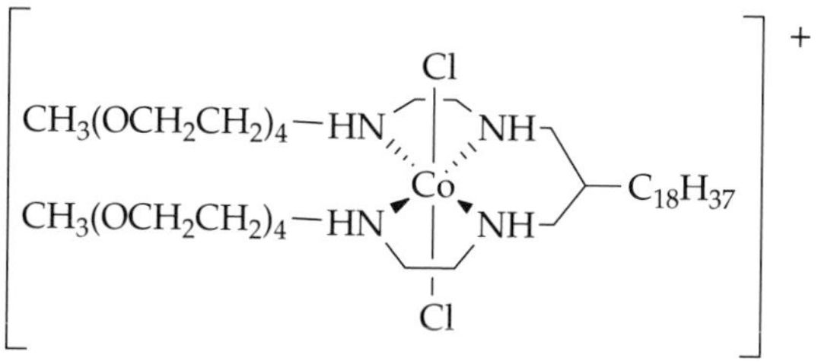

Figure 85   Co(III) annelide complex

In an effort to mimic more closely the classical structure of an amphiphile, we synthesised some iron(II) amine complexes (Figure 86) which we found to show hexagonal mesophases in water [182].

While these materials turned out to be very hydrolytically unstable, they successfully demonstrated the point that a metal complex fragment could effectively replace a more 'conventional' head group (e.g. NMe_3^+) and still generate lyomesomorphic materials.

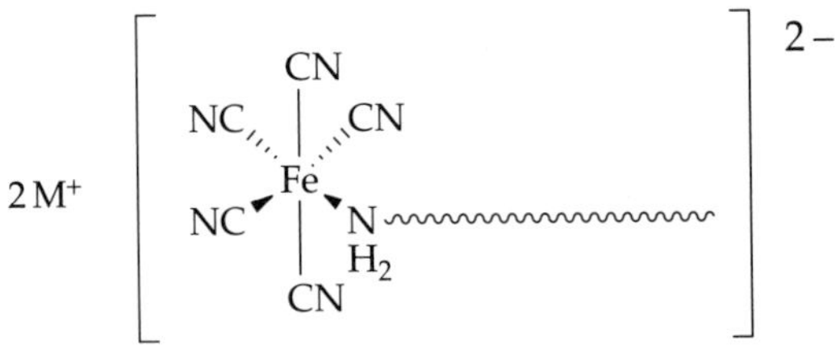

Figure 86   Amphiphilic iron(II) complexes

The way forward in this area was suggested both by the annelide work above and also by the report [183] of some surfactant Co(III) complexes (Figure 87) which were found to micellise at low concentrations (3.3×10^{-5} mol dm^{-3}) in water; no liquid crystal properties have yet been reported.

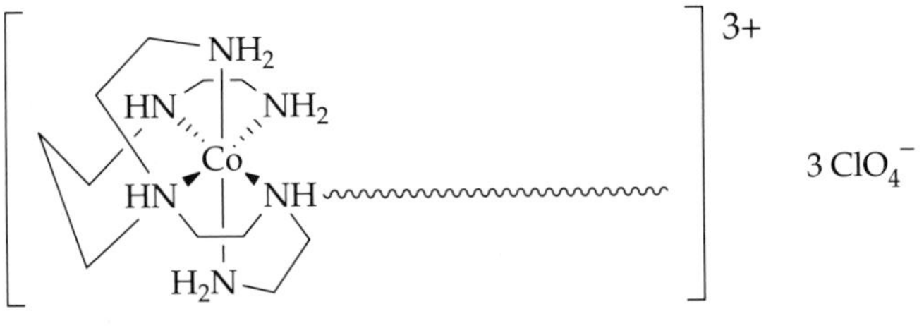

Figure 87   Surfactant Co(III) complex

Thus, the secret lay in choosing systems which were entropically-stabilised by the inclusion of chelating ligands in addition to selecting metals in oxidation states which were known to be inert. We have since shown this to be a successful approach and have re-investigated a series (Figure 88) of surfactant *tris*(bipyridyl)ruthenium(II) complexes [184] which we have shown to be mesomorphic in water [185].

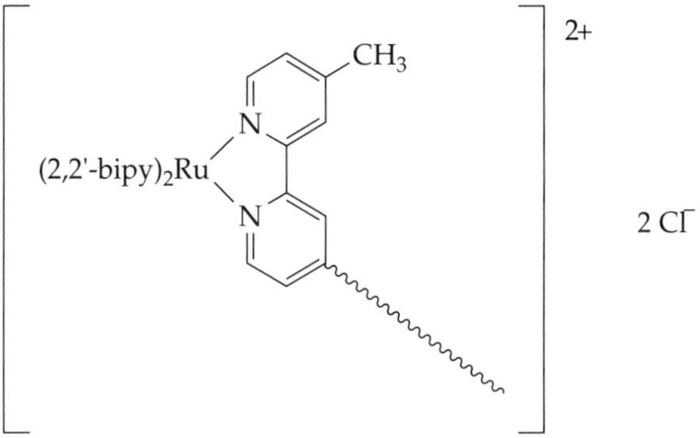

Figure 89   Lyotropic Ru(II) complexes

8.11 WHERE LIES THE FUTURE?

I feel that the area of metal-containing liquid crystals has now properly established itself and the results so far reported show something of the potential of the field. However, there is still much to do and it will probably be some time before such materials find widespread application. There are still very many metals which have not yet been used in liquid crystal systems and this is largely because the problems of generating mesomorphic materials in higher coordination number complexes have not yet been properly resolved: this is only a matter of time.

The liquid-crystalline nature of many biological systems suggests a potential rôle for metal-containing liquid crystals and there is a whole untapped area where the order of liquid crystal mesophases and the catalytic properties of metal complexes are constructively combined. The prizes are there for those who would seek them – the next decade promises to be very exciting!

8.12 ACKNOWLEDGEMENTS

I should like to thank David Dunmur (Sheffield) for valuable discussions throughout the preparation of this Chapter, Jon Rourke (Sheffield) for reading the manuscript, Julian Marsden (Sheffield) for initial help in assembling references, and finally my research group for their forebearance at various stages during the writing.

8.13 REFERENCES

1. R Virchow, *Virchows Arch*, **6**, 571 (1854); see also V Vill, *Cond Matt News*, **1**, 25 (1992) for an appreciation of some of this early work.
2. F Reinitzer, *Monatsch Chem*, **9**, 421 (1888).
3. G W Gray, K J Harrison and J A Nash, *Electron Lett*, **9**, 130 (1973).
4. J W Goodby, *Science*, **231**, 350 (1986).
5. T V Jones, *Liq Cryst Today*, **2**, 4 (1992).
6. D G McDonnell, in: *Thermotropic Liquid Crystals*, G W Gray, Ed, Wiley, Chichester (1987).
7. This classification follows that proposed in A J Leadbetter, in: *Thermotropic Liquid Crystals*, G W Gray, Ed, Wiley, Chichester (1987).
8. G W Gray and J W Goodby, *Smectic Liquid Crystals; Textures and Structures*, Leonard Hill, Glasgow (1984); D Demus and L Richter, *Textures of Smectic Liquid Crystals*, Verlag Chemie, Leipzig (1978).
9. There are however a small number of examples where the SmF phase appears out of sequence, below the SmB phase.
10. S Chandrasekhar, B K Sadashiva and K A Suresh, *Pramana*, **9**, 471 (1977).
11. See e.g. P Spegt and A Skoulios, *C R Hebd Séan Acad Sci*, **251**, 2199 (1960).
12. K Praefcke, D Singer, B Kohne, M Ebert, A Liebmann and J H Wendorff, *Liq Cryst*, **10**, 147 (1991).
13. H Bengs, O Karthaus, H Ringsdorf, C Baehr, M Ebert and J H Wendorff, *Liq Cryst*, **10**, 161 (1991).
14. K J Toyne, in: *Thermotropic Liquid Crystals*, G W Gray, Ed, Wiley, Chichester (1987).
15. D Demus, H Demus and H Zaschke, *Flüssige Kristalle in Tabellen*, Deutscher Verlag für Grundstoffindustrie, Vol I (1974) and Vol II (1984).
16. A Saupe and W Maier, *Z Naturforsch*, **16A**, 816 (1961).
17. D Demus, *Liq Cryst*, **5**, 75 (1989).
18. A C Griffin and T R Britt, *J Am Chem Soc*, **103**, 4957 (1981).
19. J W Emsley, G R Luckhurst, G N Shilstone and I Sage, *Mol Cryst, Liq Cryst*, **102**, 223 (1984).
20. R Eidenschink, F-H Kreuzer and W H de Jeu, *Liq Cryst*, **8**, 879 (1990).
21. J-M Lehn, J Malthête and A-M Levelut, *J Chem Soc, Chem Commun*, 1794 (1985).
22. I Sage, in: *Thermotropic Liquid Crystals*, G W Gray, Ed, Wiley, Chichester (1987).

23. G J T Tiddy, *Modern Trends of Colloid Science in Chemistry and Biology*, 148 (1985), Birkhauser Verlag, Basel; G J T Tiddy, *Phys Rep*, **57**, 1 (1980).

24. C Tanford, *The Hydrophobic Effect*, 2nd Edn, Wiley, New York (1980).

25. J M Seddon, *Biochem Biophys Acta*, **1031**, 1 (1990).

26. E.g. R Valenti and M L Sartirania, *Nuovo Chim*, **3D**, 104 (1984).

27. T K Attwood, J E Lydon, C Hall and G J T Tiddy, *Liq Cryst*, **7**, 657 (1990).

28. N H Hartshorne, *The Microscopy of Liquid Crystals*, The Microscope Series, **48**, Microscope Publications, London (1974).

29. A De Vries, *Mol Cryst, Liq Cryst*, **131**, 125 (1985).

30. A-M Giroud-Godquin and P M Maitlis, *Angew Chem Int Ed Engl*, **30**, 402 (1991).

31. D Vorländer, *Ber Dtsch Chem Ges*, **43**, 3120 (1910).

32. A Skoulios, *Ann Phys (Paris)*, **3**, 421 (1978).

33. D Vorländer, *Z Phys Chem Stoechiom Verwandschaftsl*, **105**, 211 (1923).

34. W R Young, I Haller and D C Green, *Mol Cryst, Liq Cryst*, **1§3**, 305 (1971).

35. J Malthête and J Billard, *Mol Cryst, Liq Cryst*, **34**, 117 (1976).

36. A-M Giroud and U T Mueller-Westerhoff, *Mol Cryst, Liq Cryst*, **41**, 11 (1977).

37. D W Bruce, E Lalinde, P Styring, D A Dunmur and P M Maitlis, *J Chem Soc, Chem Commun*, 581 (1986).

38. H Adams, N A Bailey, D W Bruce, D A Dunmur, E Lalinde, M Marcos, C Ridgway, A J Smith, P Styring and P M Maitlis, *Liq Cryst*, **2**, 381 (1987).

39. T Scheffer and J Nehring, in: *Liquid Crystals, Applications and Uses*, Vol I, B Bahadur, Ed, World Scientific, Singapore, (1990).

40. H Adams, N A Bailey, D W Bruce, R Dhillon, D A Dunmur, S E Hunt, E Lalinde, A A Maggs, R Orr, P Styring, M S Wragg and P M Maitlis, *Polyhedron*, **7**, 1861 (1988).

41. P Styring, *PhD Thesis*, University of Sheffield (1988).

42. D W Bruce, A A Maggs and M S Wragg, *unpublished work*.

43. D W Bruce and L A Green, *unpublished results*.

44. D W Bruce, D A Dunmur, E Lalinde, P M Maitlis and P Styring, *Liq Cryst*, **3**, 385 (1988).

45. J P Rourke, F P Fanizzi, N J S Salt, D W Bruce, D A Dunmur and P M Maitlis, *J Chem Soc, Chem Commun*, 229 (1990); N J S Salt, *PhD Thesis*, University of Sheffield (1990).

46. J P Rourke, F P Fanizzi, D W Bruce, D A Dunmur and P M Maitlis, *J Chem Soc, Dalton Trans*, in press.

47. D W Bruce, D A Dunmur, M A Esteruelas, S E Hunt, R Le Lagadec, P M Maitlis, J R Marsden, E Sola and J M Stacey, *J Mater Chem*, **1**, 251 (1991).
48. M A Esteruelas, E Sola, L A Oro, M B Ros and J-L Serrano, *J Chem Soc, Chem Commun*, 55 (1989).
49. M A Esteruelas, E Sola, L A Oro, M B Ros, M Marcos and J-L Serrano, *J Organomet Chem*, **387**, 103 (1990).
50. D Y Jeter and E B Fleischer, *J Coord Chem*, **4**, 107 (1974).
51. H Adams, N A Bailey, D W Bruce, S C Davis, D A Dunmur, P D Hempstead, S A Hudson and S J Thorpe, *J Mater Chem*, **2**, 395 (1992).
52. D W Bruce, D A Dunmur, S A Hudson, E Lalinde, P M Maitlis, M P McDonald, R Orr, P Styring, A S Cherodian, R M Richardson, J L Feijoo and G Ungar, *Mol Cryst, Liq Cryst*, **206**, 79 (1991).
53. M Marcos, M B Ros, J L Serrano, M A Esteruelas, E Sola, L A Oro and J Barberá, *Chem Mater*, **2**, 748 (1990).
54. D W Bruce, D A Dunmur, S A Hudson, P M Maitlis and P Styring, *Adv Mater Opt Electron*, **1**, 37 (1992).
55. D W Bruce, S C Davis, D A Dunmur, S A Hudson, P M Maitlis and P Styring, *Mol Cryst, Liq Cryst*, **215**, 1 (1992).
56. D W Bruce, D A Dunmur, P M Maitlis, P Styring and (in part) M A Esteruelas, L A Oro, M B Ros, J L Serrano and E Sola, *Chem Mater*, **1**, 479 (1991) and *idem, ibid*, **3**, 378 (1991).
57. G Etherington, A J Leadbetter, X J Wang, G W Gray and T Tajbakhsh, *Liq Cryst*, **1**, 209, (1986); G Etherington, A J Langley, A J Leadbetter and X J Wang, *Liq Cryst*, **3**, 155, (1988); D Demus, A Gloza, H Hartung, A Hawser, I Rapthel and A Wiegeleben, *Krist und Techn*, **16**, 1445, (1981); S Yano, Y Mori and S Kutsumizu, *Liq Cryst*, **9**, 907 (1991).
58. S A Hudson, *PhD Thesis*, University of Sheffield (1991).
59. J Billard, *C R Acad Sci Paris*, **305**, 843 (1987).
60. S Takahashi, M Kariya, T Yakate, K Sonogashira and N Hagihara, *Macromolecules*, **11**, 1063 (1978).
61. S Takahashi, E Murata, M Kariya, K Sonogashira and N Hagihara, *Macromolecules*, **12**, 1016 (1979).
62. S Takahashi, Y Takai, H Morimoto, K Sonogashira and N Hagihara, *Mol Cryst, Liq Cryst*, **82**, 139 (1982).
63. S Takahashi, Y Takai, H Morimoto and K Sonogashira, *J Chem Soc, Chem Commun*, 3 (1984).
64. T Kaharu, H Matsubara and S Takahashi, *J Mater Chem*, **1**, 145 (1991).
65. C Bertram, D W Bruce, D A Dunmur, S E Hunt, P M Maitlis and M McCann, *J Chem Soc, Chem Commun*, 69 (1991).
66. S E Hunt, *PhD Thesis*, University of Sheffield (1991).

67. D W Bruce, D A Dunmur, P M Maitlis, M R Manterfield and R Orr, *J Mater Chem*, **1**, 255 (1991).
68. A-M Giroud, *Ann Phys*, **3** 147 (1978).
69. U T Mueller-Westerhoff, A Nazzal, R J Cox and A-M Giroud, *Mol Cryst, Liq Cryst*, **56**, 249 (1980).
70. A-M Giroud, A Nazzal and U T Mueller-Westerhoff, *Mol Cryst, Liq Cryst*, **56**, 225 (1980).
71. M Cortrait, J Gaultier, C Polycarpe, A-M Giroud and U T Mueller-Westerhoff, *Acta Crystallogr Sect C39*, 833 (1983).
72. K W Browall and L V Interrante, *J Coord Chem*, **3**, 27 (1973).
73. M Veber, R Fugnitto and H Strzelecka, *Mol Cryst, Liq Cryst*, **96**, 221 (1983).
74. K Ohta, A Tkagi, H Muroki, I Yamamoto, K Matsuzaki, T Inabe and Y Maruyama, *J Chem Soc, Chem Commun*, 883, (1986).
75. K Ohta, A Takagi, H Muroki, I Yamamoto, K Matsuzaki, T Inabe and Y Maruyama, *Mol Cryst, Liq Cryst*, **147**, 15 (1987).
76. M Veber, P Davidson, C Jallabert, A M Levelut and H Strzelecka, *Mol Cryst, Liq Cryst Lett*, **5**, 1 (1987).
77. K Ohta, H Hasebe, H Ema, T Fujimoto and I Yamamoto, *J Chem Soc, Chem Commun*, 1610 (1989).
78. A-M Giroud-Godquin, G Sigaud, M F Archard and H Hardouin, *J Physique Lett*, **45**, L-387 (1984).
79. H Adams, A C Albeniz, N A Bailey, D W Bruce, A S Cherodian, R Dhillon, D A Dunmur, P Espinet, J L Feijoo, E Lalinde, P M Maitlis, R M Richardson and G Ungar, *J Mater Chem*, **1**, 843 (1991).
80. K Ohta, H Ema, Y Morizumi, T Watanabe, T Fujimoto and I Yamamoto, *Liq Cryst*, **8**, 311 (1990).
81. D W Bruce, R Dhillon, D Guillon, M Ibn-Elhaj and P Maldivi, *in preparation*.
82. J P Fackler Jr, J A FEtc..hin and D C Fries, *J Am Chem Soc*, **94**, 7323 (1972); J P Fackler Jr, D Coucouvanis, J A FEtc..hin and W C Seidel, *J Am Chem Soc*, **90**, 2784 (1968).
83. M Bonamico, G Mazzone, A Vaciago and L Zambonelli, *Acta Cryst*, **19**, 898 (1965); H Iwasaki, *Acta Cryst*, **29B**, 2115 (1973); H Iwasaki, M Ito and K Kobayashi, *Chem Lett*, 1399 (1978); A Domenicano, L Torelli, A Vaciago and L Zambonelli, *J Chem Soc A*, 1351 (1968).
84. C Bellitto, A Flamini, O Piovesana and P F Zanazzi, *Inorg Chem*, **19**, 3632 (1980); C Bellitto, G Dessy, V Fares and A Flamini, *J Chem Soc, Chem Commun*, 409 (1981); C Bellitto, M Bonamico, G Dessy, V Fares and A Flamini, *J Chem Soc, Dalton Trans*, 35 (1987).
85. K Ohta, H Ema, I Yamamoto and K Matsuzaki, *Liq Cryst*, **3**, 1671 (1988).
86. P E Cladis, *Mol Cryst, Liq Cryst*, **165**, 85 (1988).

87. U T Mueller-Westerhoff, A Nazzal, R J Cox and A M Giroud, *J Chem Soc, Chem Commun*, 497, (1980).
88. K L Marshall and S D Jacobs, *Mol Cryst, Liq Cryst*, **159**, 181 (1988).
89. D W Bruce, D A Dunmur, S E Hunt, P M Maitlis and R Orr, *J Mater Chem*, **1**, 857 (1991).
90. I V Ovchinnikov, Yu G Galyametdinov, G I Ivanova and L M Yagfarova, *Dokl Akad Nauk SSSR*, **276**, 126 (1984); Yu G Galyametdinov, I V Ovchinnikov, B M Bolotin, N B Etingen, G I Ivanova and L M Yagarova, *Izv Akad Nauk SSSR, Ser Khim*, 2379 (1984); I V Ovchinnikov, Yu G Galyametdinov and I G Bikchantaev, *Izv Akad Nauk SSSR Ser Fiz*, **53**, 1870 (1989); R M Galimov, I G Bikchantaev and I V Ovchinnikov, *Zh Strukt Khim*, **30**, 65 (1989).
91. Yu G Galyametdinov, D Z Zakieva and I V Ovchinnikov, *Izv Akad Nauk SSSR Ser Khim*, 491 (1986).
92. Yu G Galyametdinov, G I Ivanova and I V Ovchinnikov, *Zh Obshch Khim*, **54**, 2796 (1984).
93. Yu G Galyametdinov, I G Bikchantaev and I V Ovchinnikov, *Zh Obshch Khim*, **58**, 1326 (1988).
94. Yu G Galyametdinov, G I Ivanova and I V Ovchinnikov, *Izv Akad Nauk SSSR Ser Khim*, 1931 (1989).
95. M Marcos, P Romero, J L Serrano, C Bueno, J A Cabeza and L A Oro, *Mol Cryst, Liq Cryst*, **167**, 123 (1989).
96. M Marcos, P Romero and J L Serrano, *Chem Mater*, **2**, 495 (1990).
97. N Hoshino, H Murakami, Y Matsunaga, T Inabe and Y Maruyama, *Inorg Chem*, **29**, 1177 (1990).
98. M Marcos, P Romero, J L Serrano, J Barberá and A M Levelut, *Liq Cryst*, **7**, 251 (1990).
99. C Carfagna, U Caruso, A Roviallo and A Sirigu, *Makromol Chem Rapid Commun*, **8**, 345 (1987).
100. N Hoshino, R Hayakawa, T Shibuya and Y Matsunaga, *Inorg Chem*, **29**, 5129 (1990).
101. N Hoshino, A Kodama, T Shibuya, Y Matsunaga and S Miyajima, *Inorg Chem*, **30**, 3091 (1991).
102. U Caruso, A Roviello and A Sirigu, *Liq Cryst*, **7**, 421 (1990); **7**, 431 (1990).
103. J L Serrano, P Romero, M Marcos and P J Alonso, *J Chem Soc, Chem Commun*, 859 (1990).
104. M Marcos, P Romero and J-L Serrano, *J Chem Soc, Chem Commun*, 1641 (1989).
105. M Ghedini, S Armentano, R Bartolino, N Kirov, M Petrov and S Nenova, *J Mol Liq*, **38**, 207 (1988); G Torquati, O Francescangeli, M Ghedini, S Armentano, F P Nicoletta and R Bartolino, *Il Nuovo Cimento*, **12**, 1363 (1990).

106. A M Levelut, M Ghedini, R Bartolino, F P Nicoletta and F Rustichelli, *J Phys France*, **50**, 113 (1989).
107. G Albertini, A Guido, G Mancini, S Stizza, M Ghedini and R Bartolino, *Europhys Lett*, **12**, 629 (1990).
108. U Caruso, A Roviello and A Sirigu, *Liq Cryst*, **3**, 1515, (1988).
109. A Roviello, A Sirigu, P Ianelli and I Immirzi, *Liq Cryst*, **3**, 115, (1988).
110. J P Bayle, E Bui, F Perez and J Cortieu, *Bull Soc Chim Fr*, 532 (1989).
111. E Bui, J P Bayle, F Perez, L Liebert and J Cortieu, *Liq Cryst*, **4**, 513 (1990).
112. R Paschke, H Zaschke, A Mädicke, J R Chipperfield, A B Blake, P G Nelson and G W Gray, *Mol Cryst, Liq Cryst Lett*, **6**, 81 (1988).
113. T D Shaffer and K A Sheth, *Mol Cryst,Liq Cryst*, **172**, 27 (1989).
114. M Ghedini, M Longeri and R Bartolino, *Mol Cryst, Liq Cryst*, **84**, 207 (1982).
115. M Ghedini, S Licoccia, S Armentano and R Bartolino, *Mol Cryst, Liq Cryst*, **108**, 269 (1984).
116. M Ghedini, S Armentano and F Neve, *Inorg Chim Acta*, **134**, 23 (1987).
117. M Ghedini, S Licoccia, S Armentano and R Bartolino, *Mol Cryst, Liq Cryst*, **108**, 269 (1984).
118. C C Versace, R Bartolino, M Ghedini, F Neve, S Armentano, M Petrov and N Kirov, *Liq Cryst*, **8**, 481 (1990).
119. N Hoshino, H Hasegawa and Y Matsunaga, *Liq Cryst*, **9**, 267 (1991).
120. J Barberá, P Espinet, E Lalinde, M Marcos and J L Serrano, *Liq Cryst*, **2**, 833 (1987).
121. P Espinet, E Lalinde, M Marcos, J Pérez and J L Serrano, *Organometallics*, **9**, 555 (1990).
122. R C Elder, R D P Cruea and R F Morrison, *Inorg Chem*, **15**, 1623 (1976).
123. P Espinet, J Etxebarria, M Marcos, J Pérez, A Rémon and J L Serrano, *Angew Chem, Int Ed Engl*, **28**, 1065 (1989).
124. M Ghedini and D Pucci, *J Organomet Chem*, **395**, 105 (1990).
125. M Ghedini, D Pucci, G De Munno, D Viterbo, F Neve and S Armentano, *Chem Mater*, **3**, 65 (1991).
126. B J Bulkin, R K Rose and A Santoro, *Mol Cryst, Liq Cryst*, **43**, 53 (1977).
127. K Ohta, M Yokoyama, S Kusabayashi and H Mikawa, *J Chem Soc, Chem Commun*, 392 (1980); K Ohta, G-J Jiang, M Yokoyama, S Kusabayashi and H Mikawa, *Mol Cryst, Liq Cryst*, **66**, 283 (1981); K Ohta, M Yokoyama, S Kusabayashi and H Mikawa, *Mol Cryst, Liq Cryst*, **69**, 131 (1981); K Ohta, M Yokoyama and H Mikawa, *Mol Cryst, Liq Cryst*, **73**, 205 (1981).

128. A-M Giroud-Godquin and J Billard, *Mol Cryst, Liq Cryst*, **66**, 147 (1981).

129. A-M Levelut, *J Chim Physique*, **80**, 149 (1983).

130. A C Ribeiro, A F Martins and A-M Giroud-Godquin, *Proc 4th Portuguese Phys Conf*, Ed, Port Phys Soc, Lisbon, 1984; A C Ribeiro, A F Martins and A-M Giroud-Godquin, *Mol Cryst, Liq Cryst Lett*, **5**, 133 (1988).

131. K Ohta, A Ishii, I Yamamoto and K Matsuzaki, *J Chem Soc, Chem Commun*, 1099 (1984); K Ohta, A Ishii, I Yamamoto and K Matsuzaki, *Mol Cryst, Liq Cryst*, **116**, 299 (1985).

132. A-M Giroud-Godquin and J Billard, *Mol Cryst, Liq Cryst*, **97**, 287 (1983); K Ohta, H Muroki, A Takagi, I Yamamoto and K Matsuzaki, *Mol Cryst, Liq Cryst*, **135**, 247 (1986).

133. K Ohta, H Muroki, A Takagi, K-I Hatada, H Ema, I Yamamoto and K Matsuzaki, *Mol Cryst, Liq Cryst*, **140**, 131 (1986); H Sakashita, A Nishitani, Y Sumiya, H Terauchi, K Ohta and I Yamamoto, *Mol Cryst, Liq Cryst*, **163**, 211 (1988).

134. K Ohta, H Muroki, K-I Hatada, I Yamamoto and K Matsuzaki, *Mol Cryst, Liq Cryst*, **130**, 249 (1985); K Ohta, H Muroki, K-I Hatada, A Takagi, H Ema, I Yamamoto and K Matsuzaki, *Mol Cryst, Liq Cryst*, **140**, 163 (1986); K Ohta, H Ema, H Muroki, I Yamamoto and K Matsuzaki, *Mol Cryst, Liq Cryst*, **147**, 61 (1987).

135. K Usha and K Vijayan, *Mol Cryst, Liq Cryst*, **174**, 39 (1989).

136. S Chandrasekhar, B K Sadashiva, S Ramesha and B S Srikanta, *Pramana – J Phys*, **27**, L713 (1986).

137. S Chandrasekhar, B K Sadashiva and B S Srikanta, *Mol Cryst, Liq Cryst*, **151**, 93 (1987).

138. S Chandrasekhar, B K Sadashiva, B R Ratna and V N Raja, *Pramana – J Phys*, **30**, L491 (1988).

139. S Chandrasekhar, B R Ratna, B K Sadashiva and V N Raja, *Mol Cryst, Liq Cryst*, **165**, 123 (1988).

140. Conoscopic investigations produce optical figures which give information on the optical axiality of materials under study.

141. N J Thompson, *PhD Thesis*, University of Hull (1991).

142. B Mühlberger and W Haase, *Liq Cryst*, **5**, 251 (1989).

143. K Ohta, O Takenaka, H Hasebe, Y Morizumi, T Fujimoto and I Yamamoto, *Mol Cryst, Liq Cryst*, **195**, 135 (1991).

144. N J Thompson, G W Gray, J W Goodby and K J Toyne, *Mol Cryst, Liq Cryst*, **200**, 109 (1991).

145. P Styring, S Tantrawong, D R Beattie and J W Goodby, *Liq Cryst*, **4**, 581 (1991).

146. A-M Giroud-Godquin, M M Gauthier, G Sigaud, F Hardouin and M F Archard, *Mol Cryst, Liq Cryst*, **132**, 35 (1986).

147. R F Grant, *Can J Chem*, **42**, 951 (1964); H D Burrows and H A Ellis, *Thermochim Acta*, **52**, 121 (1982).

148. M Takekoshi, N Watanabe and B Tamamushi, *Colloid Polym Sci*, **256**, 588 (1978). This group suggested, on textural evidence, that the mesophase was SmC.

149. A-M Giroud-Godquin, J-C Marchon, D Guillon and A Skoulios, *J Phys Lett (Paris)*, **45**, 681 (1984); H Abied, D Guillon, A Skoulios, A-M Giroud-Godquin and J-C Marchon, *Liq Cryst*, **2**, 269 (1987).

150. H Abied, D Guillon, A Skoulios, H Dexpert, A-M Giroud-Godquin and J-C Marchon, *J Phys France*, **49**, 345 (1988); P Maldivi, D Guillon, A-M Giroud-Godquin, J-C Marchon, H Abied, H Dexpert and A Skoulios, *J Chim Phys*, **86**, 1651 (1989).

151. H Abied, D Guillon, A Skoulios, A-M Giroud-Godquin, P Maldivi and J-C Marchon, *Colloid Polym Sci*, **266**, 579 (1988).

152. A-M Giroud-Godquin, J-M Latour and J-C Marchon, *Inorg Chem*, **24**, 4452 (1985).

153. A-M Giroud-Godquin, J-C Marchon, D Guillon and A Skoulios, *J Phys Chem*, **90**, 5502 (1986); J-C Marchon, P Maldivi, A-M Giroud-Godquin, D Guillon, A Skoulios and D P Strommen, *Philos Trans R Soc London*, **A330**, 109 (1990).

154. R H Cayton, M H Chisholm and F D Darrington, *Angew Chem Int Ed Engl*, **29**, 1481 (1990).

155. P Maldivi, A-M Giroud-Godquin, J-C Marchon, D Guillon and A Skoulios, *Chem Phys Lett*, **157**, 552 (1989).

156. F Cukiernik, P Maldivi, A-M Giroud-Godquin, J-C Marchon, M Ibn-Elhat, D Guillon and A Skoulios, *Liq Cryst*, **9**, 903 (1991).

157. C Piechocki, J Simon, A Skoulios, D Guillon and P Weber, *J Am Chem Soc*, **104**, 5245 (1982).

158. D Guillon, A Skoulios, C Piechocki, J Simon and P Weber, *Mol Cryst, Liq Cryst*, **100**, 275 (1983).

159. D Guillon, P Weber, A Skoulios, C Piechocki and J Simon, *Mol Cryst, Liq Cryst*, **130**, 223 (1985).

160. See also: C Piechocki and J Simon, *J Chem Soc, Chem Commun*, 259 (1985); C Piechocki and J Simon, *New J Chem*, **9**, 159 (1985); D Masurel, C Sirlin and J Simon, *New J Chem*, **11**, 455 (1987); J Simon and C Sirlin, *Pure Appl Chem*, **61**, 1625 (1989); J F van der Pol, E Neeleman, J W Zwikker, R J M Nolte, W Drenth, J Aerts, R Visser and S J Picken, *Liq Cryst*, **6**, 577 (1989); M Hanack, A Beck and H Lehnmann, *Synthesis*, 703 (1987).

161. C Sirlin, L Bosio, J Simon, V Ahsen, E Yilmazer and O Bekâoğlu, *Chem Phys Lett*, **139**, 362 (1987).

162. C Sirlin, L Bosio and J Simon, *Mol Cryst, Liq Cryst*, **155**, 231 (1988).

163. C Sirlin, L Bosio and J Simon, *J Chem Soc, Chem Commun*, 379 (1987).

164. P Weber, D Guillon and A Skoulios, *J Phys Chem*, **91**, 2242 (1987).
165. C Piechocki, J Simon, J-J André, D Guillon, P Petit, A Skoulios and P Weber, *Chem Phys Lett*, **122**, 124 (1985).
166. Z Berlabi, C Sirlin, J Simon and J-J André, *J Phys Chem*, **93**, 8105 (1989).
167. M J Cook, M F Daniel, K J Harrison, N B McKeown and A J Thompson, *J Chem Soc, Chem Commun*, 1086 (1987).
168. M J Cook, A J Dunn, S D Howe, A J Thompson and K J Harrison, *J Chem Soc, Perkin Trans I*, 2453 (1988).
169. J W Goodby, P S Robinson, B-K Teo and P E Cladis, *Mol Cryst, Liq Cryst*, **56**, 303 (1980).
170. S Gaspard, P Maillard and J Billard, *Mol Cryst, Liq Cryst*, **123**, 369 (1985).
171. D W Bruce, D A Dunmur, L S Santa and M A Wali, *J Mater Chem*, **2**, 363 (1992).
172. B A Gregg, M A Fox and A J Bard, *J Am Chem Soc*, **111**, 3024 (1989); *J Phys Chem*, **93**, 4227 (1989); *J Phys Chem*, **94**, 1586 (1990).
173. B A Gregg, M A Fox and A J Bard, *J Chem Soc, Chem Commun*, 1134 (1987).
174. J Bhatt, B M Fung, K M Nicholas and C-D Poon, *J Chem Soc, Chem Commun*, 1439 (1988); M A Khan, J C Bhatt, B M Fung, K M Nicholas and E Wachtel, *Liq Cryst*, **5**, 285 (1989).
175. P Singh, M D Rausch and R W Lenz, *Liq Cryst*, **9**, 19 (1991).
176. L Ziminski and J Malthête, *J Chem Soc, Chem Commun*, 1495 (1990).
177. See e.g. T Saji and K Hoshino, *J Am Chem Soc*, **109**, 5881 (1987); M Yashiro, K Matsumoto and S Yoshikawa, *Chem Lett*, 985 (1989).
178. S Gaspard, A Hochapfel and R Viovy, *C R Acad Sci Ser C*, **289**, 387 (1979).
179. S Gaspard, A Hochapfel and R Viovy, *Springer Ser Chem Phys*, **11**, 298 (1980).
180. N V Usoltceva, V E Maizlish, V V Bykova, G A Analeva, G P Shaposhnikov and N M Kormilitsyn, *Russ J Phys Chem*, **63**, 1610 (1989); N V Usoltceva, V V Bykova, N M Kormilitsyn, G A Ananieva and V E Maizlish, *Il Nuovo Cimento*, **12D**, 1237 (1990).
181. J Le Moigne and J Simon, *J Phys Chem*, **84**, 170 (1980).
182. D W Bruce, D A Dunmur, P M Maitlis, J M Watkins and G J T Tiddy, *Liq Cryst*, **11**, 127 (1992).
183. M Yashiro, K Matsumoto and S Yoshikawa, *Chem Lett*, 985 (1989).
184. E.g. D J Cole-Hamilton and D W Bruce, in: *Comprehensive Coordination Chemistry*, G Wilkinson, R D Gillard and J A McCleverty, Eds, Pergamon, Oxford, vol 6, p 487 (1987).
185. D W Bruce, G J T Tiddy and A R Tajbakhsh, *unpublished results*.

9 Precursors for Electronic Materials

Paul O'Brien

Inorganic Materials. Edited by Duncan W Bruce and Dermot O'Hare
© 1992 John Wiley & Sons Ltd

The electronics industry uses an enormous range of inorganic materials in a diverse range of applications [1]. The present article will focus largely on the deposition of compound materials by chemical methods; the enormous and well-developed area of the processing of silicon wafers is excluded [2]. The principal method which will be discussed is the chemistry underlying the deposition of compound semiconductors by metallo-organic chemical vapour deposition (MOCVD) [3], although related methods will also be mentioned. It will be impossible to give comprehensive coverage in a short chapter but key references are given for most of the major types of materials. II/VI Materials, the area of interest in the authors own laboratory, are give a more comprehensive coverage. It seems appropriate by way of introduction to indicate the kinds of device which are at present fabricated by MOCVD and related methods.

9.1   ELECTRONIC MATERIALS

In order to appreciate the need for precursor chemistry, it is necessary to have some understanding of the kinds of functional device which are manufactured by the modern electronics industry [4]. Solid state lasers represent one of the more commonly-available electronic devices. They are classified as opto-electronic devices as they involve the generation of light [5]. However, the operation of these, and a vast number of related devices, depends on the formation of p-n junctions in semiconducting materials. In order to appreciate the challenges the chemist must meet in producing precursors for electronic materials, a basic understanding of what controls the operation of such devices is needed.

9.1.1 Semiconductors

Semiconductors are materials which contain a relatively small number of current carriers as compared to conductors such as metals. These current carriers may be either electrons (n-type materials) or electron holes (p-type materials); Figure 1 shows simple band gap diagrams for p- and n-type materials. The parameters which characterise such materials are the carrier concentration, n, (literally the concentration of carriers per cm^3) and the mobility, μ. The mobility of the carriers in a semiconductor is a more complicated function of temperature than the conductivity of a metal because, besides the temperature dependent scattering processes found in metals, the actual number of carriers (n) and their energy distribution are temperature dependent.

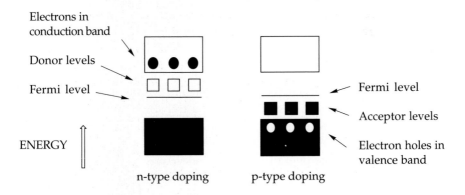

Figure 1 Schematic representation of how doping can lead to n- and p-type semiconductors; note that the exact position of the Fermi level is temperature-dependent (see reference [6])

The mobility (μ) is a measure of the ease with which the carriers can move under the influence of an electric field (the ratio of the drift velocity and the applied electric field). These parameters are generally derived from resistivity and Hall effect measurements [6].

It is interesting to think about what carrier concentrations mean in terms of the chemist's view of materials. An average quality sample of [GaAs] grown by metal-organic vapour phase epitaxy (MOVPE) might have a mobility of 130,000 $cm^2 V^{-1} s^{-1}$ and a carrier concentration of 10^{14} cm^{-3}. The carrier concentration represents 1 carrier per 2.167×10^8 gallium atom or 4×10^{-7} mole% of carriers. Carrier concentrations are influenced by a wide range of factors but the most important factors controlling carrier concentration in compound semiconductors, are non-stoichiometry or the incorporation of a dopant element. Dopants are elements related to those in the parent compound (or elemental semiconductor) which may occur as an adventitious impurity or which may be deliberately added. If the dopant element has one extra electron (e.g. substituting I for Se in $[ZnSe]_n$) then an electron will be donated to the conduction band and an n-type dopant results. However, if the dopant has one less electron (e.g. substituting Ga for Si in elemental Si) then the dopant will donate an electron hole and a p-type dopant results. Dopant concentrations tend to saturate at levels approaching 10^{19} cm^3. Some dopants are particularly detrimental to the function of semiconductors; transition metals, for example, often occupy deep levels around the centre of band gaps and can function to trap free charge carriers, reducing their lifetime considerably. Defects in the crystal lattice can also lead to levels with similar properties, one reason for an emphasis on the growth of layers of excellent crystalline morphology [7].

Another important distinction in semiconductor types is the band gap transition type. Many devices operate when a quantum of electromagnetic radiation is absorbed by, or emitted from, a transition between bands. In broad terms two kind of materials can be defined, *direct band gap* and *indirect band gap* materials. The various bands within a solid have orientations within the overall lattice and electrons within bands possess a momentum (the crystal momentum); in direct band gap materials there is no change in crystal momentum between valence and conduction bands, e.g. $[GaAs]_n$, $[CdS]_n$; in indirect band gap materials, e.g. Si or Ge, there is a change in momentum. Devices requiring the fast and efficient generation of conducting species or photons, especially lasers and other opto-electronic devices, are more effectively fabricated using direct band gap materials. Table 1 summarises some commonly-used semiconducting materials and their properties.

Table 1   Properties of some semiconductors

Semiconductor	Lattice Spacing/Å	Band-gap/eV
Element		
Si	0.5431	1.12 Indirect
Ge	0.5646	0.66 Indirect
III/V Compounds		
$[GaAs]_n$	0.5653	1.42 Direct
$[GaP]_n$	0.5451	2.26 Indirect
$[InAs]_n$	0.6058	0.35 Direct
$[InP]_n$	0.5869	1.35 Direct
$[AlAs]_n$	0.5661	2.16 Indirect
II/VI Compounds		
$[CdS]_n$	0.5832	2.42 Direct
$[CdSe]_n$	0.605	1.70 Direct
$[CdTe]_n$	0.6482	1.56 Direct
$[ZnS]_n$	0.542	3.68 Direct
$[ZnSe]_n$	0.5669	2.7 Direct
$[HgTe]_n$	0.644	———

9.1.2   p-n Junctions

The p-n junction forms the basis of the majority of electronic devices. Figure 2 shows in a schematic way what happens when p- and n- type semiconductors are placed in contact, namely that charge transfer occurs until the bands match in energy (technically until the Fermi energies are coincident). The carriers will in fact diffuse until there are equal concentrations of carriers at corresponding energies on either side of the junction in dynamic equilibrium. At the p-n junction the n-type material will be denuded of electrons, and the p-type material will have gained electrons. These regions are referred to as the depletion layers and they contain fixed equal but opposite charges [8]. The properties of this junction find numerous applications but its simplest electrical property is that when an external potential is applied, the junction behaves as a capacitor which varies as $V^{0.5}$; a variable capacitor diode or 'varactor' which can be used in frequency-locking or modulation circuits .

Another major application of p-n junctions is in photovoltaic devices. If a p-n junction were irradiated at an energy greater than the band gap then absorption of a photon would lead to the promotion of an electron from the valence band to the conduction band creating an electron-hole pair. One of this pair will be a minority carrier (i.e. the hole on the n-side or the electron on the p-side) and if the minority carrier can diffuse to the space-charge region (at the junction) before recombination takes place it will be swept across the junction. If the two sides of the junction are connected to an external circuit, a photocurrent will be observed [9]. A number of other devices are based on similar effects, in particular the converse process leads to electroluminescence, and if the device is suitably confined, solid-state lasers may be constructed (Figure 3).

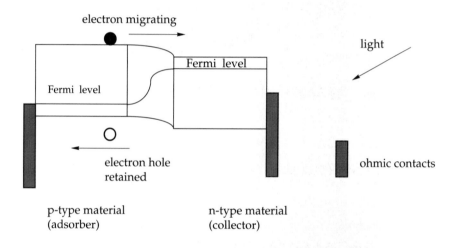

Figure 2   Schematic representation of a p-n junction in a solar cell, the spatial arrangement of the cell is shown along the x-axis, the y-axis shows relative energy levels

9.1.3   Other Materials

Microelectronic devices such as solid state lasers and complex transistors (e.g. stripe geometry laser; Figure 3) are complex two- and three-dimensional heterostructures [10]. In preparing such devices there is a need for the full range of electrical properties for various materials in the device, for instance metallisation (e.g. Au) may be needed to make contacts, insulators (e.g. $[CaF_2]_n$) are often required, and stable boundary layers to prevent interdiffusion may be prepared from silicides. Oxides have a wide range of interesting applications including infra-red detectors (bolometers) and as piezoelectric materials (Section 9.6). The still relatively-novel high T_c superconductors represent another class of materials for which chemical routes to processing may be useful.

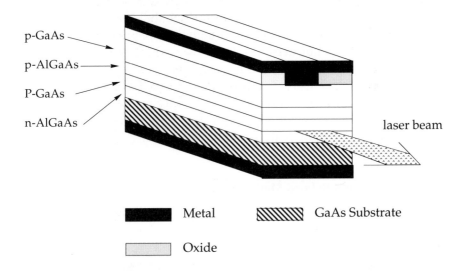

p-GaAs

p-AlGaAs

P-GaAs

n-AlGaAs

laser beam

▆▆▆▆▆ Metal   ░░░░░ GaAs Substrate

▭▭▭ Oxide

Figure 3   Stripe geometry laser in which the oxide layer isolates all but the stripe contact (stripe widths are usually in the region of 3–30 microns)

9.2   METHODS FOR CRYSTAL GROWTH

There is substantial current interest in the use of a wide range of chemical, as opposed to physical, methods for the deposition of semiconductors, and there are a number of sound reasons for such work. There are a wide range of methods for the preparation of electronic materials [11, 12]. These include growth from a melt of the material, growth from the vapour phase in sealed or flow systems and derived methods based on chemical transport (Equation (1)) in which a carrier is used to enhance the transport of one component, the zinc in this case.

$$[ZnS]_n(s) + I_2(g) \longrightarrow [ZnI_2](g) + 1/2\, S_2(g) \tag{1}$$

Such methods can be used to grow thin films of materials but have severe limitations, especially the need for relatively high temperature (typically > 800 °C) to volatilise parent material. Chemical transport can lower the optimum growth temperature, but the carriers themselves may be a problem, e.g. iodine is electrically and optically active as an impurity in II/VI materials [13]. For compound semiconductors a major problem is often that non-stoichiometry results from the high growth temperature required which is often associated with the loss of the pnictide in a III/V system (e.g. P from InP) or the chalcogen in a II/VI system (e.g. S from $[CdS]_n$), although metallic components can also be volatile (e.g. Hg in tellurides).

One approach to lowering the growth temperature is to grow from solutions, for example $[CdTe]_n$ can be grown from Te-rich solutions [14]. However, chemical methods offer the possibility of the use of very low growth temperatures combined with the possibility of chemically including a dopant potentially enabling the direct control of electrical properties.

9.2.1   MOCVD and Related Methods

Manasevit [15] pioneered the deposition of compound semiconducting, and other materials containing metals, from a reaction stream containing mixtures of volatile precursor.

The reaction of volatile precursor metal and anion species above, or at, the surface of a substrate, leading to the deposition of a solid species, is most generally termed chemical vapour deposition (CVD), or if a metal is involved, metallo organic vapour deposition (MOCVD). Other terms, for example, metal-organic vapour phase epitaxy (MOVPE), are less general and imply specific features of the system, e.g. an organometallic precursor (direct metal-to-carbon bonds in at least one of the species) and epitaxy. These terms are often used inaccurately in the literature and one reason may be that work in this area crosses the interface between materials science and conventional chemistry. An optimal deposition system will involve precursors, probably consisting of a sophisticated mixture of organometallic compounds, or a designed single precursor (chemistry), which decomposes at a solid vapour phase interface (surface science), to generate a solid phase with useful properties (materials science), in a properly-engineered reactor. A full understanding of the factors controlling the growth of solids in such systems will include: the design and characterisation of precursor molecules, the study of the adsorption and decomposition of these molecules at interfaces, and an appreciation of the kinds of solid phases of interest in materials science.

The overall chemical reactions involved may be represented in oversimplified ways, for example, as shown in Equations (2) and (3).

$$PH_3 + [Me_3In] \longrightarrow [InP]_n + 3\,CH_4 \tag{2}$$

$$H_2S + [Me_2Zn] \longrightarrow [ZnS]_n + 2\,CH_4 \tag{3}$$

However, the mechanisms of the process will be of crucial importance in determining the quality of the grown layers and when growth is optimal and epitaxial layers are grown, the process must involve a surface-controlled reaction. The surface chemistry of such systems is at present under intense study [16]. The process of layer deposition in an MOCVD reactor is represented schematically in Figure 4.

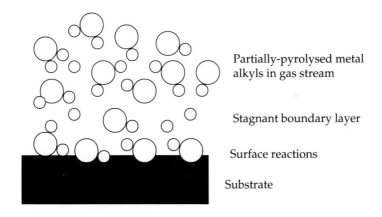

Partially-pyrolysed metal alkyls in gas stream

Stagnant boundary layer

Surface reactions

Substrate

Figure 4   Schematic representation of the MOCVD growth process

Reactor design is also important. In conventional (atmospheric pressure) MOCVD the passage of the precursors into the reactor is controlled by bubbling a carrier gas through a liquid precursor and assuming that the compound exerts its equilibrium vapour pressure; the rate of passage of the material is controlled by an up-stream mass flow controller. Solid precursors clearly present a problem in such systems but there are many cases in which solids have to be used, especially for oxide deposition. Pressurised gases (e.g. AsH_3, PH_3 or H_2Se) can be readily controlled, but represent a considerable toxic risk. A schematic representation of a conventional MOCVD reactor is shown in Figure 5.

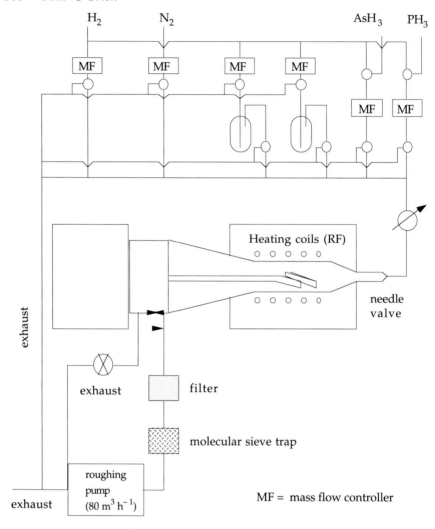

Figure 5   Schematic representation of an MOCVD reactor

9.2.2   Other Chemical Deposition Methods

9.2.2.1 *MBE and MOMBE*

The deposition of a wide range of materials can be achieved using beams
of elemental sources in high vacuum apparatus, essentially a physical

method, known as molecular beam epitaxy (MBE) [5,16]. If, instead of elemental sources, low-vapour-pressure chemical sources are used, the technique is described as metal-organic molecular beam epitaxy (MOMBE). Originally MOMBE involved the use of metal alkyl sources, however one particularly interesting area involves the deposition of compound semiconductors from single molecule precursors [16].

9.2.2.2 Solution and Related Methods

These include electrochemical methods, electroless (i.e. chemical) solution deposition, sol-gel processing and spray pyrolysis.

These methods are often important for the cheap deposition of semiconductors for use in large surface area applications such as solar cells. Many of the compounds used in such systems are dealt with in this review although these methods will not be emphasised.

Electrochemical deposition is a useful method for the deposition of CdTe for solar cell applications [17]. Electroless deposition involves the deposition of solid material from metastable solutions. In a typical procedure an alkaline solution of cadmium chloride, ammonia and thiourea is maintained above a substrate at 70–80 °C; cadmium sulphide with reasonable electrical properties can be deposited by this method [18]. The formal stoichiometry of the reaction can be written as:

$$[Cd(NH_3)_4]^{2+} + (NH_2)_2CS + OH^-$$
$$\downarrow$$
$$[CdS]_n + (NH_2)_2CO + NH_4^+ + 3\,NH_3 \tag{4}$$

The method has also been used to deposit successfully a number of ternary oxide [19] and ternary chalcogenide [20] systems.

In the spray pyrolysis method [21], an atomised jet of a solution of precursor molecules is directed toward a heated substrate. The method is used to coat glass with $[SnO_2]_n$ and has also been used to grow $[CdS]_n$. In an ideal spray pyrolysis system the solvent evaporates before reaching the surface and growth involved the interaction of vapour phase precursors with the surface, approximating to the same kinds of conditions set up in an MOCVD reactor.

Sol-gel processing [22] involves the partial hydrolysis of precursors to produce gels which are subsequently processed; this method finds extensive application in the preparation of oxides and more details are given in Section 9.6.2 which deals with oxide deposition.

9.3   REQUIREMENTS FOR PRECURSORS

The remainder of the present chapter will concentrate on describing the kinds of precursor molecule available for the deposition of electronic materials, mainly by MOCVD or MOMBE. The inorganic chemistry of the precursors is important in the control of:

1. Impurities, especially dopants, in the final material which may originate from the precursor.
2. The properties of the final material as determined by the decomposition temperature of the precursor.
3. Contamination of the material with the products of homogeneous reaction within the reactor (so-called homogeneous prereaction) of the precursors, e.g. 'snow' falling onto the substrate-limiting morphology.
4. Vapour pressure. Suitable precursors are needed for the deposition of materials under low pressure conditions (LPMOCVD), which at very low pressures approaches metal-organic molecular beam epitaxy (MOMBE) (essentially a vacuum deposition method). Some precursors, e.g. dimethylzinc, have a higher vapour pressure than is ideal for atmospheric pressure MOCVD and consequently bubblers containing such materials are cooled.

9.3.1   Classes of Molecules Used as Precursors

Chemists usually define a metal-containing species in terms of the bonds between the central metal and the surrounding, usually organic, ligating groups, and the nature of the group attached to the metal [23]. These distinctions are important as the chemical properties of the different classes of molecules are remarkably diverse, and can to a large extent be rationalised in terms of the mode of bonding. The examples have been chosen to illustrate the major classes of compound used as precursors. These include:

1. Metal-to-carbon-bonded species, e.g. the σ-bonded organometallic, dimethylzinc, [Me_2Zn].
2. Organometallic compounds containing π-bonded ligands e.g. tricarbonylmethylcyclopentadienylmanganese, [$(MeCp)Mn(CO)_3$], used in the deposition of manganese-doped materials.
3. Coordination complexes such as acetoacetonates and dithiocarbamates.

4. Adducts of organometallics (addition compounds) in which a discrete organic molecule reacts with a metal-containing compound (in this case an organometallic) to form a new species.
5. Compounds containing both organometallic and other charged groups e.g. mixed alkyl-thiolates or thiocarbamates.

9.3.2 Metal Alkyls and Adduct Purification

Metal alkyls are the most commonly-used metal precursors and are available from a wide range of commercial suppliers. The properties of some of the more commonly-used alkyls are summarised in Table 2.

Table 2   Physical properties of some metal alkyls important for semiconductor deposition*

Compound	ΔH_f	$\overline{E}$	Vapour pressure (Torr, 20 °C)
[Me$_2$Zn]	50	177	302.5
[Me$_2$Cd]	106	139	28.4
[Me$_3$Al]	−81	274	8.7
[Me$_3$Ga]	−42	247	182.3
[Me$_3$In]	173	160	1.73

* ΔH_f and mean bond energies ($\overline{E}$) in kJ mol^{-1} from ref [25]. Vapour pressure data from a compilation by Morton International.

Three aspects of the chemistry of metal alkyls are crucial to their use in MOCVD:

1. The volatility of the metal alkyl.
2. The presence of impurities in the metal alkyl and the ease of its purification.
3. The thermal and photochemical stability of the alkyl, which will be important in determining the ease of handling of the compound and its behaviour in the MOCVD reactor.

9.3.3 The Preparation of High-purity Alkyls

The preparation of the metal alkyl usually involves the reaction of an ethereal solution of a Grignard reagent with the metal chloride [23–26] (Equation (5), M = Cd, Zn; Equation (6), M = Ga, In).

$$[MCl_2] + 2\,[MeMgI] \longrightarrow [Me_2M] + [MgCl_2] \qquad (5)$$

$$2\,[MCl_3] + 6\,[MeMgCl] \longrightarrow 2\,[Me_3M] + 3\,[MgCl_2] \qquad (6)$$

The synthesis of new alkyls by this method may involve some tuning of the above reaction. For example, we have found it necessary to use [27] neopentyl bromide and cadmium iodide in order to prepare bis(neopentyl)cadmium. Other methods of preparing metal alkyls include electrochemical synthesis [28], or the use of an organometallic alkylating agent such as trimethylaluminium.

One problem with the use of metal alkyls prepared directly from such reactions may be the incorporation of halogen-containing materials such as MeI into the final precursor with the consequent incorporation of the halogen, an n-type dopant, into the final material. The purity of metal alkyls can be greatly improved by the use of suitable adducts, a technique pioneered by Bradley and Faktor, and used commercially for the preparation of InMe$_3$ from the bis(diphenylphosphino)ethane (diphos) adduct of [Me$_3$In] [29].

Two kinds of addition compounds (adducts) are of interest, solids such as [Me$_2$Cd.(2,2'-bipy)] [30] and liquid adducts such as those formed between high boiling point ethers and metal alkyls [31]. Solid adducts have the advantage that they can be purified by recrystallisation, but liquid adducts are often easier to handle on an industrial scale [31]; the chemistry of such compounds was developed in the 1960s, notably by Thiele and Coates. Another distinct advantage of such compounds is the fact that they have considerably reduced reactivity compared to the parent alkyl and are consequently easier to handle.

The metal alkyls used as precursors are in general highly-reactive, pyrophoric liquids. One important mechanism in the decomposition of many such organometallics (other than methyls) is β-elimination of hydrogen atoms to produce an alkene and a metal hydride [32] (Figure 6).

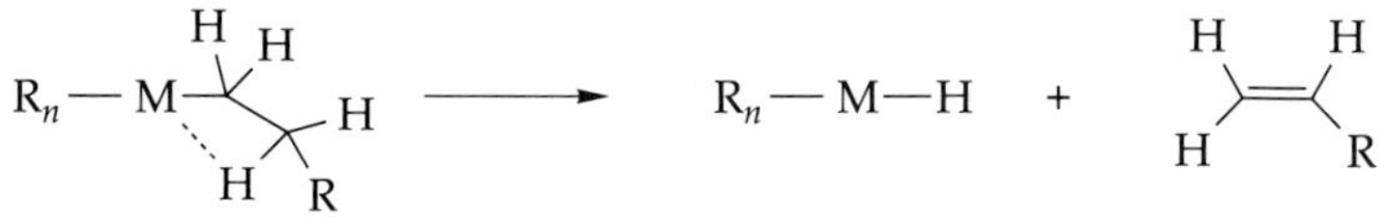

Figure 6   A representation of how β-hydrogens in alkanes can be eliminated

The elimination of alkyl groups by such mechanisms is often suggested as a significant mechanism in MOCVD reactions, often with relatively little supporting evidence. However, the reaction does provide a facile mechanism for the decomposition of higher alkyls.

9.4 PRECURSORS FOR III/V MATERIALS

9.4.1 III/V Semiconductors

The growth of thin films of III/V materials was one of the first problems to be addressed by MOCVD [33, 34] and is now a fairly mature technology with a wide range of such devices being routinely fabricated by MOCVD. These materials are particularly important in light-emitting and detecting devices such as solid state lasers (opto-electronics; e.g. $[InP]_n$); another potential application is as active electronic components operating at either higher speed or lower power dissipation than conventional group IV devices (e.g. $[GaAs]_n$). Another particular advantage of these materials is the availability of a wide range of ternary (e.g. $[AlGaAs]_n$) and quaternary (e.g. $[GaInAsP]_n$) alloys in which the band gap of the material can be tuned within a series of isomorphous alloys [35].

The materials crystallise in the zinc blende (cubic $[ZnS]_n$) structure. One particularly important result of the structure is that there are closely-spaced pairs of {111} planes (e.g. in $[GaAs]_n$ of Ga and As) separated by 0.82 Å and so two distinct {111} surfaces, {111b} (As) and {111a} (Ga), are possible. These differences in surface structure may be important in the procedures adopted for the etc.hing and growth of layers on these substrates [35].

In general, these semiconductors are grown by conventional, atmospheric-pressure MOCVD using a metal alkyl and the hydride of the pnictide [36]. The main areas of interest for the chemist are in improving the purity of precursors and controlling adventitious doping by elements such as carbon, the development of alternatives to the permanent gas sources (e.g. AsH_3 and PH_3 — both of these compounds are hazardous to handle) and in the production of single-molecule precursors for these materials (these compounds often have limited volatility and are most likely to be useful in MBE).

9.4.2 Group 13 Precursors

The metal alkyls have generally been used for the deposition of III/V materials by MOCVD and the properties of some of the more common precursors are summarised in Table 2. Many of the early problems in either the reproducibility of MOCVD growth or the quality of layers grown were eventually traced to the purity of the available metal alkyls. In early experiments in which the reaction between [Me$_3$In] and PH$_3$ was studied material precipitated from the vapour phase (a process often referred to as homogeneous prereaction); this material was believed to be [MeInPH]$_n$ [37]. The use of volatile adducts such as [Me$_3$In.NMe$_3$] was investigated [38] and such reactions were inhibited. However, the unwanted reaction is apparently catalysed by impurities in the [Me$_3$In] as no such reaction occurs when high purity metal alkyls are used [39]. Adducts have been important in the purification of metal alkyls and the purification of [Me$_3$In]. The diphos adduct is a celebrated example (Figure 7) [29] and using [Me$_3$In] dissociated from the adduct, carrier concentration of 2.6 x 10^{14} cm^{-3} and 77 K electron mobilities as high as 126,300 cm^{-2} V^{-1} s^{-1} were obtained [40].

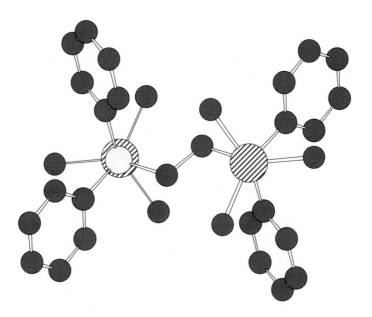

Figure 7   The diphos adduct of trimethylindium viewed down the two In–P bonds

Indium illustrates another particular problem in that trimethylindium is a solid and consequently the vapour pressure exerted by the precursor and mass transport to the reactor tend to vary during conventional growth work. Various methods are available to control the problem, such as the use of volatile adducts and specially designed delivery systems. Because of these problems, triethylindium was for a time an important precursor [41], although at present, high-purity [Me$_3$In] seems to be the precursor of choice. The search for a liquid precursor for indium led [42] to the use of dimethylethylindium. NMR studies of solutions of the compound indicate that disproportionation occurs (Equation (7)) and so the properties of this precursor, over time in a bubbler, may be suspect because of the potential for the loss of one component [42].

$$3 \, [\text{Me}_2\text{Et}_2\text{In}] \rightleftharpoons 2 \, [\text{Me}_3\text{In}] + [\text{Et}_3\text{In}] \qquad (7)$$

The alkyls of gallium are in general liquids and so there are fewer problems in their use in conventional MOCVD. Methyl-, ethyl- and t-butylgallium have all been used as precursors and the methyl is used routinely [29, 43]. Adduct purification procedures similar to those used for the indium alkyls (e.g. with diphos) are also possible.

Aluminium is particularly important for the growth of the important ternary material $[\text{Al}_x\text{Ga}_{1-x}\text{As}]_n$. In this system there are several important problems, including the control of the mole percentage of aluminium in the final product and carbon incorporation [44]. Aluminium alkyls have been extensively used as precursors but recently the use of alane ([AlH$_3$]) has become important. Adducts of alane and galane have been used to grow [AlGaAs]$_n$ with low carbon incorporation [45].

Another class of molecules which has been used in the deposition of III/V materials is the chelate supported organometallics (Figure 8). These molecules have improved coordinative saturation compared to the simple alkyls and are consequently much easier to handle. There is apparently little or no problem with the transfer of nitrogen from these precursors to the deposited layer and when such precursors are available in high states of purity, they may become useful precursors for the routine deposition of III/V materials [46, 47].

Figure 8   Some chelate supported group 13 metal alkyls used in materials deposition; typically R and R' are Me or Et groups and M is Al, In or Ga

9.4.3   Alternative Group 15 Precursors

The extreme toxicity of both arsine and phosphine has led to a significant effort to develop alternate precursors for the group 15 component. The *tris* methyls or ethyls of both arsenic and phosphorus have high pyrolysis temperatures and are not promising precursors, giving rise to significant carbon contamination [48, 49]. The use of higher alkyls such as the tributyl derivatives, especially in the presence of one or more hydrogens [50], leads to lower pyrolysis temperatures; these compounds are potentially important alternate precursors for the group 15 element. The precise mechanisms involved in the decomposition of these compounds are important in understanding carbon incorporation into the films. Sources containing at least one hydrogen atom, e.g. $But AsH_2$, can be used in the growth of low-C GaAs. Various decomposition pathways have been suggested as being important [51] including intramolecular coupling (Equation (8)) and β-elimination (Equation (9)), whose importance increases with temperature.

$$[As(t\text{-}C_4H_9)H_2] \longrightarrow [AsH] + i\text{-}C_4H_{10} \qquad (8)$$

$$[As(t\text{-}C_4H_9)H_2] \longrightarrow [AsH_3] + i\text{-}C_4H_8 \qquad (9)$$

9.4.4   Single-molecule Precursors

Such molecules are potentially excellent precursors for the deposition of semiconductors by MOMBE. There has been considerable recent interest in metal-organic precursors for the deposition of III/V semiconductors such as gallium arsenide and indium phosphide, in which both the metal ion and group 13 element are contained in a single molecule. One successful approach has involved the use of dimers, first described by Coates [52], of the general formula shown below (Figure 9).

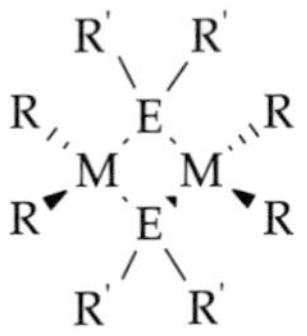

Figure 9   Parent structural type for dimeric single molecule III/V precursors; a wide range of alkyl substituents for both R and R' are known

Notable examples of precursors in this class include [Me$_2$Ga(μ-t-Bu$_2$As)]$_2$ as described by Cowley and Jones [53], and [Me$_2$In(μ-t-Bu$_2$P)]$_2$ developed by Bradley and co-workers [54]. The advantages of such molecules as precursors include safer handling and the potential for low-temperature deposition. The use of such molecules for the deposition of III/V materials has been reviewed [55].

Examples of more growth-orientated work include the deposition of good-quality GaAs from [Me$_2$Ga(μ-t-Bu$_2$As)]$_2$, at 10^{-4} Torr in H$_2$ gas [56]. Carbon contamination was below the levels which could be detected by secondary ion mass spectrometry (at least ppm sensitivity). The mechanisms responsible for the clean elimination of the ligand were suggested to be as shown in Figure 10 on the evidence of effluent gas analysis.

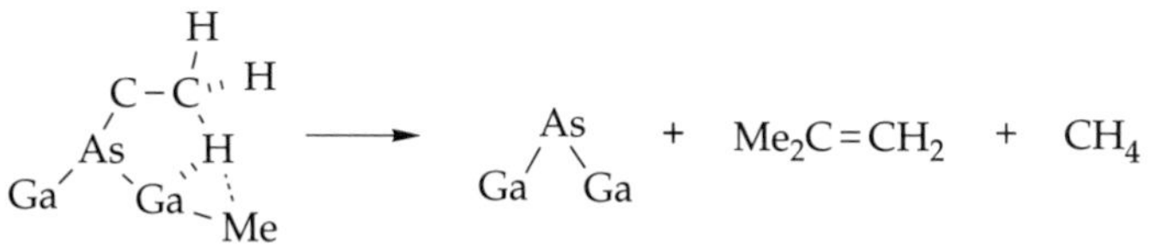

Figure 10   A possible mechanism for the surface decomposition of [Me$_2$Ga(μ-t-Bu$_2$As)]$_2$

Recent pyrolysis studies on related trimeric precursors, such as [Me$_2$Ga(μ-As-iPr)]$_3$, show that these compounds react at 150 °C to produce diarsines; further studies of dimers indicate the importance of β-elimination and intramolecular coupling reactions [57].

This kind of precursor chemistry has been extended to more complex cage structures involving RM and R'E fragments [58, 59]. The cubanes, [t-BuGa(μ_3-E)]$_4$, (E = S, Se or Te) have also been reported; related species containing aluminium have a dimeric compounds with μ_2-bridging chalcogens [60]. The cubane [t-BuGa(μ_3-E)]$_4$ can be used to deposit a novel cubic form of GaS by low-pressure MOCVD [61]. This work emphasises the important fact that growth by MOCVD often takes place well away from equilibrium conditions and under kinetic control with the consequence that metastable phases (e.g. cubic [GaS]$_n$ or high temperature forms) can be produced under relatively mild conditions.

9.5   PRECURSORS FOR II/VI MATERIALS

9.5.1   II/VI Materials

There are at least three main reasons for the substantial interest in the deposition of II/VI materials:

1. High-quality $[ZnSe]_n$ (and $[ZnS]_n$) in which doping could be reproducibly controlled to produce p- and n-type material could form the basis of a range of opto-electronic devices operating in the blue region of the visible spectrum [62].
2. The related materials, $[CdS]_n$ and $[CdTe]_n$, form the basis of important solar cell technologies [63].
3. The narrow-band-gap materials, particularly cadmium mercury telluride (CMT, $[Cd_xHg_{1-x}Te]_n$) are important in infrared opto-electronic devices [64]. The properties of the alloy can be tuned but compositions $x = 0.3$ and $x = 0.2$ are particularly important as they match with the 3–5 µm and 8–14 µm near infrared windows in the terrestrial atmosphere. One clever way in which CMT has been grown is the interdiffused multilayer process (IMP) [65] in which very thin layers (*ca* 0.1 µm) of $[CdTe]_n$ and $[HgTe]_n$ are grown on top of each other under close-to-optimal conditions for the deposition of each binary material. The layers are then annealed and interdiffusion leads to CMT of good quality.

9.5.2   Metal Precursors for II/VI Deposition

Deposition studies are dominated by the use of the methyls of cadmium or zinc, or an adduct of these compounds, while metallic mercury tends to be the precursor of choice for the deposition of CMT and related materials [66]. It is relatively rare to see the use of metal sources other than dimethylzinc or dimethylcadmium, even though these may have distinct advantages. Jones and co-workers reported [67] the use of diethylcadmium for the deposition of CMT. An advantage of this alkyl may be a lower decomposition temperature, better matched to the newer tellurium precursors (Section 9.5.3) than dimethylcadmium. It should be noted that the higher alkyls of cadmium are particularly prone to photochemical decomposition and are particularly hazardous materials, even by the standards of metal alkyls. The reasons for this are probably particularly stable radical species, generated by β-elimination reactions, during the decomposition of such alkyls.

Recently bis(neopentyl)cadmium(II), a low-melting point solid (mpt *ca* 40 °C) of remarkable thermal and photolytic stability, has been successfully used in combination with H_2S to grow films of $[CdS]_n$ on $[GaAs]_n$ *{111a}* at 350 °C. The growth rate was low at *ca* 0.06 μm h^{-1} and there was some homogeneous reaction between the alkyl and hydrogen sulphide; the films showed hexagonal features. The low growth rate obtained with the source at room temperature suggests that the alkyl is unsuitable for atmospheric pressure MOCVD. Bis(neopentyl)cadmium(II) may have an application as a p-type dopant for $[InP]_n$ and related III/V alloys, an alternative to controlling the vapour pressure of dimethylzinc by the use of an adduct [27]. The corresponding manganese(II) species may prove to be a good precursor for manganese. At present the commercially available source is tricarbonylmethylcyclopentadienylmanganese, although alternative species with better decomposition properties are now appearing; the use of such compounds has been the subject of an excellent review [68].

Dimethylmercury has been used as a precursor for the deposition of CMT, but it is expensive and represents a persistant toxic hazard [69,70].

9.5.2.1 Adducts as Precursors

Adducts of the alkyls of zinc and cadmium also find application as precursors for the growth of wide-band-gap II/VI semiconductors by MOCVD. The dioxan [71, 72, 73], thioxan [73], triethylamine [74–76] and triazine [75, 77, 78] adducts of dimethylzinc successfully inhibit homogeneous prereaction in the growth of $[ZnSe]_n$ and related alloys by MOCVD. Adducts of dimethylcadmium have been used to grow $[CdSe]_n$ and $[CdS]_n$ [79, 80]. The use of such adducts has several potential advantages:

1. The vapour pressure of the metal alkyl is effectively reduced, thus eliminating the necessity to cool bubblers containing dimethylzinc.
2. Reaction occurs before the precursors reach the hot zone of the reactor, hence the so-called 'homogeneous prereaction' may be considerably limited.
3. Layers may have improved electrical properties due to the purification of the alkyl during the preparation of the adduct.

Despite the increasing use of such compounds in MOCVD, little is known about the nature of the chemical species present in the gas phase in the MOCVD reactor.

Recent infrared studies indicate that many of the adducts of dimethylcadmium and dimethylzinc are nearly fully dissociated in the vapour phase [81]. Totally-dissociative vapourisation is common for organometallic species and one of the simplest examples is trimethylindium which is a tetramer in the solid state [82] and a monomer in the vapour phase [83]. The adducts of dimethycadmium with the chelating ligand tetramethylethylenediamine, show some evidence of association at room temperature; under MOCVD conditions any association is likely to be extremely limited [84].

9.5.3   Nature of the Effect of Adducts on Growth

Growth results (Table 3) indicate that there is a marked difference in the extent to which each adduct inhibits 'homogeneous prereaction' with the group 16 hydride during growth. This is a surprising observation in view of the similar, and largely dissociated nature of the adducts in the vapour phase. The triethylamine and trimethyltrazine adduct of dimethylzinc have been used to grow high-quality $[ZnSe]_n$ (Figure 11).

Table 3   Typical growth results indicating the effect of adduct precursors on the quality of layers of $[ZnSe]_n$

Precursor	Material Grown	Growth Temp/°C	$n_{77 K}$ (cm^{-3})	$\mu_{77 K}$ $(cm^2 V^{-1} s^{-1})$	Comments
$[Me_2Zn]$	$[ZnSe]_n$	325	3.2×10^{15}	1708	Prereaction [73]
$[Me_2Zn.dioxan]$	$[ZnSe]_n$	325	9.1×10^{15}	630	Some inhibition of prereaction, reaction in hot zone only [73]
$[Me_2Zn.thioxan]$	$[ZnSe]_n$	325	2.9×10^{15}	832	Some inhibition of prereaction, reaction in hot zone only [73]
$[Me_2Zn.NEt_3]$	$[ZnSe]_n$	350	3.6×10^{14}	6455	No prereaction [75]
$[Me_2Zn.tri^*]$	$[ZnSe]_n$	350	layers depleted		No prereaction [75]
$[Me_2Cd.dioxan]$	$[CdSe]_n$	300–500	$10^{16}–10^{17}$	–	Some inhibition of prereaction [75]
$[Me_2Cd.tht^*]$	$[CdSe]_n$	300–500	$10^{16}–10^{17}$	–	More inhibition of prereaction [75]

*   tri = hexahydro-1,3,5-trimethyl-1,3,5-triazine; tht = tetrahydrothiophene

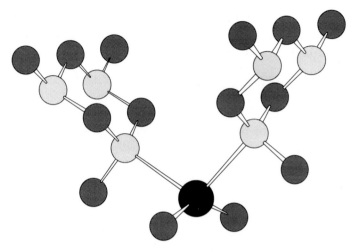

Figure 11   The triazine adduct of dimethylzinc

The effectiveness of the adducts in inhibiting homogeneous reaction in MOCVD reactors had been tacitly interpreted (e.g. [75]) in terms of the 'blocking' of the reaction by adduct formation in the vapour phase. The implication being that homogeneous reaction proceeds via an initial step involving the formation of an addition compound (adduct) between the group 12 source and the metal alkyl (a 'pre-equilibrium' process (Equation (10)), in chemical terminology an associative reaction) which is inhibited by excess of a 'stronger' Lewis base (Equation (12)). The kind of scheme envisaged is shown in Equations (10)–(12); X = chalcogen, L = Lewis base).

$$[Me_2Zn] + H_2X \rightleftharpoons [Me_2Zn.XH_2] \qquad (10)$$

$$[Me_2Zn.XH_2] \longrightarrow ZnX + \text{volatile products (CH}_4 \text{ etc.)} \qquad (11)$$

$$[Me_2Zn] + L \rightleftharpoons [Me_2Zn.L] \qquad (12)$$

As adducts formed between strong Lewis bases and group 12 alkyls are effectively totally dissociated in the vapour phase, then inhibition of homogeneous reaction ('pre-reaction') must occur by some other mechanism.

The extreme alternative mechanism to an associative process, as described above, is a dissociative mechanism. Free radicals (CH$_3$•, etc.) are now well-established components of the reactive vapour phase in

MOCVD reactors containing group 13 precursors [85]. An alternative mechanism for the homogeneous reaction could invoke metal–carbon bond homolysis (Equation (13)).

$$[Me-M-Me] \longrightarrow Me^{\bullet} + [^{\bullet}M-Me] \tag{13}$$

The highly-reactive intermediates thus formed could react with other molecules in the vapour phase. In the absence of a better electron donor, the metal intermediates are likely to react with the chalcogen-containing precursor leading to homogeneous reaction and 'snow'. However, in the case of adduct precursors the vapour will contain a considerable (often stoichiometric or greater) quantity of the Lewis base. The bases are much better ligands than the chalcogen precursors and are likely to act as 'traps' for the reactive metal-containing intermediates. Thus, these reactive intermediates could pass through the hot zone of the reactor without the reaction proceeding far enough for the nucleation and growth of a particulate material.

It is interesting to consider the recent conclusions of Jackson concerning the homogeneous decomposition of the group 12 metal alkyls [86]. At the temperatures typically used for the preparation of II/VI materials, 400–500 °C, simple bond homolysis does not seem likely. However, radical and/or Lewis acid species could be generated during the decomposition of the initial complex formed on the association of the alkyl and the chalcogen (Equation (14)).

$$[Me_2Zn] + H_2X \longrightarrow [Me_2Zn.XH_2] + CH_4 +$$

$$[MeZnXH] + Me^{\bullet} + [ZnHX] \tag{14}$$

This scheme combines features of both of the earlier mechanisms. Finally if the first step in the homogeneous decomposition is taken as being a bimolecular collision between the alkyl and chalcogen containing materials; such interactions will be less likely at lower partial pressures of the alkyl and in the presence of a third body, the Lewis base, leading to no orereaction.

In summary, there are a number of ways in which adducts could be operating to control unwanted reactions; these include:

1. The inhibition of surface-, possibly acid-catalysed, reactions in the cool zone of the reactor.
2. Unwanted reactions are often less apparent with high purity reagents, so the improved purity of adducted metal alkyls may be important.

3. With highly-volatile precursors such as dimethylzinc, the adducts provide for better control of mass flow. Under equivalent flow conditions a lower partial pressure of the metal alkyl in the MOCVD reactor will be supported by an adduct as compared to the pure alkyl. Homogeneous reactions are likely to proceed at rates proportional to the partial pressure of metal alkyls, a lower partial pressure will lead to a lower rate of homogeneous reaction.

4. The Lewis bases from which adducts are formed may play a rôle in trapping highly-reactive intermediates generated during the decomposition of metal alkyls.

In considering factors 3 and 4 above, it is important to note that adducts do not need to stop totally the formation of precipitates from the vapour phase but only to sufficiently inhibit their formation so that they pass through the hot zone of the reactor before forming particulate material. As such reactions probably proceed by a process of nucleation and growth, inhibition of either of these steps could stop significant homogeneous reaction occurring as the growing nuclei of metal-chalcogen material are likely to contain highly-reactive (Lewis acidic) sites which could be blocked by the Lewis base.

The lack of significant association of such adducts in the vapour phase rationalises early failures to grow $[ZnO]_n$ directly from the dioxan adduct of dimethylzinc [75]. Other than in the presence of a large excess of dioxan, films of metallic zinc were obtained. There are clearly at least two distinct problems in growing directly from such adducts:

1. A high temperature is required to crack the organic ether.
2. The lack of a strongly-bonded adduct in the gas phase means that the concentration of species with proximal oxygen and zinc atoms will be low.

The reaction is unlikely to proceed via the surface decomposition of an adduct (or the ligand) in the absence of an excess of the Lewis base.

Much needs to be learned about the properties of the adducts of group 12 metals and the way they affect the quality of grown layers. More information concerning vapour pressures, properties in the vapour phase and, above all, the mechanisms of homogeneous decomposition is needed.

9.5.4 Alternative Group 16 Sources

A number of papers have reported the use of alternate group 16 sources. The aims of such work include: the elimination of homogeneous reaction, lower growth temperatures and photochemically-active systems. It seems appropriate to indicate the main directions such work has taken, although the topic is not a major one in the present article.

In an effort to avoid the problem of prereaction, Jones and Sritharan reported the use of diethylselenium as a source of Se in the growth of $[ZnSe]_n$, films of excellent crystallinity were grown at 450 °C [87]. Similarly [88, 89] thiophene, selenophene and furan were used to grow films of zinc and cadmium chalcogens or oxides; the utility of these heterocycles as precursors is somewhat limited by the high temperatures (> 400 °C for $[ZnS]_n$ and $[ZnSe]_n$ and >300 °C for $[ZnO]_n$) required to crack the heterocycle; homogeneous reaction was however eliminated. Alkyl chalcogens and their derivatives, such as methylselenol, have been used by Isemura and others to grow a wide range of II/VI materials [90–92]; notably, using MeSeH, epitaxial $[ZnSe]_n$ was grown at 300 °C [92]. Takata *et al.* have used gaseous CS_2 as a source for the growth of $[ZnS]_n$ at 400 °C [93].

In related work, the synthesis and purification of a number of organotellurium compounds has been reported [94]. These include [Et$_2$Te] and [^{i}Pr$_2$Te]. Such compounds have been used to lower the growth temperature needed for CMT deposition, often with photochemical assistance [95]. Allyl tellurides, although of lower volatility, provide another possibility [96]. It should be remembered that lowering the growth temperature for CMT is particularly important because of the volatility of mercury.

9.5.5 Single-molecule Precursors

The chemistry of cadmium and zinc with the chalcogens is typified by the formation of polymeric structures with tetrahedral metal ions, and the adamantyl structure of many thiolates with zinc and cadmium provide well-characterised examples of such behaviour [97–99] as do the chain complexes formed by pyridinethione and 2,3-mercaptobenzthiazole with cadmium and zinc [100, 101] and mercury [102]. Mercury, as is often the case, proves exceptional and many years ago, Bradley [103] showed that the mercury complex of t-butylthiol has only weak interactions between essentially molecular units in the solid state.

A number of approaches can be taken to increase the volatility of thiolate complexes of group 12 metals. Perhaps the simplest is to increase the bulk of the thiolate so limiting polymerisation and hopefully increasing the volatility of the compound. Work on bulky thiolates was initially pioneered by Dilworth [104], although Bochmann [105–107] has extended and developed this chemistry to produce a range of molecules based on 2,4,6-tri-t-butylbenzene. Mixed alkyl thiolates with this ligand have recently been reported [107]. These compounds have been used to deposit thin films of the metal chalcogen in preliminary low pressure growth experiments. The problem with such ligands is that steric bulk is achieved by the incorporation of large numbers of carbon atoms, because of which carbon incorporation into thin films grown from such precursors seems a distinct possibility. In related work, the bulky silicon-based systems (e.g. $[MESi(SiCH_3)_3]$ M = Zn, Cd, Hg and E = Se or Te) have been used by Arnold to deposit a range of chalcogenides [108].

Many metal chalcogens have been shown to decompose to the corresponding II/VI material and an alternate approach to the preparation of precursors is to modify the properties of the thiolate by forming adducts. A novel series of precursors for the deposition of II/VI materials have been reported by Steigerwald *et al.* [109, 110]. Adducts of the phenyl and t-butyl chalcogens $[M(ER)_2]$ (M = Zn, Cd, Hg; E = S, Se, Te) have been prepared with 1,2-bis(diethylphosphino)ethane (depe). Complexes containing one and two mole equivalents of the phosphine have been isolated. The 1:2 species are polymeric and the 1:1 complexes dimers; crystal structures have been reported for the complexes $[Cd_2(SeC_6H_5)_4.depe]_n$ and $[Hg(SeC_6H_5)_2.depe]_2$.

The decomposition of such compounds in high-boiling-point solvents such as 4-ethylpyridine leads to the deposition of sub-nanometer clusters of, for example, $[CdSe]_n$ from $[Cd_2(SeC_6H_5)_4.(depe)]_n$; $[CdSe]_n$ was similarly obtained as the result of sealed-tube pyrolysis. Similarly the decomposition of bis(chalocogens) [111, 112] gives rise to metal-chalcogen-containing phases.

Other classes of molecule which have proved useful for the deposition of sulphide films include coordination complexes such as dithiocarbamates and thiophosphinates. The diethyldithiocarbamates are air-stable solids, with dimeric metal-containing units [113, 114], but are apparently monomeric in the vapour phase (as judged from mass spectra). In this case, each ligand atom provides two (initially) equivalent donor sulphur atoms, and a reasonable degree of coordinative saturation (typically 5-coordinate trigonal bipyramidal) can be achieved in a dimeric rather than a polymeric structure. Thin films of cadmium and cadmium zinc sulphide have been grown by low-pressure MOCVD (10^{-4} Torr, 370–420 °C) using the cadmium and zinc complexes of diethyldithiocarbamic acid [115]. The films were typically: specular, hexagonal (from RHEED (Reflection High-energy Electron Diffraction) measurements) and with optical band gaps in the region of 2.4 eV for $[CdS]_n$.

Takahashi *et al.* [116] reported the growth of cadmium sulphide using dimethylthiophosphinates $[M(S_2PMe_2)_2]$ (M = Cd or Zn); they had commented that the dithiocarbamates would be unsuitable as precursors, an opinion based solely on their preliminary sublimation data. However, the use of compounds containing phosphorus is itself undesirable; cadmium sulphide is normally n-type due to non-stoichiometry, and doping with phosphorus would lead to highly-compensated, semi-insulating materials. In a subsequent paper, Williams demonstrated [117] that highly-orientated sulphide films could be grown using the dimethylthiophosphinates as precursors. There is also a recent conference report of the growth of sulphide films from zinc and cadmium thiocarbamates [118]. Such compounds may open up the possibility of the cheap vacuum deposition of such materials for coating and photovoltaic applications.

Derivatives of these complexes have recently been prepared and complexes of the stoichiometry $[RME_2CNR'_2]$ (M = Zn, R = Me or Et, R' = Me or Et, E = S; M = Zn, R = Me or Et, R' = Me, E = Se and M = Cd, R = Me , R' = Et; E = S or Se) are known [119–121]. The X-ray single crystal structure of $[MeZnS_2CNEt_2]$ is shown in Figure 12.

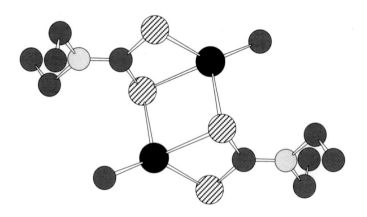

Figure 12   Single crystal X-ray structure of $[MeZnS_2CNEt_2]$

These compounds decompose cleanly under vacuum (10^{-2} Torr), to thin films of the metal chalcogen (typically ZnSe from R = R' = Et, 450 °C). As judged by mass spectral results, the dimers remain intact in the vapour phase and detailed studies of the decomposition of these complexes are in hand.

Finally, Jones and Cowley [122] have used the phosphiding agent $[Ga(t-Bu_2P)_3]$ to prepare the unusual cadmium and zinc complexes $[MeM(\mu-t-Bu_2P]_3$; the cadmium complex contains a Cd_3P_3 ring. A bis complex with mercury was also reported. These complexes are potential precursors for II/V materials such as $[Cd_3P_2]_n$.

9.6 PRECURSORS FOR OXIDES

9.6.1 High T_c Superconductors

The still relatively-novel 'high T_c' superconductors were initially studied as bulk materials [123], however many of the potential applications of such materials will necessarily involve the use of thin films [124]. As for semiconductors there are many potential advantages in the deposition of such films by MOCVD and these include relatively mild processing conditions together with the possibility of good control of structure and morphology and an established potential for large scale processing.

The most extensively studied of the high T_c materials are the so-called 123 compounds, $[MBa_2Cu_3O_{7-x}]_n$ (M = Y, Nd, Sm, Eu, Gd, Dy, Ho, Er, Tm, Yb or Lu) [124]. There are now several reports of the deposition of such materials by MOCVD [125]. The precursors used for the deposition of these compounds need to have sufficient volatility and stability to give reproducible growth results. In contrast to the growth of compound semiconductors where the use of metal alkyls predominates, oxide depositions tend to use precursors based on β-diketonates. The metal complexes of these ligands have reasonable vapour pressures but tend to suffer from problems associated with long term stability. Consequently, there have been several studies of the chemistry of such compounds in recent years.

The parent 2,4-pentandioanato complexes do not generally show sufficient volatility for use in MOCVD, although bis(2,4-pentandionato)copper(II) ($[Cu(acac)_2]$) has been used for the deposition of 123 compounds [126]. The volatility of metal complexes is in part controlled by interactions between individual metal complexes in the solid state. Such interactions can be reduced by increasing the bulk of the ligand and so the introduction of t-butyl groups as in 2,2,6,6-tetramethyl-3,5-heptanedione (Htmhd) (Figure 13) leads to increased volatility and indeed may be required, for example in the case of yttrium [127].

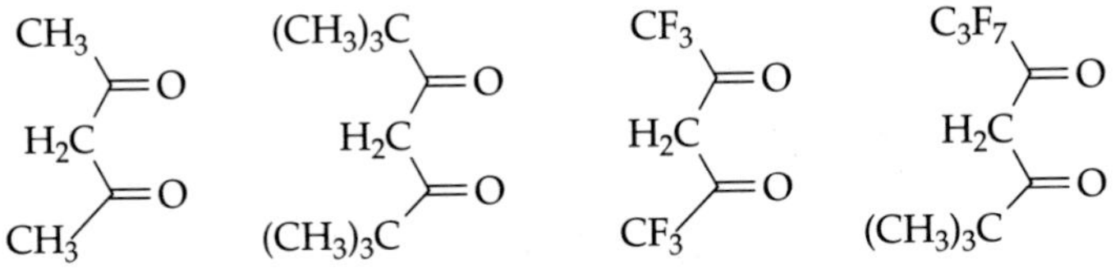

Hacac	Htmhd	Hfacac	Hfod

Hacac   2,4-pentandione (acetoacetone)
Htmhd   2,2,6,6-tetramethyl-3,5-heptanedione
Hfacac hexafluroacetoacetone
Hfod   1,1,1,2,2,3,3-heptafluoro-7,7-dimethyloctane-3,5-dione

Figure 13   Some β-diketonate ligands commonly used in preparing precursors for oxides

Copper and yttrium complexes of 2,2,6,6-tetramethyl-3,5-heptanedione sublime at temperatures between 125 and 160 °C at atmospheric pressures, but much higher temperatures are needed to sublime the corresponding barium complex. Barium complexes tend to decompose at the elevated source temperatures required and the vapour pressures of the decomposition products are different from those of the original material which leads to a variation of the barium content of the reactor and inhomogeneous growth [128]. Consequently much of the work on precursors for 123 compounds has concentrated on developing new, more reliable sources for barium. Similar problems are also encountered for calcium and strontium [129]. Crystal structures are now available for several of such complexes (e.g. Figure 14) [129–132].

One approach to improving the volatility of β-diketonates is the introduction of fluorinated groups into the β-diketonate. In the case of 123 compounds grown from such precursors, barium fluoride contamination is common. The principal fluorinated ligands which have been studied are trifluoroacetoacetonate, hexafluoroacetoacetonate (Hhfa) and 1,1,1,2,2,3,3-heptafluro-7,7-dimethyloctane-3,5-dione (Hfod) [130,133]. One problem with such precursors is that barium fluoride tends to be incorporated into the films. Although these sublime at lower temperatures than the hydrocarbon analogues, their decomposition temperatures are still close to the temperature required for sublimation and significant decomposition of the precursor occurs in the bubbler during decomposition.

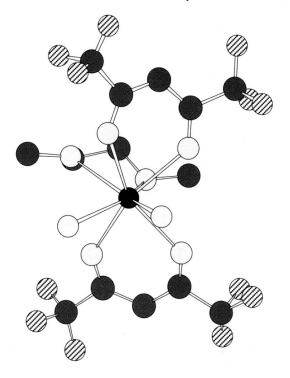

Figure 14 Single crystal X-ray structure of [Ca(hfac)$_2$(H$_2$O)$_2$(MeOCH$_2$CH$_2$OMe]

One approach to stabilising the precursor may be to add a ligand to the system. It can be argued that decomposition occurs in these systems because it is relatively hard to saturate coordinatively the metal ion making hydrolysis facile. In the case of [Ba(tmhd)$_2$] an excess of the ligand Htmhd passed over the precursor in the bubbler markedly improves the performance of the precursor [133]. This kind of approach has been taken in studies of such compounds by gas chromatography [130]. A second approach involves the use of a neutral coordinating ligand; for example, [Ba(hfa)$_2$] exhibits greater stability and enhanced volatility when coordinated to polyethers (CH$_3$O(CH$_2$CH$_2$O)$_n$CH$_3$; n = 2–4)[134]. Finally, Barron and co-workers have demonstrated that the addition of a nitrogen-containing base to (e.g. NH$_3$, NEt$_3$ or tetramethyl-ethylenediamine) leads to the formation of liquids in the barium bubbler and the reproducible transport of the barium-containing precursor [Ba(tmhd)$_2$]; similar results have been obtained with calcium and strontium precursors [135].

9.6.2   Other Oxides

Oxides have a wide range of applications [136] in the electronics industry other than in superconductors; some typical materials and their uses are summarised below.

Table 4   Some oxides and their electrical properties

Material	Application
$[BaTiO_3]_n$	Dielectric
$[LiNbO_3, AlPO_4]_n$	Piezoelectric
$\beta\text{-}[Al_2O_3]_n$	Superionic conductors
$[MgFe_2O_4, CoFe_2O_4]_n$	Thermistors
$[ZnO]_n$	Varistors
$[KTaO_3]_n$	Ferroelectrics

The chemical preparation of such materials tends to be dominated by the use of acetoacetonates as discussed above, and metal alkoxides. The chemistry of homoleptic metal alkoxides was pioneered by Bradley and a book [137] and several reviews detail this important area of chemistry [138–140].

Metal alkoxides are prepared by a number of methods including the reaction of an alkali metal alkoxide with the chloride of a higher valency metal (Equation (15)).

$$[TiCl_4] + 4\,NaOEt \longrightarrow [Ti(OEt)_4] + 4\,NaCl \qquad (15)$$

This method is not suitable for many important metals (e.g. Zr, Hf, Nb and Ta) for which stable heterometallic alkoxides are formed (e.g. $NaZr_2(OR)_9$); in such cases anhydrous ammonia can be used to dechlorinate the metal (Equation (16)).

$$[ZrCl_4] + 4\,NH_3 + 4\,ROH \longrightarrow [Zr(OR)_4] + 4\,NH_4Cl \qquad (16)$$

There are various other methods for the synthesis of metal alkoxides and these can be found in various reviews [137–140].

One of the most characteristic reactions of metal alkoxides is the ease with which they undergo hydrolysis to the metal oxide (Equation (17)).

$$[M(OR)_4] + 2\,H_2O \longrightarrow [MO_2]_n + 4ROH \qquad (17)$$

Obviously, in the synthesis and characterisation of metal alkoxides, the rigorous exclusion of water is important. The hydrolysis and partial hydrolysis of such metal alkoxides forms the basis of the sol-gel process [22]. It is, however, important to note that the precise nature of the species forming during the hydrolysis of such materials is far from clear.

Volatile alkoxides are clearly potentially useful for the deposition of metal oxides by MOCVD. The decomposition of such compounds was first studied by Bradley and Faktor in the 1950s when they discovered that the decomposition of zirconium t-butoxide at 200–250 °C involved the dehydration of the tertiary alcohol on the surface of the glass reactor [141] (Eqns. (18–20)).

$$[Zr(OBu^t)_4 + H_2O \longrightarrow [ZrO(OBu^t)_2] + 2\,Bu^tOH \qquad (18)$$

$$2\,Bu^tOH \longrightarrow 2\,H_2C{=}CMe_2 + H_2O \qquad (19)$$

The overall reaction may be represented as shown in Equation (20).

$$[Zr(OBu^t)_4] \longrightarrow ZrO_2 + 4CH_2{=}CMe_2 + H_2O \qquad (20)$$

Such a hydrolytic, thermal decomposition, leading to the metal oxide and volatile products suggests, as has been proved in practise, that such compounds will make good precursors for oxides.

The growth of more complex oxides such as ferroelectrics and nonlinear optical materials may require a more sophisticated approach, and matching the volatilities of precursors can be a problem. Lithium niobate has been grown by using a lithium β-diketonate, [Lithd], in combination with $[Nb(OMe)_5]$; $[LiNbO_3]_n$ was grown in a stream of argon at 450 °C [142].

A number of lead-containing oxides are important materials and their deposition has usually involved $[Pb(OBu^t)_2]$ or $[Pb(fod)_2]$. The ultimate aim is to produce a ferroelectric phase such as $[PbSc_{0.5}Ta_{0.5}O_3]_n$ (PST) as the perovskite [143, 144]. Unlike some of the other fluorinated acetoacetonates discussed, $[Pb(fod)_2]$ seems relatively thermally stable and exists, in the solid state (Figure 15), as a dimer involving bridging fluorine atoms [145].

524   *Paul O'Brien*

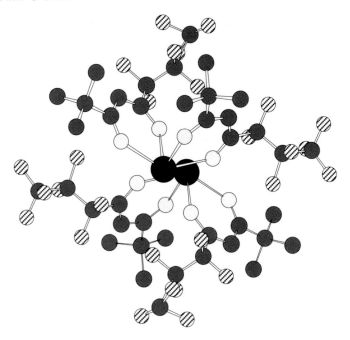

Figure 15   Structure of [Pb(fod)₂]

9.6.2.1 Sol-gel Processing

The sol-gel process (Figure 16) is an important chemical method for processing alkoxides and involves the controlled hydrolysis of a mixture [22].

A sol is a suspension or dispersion of discrete, colloid particles (10–1000 Å) in a liquid which is destabilised by concentrating to give a gel, an amorphous colloidal solid containing a fluid component dispersed in a three dimensional network. There are obviously many stages at which the process can be controlled and some of the advantages of the method include the availability of high-purity metal precursors, the homogeneity of the components at the molecular level and the low processing temperature required for the final step.

Examples of the use of the sol-gel process include the deposition of ferroelectric oxides such as barium strontium or lead lanthanum zirconium titanates [146]. Other applications include the preparation of $[TiO_2]_n$ a t low temperature and the use of mixtures of sodium, zirconium, silicon and phosphorus alkoxides to produce an amorphous form of the superionic conductor $[Na_3Zr_2Si_2PO_{12}]_n$ (NASICON) at 1,000 ° C [147].

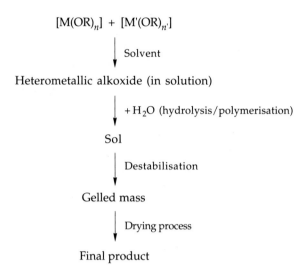

$$[M(OR)_n] + [M'(OR)_{n'}]$$

$\downarrow$ Solvent

Heterometallic alkoxide (in solution)

$\downarrow$ $+ H_2O$ (hydrolysis/polymerisation)

Sol

$\downarrow$ Destabilisation

Gelled mass

$\downarrow$ Drying process

Final product

Figure 16   Schematic representation of sol-gel processing

9.7   OTHER MATERIALS IMPORTANT IN SEMICONDUCTOR FABRICATION

9.7.1   Metallisation

Clearly the fabrication of many microelectronic devices requires the formation of contacts to conducting metals. The metals used for contacts tend to be noble metals such as gold or silver (which are sometimes deposited by evaporation methods), platinum group metals and copper. In many applications copper would make an ideal contact because of its ready availability and low cost, however until quite recently there have been no really good precursors for the deposition of thin films of copper. Aluminium is another important metal used as a contact and deposition studies of aluminium using alane seem promising [148].

9.7.1.1   Platinum Metals and Noble Metals

A wide range of complexes of platinum has been used to deposit thin films of this metal. However, because of the tendency of this metal to form fairly strong bonds to various elements, especially carbon, contamination of the films can be a major problem.

The earliest work used β-diketonate ligands and the parent [Pt(acac)$_2$] led to films which were very heavily contaminated with carbon [149]. Puddephatt has reported the use of a range of organometallic complexes of platinum to deposit the metal in vacuo and in a hydrogen atmosphere [150–152]. The use of H$_2$ greatly reduces the contamination of the films with carbon from around 50% to 5% as methyl groups are hydrogenated to methane and eliminated. Some of the compounds used for the deposition of platinum are summarised in Table 5.

Table 5   Some complexes used for the deposition of platinum metal

Compound	Deposition Gas	Carrier	Impurities	Ref
Temp/°C				
[Pt(acac)$_2$]	555	vacuum	50%C	[149]
[(COD)PtMe$_2$]	250	H$_2$	4%C, 2%O	[150–151]
[CpPtMe(CO)]	250	vacuum	27%C, 2%O	[150–51]
[CpPtMe$_3$]	95	H$_2$	<1%C	[153]

A recent study of the deposition of platinum from cyclopentadienyl- or methylcyclopentadienyl-trimethylplatinum in a hydrogen atmosphere (90–180 °C) led to platinum films containing less than 1 atom% of carbon and oxygen and no other impurities [153]. The reaction is autocatalysed by the fresh platinum deposit and both the initial deposition of platinum and the subsequent catalysed deposition are heterogeneous processes. The nature of the species controlling the autocatalytic step is as yet unknown.

Essentially similar approaches have been taken to the deposition of gold and platinum metal films. Trimethylphosphine adducts of gold(I) or gold(III) methyls have been used for the deposition of gold [154]. The mixed-methyl β-diketonate, [AuMe$_2$(hfa)], has been used for the laser-induced deposition of gold [155].

9.7.1.2 Copper

There is considerable interest at the moment in the deposition of copper as an interconnect in microelectric circuits. Originally, workers used the β-diketonates, a hydrogen atmosphere often being needed for the deposition of good quality films [156]. Another line of work involved alkoxides such as [CuO^{t}Bu]$_4$; these precursors led to films with very low levels of carbon incorporation but with oxygen contamination at about 5% [157].

Cyclopentadienyl complexes (e.g. [CpCuPR$_3$]; R is various alkyl groups) have been prepared and in, for example, the trimethylphosphine adduct, carbon contamination varied between 2 and 16% depending on the exact conditions for the deposition. Analysis of the reaction products led to the suggestion that the predominant reaction could be represented as shown in (Equation (21)) [158].

$$[CpCu(PMe_3)] \longrightarrow Cu + PMe_3 + CpH \text{ (trace)}$$
$$+ \text{ 1,5-dihydrofulvalene } + \text{ others} \qquad (21)$$

Perhaps the most exciting development in the deposition of copper has been the realisation that the monomeric adducts of β-diketonates of copper(I) (Figure 17) can decompose very cleanly, predominantly via a disproportionation mechanism, to copper metal and the corresponding copper(II) β-diketonate (Equation (22)) [159, 160] .

$$2 [(hfa)Cu(PMe_3)] \longrightarrow Cu + [Cu(II)(hfa)_2] + 2PMe_3 \qquad (22)$$

Compounds are now known with a range of phosphines and β-diketonates. Other adducting groups have also been used e.g. copper metal of excellent quality has also be deposited for the μ^2-complex of 2-butyne and copper(I) hexafluoracetoacetonate [161].

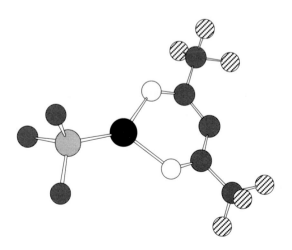

Figure 17 Structure of [(hfa)Cu(PMe$_3$)]

Excellent copper films can be grown from such precursors at rates from $10->1000$ Å min^{-1}, resistivity of the films can be near bulk e.g. 1.7 $\mu\Omega$ cm.

9.7.2 Silicides

These have a number of applications in silicon-based semiconductor fabrication technology including [162]:

1. As barrier materials between metal conductors and silicon, to prevent interdiffusion of silicon and metal atoms.
2. As active layers in Schottky and multiple-quantum-well devices.

Volatile metal complexes containing a silicon-metal bond are the logical precursors for such materials and typical deposition reactions are as carried out in flow systems with an excess of He carrier gas (Equations (23–25)).

$$[H_3SiCo(CO)_4] \longrightarrow CoSi \qquad (23)$$

$$[H_3SiM(CO))_5] \longrightarrow M_5Si_3 + MSi_x \; (x \; ca \; 1.25; M = Mn \; or \; Re) \quad (24)$$

$$[(H_3Si)_2Fe(CO)_4] \longrightarrow \beta\text{-}[FeSi_2]_n \qquad (25)$$

9.8   ACKNOWLEDGEMENTS

I am indebted to my colleagues at Queen Mary and Westfield College especially Professors Bradley and Aylett for numerous stimulating discussions and for providing selections of reprints. Much of our own work has been carried out in collaboration with Dr Tony Jones of Epichem whose help and support is gratefully acknowledged. Finally, I would like to thank the editors for the invitation to write the present chapter and for their forebearance.

9.9   REFERENCES

1. For a detailed survey see for example, *The Chemistry of the Semiconductor Industry*, S J Moss and A Ledwith, Eds, Blackie, Glasgow (1987).
2. P John, in: *The Chemistry of the Semiconductor Industry*, S J Moss and A Ledwith, Eds, Blackie, Glasgow, Chapter 5, p 98 (1987).

3. G B Stringfellow, *Organometallic Vapour Phase Epitaxy: Theory and Practice*, Academic Press, New York (1989).
4. For a readable, contemporary account, see C R M Grovenor, *Microelectronic Materials*, Adam Hilger, Bristol (1989).
5. For a readable review of materials applications see *Scientific American*, **235(4)** (1986).
6. H M Rosenburg, *The Solid State*, 2nd Edn, Clarendon Press, Oxford, p 150 et seq (1978).
7. H M Rosenburg, *The Solid State*, 2nd Edn, Clarendon Press, Oxford, p 30 (1978).
8. H M Rosenburg, *The Solid State*, 2nd Edn, Clarendon Press, Oxford, p 160 (1978).
9. S M Sze, *Semiconductor Devices: Physics and Technology*, Wiley, Chichester, p 289 et seq (1985): ref 4 , Chapter 9, p 431.
10. S M Sze, *Semiconductor Devices: Physics and Technology*, Wiley, Chichester, 1985 p 69.
11. I R Grant, in: *The Chemistry of the Semiconductor Industry*, S J Moss and A Ledwith, Eds, Blackie, Glasgow, Chapter 4, p 41 (1987).
12. J Woods, in: *The Chemistry of the Semiconductor Industry*, S J Moss and A Ledwith, Eds, Blackie, Glasgow, Chapter 4, p 64 (1987).
13. See ref 12 and e.g. S Fujita, H Mumoto, H Takebe and T Nooguchi, *J Crystal Growth*, **49**, 326 (1979).
14. T Taguchi, J Shirafugi and Y Inuishi, *Rev Phys Appl*, **12**, 117 (1977), and references therein.
15. H M Manasevit, *J Crystal Growth*, **55**, 1 (1981).
16. G J Davies and D A Andrews, *Chemtronics*, **3**, 3 (1988).
17. D Ham, K K Mishra, A Weiss and K Rajeshwar, *Chem Mater*, **1**, 619 (1989).
18. I Kaur, D K Pandya and K L Chopra, *J Electrochem Soc*, **127**, 943 (1980).
19. N C Sharma, R C Kainthla, D K Pandya and K L Chopra, *Thin Solid Films*, **60**, 55 (1979).
20. D Raviendra and J K Sharma, *J Appl Phys*, **58**, 838 (1985).
21. J B Mooney and S B Radding, *Ann Rev Mater Sci*, 12 (1982).
22. C J Brinker and G Scherrer, *Sol-Gel Science*, Academic Press, London, (1990).
23. For an excellent introduction see: Ch Elschenbroich and A Salzer, *Organometallics*, Verlag Chemie, Weinheim (1989).
24. A C Jones, *Chemtronics*, **4**, 15 (1989).
25. For comprehensive review information on synthesis see: *Comprehensive Organometallic Chemistry*, G Wilkinson, F G A Stone and E W Abel, Eds, Pergammon, Oxford (1982).
26. G E Coates and K Wade, *Organometallic Compounds: Vol 1*, Methuen, London (1967).

27. P O'Brien, J R Walsh, A C Jones and S A Rushworth, *Polyhedron*, **9**, 1483 (1990).

28. J B Habeeb and D G Tuck, *J Organomet Chem*, **146**, 213 (1978).

29. D C Bradley, H Chudzynska, M M Faktor, D M Frigo, M B Hursthouse, B Hussain and L M Smith, *Polyhedron*, **7**, 1289 (1988).

30. D V Shenai-Khatkhate, E D Orrell, J B Mullin, D C Cupertino and D J Cole-Hamilton, *J Crystal Growth*, **77**, 27 (1986).

31. D V Shenai-Khatkhate, M B Parker, E D McQueen, J B Mullin and D J Cole-Hamilton, *Phil Trans R Soc Lond*, **A330**, 101 (1990).

32. A W Parkins and R C Poller, *An Introduction to Organometallic Chemistry*, Macmillan, London p 19 (1986).

33. H Manasevit and W I Simpson, *J Electrochem Soc*, **116**, 1725 (1969).

34. H Manasevit and W I Simpson, *J Electrochem Soc*, **120**, 135 (1973).

35. I R Grant, *The Chemistry of the Semiconductor Industry*, J A Moss and A Ledwith, Eds, Blackie, Glasgow, Chapter 4, p 43 (1987).

36. For an excellent review emphasising mechanistic aspects see: P Zanella, G Rossetto, N Brianese, F Ossola, M Porchia and J O Williams, *Chem Mater*, **3**, 225 (1991).

37. R Didchenko, J E Alix and R H Toeniskoetter, *J Inorg Nucl Chem*, **35**, 35 (1960).

38. A K Chatterjee, M M Faktor, R H Moss and E A D White, *J de Physique*, **43**, C5 (1982).

39. This suggestion seems to have gained general acceptance, see refs 24, 30 and 36.

40. A H Moore, M D Scott, J J Davies, D C Bradley, M M Faktor and H Chudzynska, *J Crystal Growth*, **77**, 19 (1988).

41. M Lopez Coronado, E J Abril and M Aguilar, *Abstract of the 3rd European Workshop on MOVPE Montpellier*, p 70 (1989).

42. D C Bradley, H Chudzynska and D M Frigo, *Chemtronics*, **3**, 159 (1988).

43. K J Monserrat, J N Tothill, J Haigh, R H Moss, C S Baxter and W M Stobbs, *J Crystal Growth*, **93**, 466 (1988).

44. J S Foord, A J Murrell, D O'Hare, N K Singh, A T S Wee and T J Whitaker, *Chemtronics*, **4**, 262 (1989).

45. D A Bohling, G T Muhr, C R Abernathy, A S Jordan, S J Pearton and W S Hobson, *J Crystal Growth*, **107**, 1068 (1991).

46. A Molassioti, M Moser, A Stapor, F Scholz, M Holstalek and L Pohl, *Appl Phys Lett*, **54**, 857 (1989).

47. L Pohl, M Hostalek, H Schumann, U Hartmann, W Wasserman, A Brauers, G K Regel, R Hovel, P Balk and F Scholz, *J Crystal Growth*, **107**, 309 (1991).

48. G B Stringfellow, *J Electron Mater*, **17**, 327 (1988).

49. L M Fraas, P S McLeod, R E Weiss, L Partain and J A Cape, *J Appl Phys*, **62**, 299 (1987).

50. For a full discussion see ref 36.
51. S H Li, N I Buchan and G B Stringfellow, *J Crystal Growth*, **98**, 309 (1989).
52. O T Beachley and G E Coates, *J Chem Soc*, 3241 (1965).
53. A H Cowley, B L Benac, J G Ekerdt, R A Jones, K B Kidd, J Y Lee and J E Miller, *J Am Chem Soc*, **110**, 6248 (1988).
54 . D A Andrews, G A Davies, D C Bradley, M M Faktor, D M Frigo and E D White, *Semicond Sci Technol*, **3**, 1053 (1988).
55. A H Cowley and R A Jones, *Angew Chem, Int Ed Engl*, **28**, 1208 (1989).
56. J E Miller, K B Kidd, A H Cowley, R A Jones, J G Ekerdt, H J Gysling, A A Wernberg and T N Blanton, *Chem Mater*, **2**, 589 (1990).
57. J E Miller, M M Mardones, J W Nail, A H Cowley, R A Jones and J G Ekerdt, *Chem Mater*, **4**, 447 (1992).
58. A H Cowley, R A Jones, M A Mardones, J L Atwood and S G Bott, *Angew Chem, Int Ed Engl*, **30**, 1141 (1991).
59. H Hope, D C Pestana and P P Power, *Angew Chem Int Ed Eng*, **30**, 726 (1991).
60. A H Cowley, R A Jones, P R Harris, D A Atwood, L P Contreras and C J Burek, *Angew Chem Int Ed Engl*, **30**, 1143 (1991).
61. A N MacInnes, M B Power and A R Barron, *Chem Mater*, **4**, 11 (1992).
62. H Kukimoto, *J Crystal Growth*, **101**, 953 (1990).
63. H J Hovel, *Semiconductors and Semimetals, Vol II: Solar Cells*, Academic Press, New York (1975).
64. L M Smith and J Thompson, *Chemtronics*, **4**, 60 (1989).
65. J Tunnicliffe, S J C Irvine, O D Dosser and J B Mullin, *J Crystal Growth*, **68**, 245 (1984).
66. For reviews see: S J C Irvine, J B Mullin, J Geiss, J Gough and A Royle, *J Crystal Growth*, **93**, 732 (1988): J B Mullin, D J Cole-Hamilton, S J C Irvine, J E Hails, J Geiss and J S Gough, *J Crystal Growth*, **101**, 1 (1990).
67. L M Smith, J Thompson, A C Jones and P R Jacobs, *Mater Lett*, 722 (1988).
68. G N Pain, G I Christiansz, R S Dickinson, G B Deacon, B O West, K McGregor and R S Rowe, *Polyhedron*, **9**, 921 (1990).
69. J L Schmit, *J Vac Sci Technol*, **89**, A3 (1985).
70. C-H Wang, P-Y Lu and L-M Williams, *Appl Phys Lett*, **49**, 1372 (1986).
71. P J Wright, B Cockayne,A J Williams, A C Jones and E D Orrell, *J Crystal Growth*, **84**, 552 (1987).
72. P J Wright, B Cockayne and A C Jones, *Chemtronics*, **3**, 35 (1988).
73. B Cockayne, P J Wright, A J Armstrong, A C Jones and E D Orrell, *J Crystal Growth*, **91**, 57 (1988).

532 *Paul O'Brien*

74. P J Wright, P J Parbrook, B Cockayne, A C Jones, E D Orrell, H P O'Donnell and B Henderson, *J Crystal Growth*, **94**, 441 (1989).
75. P J Wright, P J Parbrook, B Cockayne, A C Jones, P O'Brien and J R Walsh, *J Crystal Growth*, **104**, 601 (1990).
76. P J Parbrook, P J Wright, B Cockayne, A G Cullis, B Henderson and K O'Donnell, *J Crystal Growth*, **106**, 503 (1990).
77. M B Hursthouse, M Motevalli, P O'Brien, J R Walsh and A C Jones, *J Materials Chem*, **1**, 139 (1991).
78. M B Hursthouse, M Motevalli, P O'Brien, J R Walsh and A C Jones, *Organometallics*, **10**, 3192 (1991).
79. S P J Wright, B Cockayne, A C Jones, E D Orrell, P O'Brien and O F Z Khan, *J Crystal Growth*, **94**, 97 (1989).
80. A C Jones, S A Rushworth, P J Wright, B Cockayne, P O'Brien and J R Walsh, *J Crystal Growth*, **97**, 537 (1989).
81. O F Z Khan, P O'Brien, P A Hamilton, J R Walsh and A C Jones, *Chemtronics*, **4**, 244 (1989).
82. E I Amma and R E Rundle, *J Am Chem Soc*, **80**, 4141 (1958).
83. G E Coates and R A Whitcombe, *J Chem Soc*, 3351 (1956).
84. M J Almond, M P Beer, K Hagen, D A Rice and P J Wright, *J Mater Chem*, **1**, 1065 (1991).
85. J E Butler, N Bottka, R S Sillmon and D K Gaskill, *J Crystal Growth*, **77**, 163 (1986).
86. R L Jackson, *Chem Phys Lett*, **163**, 315 (1989).
87. S Sritharan and S A Jones, *J Crystal Growth*, **66**, 231 (1984).
88. P J Wright, R J M Griffiths and B Cockayne, *J Crystal Growth*, **66**, 26 (1984).
89. H Mitsuhashi, I Mitsuishi, M Mizuta and H Kukimoto, *Jap J Appl Phys*, **24**, L578 (1985).
90. S Fujita, M Isemura, T Sakamoto and N Yoshimura, *J Crystal Growth*, **86**, 263 (1988).
91. M Mitsuhashi, I Mitsuhashi and H Kukimoto, *Jap J Appl Phys*, **24**, L864 (1985).
92. S Fujita, T Sakamoto and M Isemura, *J Crystal Growth*, **87**, 581 (1988).
93. S Takata, T Minami, Y T Miyata and H Nanto, *J Crystal Growth*, **86**, 257 (1988).
94. D V Shenai-Khatkhate, P Webb, D J Cole-Hamilton, G W Blackmore and J B Mullin, *J Crystal Growth*, **93**, 744 (1988).
95. S J C Irvin, H Hill, O D Dosser, J E Hails, J B Mullin, D V Shenai-Khatkhate and D J Cole-Hamilton, *Materials Lett*, **7**, 25 (1988).
96. For a review on the use of allyl tellurides see: J E Hails, S J C Irvine, J B Mullin, D V Shenai-Khatkhate and D J Cole-Hamilton, *Mat Res Soc Symp Proc*, **131**, 75 (1989).

97. I G Dance, R G Garbutt, D C Craig and M L Scudder, *Inorg Chem*, **26**, 3732 (1987).
98. I G Dance, R G Garbutt, D C Craig and M L Scudder, *Inorg Chem*, **26**, 4057 (1987).
99. I G Dance, R G Garbutt and M L Scudder, *Inorg Chem*, **29**, 1571 (1990).
100. M B Hursthouse, O F Z Khan, M Mazid, M Motevalli and P O'Brien, *Polyhedron*, **9**, 541 (1990).
101. O F Z Khan and P O'Brien, *Polyhedron*, **10**, 325 (1991).
102. S Wang and J P Fackler, *Inorg Chem*, **28**, 2615 (1989).
103. D C Bradley and D N Kunchur, *J Chem Phys*, **8**, 2258 (1964).
104. E S Gruff and S A Koch, *J Am Chem Soc*, **112**, 1245 (1990).
105. M Bochmann, K Webb, M Harman and M B Hursthouse, *Angew Chem Int, Ed Eng*, **29**, 638 (1990).
106. P J Blower and J R Dilworth, *Coord Chem Rev*, **76**, 121 (1987).
107. M Bochmann, A P Coleman and A K Powell, *Polyhedron*, **11**, 507 (1992).
108. B O Dabbousi, P J Bonasia and J Arnold, *J Am Chem Soc*, **113**, 3186 (1991); P J Bonasia and J Arnold, *Inorg Chem*, **31**, 2508 (1992).
109. J G Brennan, T Segrist, P J Carroll, S M Stuczynski, P Reynders, L E Brus and M L Steigerwald, *J Am Chem Soc*, **111**, 4141 (1989).
110. J G Brennan, T Segrist, P J Carroll, S M Stuczynski, P Reynders, L E Brus and M L Steigerwald, *Chem Mater*, **2**, 403 (1990).
111. K Osakada and T Yamamoto, *J Chem Soc, Chem Commun*, 117 (1987).
112. M L Steigerwald and C R Sprinkle, *J Am Chem Soc*, **109**, 7200 (1987).
113. M Bonamico, A Vaciago and L Zambonelli, *Acta Cryst*, **19**, 898 (1965).
114. M Bonamico and G Dessy, *J Chem Soc* (A), 264 (1971).
115. D M Frigo, O F Z Khan and P O'Brien, *J Crystal Growth*, **6**, 989 (1989).
116. Y Takahashi, R Yuki, M Sugiura, S Motojima and K Sugiyama, *J Crystal Growth*, **50**, 491 (1980).
117. M A H Evans and J O Williams, *Thin Solid Films*, **87**, L1 (1982).
118. A Saunders, A Vecht and G Tyrell, *Ternary Multiary Cmpd Proc 7th Int Conf*, 1986 (publ 1987) CA108:66226h (1987)).
119. M B Hursthouse, M A Malik, M Motevalli and P O'Brien, *Organometallics*, **10**, 730 (1991).
120. M A Malik and P O'Brien, *Chem Mater*, **3**, 999 (1991).
121. P O'Brien, M A Malik, M B Hursthouse and M Motevalli, *Polyhedron*, **11**, 45 (1992).
122. B I Benac, A H Cowley, R A Jones, C M Nunn and T C Wright, *J Am Chem Soc*, **111**, 4986 (1989).
123. J G Bedorz and K A Muller, *Z Phys B*, **64**, 189 (1986).

534   *Paul O'Brien*

124. *The Science and Technology of Thin Film Superconductors*, R D McConnell and S A Wolf, Eds, Plenum Press, New York (1989).
125. K Watanabe, H Yamane, H Kuroosawa, T Hirai, N Kobayashi, K Noto and Y Muto, *Appl Phys Lett*, **54**, 575 (1989).
126. D S Richeson, L M Tonge, J Zhao, J Zhang, H O Marcy, T J Marks, B W Wessels and C R Kannewurf, *Appl Phys Lett*, **54**, 2154 (1989).
127. E W Berg and J J Acosta, *Anal Chim Acta*, **40**, 101 (1968).
128. T Nakomori, H Abe, T Kanamori and S Shibata, *Jap J Appl Phys*, **27**, L1265 (1988).
129. A P Purdy, A D Berry, R T Holm, M Fatemi and K K Gaskill, *Inorg Chem*, **28**, 2799 (1989).
130. S B Turnipseed, R M Barkley, and R E Sievers, *Inorg Chem*, **30**, 1164 (1991).
131. D C Bradley, M Hasan, M B Hursthouse, M Motevalli, O F Z Khan, R G Pritchard and J O Wiliams, *J Chem Soc, Chem Commun*, 575 (1992).
132. G Rossetto, A Polo, F Benetollo, M Porchia and P Zanella, *Polyhedron*, **11**, 979 (1992).
133. P H Dickinson, T Geballe, A Sanjurjo, D Hildebrand, G Craig, J Collman, S A Banning and R E Sievers, *J Appl Phys*, **66**, 444 (1989).
134. R Gardiner, D W Brown, P S Kirlin and A L Rheingold, *Chem Mater*, **3**, 1053 (1991).
135. A R Barron, J M Buriak, L Cheatham and R Gordon, *The Electrochemical Society 177th Meeting Abstract 943 HTS Montreal Canada* (1990).
136. L G Hubert-Pfalzgraf, *New J Chem*, **11**, 663 (1987).
137. D C Bradley, R C Mehrotra and P D Gaur, *Metal Alkoxides*, Academic Press, London (1978).
138. D C Bradley, *Chem Rev*, **89**, 1317 (1989).
139. R C Mehrotra, *Advances in Inorganic and Radiochemistry*, Academic Press, London, vol XXVI, 269 (1989).
140. M H Chisholm, in: *Inorganic Chemistry Towards the 21st Century*, American Chemical Society, Washington DC, Chapter 16, p 243 (1983).
141. D C Bradley and M M Faktor, *Trans Faraday Soc*, **55**, 2117 (1959).
142. B J Curtis and H R Brunner, *Mater Res Bull*, **10**, 515 (1975).
143. L M Brown and K S Mazdiyasni, *J Am Chem Soc*, **53**, 91 (1970).
144. C J Brierley, C Trundle, L Considine, R M Whatmore and F W Ainger, *Ferroelectrics*, **91**, 181 (1989).
145. L A Khan, M A Malik, M Motevalli and P O'Brien, *J Chem Soc, Chem Commun*, in press (1992).
146. D R Ulrich, *J Non Crystalline Solids*, **100**, 74 (1988).
147. J P Bouilot, *Ann Chim Science des Materiaux*, **10**, 305 (1985).

148. W L Gladfelter, D C Boyd and K F Jensen, *Chem Mater*, **1**, 339 (1989).
149. M Rand, *J Electrochem Soc*, **120**, 686 (1973).
150. R Kumar, S Roy, M Rashidi and R J Puddephatt, *Polyhedron*, **8**, 551 (1989).
151. B Nixon, P R Norton, E C Ou and R J Puddephatt, *Chem Mater*, **3**, 222 (1991).
152. N H Dryden, R Kumar, E C Ou, M Rashidi, S Roy, P R Norton and R J Puddephatt, *Chem Mater*, **3**, 677 (1991).
153. Z Xue, H Thridandam, H D Kaesz and R F Hicks, *Chem Mater*, **4**, 162 (1992).
154. R J Puddephatt and I Treuernicht, *J Organomet Chem*, **319**, 129 (1987).
155. T T Kodas, T H Baum and P B Comita, *J Crystal Growth*, **87**, 378 (1988).
156. R L Van Hemert, L B Spendlove and R E Sievers, *J Electrochem Soc*, **112**, 1123 (1965).
157. P M Jeffries and G S Girolami, *Chem Mater*, **1**, 8 (1989).
158. M J Hampden-Smith, T T Kodas, M Paffett, J D Farr and H-K Shin, *Chem Mater*, **2**, 636 (1990).
159. H-K Shin, K-M Chi, M J Hampden-Smith, T T Kodas, J D Farr and M Paffett, *Adv Mater*, **3**, 246 (1991).
160. H-K Shin, M J Hampden-Smith, E N Duesler and T T Kodas, *Polyhedron*, **10**, 645 (1991).
161. T H Baum and C E Larse, *Chem Mater*, **4**, 365 (1992).
162. B J Aylett and A A Tannahill, *Vacuum*, **35**, 435 (1985).

Index